BUREAU OF FLORA AND FAUNA, CANBERRA
AND
DEPARTMENT OF CONSERVATION AND LAND MANAGEMENT, WESTERN AUSTRALIA

ABRS

AUSTRALIAN FLORA AND FAUNA SERIES
NUMBER 8

THE BANKSIA ATLAS

Anne Taylor and Stephen Hopper

Australia
1788-1988

Australian Government Publishing Service
Canberra

Bureau of Flora and Fauna
G.P.O. Box 1383, Canberra, A.C.T.
2601, Australia

Department of Conservation and Land Management, Western Australia,
Western Australian Wildlife Research Centre,
P.O.Box 51 Wanneroo, W.A. 6065

ISSN 0813-6726
ISBN 0 644 07124 9

Cover photographs
Honey possum on *Banksia grandis* (S.D.Hopper)
Inset: Contributors looking at maps (A.Taylor)

Printed in Australia by Canberra Publishing and Printing Co., Fyshwick, A.C.T.

FOREWORD

Australia has a vast land area, and a low population concentrated in near-coastal regions of the southern half.

These geographical factors have been used as arguments against the use of volunteer biologists in surveying Australia's biodiversity. Ten years ago the Royal Australasian Ornithologist's Union (RAOU) was active in promoting the 'Atlas of Australian Birds' project. The success of the project defied the doomsayers and those who insisted it "couldn't be done". The resultant publication now ranks as one of the best distribution Atlases, and has won praise and acclaim around the world.

While the RAOU project was in full swing Stephen Hopper in Western Australia became interested in similar projects involving flora. His work with that of others resulted in the FLORAPLOT system being established to store information on and map the distribution of Western Australian Flora.

In 1983 the Australian Biological Resources Study (ABRS) became interested in establishing a nationwide distribution study of a significant plant genus. Fortuitous interaction with Stephen Hopper's group in Western Australia resulted in the "Banksia Atlas" program - run by the Western Australian Department of Conservation and Land Management, with funding support from ABRS. As Western Australia had the lion's share of Banksia species it made sense for the project and the national coordinator, Anne Taylor, to be based in Perth, with substantial support in eastern Australia provided by Roger Hnatiuk in the Bureau of Flora and Fauna.

The result of the three-year study is this book "The Banksia Atlas". It represents a significant effort by many members of the community led by the enthusiasm of Stephen Hopper and Anne Taylor. The Atlas has vastly improved our detailed knowledge of Banksia, and helped the discovery of two new species and a range of infra-specific taxa. I hope it will be the first of many such studies and publications. Certainly our ability to understand and therefore manage adequately these beautiful plants has been tremendously improved. "The Banksia Atlas" is more than just a fascinating book - it is a conservation tool of great power.

All concerned with its production, from national coordinators to state leaders and, most importantly, individual contibutors, deserve not only praise but the thanks of the community for helping in the mapping and thus production of this volume.

Peter Bridgewater
Bureau of Flora and Fauna
December 1987

CONTENTS

INTRODUCTION

Where do the 75 species of Australian banksias grow in the wild? This book reports on a project aimed at addressing the question in a novel way for Australian plants. Rather than attempting to map each species from exhaustive field surveys, the project sought the assistance of volunteers in all States and Territories. The Department of Conservation and Land Management provided a central computerised data processing facility and supporting materials for recording observations in the field. The volunteers provided the labour, contributing observations made in the wild at their own expense.

The outcome of this project, The Banksia Atlas, attests to the enthusiasm, diligence and competence of the volunteer contributors. In the short space of three years (1984-1986), the banksias were mapped and studied at a level of detail surpassed in few groups of Australian plants. Hundreds of people became interested in banksias and new species were identified. Banksias turned up in unexpected places. Significant contributions were made to knowledge of their response to fire, to the timing of flowering and new shoot growth, and on the habitats in which they grow. Many conservation problems were elucidated and promising areas for further research have been identified.

Thus, the Atlas complements existing literature (Holliday and Watton 1975; George 1981, 1984) in providing accurate maps and a compendium of ecological and reproductive information on banksias as recorded by contributors. No attempt was made to review the literature on banksias comprehensively. This was considered beyond the scope of the present work.

We hope that this Atlas is the forerunner of many. The project has demonstrated that public participation in plant conservation research does work, and work well. There is a rich field here for future collaboration between volunteers and those professional botanists willing to venture into this stimulating and rewarding arena of inquiry.

Why banksias?

In a national project such as this, the choice of plants to be mapped is of fundamental importance. It is desirable that the plant group occurs in all States and Territories. It should consist of a manageable number of species whose distinguishing features are well documented and easily learnt by interested volunteers who may lack formal botanical training. A well-written inexpensive field guide supported by a detailed modern taxonomic revision is highly desirable. And lastly, a strong aesthetic appeal to people of all ages and walks of life is a central ingredient to the success of an Atlas dependent on volunteer contributions.

In view of these points, banksias were a natural choice for the Atlas. They rank alongside eucalypts and wattles as among the most distinctively Australian native plants known to the public. The majority of banksias have characteristic leaves and big colourful flower spikes that make each species easy to recognise. A recent taxonomic treatment (George 1981) and books (Holliday & Watton 1975; George 1984) were available during the Atlas as aids in learning to identify species.

Banksias have wide appeal among horticulturists and wildflower enthusiasts. Their showy flowers and rich nectar production attract birds and mammals into home gardens. A thriving cut-flower export trade exists in many banksias, which are grown increasingly on farms as a crop to supply markets, mostly in Europe (Burgman & Hopper 1982).

A growing scientific literature has highlighted the importance of banksias in natural ecosystems. Many species are dominant trees or shrubs in vegetation types across Australia. They play a central role in providing food and shelter to a wide range of native animals and plants. The banksias themselves are masters of deriving food from nutrient-deficient sandy soils. They use a dense mat of fine proteoid roots near the soil surface to accomplish this.

Thus, the conservation of banksias in the wild is important for aesthetic, economic, scientific and biological reasons. Accurately mapping their distribution is a vital first step in a conservation program. Armed with such knowledge, it is then possible to determine which species occur within protected areas such as national parks, and which do not. Rare and endangered species can be identified and appropriate management actions planned to ensure against their extinction.

We note with particular concern that many of the most beautiful Western Australian banksias are highly susceptible to the dieback fungal pathogen *Phytophthora cinnamomi* (Fig. 12). In Western Australia, this fungus has infected many banksia communities along the south coast and in the forest regions and continues to spread.

An essential tool in monitoring, and hopefully controlling, the spread of the dieback fungus is a comprehensive data bank and set of maps for the species most susceptible to its effects. This was one of the many potential uses of the *Atlas* that provided an additional stimulus to the choice of the banksias as a suitable group to map.

Reliability of the Atlas data

Before describing in detail how the *Atlas* was organised (Chapter 1), and its major achievements and findings (Chapter 3), we wish to comment briefly on a frequently-asked question "How reliable are the *Atlas* maps and associated data?". This concern is often raised by botanists who consider that voucher herbarium specimens need to be collected so that any future queries about incorrect identifications can be answered.

Clearly, this course of action was not possible for *The Banksia Atlas*. An influx of 24 781 specimens, one for each mapping record, over a two year period would have strained the resources of existing Australian herbaria to the limit, especially given the size and bulk of herbarium specimens of *Banksia*. Moreover, with a modern taxonomic revision and well-illustrated guides available, identification problems exist for only a few groups of species in *Banksia*.

Thus, it was considered to be inappropriate to request voucher specimens from contributors as a routine exercise. Nevertheless, as described in detail in Chapter 1, sight records were subjected to rigorous examination, checking and rechecking by the State and national coordinators and by computer validation routines (e.g. checking if the latitude and longitude of a record fell within the coordinates of the shire recorded on the sheet). Moreover, specimens were requested when an identification problem arose or a significant range extension was reported. Having mapped herbarium specimen collection sites for each *Banksia* to compare with the *Atlas* data (Chapter 5), we are confident that the *Atlas* maps of sight records contain less errors than localities given on herbarium sheets.

We are also sure that this _Atlas_ is not the last word on _Banksia_ distributions. Contributors recorded banksia areas with varying thoroughness depending on accessability and other constraints. Also, the distribution of a species is dynamic, changing through time as populations become extinct or new colonies establish. The present _Atlas_ should be viewed, therefore, as a slice in time, accurately mapping a sample of populations of each banksia. Future workers will, no doubt, correct errors of omission, and document changes in the status of populations recorded here.

In the case of the associated _Atlas_ data on habitat, fire response, pollination etc., there were some problems (described in Chapters 2 and 5) for some fields of information, and we caution readers where appropriate. However, most such information recorded by contributors has been carefully checked and appears to be accurate.

Acknowledgments

We are grateful to all who have contributed to this project. In addition to persons listed below as contributors, we wish to record our thanks to Norman Hall and Wilf Lehre for early developmental work on computer programs, to Andrew Brown for assistance in compiling camera-ready copy for publication, and to Raelene Hick for typing the manuscript. The Australian Biological Resources Study (ABRS) generously supported a pilot project in 1979-80 that made scaling up to the national _Banksia Atlas_ possible. The continued support from ABRS for _The Banksia Atlas_ was central to the successful completion of the project. The Curators and staff of all major Australian herbaria were most helpful in assisting contributors and coordinating staff.

We have attempted to present the observations of contributors as accurately as possible. However, for those errors or omissions that may have escaped our attention in the preparation of the book, we accept responsibility.

CONTRIBUTORS

Project Director
Stephen Hopper

National Coordinator
Anne Taylor

Computer Programmer
Paul Gioia

Part-time Assistants to the National Coordinator
Scott Wilson
Karin Palmer
Anna Napier
Robin Andersson

Administrative and Mailing Support, Bureau of Flora and Fauna, Canberra
Roger Hnatiuk

Taxonomic Adviser
Alex George

State and Territory Coordinators

Australian Capital Territory
Alex George

New South Wales
Ross Goldingay
Paul Spratt

Northern Territory
Anne Taylor

Queensland
Paul Taylor

South Australia
Les Gray

Tasmania
Bruce Champion

Victoria
Alf Salkin

Western Australia
Anne Taylor

Botanical Illustrators

	Pages
Gwen Collison	144, 146, 148, 154, 164, 228
Ross Goldingay	196, 224
David Hutchison	72, 74, 90, 220
Rosemary Opala	62, 186, 214
Sue Patrick	60, 66, 76, 78, 86, 88, 92, 94, 96, 98, 102, 106, 108, 110, 112, 114, 116, 118, 120, 122, 124, 126, 132, 134, 136, 138, 140, 142, 150, 152, 156, 160, 168, 170, 172, 174, 176, 178, 180, 182, 184, 188, 190, 194, 202, 204, 206, 210, 212, 216, 218, 222, 230, 232, 234, 236, 238, 240, 242, 244, 246, 248, 256, 252
Margaret Pieroni	64, 68, 70, 80, 82, 84, 100, 104, 128, 130, 158, 162, 192, 198, 200, 208, 226, 254, 256

Photographers

Andrew Brown
John Falconer
Alex George
Stephen Hopper
Tom Hill
Greg Keighery
George Schmidt
Anne Taylor
Peter Taylor
Graham Velterop

Sight Record Sheet Contributors

The following list includes the names of all known *Atlas* contributors. There were many more who assisted these listed contributors but did not identify themselves as having filled in Sight Record Sheets. We apologise to any contributor whose name has been omitted and thank all who helped. (Number of sheets is in brackets, followed by State in which contributor was resident.)

John Adams	(10)	WA
Joan & Laurie Adams	(120)	NSW
Elizabeth Addario	(1)	NSW
Norma Ali	(53)	TAS
Jennie Allen	(394)	WA
Mr & Mrs S. Allen	(22)	WA
Terry & Helen Allen	(32)	WA
Margaret Allan	(13)	TAS
Eric Anderson	(5)	QLD
Robin & Denis Andersson	(129)	WA
APPM Forest Products	(5)	TAS
Paul Armstrong	(13)	WA
M. Asa	(2)	ACT
Beth Atkins	(7)	WA
Tony Atkinson	(52)	WA
Geoff & Beryl Austin	(20)	WA
P. Austin	(6)	NSW
John Baas	(2)	WA
Bob & Barbara Backhouse	(54)	WA
Gary Backhouse	(37)	VIC
Diane Baker	(4)	WA
Betty Ballingall	(62)	QLD
Heather Bambrick	(2)	QLD
Judith Barker	(5)	VIC
Greg Barrett	(23)	WA
Mirranie Barker	(43)	WA
Mr & Mrs P. Baxter	(1)	NSW
Tony Bean	(11)	QLD
Jeff Beard	(10)	WA
Ted Belcher	(3)	NSW
Don & Barbara Bellairs	(18)	WA
David Bennett	(2)	WA
Graham Bennett	(32)	QLD
Jo Benyon	(27)	ACT
Arnold Berkin	(21)	VIC
Chris Berry	(2)	WA
Rowland Bingham	(1)	NSW
Sue Birney	(2)	WA
Elsie Bishop	(9)	WA
Trevor Blake	(16)	VIC
Doris & Vic Blampey	(386)	WA
Tony Bloomfield	(2)	TAS
John Blyth	(6)	QLD

Beryl & Fred Boase	(5)	WA
Jean Bold	(6)	WA
R. Boulton	(1)	ACT
R.A. & J.A. Boyd	(7)	NSW
Grace Boyd	(2)	QLD
John Boyle	(57)	WA
Keith Bradby	(77)	WA
Wendy Bradshaw	(4)	WA
George Brandt	(1)	ACT
Mary Bremner	(85)	WA
Reg & Ruth Brian Davis	(5)	WA
Peter Bridgewater	(13)	ACT
Lola Broadhurst	(19)	WA
Andrew Brown	(3)	WA
Judith Brown	(26)	WA
Clive Brownsea	(18)	VIC
Allan Burbidge	(11)	WA
Dick Burns	(52)	TAS
Neil Burrows	(12)	WA
Keith & Ann Cairncross	(11)	NSW
Peter Cale	(6)	WA
Clem Campbell	(2)	QLD
Jason Campbell	(1)	NSW
Jim Carpenter	(111)	TAS
Dan Carter	(21)	WA
Barry Carter	(5)	ACT
J. Carter	(5)	ACT
Racquel Cavallaro	(2)	WA
Tony Cavanagh	(4)	VIC
Ian Chamberlain	(1)	QLD
Bruce Champion	(80)	TAS
Arthur & Kathleen Charlton	(7)	WA
John Chilvers	(51)	WA
H. Chinchen	(3)	NSW
Ruth Clark	(1)	WA
Bronwyn Clarke	(3)	ACT
Lyn Clarke	(9)	WA
Mr & Mrs G. Clarke	(36)	ACT
Wayne & Desrae Clarke	(6)	WA
Steve Clemesha	(43)	NSW
Alison Clifton	(1)	WA
Jeanette Closs	(10)	TAS
Phil Collier	(61)	TAS
Kevin & Kathy Collins	(199)	WA
Gwen Collison	(1)	NSW
John Conran	(23)	QLD
Sam Cook	(3)	WA
Ron & Margaret Cooper	(114)	NSW
Diana Cordiner	(5)	NSW
Colin & Verna Cornford	(11)	QLD
Doug Coughran	(14)	WA
Ron & Betty Cox	(14)	NSW
Jenny Craigen	(5)	NSW
Peter & Lorraine Crammond	(7)	SA
Ray Cranfield	(2)	WA
Chris Creighton	(52)	WA
Gay Crowley	(3)	QLD
Val Crowley	(43)	WA
Eileen Croxford	(69)	WA
Audrey Crutchfield	(5)	NSW
Rosemary & Jeff Cugley	(13)	WA
Robyn Cummings	(11)	NSW
Paul & Thelma Daniell	(8)	WA
Alan Danks	(33)	WA
John Davey	(2)	WA
Doreen Davidson	(34)	WA
Jenny Davies	(2)	WA
Steve Dawson	(19)	WA
Louisa De Braganca	(3)	WA
Roslyn Delecourt	(3)	WA
Harry De Jong	(33)	WA
Denmark Wildflower Group	(72)	WA
Joan Dewhurst	(7)	SA
Peter Donan	(2)	QLD
Doris Dove	(10)	QLD
Greg Drake & Ann Knight	(23)	NSW
Morris Duggan	(1)	ACT
W.T. Dullard	(4)	VIC
Jennifer Duwing	(13)	WA
Linn Dyason	(15)	QLD
Greg Ebworth	(4)	VIC
Graham Edwards	(123)	WA
Robin Elkin	(13)	NSW
Peter Engler	(73)	WA
Ted Errey	(8)	VIC
Barbara & Margaret Evans	(110)	WA
Lalage & John Falconer	(5)	WA
Ian & Gwynneth Fardon	(28)	NSW
Ken Felton	(13)	TAS
Charles Felton	(6)	WA
Lee Fernie	(8)	WA
Peter & Anita Fielding	(1)	NSW
Keith & Lindsay Fisher	(4)	NT
Ian Fowler	(20)	NSW
Lawrence Frizzell	(6)	WA
Janice & John Fuss	(22)	SA
Mary Gandini	(22)	QLD
Ray Garstone	(11)	WA
Stephen Gee	(33)	WA
Elizabeth George	(4)	WA
Howard & Dorrie Gibbs	(16)	WA
Anne Giblin	(8)	NSW
Grace Gibson	(17)	NSW
John Gibson	(168)	ACT
Lloyd Gibson	(40)	ACT
Brendon Gill	(1)	NSW
Alan Gillanders	(14)	QLD
Ross Goldingay	(3)	NSW
Peter Golos	(6)	WA
Robert Goodman	(14)	WA
Chris Goodsell	(24)	WA
Allan Goodwin	(8)	VIC
Mal Graham	(56)	WA
Les Gray	(706)	SA
Jim Green	(33)	WA
Olga Green	(10)	WA
Kay Greeves	(9)	TAS
R. Gregory-Smith	(6)	ACT
Joy Grieg	(2)	VIC
Lois Grover	(4)	WA

Linley Gueho	(6)	WA
Mark Gunson	(1)	WA
Murray Haby	(373)	VIC
Kate & Mark Hackling	(11)	WA
Roger Hall	(25)	WA
Annette Hallpike	(125)	VIC
Stephen Hamilton	(13)	VIC
Brenda Hammersley	(40)	WA
David Handscombe	(3)	VIC
M.V. Hannah	(3)	NSW
Gary Hardy	(37)	QLD
Paul Harris	(7)	QLD
Joy Harrop	(9)	WA
Janet Hauser	(2)	QLD
Jan Hay	(4)	WA
K. Haynes	(14)	ACT
Frank Hebden	(7)	NSW
Joan & Ray Hegarty	(18)	ACT
Cyril Henshaw	(463)	VIC
Grant Hewett	(16)	WA
Nigel Hewett	(84)	WA
Lynne Heywood	(12)	NSW
Graeme Hill	(6)	NSW
Joyce Hill	(1)	QLD
Sheila Hill	(74)	WA
Maria Hitchcock	(52)	NSW
Meri Hitchins	(2)	WA
Roger Hnatiuk	(11)	ACT
Margaret Holmes	(9)	NSW
Harald Hoffman	(26)	WA
Peter Hook	(26)	WA
Jean Hooper	(3)	WA
Stephen Hopper	(1)	WA
John Howard	(6)	NSW
Wade Howlett	(4)	VIC
Bert & Alice Humphreys	(3)	WA
David Hutchison	(8)	WA
Ida Jackson	(93)	SA
Robert Jago	(9)	QLD
David James	(26)	WA
Rhoda & Harry Jeavons	(4)	NSW
Laurence Johnson	(18)	NSW
Brendan Johnston	(54)	NSW
Chris Johnston	(12)	NSW
Pat Jordon	(4)	NSW
Wendy Kappelle	(43)	WA
Morton Kaveney	(1)	NSW
Ian Kealley	(91)	WA
R. Kefford	(33)	ACT
Bronwen Keighery	(5)	WA
Margaret Kelly	(3)	QLD
Claire Kennedy	(11)	NSW
Anne Kerr	(15)	VIC
Sandy King	(9)	WA
Linda King	(1)	NSW
Ian Kirwan	(21)	VIC
Matthew Kirwan	(23)	VIC
Horace Knox	(2)	WA
Kowree Field Naturalists Club	(26)	VIC
Raymond Lacken	(4)	QLD
Greg Lamont	(5)	NSW
Maryanne Larkin	(44)	NSW
Stephen & Meg Le Fanu	(8)	WA
Carol Leggett	(19)	QLD
Pattie Leighton	(30)	WA
Penny Leighton	(19)	WA
Malcolm Lewis	(9)	WA
Brett Ley	(8)	NSW
Lenore Lindsay	(21)	QLD
Claire Lithgow	(115)	SA
Robert & Lorraine Little	(28)	WA
Sandra Lloyd & Rod Randall	(2)	WA
Shirley Loney	(6)	WA
Roger Lord	(1)	QLD
Enid Lowe	(20)	WA
Tim Lowe	(8)	WA
Noni Mammatt	(2)	WA
Philip Manley	(14)	QLD
Eric Manning	(3)	NSW
Pauline March	(4)	NSW
Michael Marmach	(18)	VIC
Marrie Family	(2)	QLD
Neil Marriott	(15)	VIC
Cyril Marhsall	(22)	NSW
Sheelagh Marshall	(7)	WA
Bill Martin	(12)	QLD
Debbie Martin	(5)	WA
Gary Martin	(12)	WA
Peter Mawson	(53)	WA
Neil Maxted	(6)	WA
Marjorie May	(37)	ACT
Ian McAllan	(35)	NSW
Eric & Margaret McCrum	(7)	WA
Ian McDonald	(7)	NSW
Donna McDuff	(19)	QLD
Shaun McGowan	(16)	VIC
Ross & Bev McGuinness	(50)	WA
Duncan McKay	(7)	NSW
Helen McKechnie	(10)	NSW
Inez Mcloughlin	(5)	VIC
Robert McLure	(12)	VIC
F.C. McNamara	(2)	WA
Graham Mee	(23)	NSW
M. Melkman	(2)	QLD
Simon Merewether	(5)	WA
Michael Merrony	(15)	ACT
Col Middleton	(6)	QLD
Phillipa Middleton	(3)	SA
Kevin Mills	(632)	NSW
Alan Moore	(49)	WA
Noel Moore	(2)	NSW
Irene Morcombe	(1)	WA
Joan Moriarty	(2)	NSW
Gary Muir	(9)	WA
Jane Muir	(7)	WA
Dick Mumford	(95)	WA
Lorna Murray	(22)	QLD
Bob Mylius	(3)	VIC
Anna Napier	(23)	WA

Clive & Wendy Napier	(14)	WA
National Park & Wildlife Service	(35)	NSW
Ken Newbey	(152)	WA
Flora Nichols	(4)	WA
Alan Notley	(1)	WA
Wynne O'Brien	(3)	NSW
Michael O'Keefe	(2)	NSW
Rosemary Opala & Evelyn Peacock	(153)	QLD
Ray Page	(6)	NSW
Andrew Paget	(2)	VIC
Karin & Michael Palmer	(29)	WA
Sue Palmer	(7)	WA
Rae & Rodney Papenfus	(24)	WA
Margaret Parris	(27)	NSW
I.G. Paterson	(1)	QLD
Jim Paterson	(7)	TAS
Marlene Paterson	(13)	WA
Sue Patrick	(3)	WA
Grant Pearson	(8)	WA
Peter Penney	(1)	SA
Jan & Don Penny	(6)	QLD
Patricia Perrin	(5)	NSW
Maria Pesavento	(6)	QLD
Tony Phillips	(5)	VIC
Frank & Joy Phillips	(3)	WA
E. Pickering	(39)	ACT
Margaret Pieroni	(2)	WA
Pat Plozza	(1)	WA
James Plummer	(3)	SA
Terry Powell	(6)	WA
Max & June Preece	(20)	VIC
Lois Prictor	(21)	VIC
Peter & Ann Radke	(44)	QLD
Beryl Rainer	(3)	QLD
Frances Reay	(34)	SA
James Redmond	(6)	ACT
Stephanie Renwick	(4)	VIC
Barry Rick	(3)	WA
Paul & Thelma Roach	(197)	SA
Martin Roberts	(13)	WA
Jill Roberts	(9)	TAS
William Roberts	(6)	NSW
Claire Robertson	(1)	VIC
Victor Robertson	(21)	QLD
Ed Robinson	(17)	WA
Jim & Pearl Rogers	(5)	WA
Gary Rowe	(2)	NSW
Alf & Esme Salkin	(1205)	VIC
Pam Sanderson	(3)	WA
Paulette Savage	(4)	WA
George Schmidt	(66)	WA
Mr & Mrs J. Scott	(2)	WA
R. Scott	(1)	WA
Hugh Seeds	(5)	WA
Janet Semmens	(24)	VIC
Nick Shepherd	(1)	NSW
Mary Sherwood	(18)	WA
R.K. Shoosmith	(15)	WA
Dave & Dot Sieber	(117)	WA
John Simpson	(2)	QLD
Marion & John Simmons	(4)	TAS
Ed Smidt	(10)	WA
Basil & Mary Smith	(17)	WA
Louise Smith	(5)	WA
Margaret Smith	(2)	NSW
Peter Smith	(7)	WA
Ralph Smith	(4)	WA
Andrew & Lois Sourry	(70)	NSW
Society for Growing Australian Plants		
Central Highlands Group	(25)	VIC
Geraldton Branch	(5)	WA
Maryborough Branch	(11)	QLD
Murray Bridge Group	(2)	SA
Southern Eyre Group	(7)	SA
Paul Spratt	(2)	NSW
Barb St John	(49)	SA
Don Stallard	(5)	NT
Helen & Mick Statham	(22)	TAS
Jim Steenson	(6)	NSW
Norman & Jane Stevens	(11)	WA
Trevor Stoneman	(8)	WA
Philip Strong	(27)	NSW
Margery & Graham Stutchbury	(26)	QLD
Donna & Peter Summers	(23)	WA
Ted Summers	(3)	SA
Betty & Jim Swainson	(30)	WA
Katie Syme	(4)	WA
Janet Tallon	(2)	QLD
Tony Tapper	(80)	WA
Anne Taylor	(100)	WA
Helen Taylor	(31)	WA
Paul Taylor	(21)	QLD
Trevor Thomas	(40)	SA
Lyndal Thorburn	(2)	ACT
Rosemary & Neville Thorn	(17)	WA
Ross Thompson	(7)	NSW
Hartley Tobin	(9)	VIC
David & Helen Tranter	(11)	TAS
Bill Turner	(7)	NSW
Doug Twaits	(45)	VIC
R. Vallak	(11)	ACT
Stephen Van Leeuwen	(26)	WA
Mr & Mrs Van Rijnswoud	(1)	WA
Graham Velterop	(71)	WA
Ken Veness	(10)	NSW
Honor Venning	(8)	WA
Colleen Vigar	(16)	SA
Graham Vincent	(14)	SA
Don Voigt	(17)	WA
W.A. Dept Conservation & Land Management, Dwellingup	(185)	WA
Marjorie & Len Wall	(125)	TAS
Ken Wallace	(2)	WA
Paul & Robin Walker	(5)	NSW
Brian Walters	(153)	NSW

Mr & Mrs B. Ward	(10)	NSW
Peter & Carolyn Wardle	(12)	WA
Ross Weaver	(5)	WA
Rauleigh Webb	(29)	WA
Stan Webster	(19)	WA
Bob Weston	(11)	ACT
A. Wheeler	(7)	ACT
Rob Whelan	(6)	NSW
Jim Whelan	(2)	VIC
John & Joyce White	(14)	WA
Beth Williams	(124)	NSW
Don Williams	(31)	WA
Keith Williams	(5)	WA
Pip Williams	(5)	SA
Eleanor Williams	(3)	WA
Eric Williamson	(30)	QLD
Jim Williamson	(16)	WA
Alan Willis	(3)	VIC
Mae Willmot	(5)	NSW
Ian Wilson	(49)	WA
Scott Wilson	(4)	WA
Bob Wiltshire	(2)	WA
Meri Wood	(2)	NSW
Fay Woods	(3)	NSW
Maxine & Jack Woodhouse	(4)	WA
Tony Woolford	(40)	TAS
John Wrigley	(4)	NSW
Person unknown (ident. code CBA)	(27)	NSW

1. HOW THE ATLAS PROJECT WAS ORGANISED

This chapter provides information on how The Banksia Atlas originated and how it was organised. Reasonable attention to detail is provided so that the lessons learnt in this first national plant atlas can benefit the development of similar projects in the future.

BACKGROUND

In a country the size of Australia with only a relatively small number of professional botanists located mainly in capital cities, the potential for volunteers to assist in the collection of scientific data is enormous. The Banksia Atlas was a three-year project (1984-1986) which sought to explore this potential.

Previously, there had been plant recording schemes at the State level, but as an Australia-wide mapping project relying entirely on volunteer contributors, The Banksia Atlas was the first of its kind.

This approach had been used successfully in other countries, particularly the United Kingdom, whose Atlas of the British Flora (Perring & Walters 1962) was published on the basis of records contributed by amateur plant enthusiasts throughout the country. In Australia, the Royal Australasian Ornithologists' Union (RAOU) enlisted 3 000 volunteers over a five-year period (1976-81) to provide records for The Atlas of Australian Birds (Blakers *et al.* 1984). Its success prompted botanists to consider whether the same thing could be tried with plants.

Between 1979 and 1983 S.D. Hopper from the Western Australian Department of Fisheries and Wildlife (now the Department of Conservation and Land Management) selected kangaroo paws (*Anigozanthos*) and orchids as the subjects for two pilot projects to test the feasibility of computer-based flora mapping and of using volunteers for field recording. A special computer program, FLORAPLOT, was developed, which allowed point plotting of locations according to a number of specified criteria, e.g. by species, flowering time, geographic area.

During 1983 discussions took place between the Australian Biological Resources Study (ABRS) and the Department of Fisheries and Wildlife over setting up the first national plant atlas. It was decided to focus on one high-profile group of plants and the unanimous choice was banksias. At that time the distribution of banksias was known largely on the basis of herbarium specimens, some dating back to 1770, which were plotted on maps in a revision of *Banksia* by George (1981). Two other publications, A Field Guide to Banksias by I. Holliday & G. Watton (1975) and The Banksia Book by A.S. George (1984), had shown the general distribution of each species as blacked-in areas but these maps did not show all isolated pockets or gaps in distribution. It was considered useful to see how the present situation compared and to see what degree of coverage could be achieved by a large number of plant enthusiasts.

Approval for The Banksia Atlas was given and the project began in February 1984. Joint funding by ABRS and the Department of Conservation and Land Management allowed the appointment of a full-time national coordinator based in Perth, Western Australia. Five months to design, test and produce the final recording kit allowed the project to go public in July 1984. Two years later (31 August

1986), the recording phase ended. 1 017 recording kits had been sent out and approximately 13 000 record sheets received, representing some 25 000 records of taxa.

ADMINISTRATION

The national coordinator (Anne Taylor) was based at the Western Australian Wildlife Research Centre. It was appropriate to base The Banksia Atlas in Western Australia since 57 of the 72 then known species occur in that State. Moreover, staff at the Wildlife Research Centre had six years' experience in running pilot plant atlas projects on kangaroo paws and orchids.

The national coordinator was responsible for the day to day running of the project. Recruiting and encouraging recorders, compiling and sending out recording kits, organising workshops and field trips, giving talks to interested groups, answering queries, checking and double checking record sheets, producing newsletters and interim maps, doing banksia field work and filling in record sheets, and a mountain of clerical work all fell within the domain of the national coordinator. Throughout the project, administration remained largely centralised. Although producing a heavy workload, the undoubted benefit was an overall standardisation of procedures. Potential problems relating to variable interpretation of requirements were reduced to a minimum.

State Coordinators

A volunteer coordinator was recruited from each State and Territory excepting Western Australia and the Northern Territory which were covered by the national coordinator. This allowed the initial checking of record sheets to be carried out within that State. A State coordinator's other duties included help with banksia identification or problems with record sheets, organisation of field trips, giving talks to interested groups, identifying areas in need of special coverage and directing volunteer's recording efforts accordingly. In practice, most State coordinators found the time taken to edit record sheets and advise on recording problems was as much as they could cope with. Several volunteers commented that they would have liked their State coordinator to take a more active role particularly in organising field trips. Whilst State coordinators agreed that this would be good in theory, all but one said that they hadn't had time to do it.

In order to check the locations specified on record sheets, State coordinators required detailed maps of their State. Some already had a comprehensive collection and were prepared to purchase those they needed to fill in the gaps. Others had no maps at all. For them to have purchased an entire collection for their State would have been prohibitively expensive. By supplying them with a few `key' maps of banksia-rich areas, the problem was partly overcome. When record sheets came in from other areas they used the Reader's Digest Atlas of Australia (scale 1:1 000 000). Although adequate for checking approximate locations, it was not good enough when people were recording to the nearest 30" (900 m) or 10" (300 m). Inability to accurately check a location first time round (i.e. at State coordinator level) meant that the national coordinator had to write back to the original recorder for confirmation. This increased the administrative load considerably. Access to detailed maps is an essential part of coordinating a project like The Banksia Atlas. Future projects should always allow for sufficient funds to purchase detailed maps whenever they are needed.

Personal Communication and Feedback

Personal communication was given a high priority throughout the project. All letters and queries were answered personally and "standardised" replies avoided. When a volunteer sent in their first record sheets, a letter of thanks was always sent. If necessary, recording errors were pointed out and advice given on correct recording techniques. If they had sufficient time, State coordinators took on this additional task.

Towards the end of the recording phase a questionnaire was sent to all registered volunteers with the aim of obtaining feedback on the project. In their returns, volunteers expressed a very high level of satisfaction with the communication they had had with both national (99%) and State (93%) coordinators. Many also commented on the quality of the personal attention they had received, saying that it had been a strong inducement to maintain their recording activities.

Constructive, positive feedback from coordinators is an essential part of a project like *The Banksia Atlas*. Undoubtedly there are some "hard core" volunteers who would continue recording with or without any feedback but for the majority the more feedback provided the greater their recording activities.

Volume of clerical work

The amount of clerical work generated throughout the project was very large and should not be underestimated for future projects. Without the help of temporary part-time assistants the project was in danger of floundering under a mountain of paperwork. Successive attempts were made to attract voluntary help to assist at office headquarters, but few offers were forthcoming. Compared with other Atlasses and their administration, this reluctance was surprising. It was probably due to a number of factors including (i) the location of the Wildlife Research Centre some way out of Perth, (ii) association of the project with a government department and therefore a feeling that paid employment should be forthcoming, (iii) the lack of an already existing group to identify with (in contrast to *The Bird Atlas* and RAOU), (iv) the reluctance of most plant enthusiasts to give up more than a certain amount of their time to banksias, when other plants were also demanding their attention.

RECRUITING

For the first six months of the project (July - December 1984) the main recruiting effort was in Western Australia. From January 1985 to September 1985, attention was focused on the eastern States. All active recruiting ceased in September 1985. A few volunteers continued to ask for and be sent kits up until June 1986, two months before the recording phase ended.

To begin with, anyone expressing interest in the project was sent a recording kit. However, it was soon realised that many enquiries were based on fairly inaccurate information about the project. Also the number of volunteers joining the project was outstripping the national coordinator's ability to cope with the paperwork. Recruitment was accordingly scaled down and became more selective. It was decided to first send an information sheet about the *Atlas* to any interested person, followed by a recording kit if the enquirer asked for one. Such an approach is recommended for any future projects.

The main method of recruitment was to organise workshop/seminars and field trips among special interest groups, e.g. Society for Growing Australian Plants, Wildflower groups, Naturalists' Clubs, "Friends" of National Parks, etc. This proved a very effective method of advertising the project. Presented with the immediacy of the task and its relevance to the local situation, volunteers responded with enthusiasm. Personal contacts were established between volunteers and the coordinator and between volunteers within a group. The latter could provide a social framework for future recording activities, and could also act as a "self-help" group if problems arose. The emphasis in both talks and field trips was on training volunteers to become competent recorders. This meant learning to identify the various local banksias as well as practising recording techniques. Taking part in such training sessions gave volunteers a greater sense of confidence in their recording abilities.

During 1985, the national coordinator spent almost four months away from Western Australia recruiting and meeting volunteers in all other States and the Australian Capital Territory. By the end of the project a total of 36 workshop/seminars and 27 field trips had been held with an estimated 1 000 people attending. Additional meetings/field trips etc. had been organised by at least one of the State coordinators.

Other methods of advertising included the production of an audio visual, articles in magazines/journals etc., newspaper reports, and talks on radio. A few direct approaches to certain people were also made, this method always proving successful.

In the questionnaire, volunteers were asked how they had heard about the Atlas. Results are summarised in Table 1. It should be noted that volunteers could tick more than one box and therefore percentages add up to more than 100. The results clearly reflect the way in which the project was publicised. In second place, "word of mouth" suggests that a project will largely advertise itself where people see clear benefits, and also enjoy taking part in the work.

Table 1. How volunteers heard about the Atlas.

Publicity	Occurrences	
Magazine	43	(10%)
Local newspaper	46	(11%)
National newspaper	24	(6%)
Radio	12	(3%)
Word of mouth	99	(24%)
Professional network	46	(11%)
Special interest group	215	(52%)
Other	11	(3%)
Total no. of returns	410	(100%)

Advertising the Atlas was in no way exhaustive because, as already explained, saturation of the national coordinator's resources was reached at an early stage. Also, it was considered more important to maintain personal contact with existing recorders, to train them to a high standard of recording, and to provide them with as much feedback as possible so that they could carry on their recording activities most efficiently, than to give greater attention to attracting more people to the project. With hindsight, this decision proved correct.

It is also clear that not enough recruiting effort was directed at the farming community. Those farming families who did contribute provided records which were invaluable, coming largely from privately owned areas of natural vegetation, which are generally inaccessible to passing recorders. Articles were planned for relevant farming magazines e.g. The Countryman, The Western Farmer, and also visits to Country Women's Association groups and country schools were intended, but none eventuated due to the lack of time and resources.

A small number of school children expressed interest in the project and were sent recording kits. Their initial enthusiasm, however, soon disappeared. It was evident that the record sheet was too complex for most of them. Also they required the support and enthusiasm of a school teacher or interested parent to assist with their recording. It was beyond the resources of the national coordinator to provide the frequent reminding and reinforcement that children often require.

Several professional botanists or biologists showed interest in the project and were sent recording kits. However, the number of completed record sheets they submitted was very low, most expressing that lack of time prevented them contributing to the project. However, they did serve to advertise the project, e.g. among students and colleagues.

THE RECORDING KIT

The following components made up the recording kit. They are illustrated in Figs 1-5.

- An introductory letter (Fig.1).
- An **Instruction Booklet** and **Supplementary Field Guide** (Fig. 2). Detailed instructions and worked examples showing how to fill in the record sheets formed the first half of the booklet. The second half comprised a field guide to supplement and update A Field Guide to Banksias by I. Holliday and G. Watton. At the start of the project this was the only identification text available. A recent revision of banksias (George, 1981) meant that, by itself, this text was no longer adequate.
- A pad of 100 computer-based sight record sheets with all information coded alphabetically or numerically (Figs 3 & 4).
- A pocket-sized field notebook designed to provide reminders of data to be noted in the field for later transferral to sight record sheets (Fig. 5).
- For W.A. recorders only, a 1:1 000 000, scale map showing National Parks and Nature Reserves, and lines of latitude and longitude. Similar maps were not available for the other States.
- Copies of all past and present newsletters.

DATA COLLECTION

Only banksias growing in the wild were to be recorded. Generally, the status (wild or cultivated) of any banksias was obvious. Whenever there was any doubt, inquiries amongst local people usually solved the problem.

Any contributor could send in record sheets from any part of Australia. Some began by recording the banksias of their home area, but many did not, apparently thinking that they were either not relevant or interesting enough, or that they must have already been recorded. For these people, field trips to unknown territory was what was required. Many non-contributors apparently had the same attitude. In the returned questionnaires, "unable to get away on field trips" was the main reason specified for not contributing to the project. This attitude towards recording local plants may occur in future projects. Every effort should be made to overcome it, as local people have

BANKSIA ATLAS

A National Computer-based Flora Survey

Anne Taylor (Co-ordinator)
Wildlife Research Centre
P.O. Box 51
Wanneroo, W.A. 6065
Telephone (09) 405 1555

Many thanks for offering to participate in the Banksia Atlas.

You will find enclosed the following items:

1. A pad of Sight Record Sheets (SRS)
2. A Field Notebook (please note errata)
3. Instruction Booklet and Supplementary Field Guide

Please read pages 1-8 of the Instruction Booklet carefully. This describes the background to the Banksia Atlas and outlines general recording instructions.

To complete your Volunteer's Kit you will also need a copy of one of the following field guides:

"The Bankia Book" by Alex George, publ. Kangaroo Press. Available as from October 1984.

"A Field Guide to Banksias" by Ivan Holliday and Geoffrey Watton, publ. Rigby 1975. Now out of print, but should be available in some bookshops. Please note carefully the amendments to this book listed on page 27 of the Banksia Atlas Instruction Booklet.

Now its largely up to you! Although there will be group field trips to go on, the bulk of information is likely to come from individuals recording banksias in the areas around their homes and from their own weekend/holiday excursions.

Only records of naturally occurring banksias are required. Ignore any plants that have been deliberately planted or are growing under cultivation. Please record both the types you regard as "common" as well as the more unusual. We want every record that you can send, nothing will be wasted.

If you have any previously recorded banksia information that includes at least an accurate locality, date of observation and species present, it may also be entered onto a sight record sheets(s) and sent in.

-2-

Additional Tips on Field Recording

1. If you are filling in SRS directly (i.e. in the field) I suggest you use a pencil. On returning home you can go over the records again in biro, using the carbon paper provided if you wish to keep a duplicate copy for yourself.

2. Take a compass - both for your own safety and also for greater accuracy when filling in the SRS.

3. If you have access to larger scale maps (eg, 1:100 000 series) than the one provided, please use them - the larger the scale of the map, the more accurate is your locality information likely to be.

N.B. Completed sight record sheets should be sent to your State Co-ordinator, (see below). If you live in a state which does not yet have a co-ordinator please send record sheets directly to myself.

Please contact me if you have any queries or need additional materials.

Wishing you good Banksia hunting!

Yours sincerely

Anne Taylor

STATE CO-ORDINATORS

N.S.W.	**VICTORIA**	**NORTHERN TERRITORY**
Paul Spratt 10 Stelling Avenue KANWAL 2259	Alf Salkin 38 Pinewood Drive MT WAVERLEY 3149	
SOUTH AUSTRALIA	**TASMANIA**	**QUEENSLAND**
Leslie Gray 19 New York Road ABERFOYLE PARK 5159	Bruce Champion PO Box 1 ROSNY PARK 7018	Paul Taylor 7 Miles Street BUNDABERG 4670
WESTERN AUSTRALIA	**A.C.T**	
Anne Taylor (as above)	Alex George 13 Hawkesbury Crescent FARRER 2607	

Fig. 1 Introductory letter sent to intending contributors to *The Banksia Atlas*.

BANKSIA BLECHNIFOLIA F. Muell.

From the Latin, *folium*, a leaf, and *Blechnum*, a genus of ferns, in reference to the shape of the leaves.

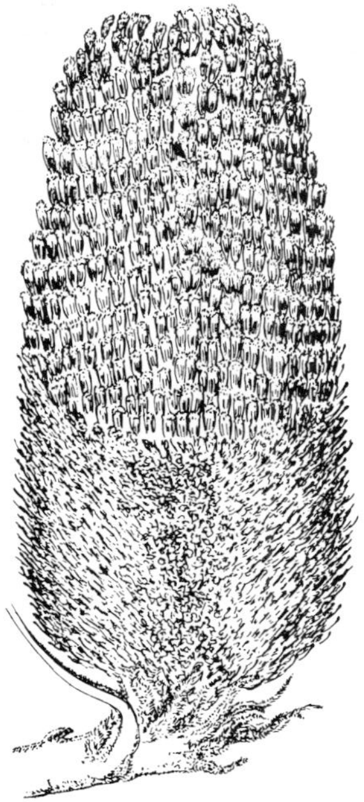

Flower spike (x½)

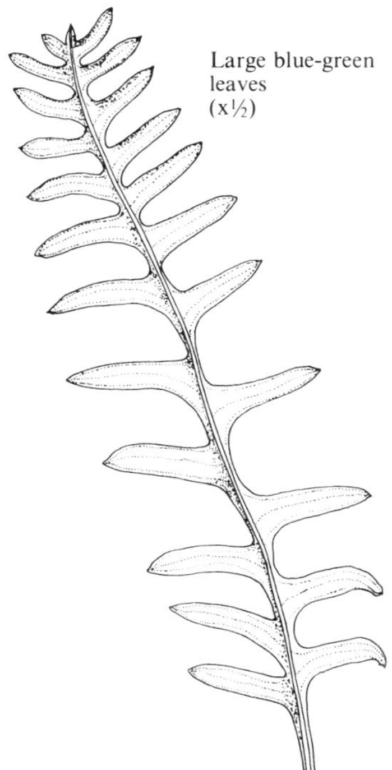

Large blue-green leaves (x½)

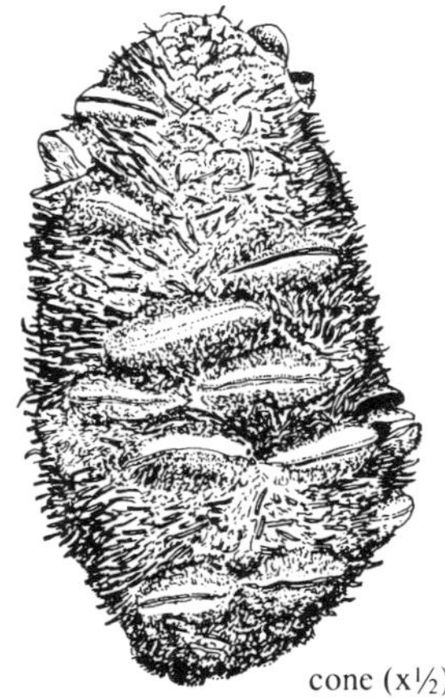

cone (x½)

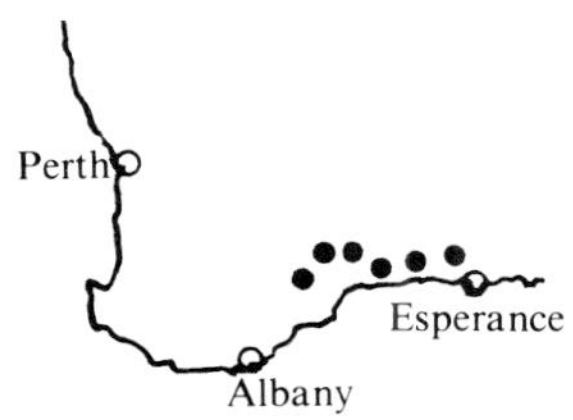

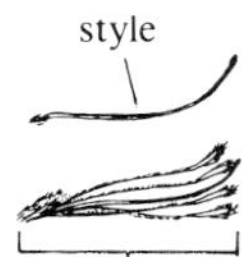

A single flower (x½)

Distinctive characteristics: Easily distinguished from the other prostrate species by its large blue green leaves divided almost to the mid-rib by entire lobes 2-5 cm in length. Flower colour is also different, being dominantly reddish-pink, turning pale brown then grey after flowering.

Other characteristics: A low shrub with horizontal branches at ground level or just below. Young growth is bronze-red and velvety. Flower spikes are 9-19 cm long and borne erect at the tips of the stems. The follicles on the fruiting cone are often almost obscured by the persistent old flowers.

Flowering period: Late September - mid November.

Distribution and habitat: S.W. Western Australia between Jerramungup, Gibson and Lake King, not recorded within 10 km of the coast. Occurs in sands, amongst small shrubs and mallee eucalypts.

Direction from Nearest Place

The direction is FROM the nearest named place TO the record locality.

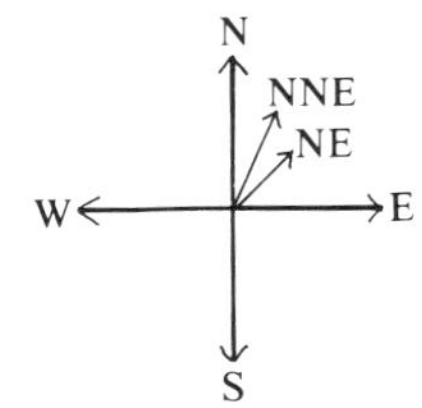

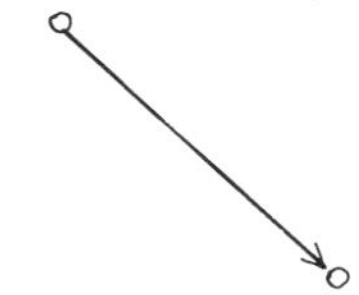

When entering this information, the standard form for the sixteen major compass points should be used. They are N, NNE, NE, ENE, E, ESE, SE, SSE, S, SSW, SW, WSW, W, WNW, NW, NNW and should be entered from left to right as follows:

N   NNE   NE

Further Details of Location

Write any additional information which would assist in relocating the Record Locality. This may be a plain word description e.g.

LO JUNCTION OF YORK ROAD AND INKPEN R
OAD
LO SOUTH SIDE OF RESERVE NEAR GRAVEL
PIT

If distances are included, it is important to distinguish between straight line distance, and distance along a road (or river). To avoid confusion the following rules regarding recording of distance should be followed. Where straight line distance is used, the distance should be entered first, followed by direction, e.g.

LO 31KM N OF ENEABBA WHERE SAMPSON RD
APPROACHES SMITH RD

Where distance along a road or river is used, the direction should be entered first, followed by distance, then the road or river name. In this case the use of the word "along" is strongly recommended, e.g.

LO SE 11KM FROM PERTH ALONG BROOKTON
HIGHWAY

When measuring distance from any town or village, the post office is assumed to be the starting point unless otherwise stated.

Fig. 2 Sample pages from the instruction booklet and supplementary field guide provided to *Atlas* contributors.

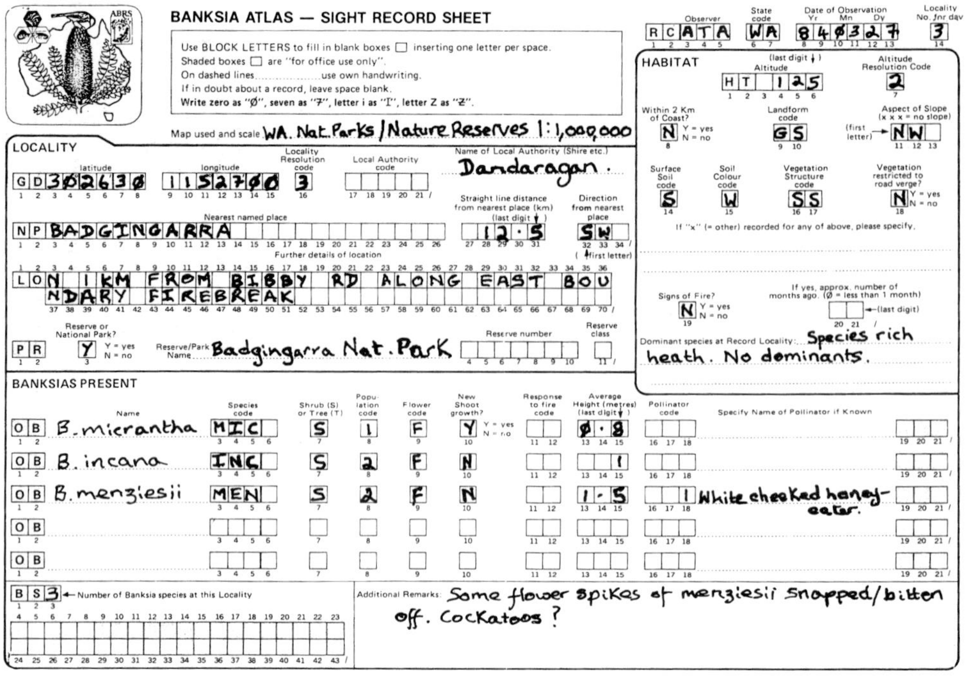

BANKSIA ATLAS — SIGHT RECORD SHEET

Use BLOCK LETTERS to fill in blank boxes ☐ inserting one letter per space.
Shaded boxes ☐ are "for office use only".
On dashed lines.................use own handwriting.
If in doubt about a record, leave space blank.
Write zero as "Ø", seven as "7", letter i as "I", letter Z as "Z".

Map used and scale WA. Nat. Parks / Nature Reserves 1:1,000,000

Observer RC ATA | State code WA | Date of Observation Yr Mn Dy 84 Ø3 27 | Locality No. for day 3

LOCALITY

latitude GD 3Ø2630 | longitude 1152700 | Locality Resolution code 3 | Local Authority code | Name of Local Authority (Shire etc.) Dandaragan.

Nearest named place NP BADGINGARRA | Straight line distance from nearest place (km) (last digit) 12·5 | Direction from nearest place SW (first letter)

Further details of location: LON 1KM FROM BIBBY RD ALONG EAST BOUNDARY FIREBREAK

Reserve or National Park? PR Y (Y = yes N = no) | Reserve/Park Name Badgingarra Nat. Park | Reserve number | Reserve class

HABITAT

Altitude (last digit) HT 125 | Altitude Resolution Code 2

Within 2 Km of Coast? N (Y = yes N = no) | Landform code GS | Aspect of Slope (x x x = no slope) (first letter) NW

Surface Soil code S | Soil Colour code W | Vegetation Structure code SS | Vegetation restricted to road verge? N (Y = yes N = no)

If "x" (= other) recorded for any of above, please specify.

Signs of Fire? N (Y = yes N = no) | If yes, approx. number of months ago. (Ø = less than 1 month) (last digit)

Dominant species at Record Locality: Species rich heath. No dominants.

BANKSIAS PRESENT

	Name	Species code	Shrub (S) or Tree (T)	Population code	Flower code	New Shoot growth?	Response to fire code	Average Height (metres) (last digit)	Pollinator code	Specify Name of Pollinator if Known
OB	B. micrantha	MIC	S	1	F	Y		Ø·8		
OB	B. incana	INC	S	2	F	N		1		
OB	B. menziesii	MEN	S	2	F	N		1·5	1	White cheeked honey-eater.
OB										
OB										

BS 3 ← Number of Banksia species at this Locality

Additional Remarks: Some flower spikes of menziesii snapped/bitten off. Cockatoos?

Fig. 3 A completed Sight Record Sheet

State Codes

AC = ACT
NS = NSW
NT = N. Terr.
TA = Tas.
QU = Qld.
SA = Sth. Aust.
VI = Vic.
WA = West. Aust.

Locality Resolution Codes

1 = 1'' 2 = 10'' 3 = 30''
4 = 1' 5 = 10' 6 = 30'
7 = 1°

Altitude Resolution Codes

1 = 5m 2 = 20m
3 = 50m 4 = 100m
5 = 250m 6 = 500m

Landform Codes

HT = Hill/mountain top
SS = Steep slope (> 20°)
GS = Gradual slope (<20°)
VA = Valley bottom or swale
RK = Rock outcrop
FL = Flat
SW = Seasonally wet swamp
LE = Lake edge
RB = Riverbank or dry bed
XX = Other

Soil Type Codes

S = sandy
L = loamy
C = clayish
G = gravel
R = rocky
T = laterite
P = peaty
N = sand over laterite
V = sand over rock
X = other

Soil Colour Codes

B = brown
G = grey
K = black
R = red
W = white
Y = yellow
N = orange
X = other

Vegetation Structure Codes

GR = Grassland
SS = Small shrubs (<2m)
LS = Large shrubs (>2m)
MA = Mallee
WL = Woodland
FT = Forest
RF = Rainforest
CF = Cleared farmland/plantations, orchards etc.
XX = Other

Population Size Codes

1 = 1 - 10
2 = 10 - 100
3 = more than 100

Flower Codes

B = majority in bud
F = majority in full flower
A = mainly old flowers and/or immature fruiting cones
N = no flowers

Response to Fire Codes

K = Killed by fire — no seedlings present
S = Killed by fire — seedlings present
L = Not killed — resprouting from below ground
T = Not killed — resprouting from trunk

Pollinator Codes

1 = Bird
2 = Mammal
3 = Butterflys/Moths (Lepidoptera)
4 = Bees, Wasps, Ants (Hymenoptera)
5 = Beetles (Coleoptera)
6 = Flies (Diptera)
X = Other
N/ = No pollinator observed

Species Codes * First 3 letters of a species' name = species code EXCEPT B.candolleana = CAD , B.canei = CAE

ACU	aculeata	GRA	grandis	OCC	occidentalis
AEM	aemula	GRO	grossa	ORE	oreophila
ASH	ashbyi	HOO	hookerana	ORN	ornata
ATT	attenuata	ILL	ilicifolia	PAL	paludosa
AUD	audax	INC	incana	PET	petiolaris
BAU	baueri	INT	integrifolia	PIL	pilostylis
BAX	baxteri	INTI	" var. integrifolia	PLA	plagiocarpa
BEN	benthamiana	INTC	" var. compar	PRA	praemorsa
BLE	blechnifolia	INTA	" var. aquilonia	PRI	prionotes
BRO	brownii	LAE	laevigata	PUL	pulchella
BUR	burdettii	LAEL	" subsp. laevigata	QUE	quercifolia
CAL	caleyi	LAEF	" subsp. fuscolutea	REP	repens
CAD	candolleana*	LAN	lanata	ROB	robur
CAE	canei*	LAR	laricina	SAX	saxicola
CHA	chamaephyton	LEM	lemanniana	SCA	scabrella
COC	coccinea	LEP	leptophylla	SCE	sceptrum
CON	conferta	LIN	lindleyana	SER	serrata
CONC	" var. conferta	LIT	littoralis	SOL	solandri
CONP	" var. penicillata	LITL	" var. littoralis	SPE	speciosa
CUN	cuneata	LITS	" var. seminuda	SPH	sphaerocarpa
DEN	dentata	LUL	lullfitzii	SPHS	" var. sphaerocarpa
DRY	dryandroides	MAR	marginata	SPHC	" var. caesia
ELD	elderana	MED	media	SPHD	" var. dolichostyla
ELE	elegans	MEI	meisneri	SPI	spinulosa
ERI	ericifolia	MEIM	" var. meisneri	SPIS	" var. spinulosa
ERIE	" var. ericifolia	MEIA	" var. ascendens	SPIO	" var. collina
ERIM	" var. macrantha	MEN	menziesii	SPIU	" var. cunninghamii
GAR	gardneri	MIC	micrantha	TEL	telmatiaea
GARG	" var. gardneri	NUT	nutans	TRI	tricuspis
GARB	" var. brevidentata	NUTN	" var. nutans	VER	verticillata
GARH	" var. hiemalis	NUTC	" var. cernuella	VIC	victoriae
GOO	goodii	OBL	oblongifolia	VIO	violacea

Fig. 4 Computer codes for *Banksia* species, habitat and other data as they appeared on the reverse side of each Sight Record Sheet.

the best knowledge of their local flora and are best suited to record it.

Duplication of recording for any particular area was an initial concern expressed by many volunteers but in fact it rarely happened. In any case, a duplicate record would more than likely have occurred at a different time of year and would provide new data on aspects such as flowering, new shoot growth and pollinators.

Interim Maps and Newsletters

Two sets of interim distribution maps were sent to all contributors in July 1985 and December 1985 and one extra set was sent in October, 1985, to contributors from the eastern States. These three sets of maps served to illustrate areas and species that were already well covered and so indicated to volunteers where their future efforts were most needed. The extra set to the eastern States showed localities known from herbarium specimens as well as a prediction of potential areas based on the BIOCLIM method (Busby, 1984, Nix, 1986). Their other important role was to demonstrate that the project was producing useful results. Fig. 6 illustrates *Banksia benthamiana* maps as they appeared in *Interim Maps I* and *II*.

More detailed distribution maps for each State were sent to the relevant State coordinator and to any special interest groups who had been visited by the national coordinator. Volunteers were also advised that detailed distribution maps of any specified area were available on request. Towards the end of the project, Western Australia was studied in detail to determine which areas, if any, remained poorly covered. Areas so identified were given special attention with letters written to both registered and prospective volunteers in that area and copies of the relevant distribution maps sent out. A few keen Perth volunteers also made lightning visits to such areas. Some other States received similar treatment, others not so, depending on the

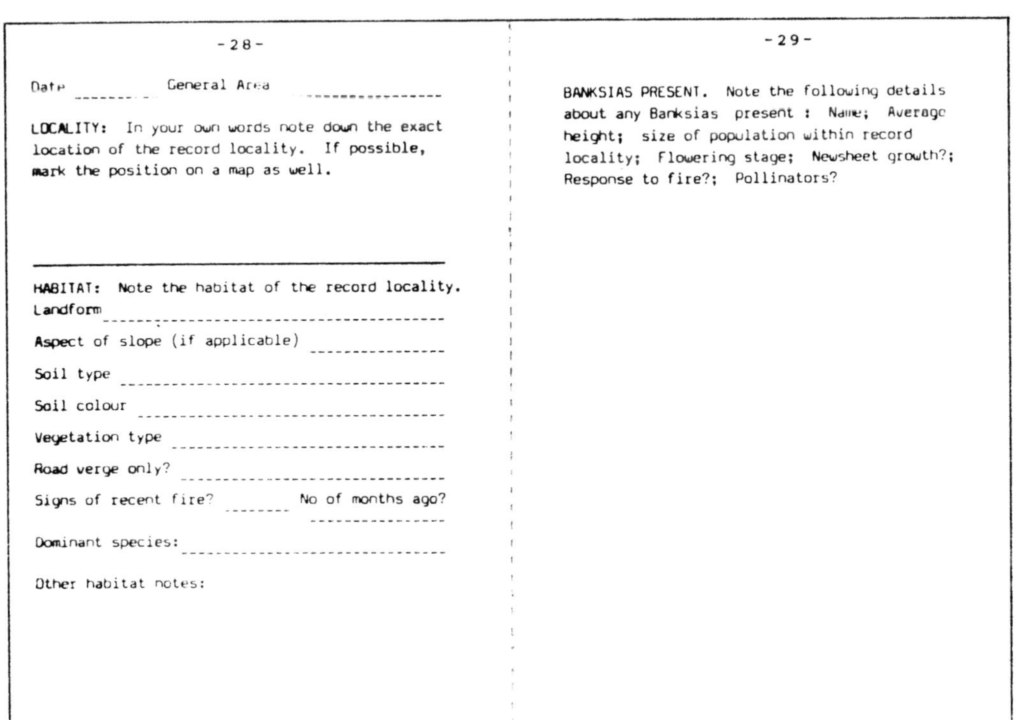

-28-

Date ________ General Area ________________

LOCALITY: In your own words note down the exact location of the record locality. If possible, mark the position on a map as well.

HABITAT: Note the habitat of the record locality.

Landform ________________

Aspect of slope (if applicable) ________________

Soil type ________________

Soil colour ________________

Vegetation type ________________

Road verge only? ________________

Signs of recent fire? ______ No of months ago? ______

Dominant species: ________________

Other habitat notes:

-29-

BANKSIAS PRESENT. Note the following details about any Banksias present : Name; Average height; size of population within record locality; Flowering stage; Newsheet growth?; Response to fire?; Pollinators?

Fig. 5 Sample pages from the field notebook provided to Atlas contributors.

availability of the State coordinator's time.

Whereas distribution maps were sent only to volunteers who had sent in at least one record sheet, newsletters went to all persons who had received a recording kit. It was hoped that continued receipt of the newsletters might stimulate into action those volunteers who had not yet started recording. In a number of cases this did occur, though by the end of the project 60% of those people who had been sent recording kits had not contributed a record sheet. The RAOU had found a similar level of non-participation. In projects of this nature it seems that a fairly high drop-out rate is likely and should be planned for accordingly.

The importance of the five newsletters in the coordination of the project cannot be overemphasised. They acted as a link between recorders all over Australia, informing them of progress and enabling them to learn of the trials and triumphs of their colleagues. They also served simply as a reminder about the Atlas, causing records lying dormant in volunteers' field notebooks to be hastily transferred to record sheets. Following publication of each newsletter there was a noticeable increase in records received.

The newsletters included reports and illustrations of unusual banksias. There were also articles covering identification of difficult species. Forthcoming meetings and field trips were advertised, and a brief account given of the national coordinator's activities during the previous few months. Those parts of the record sheet that were causing particular problems were highlighted and additional instructions provided. Volunteers were encouraged to contribute both articles and illustrations for the newsletter and did so with enthusiasm. They varied from the serious, e.g. "Some notes on leaf variation in the *Banksia spinulosa* complex in N.S.W.", to the not so serious, e.g. Fig. 7.

Use of Sight Record Sheets and Field Notebooks

When recording in the field, volunteers had the option of using record sheets directly or making notes in the field notebook and at a later stage transferring these data onto the record sheets. Most people opted for the latter. Although initially it seemed to be a duplication of effort, in fact it was much easier to use this method than to be continually altering data which had already been entered on the record sheet. There was also the convenience of using a pocket sized notebook in the field compared to a pad of A4 sized record sheets.

Volunteers were asked to retain a carbon copy of every record sheet they submitted and a sheet of carbon paper was included in the kit. Not only could this act as a record of their own achievements, but if a record had to be queried it would simplify the process of investigation.

Use of Maps

There were no set maps which had to be used for working out details of locations. Volunteers could use whatever maps were available to them. This was the only practical option in a country as large as Australia where volunteers could not be expected to purchase a whole range of one particular type of map. Also, although the 1:100 000 Natmap series covered most parts of Australia where banksias occur, it was not readily available throughout the country. When volunteers in country areas found it hard to obtain a certain map, then there was little point in insisting on its usage. In

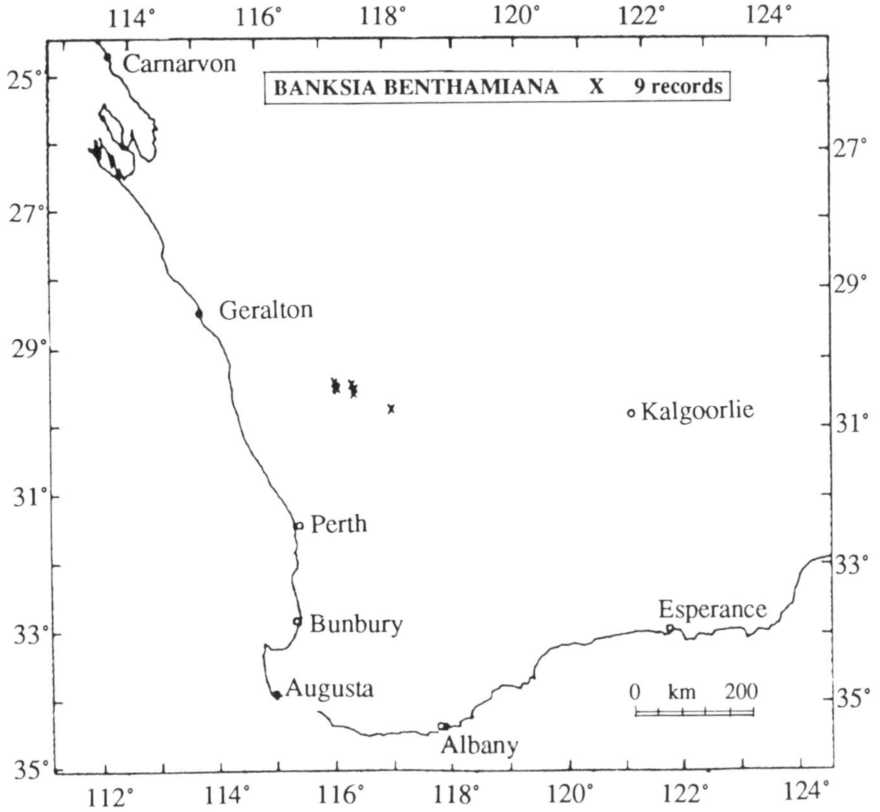

B. benthamiana - 8 of the 9 sight records are around Wubin (Mr and Mrs Smith). The ninth is from near Kulja (H. Hoffman). Herbarium records indicate another population based around Mullewa - volunteers please investigate.

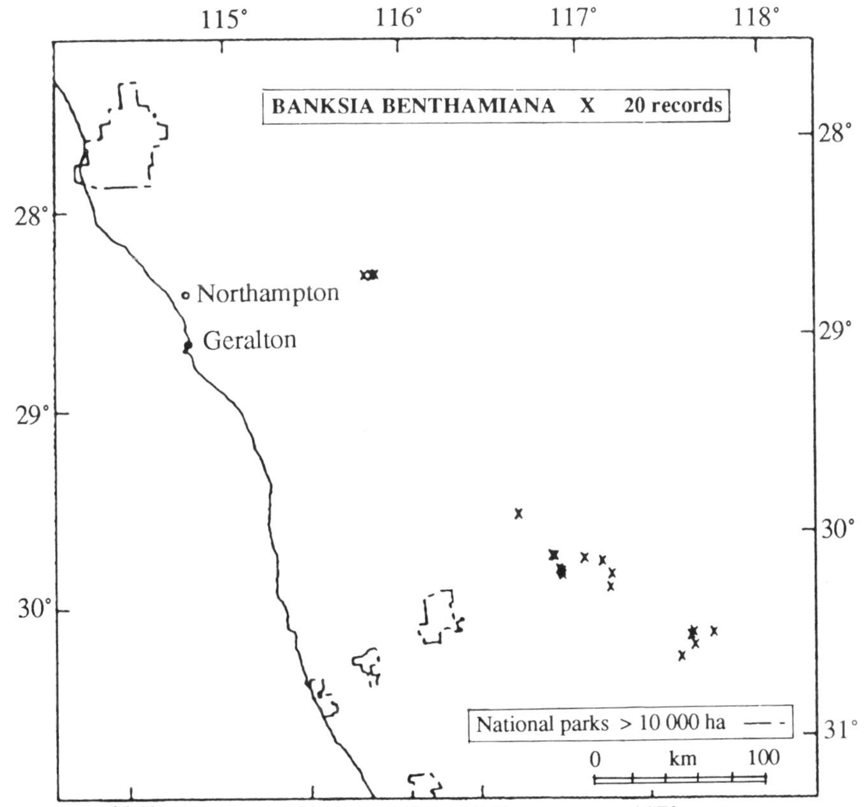

B. benthamiana - the missing population from east of Mullewa has been found by P. Engler. Volunteers please investigate areas between this population and the one around Wubin for new sightings.

Fig. 6 Maps of *B. benthamiana* produced for the *Interim Map Series I* [top] and *II* [bottom].

"It seems to have started since I joined the Banksia Atlas Programme."

Fig. 7 Newsletter contributions. Drawn by Rosemary Opala.

practice, most people used either Natmaps (scale 1:100 000), the Reader's Digest Atlas of Australia (scale 1:1 000 000) or in Western Australia, the map that had been supplied in the recording kit (scale 1:1 000 000). The 1:50 000 Forests Department Maps in Western Australia were also used frequently.

Problems with Map Work and the Record Sheet in General

For many volunteers, map work presented the biggest problem. Errors in recording latitude and longitude were numerous, especially in the early stages of the project. Gradually, volunteers became more proficient and errors were reduced.

Additional difficulties experienced with the record sheet are considered in detail in Chapter 2. Most problems arose from (i) the cut and dried nature of categories on the sheet, which often seemed to need qualifying (not acceptable as the sheet was designed to provide computer input), (ii) the difficulty of establishing a universally agreed and remembered meaning for such terms as "woodland", "forest", "rainforest", "sandy", "loamy" etc. To an extent these problems were resolved during the course of the project and solutions published in newsletters. However, it should be stated that, in the species' reports, not every category meant precisely the same thing to all recorders at every stage of the survey.

Use of Data From Other Surveys

Records from other surveys were occasionally used, if a volunteer was prepared to transfer the data onto the *Atlas* record sheets. There was no time available for the national coordinator to take on this additional task. However, this approach was not encouraged as inaccuracies could easily result from differing interpretations of data.

Identification

Compared with other groups of native plants, most banksias are easy to identify. Volunteers with little previous experience in identifying plants found that, with the aid of a field guide, they could sort out most taxa. A few banksias caused problems throughout the project. These are listed below and are discussed in more detail in Chapter 3.

- *B. spinulosa* - the three varieties often confusing;
- *B. integrifolia* - the three varieties often confusing;
- *B. micrantha* - easily mistaken for *B. sphaerocarpa* var. *sphaerocarpa*;
- *B. sphaerocarpa* var. *sphaerocarpa* - easily mistaken for *B. sphaerocarpa* var. *caesia*;
- *B. gardneri* var. *gardneri* - variability in leaf shape caused confusion, and young leaves of *B. repens* very similar to *B. gardneri* var. *gardneri*;
- *B. gardneri* var. *gardneri* and var. *hiemalis* - impossible to differentiate if flowers absent;
- *B. nutans* - the two varieties often confusing.

Initially there was a marked conservatism amongst volunteers to record anything unusual or different. As confidence in identification and recording abilities grew, this attitude diminished. An advantage of this conservatism was that it undoubtedly reduced the number of incorrect sightings. A disadvantage was that potentially interesting or unusual records were possibly lost through the recorder deciding to "play safe" by classifying the banksia under the most likely taxon. Since for many volunteers this was their first attempt at systematic plant recording, this attitude was to be expected. Projects like The Banksia Atlas help to increase volunteers' confidence in their own abilities. Future atlasses will benefit accordingly.

Since the project was based on sight records, herbarium specimens were generally not required. However, if in doubt about the true nature of a sighting, volunteers were instructed to send a photograph and/or leaf specimen to their State coordinator. A description of growth habit, flower colour, associated vegetation, or any other features of note were to accompany the specimen. Special permits were required by all States when collecting from Crown Land. The names and addresses of relevant authorities were provided.

Whenever an unusual or particularly interesting subject was located, a full voucher specimen (leaves, flower, fruit) was obtained where possible. All voucher specimens were deposited in the appropriate State herbarium or the Australian National Herbarium. Some specimens were confirmed as new taxa in time to be included in The Atlas, namely B. oligantha, B. epica and B. spinulosa var. neoanglica described by A.S. George (1988). Their distributions, habitat and biological summaries appear in The Banksia Atlas, pages 108, 190 and 241. Full illustrations of their leaves, flowers and fruits are on pages 29, 31 and 244. More recent research has led to the recognition of several new varieties and subspecies which were identified too late in The Atlas to be treated as separate taxa. These include B. leptophylla var. melletica (George 1988), B. oblongifolia var. minor (Conran & Clifford 1987), B. seminuda subsp. remanens (Hopper ,in press) and B. occidentalis subsp. formosa (Hopper, in press). Other specimens raise many questions which have not yet been answered. They await the detailed attention of future botanical researchers.

Geographical Sampling Problems

Geographically biassed data collection is an inevitable problem in an atlas project, because volunteer recorders are not uniformly active throughout the survey area. The geographical bias favours the most popular holiday locations and also the domicile of the most active recorders. Much of the south coast of Western Australia was well covered at an early stage, as were popular national parks such as Kalbarri National Park. Sometimes, however, human psychology can work in perverse ways. The Stirling Range National Park in Western Australia is a very popular reserve, especially amongst people interested in the flora of that State. Being something of a botanists' mecca, the national coordinator had assumed that there would be no problems in coverage. Apparently volunteers had made the same assumption and thought that "somebody else must have recorded there". A few months before the recording phase ended it was realised that in fact the area was quite poorly covered. Local volunteers were contacted and successfully redressed the balance.

Coverage

By the end of the project the general coverage achieved throughout Australia was excellent (see Fig. 8). There remained only a few areas where coverage could be considered inadequate. A longer recording phase (more than two years) would have allowed such areas to be given special attention. As it was, they became apparent too late in the project.

DATA CHECKING AND PROCESSING

Both manual checking and computer validation techniques were used to check record sheets. All record sheets went through at least three or four checking processes - State coordinator, national coordinator, computer validation, and visual assessment of maps.

The first sheets submitted by any volunteer were checked thoroughly and, depending on their accuracy, subsequent sheets were checked accordingly. Even with very competent recorders, spot checks continued throughout the project. Obvious errors could be corrected easily, but often it was necessary to refer back to the volunteer who was asked to verify the record. More often than not, it was the locality details that were queried. If it was an unusual sighting, a voucher specimen was requested. If it was impossible for the original recorder to supply this (e.g. records made when a long way from home), attempts were made to contact volunteers in the relevant area to check the doubtful record. If all sources of enquiry failed, the record was omitted. In practice many considerations were taken into account when deciding whether to accept or reject a record, e.g. the recorder's ability, the species or variety of Banksia in question (some are easier to identify than others) and the conformity of the record (the presence of a species in a typical habitat in the right part of the country). The more a coordinator learnt about the abilities of individual recorders, the more efficiently they were able to check their records.

It is important that the time taken between first receiving a sheet and its subsequent checking and processing is as short as possible. Any query back to the original recorder is much easier to answer if it is still fresh in their memory. Also, apparent delays in the processing of record sheets can easily lead to contributors feeling that their efforts are being wasted. However, delays were occasionally inevitable, particularly since the national coordinator spent almost four months of 1985 away on interstate visits. Also State coordinators were working voluntarily in whatever spare time they had available.

Not all State coordinators were equally thorough in checking record sheets. Record sheets from those who were known to be less thorough were given greater attention by the national coordinator. Any problems which State coordinators expressed in checking record sheets were dealt with as they arose. State coordinators would undoubtedly have benefited from a list of "checking guidelines", updated and modified whenever necessary. This would have led to a greater uniformity in checking techniques.

In the office of the national coordinator, the number of record sheets submitted by each observer was recorded on his or her address card, and a separate tally of contributors and their number of completed record sheets also maintained. The current list was published in each newsletter.

The combination of observer code, State, date of recording, and locality number for day (i.e. all the information in the top right hand corner of sheet) provided the means for identifying an individual sheet and was used for all subsequent collating and filing of sheets. The system permitted any particular sheet to be retrieved within seconds.

Every few weeks, information on the sheets was punched onto a magnetic tape which was then loaded onto a Cyber mainframe computer and added to the existing data base. Verification checks produced a punching accuracy close to 100%.

The computer was programmed to indicate major inconsistencies in the data, e.g. impossible dates, lat./long. coordinates not conforming with local authority boundaries, incorrect codes, cases where the 'number of species' written on the sheet did not tally with the actual number of recordings made. The errors that had given rise to these inconsistencies were corrected and the batch of new data combined with that already in the data bank.

In order to plot the distribution maps a subset of data was extracted and stored on a

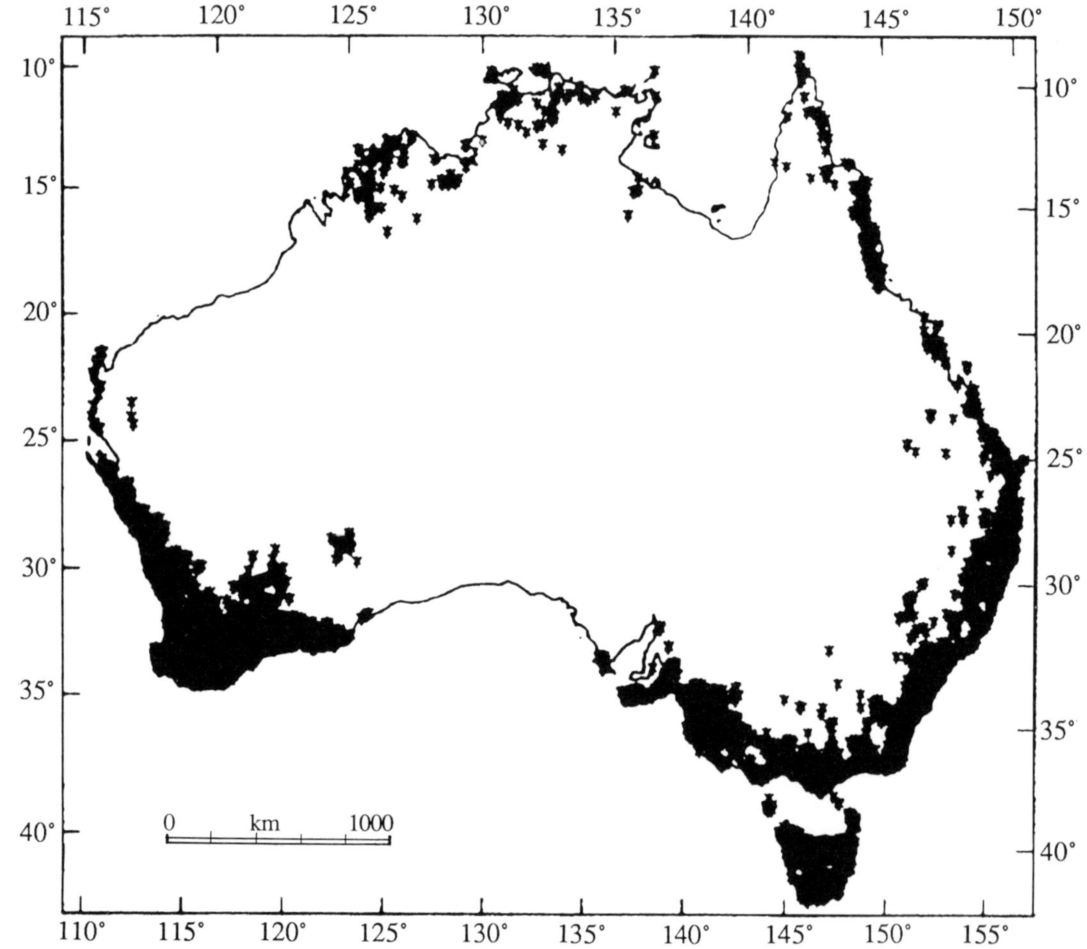

Fig. 8 Distribution of *Banksia* in Australia, based on 24 781 records made for Atlas.

temporary file which was then transferred to a Tektronix 4054 Graphics microcomputer. Indexing of these subset data was by species, enabling the plotting of distribution maps using the FLORAPLOT program.

Once plotted, the maps were scanned for possible errors. Records identified in this way were checked for both species and locality information in the manner already described. Any resultant corrections or deletions were incorporated in the main data base on the Cyber mainframe.

In the final plotting of distribution maps a few species had such a concentration of recordings in certain areas that the effect was for the plotter to tear a hole in the paper! This could, of course, have been overcome by a series of smaller scale maps on which the crosses, representing record localities, would become more separated. However, limited space in this *Atlas* publication did not allow for more than a single distribution map for each species. The only alternative was to filter the original data, allowing only one record to appear in each 30 second latitude/longitude cell. This was done for *B. ericifolia*, *B. ericifolia* var. *ericifolia*, *B. integrifolia* var. *integrifolia*, *B. robur*, *B. spinulosa* var. *spinulosa*. For *B. marginata*, *B. integrifolia* and *B. serrata* a one minute cell was used.

This filtering process was used only in the presentation of the distribution maps. All records of the above species were used in the tabulations and report summaries for habitat, fire response, pollinators etc.

Towards the end of the project the entire data set was transferred from the Cyber mainframe to an IBM microcomputer. Using a DBase III program, the data were sorted and analysed for each species according to the criteria that appeared on the original record sheet, e.g. habitat, flowering times, response to fire, conservation status. These results together with a distribution map for each species are presented in Chapter 5.

FLORAPLOT is planned to be upgraded to either a mainframe or personal computer. It will then be transferable to other users in other States or countries.

Summary of Points to Consider for Future Atlasses

- Consider first sending information sheets to prospective volunteers followed by recording-kits when they continue to show interest.
- Choose State coordinators carefully. They should be enthusiastic, good organisers and preferably have a good deal of free time. They are the key person in their State and on them the success or failure of the project will depend.
- Conduct as many field trips as possible. Make sure they are seen as enjoyable excursions and not just hard work.
- Supply State coordinators with maps if necessary.
- Do as much personal letter writing as possible.
- Maximise feedback, e.g. responding to record sheets/specimens sent in; producing newsletters and maps.
- For national projects, it is essential for the coordinator to be away from head office for substantial periods. Funds should be provided for at least part-time replacement.

- For national projects consider the employment of two national coordinators, one in the east and the other in the west. (In The Banksia Atlas, most of the national coordinator's time was spent on correspondence/liaison with eastern States contributors, to the detriment of ensuring comprehensive coverage of areas in Western Australia where many more species occurred.)

- Pay greater attention to recruiting among the farming community.

- Paid (professional) botanists/biologists etc. provide very few results. Many show initial interest but generally do not have the time to fill in record sheets.

- School children show initial enthusiasm but need constant reminding and reinforcement. For any chance of success, they must have the support and enthusiasm of their school teacher.

- It may be better to go for quality (rather than quantity) of volunteers. A large number of poorly trained volunteers are a drain on the coordinator's time. Most significant results generally come from a small number of very competent and dedicated recorders.

2. EVALUATION OF THE BANKSIA ATLAS SIGHT RECORD SHEET

The aim of this chapter is to evaluate the Sight Record Sheet, both in terms of problems experienced by recorders in using the sheet, and in the usefulness of the results obtained.

The Banksia Atlas Sight Record Sheet (Fig. 3) was probably one of the most complex ever presented to volunteers. It requested habitat and biological data as well as locality information. Its appearance undoubtedly frightened off some potential contributors altogether. This at least had the merit of allowing the national coordinator more time to concentrate on the active contributors.

From returned questionnaires, 66% of contributors experienced problems with the record sheet. This usually meant that they had found a section difficult and/or time consuming to work out rather than that they were unable to complete it at all. The main problem areas expressed by contributors were as follows*: altitude (43%)1; latitude/longitude (37%); pollinators (24%); soil type (23%); identification of species (15%); fire response (13%). All other items were less than 10%.

In terms of results obtained, the least successful areas were fire response and pollinators. Inclusion of such topics in any future project will require considerable revision of the recording techniques as used in The Banksia Atlas.

* Contributors could express as many items as they wished. Therefore the sum of the percentages exceeds 100%.

Design and Coding

The sheet was designed to allow recording on the upperside with all computer codes listed on the reverse. In the form of a pad bound along the top edge, this allowed the codes from the previous sheet to be read whilst recording on the current sheet. This design proved satisfactory.

Codes were chosen to be readily associated with whatever they represented and thus be easily remembered. Banksia codes consisted of the first three letters of a species name, and a fourth letter if a variety was specified, e.g. B. spinulosa var. spinulosa = SPIS. Other codes followed suit, e.g. forest recorded as FT, woodland as WL and so on. The use of easily remembered codes was successful with completed record sheets showing very few errors. With species names for example, there were virtually no errors.

Core Data/Optional Data

The sheer amount of information required was daunting to some people. After all, the instructions had advised them to "Put maximum effort into completing each record sheet accurately and thoroughly" and to "Fill in as much of the sheet as you can - don't miss out difficult sections because they take a bit more time to complete". To some, this implied that an incomplete record sheet was not acceptable. If they had noted a banksia's occurrence but had insufficient time to do a thorough habitat survey, they often did not fill in a sheet. By the middle of the project it was realised that many useful contributions were being lost in this way. New instructions were issued stating that it was not essential to fill in the entire sheet though certain parts such as locality, name of banksia, observer code and date could not be omitted. This concept of "core data" and "optional data" is recommended for any future projects. There will always be some people who have either greater knowledge or time or enthusiasm and who will want to fill in the whole sheet. Others, who for whatever reason are unable to complete the entire sheet, need to be informed which are essential data and which are not.

Size of Record Locality

An area of 500 m x 500 m was defined as the maximum size for a record locality. This was a compromise between (i) the area that can realistically be covered on foot (ii) the fact that smaller record localities would have made coverage of the whole of Australia unlikely.

The main problem arising from size of record locality was in situations where vast areas are covered with uniform vegetation and a uniform assemblage of banksias. Recorders were reluctant to fill in numerous repeat record sheets. It was suggested, therefore, that they define the area concerned and complete a sample record sheet which would be representative of the entire area. On the basis of this sample, repeat record sheets were filled in at head office, a time-consuming process.

It is possible that 'in-between' locations could be generated artificially by computer. The acceptability of such records would depend largely on the thoroughness of the original recorder's observations.

Latitude/longitude v. Grid reference

Atlasses in Britain use grid references for recording locations. Their distribution maps show species as either present or absent within a 10 km square. The Banksia Atlas followed the recommendations of ABRS in using latitude and longitude and in the distribution maps showing the actual point locations of recordings. The superimposition of a grid could be carried out at any future stage if the need arose, e.g. for detailed phytogeographical or ecological studies.

In Australia, the advantage of latitude and longitude is that it is universal and is displayed on most maps whereas the national grid is missing from many maps. The maps most commonly used were the 1:100 000 NATMAPS followed by the Reader's Digest Atlas of Australia, and Western Australian Department of Fisheries and Wildlife map (both scale 1:1 000 000). All these maps show latitude and longitude. Only the NATMAPS show the national grid. A disadvantage of latitude/longitude is that the method of recording is not well known and even when understood, is prone to error. On many NATMAPS, latitude and longitude are indicated only at the 10 minute divisions, making accurate point recording difficult and time consuming to work out. In comparison, the specification of a grid reference is both quick and easy. Many recorders stated that they would have preferred to use grid references. Certainly there would probably have been fewer errors, reducing considerably the time spent on editing record sheets. Although most banksia areas are covered by NATMAPS, their availability varies from State to State. In Western Australia, they are not well known and are difficult to purchase. In this State most recorders used either the map supplied in the kit, the Travellers Atlas of Western Australia (Department of Lands and Surveys), or the Forests Department maps (scale 1:50 000), none of which shows the national grid. In New South Wales, the 1:25 000 topographic maps were the ones most commonly used.

The best approach is to recognise the diversity of maps currently being used throughout Australia and to allow for either grid reference or latitude/longitude to be specified, whichever is the easier. Computer programs are now available which can readily convert grid references to latitude/longitude and vice versa.

Locality Resolution Code

Locality resolution codes are essential, especially when recorders are using maps ranging from scale 1:4 000 to 1:2 500 000. The resolution code reflects the degree of accuracy for the specified latitude/longitude. It will depend largely on the scale of the map being used and also on the recorder's ability to work out where they are. This is not always an easy matter in areas of flat land and uniform vegetation with inadequate maps. In practice nearly all sight records were submitted at resolution code 2, 3, or 4 (10", 30", 1'). Very occasionally a code 5 (10') appeared. It was not necessary to include the other 3 codes. There were marked interstate differences. Tasmanian volunteers generally recorded to the nearest 10" (300 m), as did many New South Wales volunteers. In Western Australia, recording to the nearest 1' (1.8 km) was usual.

The supporting locality information, e.g. distance and direction from nearest named place, was essential for correcting erroneous latitudes and longitudes. It also pinpoints the actual locality far more accurately than does a specified latitude/longitude. Therefore, it should never be omitted.

Reserve or National Park?

These were defined as being areas of reserved Crown land where native flora and fauna are protected. In Western Australia, the situation was relatively clear-cut and only nature reserves and national parks (both managed by the Department of Conservation and Land Management) were to be included.

In other States, the situation was more complex. Victoria, for example, has national parks, State parks, coastal parks, historic parks and wildlife reserves. It also has a large number of forest reserves set up specifically for the conservation of flora. All are managed by the Department of Conservation, Forests and Lands. To be consistent with Western Australia, it was decided to omit the forest reserves. The average recorder, however, saw no distinction between these various categories and continued to include all as reserves.

In New South Wales there was the problem of State recreation areas. These areas, which are managed by the National Parks and Wildlife Service, have a greater emphasis on recreation but still afford considerable protection to native flora. Should they or should they not be included?

The overall problem was compounded as the design of the record sheet allowed only the reserve or park number to be recorded. Whilst this is readily available in Western Australia, it is not so in other States. To create a new numbering system specifically for the Atlas required strict definitions of what was and what was not to be included.

The system adopted involved the acquisition of a list of reserves and parks from each of the State's conservation authorities and the assignment of an identifying number to each listing. Forest reserves and local reserves administered by shires, councils, etc. were not included. The State recreation areas in New South Wales were included. A better design would have allowed either the name or number to be recorded. This would have allowed flexibility in definitions and would have meant that all data were recorded whether a number could be assigned or not.

Local Authority

The specification of the local authority in which the banksia record occurred was intended to serve one main purpose. It would be used to validate the specified latitude/longitude and would highlight any discrepancies. Subsequent investigations could lead to either the local authority code or the lat./long. being corrected. For such a computerised validation to take place it was necessary to have defined the local authority boundaries in terms of their lat./long. coordinates. In Western Australia this was achieved at an early stage in the project. Other States and Territories were planned to follow. However, the amount of work involved in this task had been underestimated and it was never completed. Western Australia remained the only State where this computer validation technique was used.

Volunteers in Western Australia were provided with a map showing local authority boundaries thus making their recording easier. This was not the case in the rest of Australia, where maps had not been provided and volunteers often experienced considerable difficulty identifying the relevant local authority. It is suspected that 'guesses' were often made. It is recommended that future projects should ask for such information only if it is definitely needed, and then only if relevant maps can be provided.

Altitude

This was the main difficulty experienced with the record sheets. There were interesting interstate differences. For example, altitude was no problem to Tasmanian or Australian Capital Territory recorders but a considerable problem existed for Western Australians. This is no doubt partly due to the types of maps used. In Western Australia these were mainly forestry maps, Travellers Atlas, and the Fisheries and Wildlife Map supplied in the recording kit. None of these maps include altitude information. However, even when using maps which did include altitude (e.g. NATMAPS) there was considerable inability and/or reluctance to record this information.

It is recommended that any future projects pay particular attention to the difficulties volunteers experience in recording both locality and altitude information. A small leaflet would be useful to train volunteers to become proficient in these techniques.

With altitude resolution codes, only codes 2, 3 and 4 (20 m, 50 m, 100 m) were necessary. The rest were scarcely used and could have been omitted.

Landform

Initial difficulties were experienced in the assignment of suitable codes to "sand dunes", "undulating country", and "coastal cliffs". It was decided to include the first two in the category "gradual slope", and the latter in "hilltop". Volunteers were notified accordingly.

Another problem involved variations in landform over small distances. Whenever any aspect of the habitat changed volunteers had been instructed to fill in a new sheet. For example, a lake edge group of *B. littoralis* should have been recorded separately from a population of B. grandis only metres away but on a gradual slope. Understandably, there was reluctance to repeat record sheets in this way. Generally, both records were included on the same sheet and mention made of the differing habitats under "Additional Remarks". At head office this meant the re-writing of two new sheets in the correct manner.

Aspect of Slope

Wherever a uniform slope existed, this was no problem. With undulating or changeable country it was impossible to generalise and recorders were instructed to leave this box blank. The usefulness of the records received is debatable. It may be advisable to omit this item from future surveys.

Soils - Type and Colour

Because contributors to the unpublished W.A. Orchid Atlas pilot project had difficulties in recording complex soil types, classification for The Banksia Atlas was kept deliberately simple. As with other codes, if volunteers were unable to match their observed soil with any of the listed types, they could enter 'X' in the ascribed box and write the actual soil type underneath. In head office, a new code was assigned to each addition and from then on was included on any relevant record sheets.

Uniform recording of soil type and colour is difficult to achieve over the whole of Australia. A person's assessment of soil type is often based on the type of soil with which they are most familiar. A Perth recorder who is used to soils of almost pure sand may classify a particular soil as loamy when a Tasmanian recorder (used to heavier soils) may refer to the same soil as sandy.

There were other specific problems expressed by contributors. Many commented on the artificiality of recording surface soils when, beneath the surface, soil type and colour were often completely different. In the more mountainous areas, rock type was seen to be just as important, yet there was no place to record this. Another problem expressed was the frequent changes in soil type over small distances which could not be conveyed adequately on a record sheet.

In view of all these difficulties it came as something of a surprise to find that in the final results soils had in fact been expressed very clearly with specific banksias showing marked preferences for certain soil types.

Vegetation Structure

This generally worked well. The distinctions between "woodland" and "forest" and between "rainforest" and "forest" were hazy in the minds of some recorders. Although each category had been defined and illustrated in the Instruction Booklet, apparently greater emphasis on definitions is needed.

Fire

The number of months since a fire was often difficult to determine and figures should be regarded as approximate only.

Intensity is an important component of fire, affecting the way in which plants respond. B. ornata, for example, normally is killed by fire and regenerates from seed. A cool fire may, however, have no effect, whilst a hot fire may incinerate both plants and seed. Usually B. aemula is not killed by fire, the mature plants regenerating from both trunk and lignotuber resprouts. However, a hot fire may kill the mature plants and regeneration is from seed.

Even a single fire is not uniformly hot. A few degrees' difference in temperature between the centre of a burn and its edge can markedly alter the response. Fires can also be patchy, burning some areas and leaving nearby ones untouched. Separate record sheets should have been filled in for the burnt and unburnt areas. However, contributors were reluctant to repeat record sheets in this way. The result was inaccurate recording producing the sorts of problems discussed in Chapter 5. The complexity of fire and fire response was difficult to incorporate within the confines of the standard record sheet. An additional specialist sheet may be worthwhile.

Many recorders found fire to be a fascinating subject judging by the number of additional comments added to their coded data. They have gained new insights into the relationships between fire and native vegetation in Australia. In terms of results, however, the methods used to record fire response need to be considerably revised.

Shrub/tree?

Some banksias (e.g. *B. marginata*, *B. integrifolia*) grow as both shrubs and trees, sometimes within the same area. The single box did not allow these to be recorded.

Population Size

The population size of a species is essential information when assessing its conservation status. A species may be widespread in distribution and yet at each location there may be only one or two plants. Conversely, it may exist as large populations at only a few sites. If the latter are all on private property or road verges and the species has only small numbers in conservation areas, it is clearly at risk.

There were few problems in recording population size once contributors realised that estimates rather than actual counts were required. For the purpose of assessing conservation status the following recommendations are suggested: (i) population categories modified (e.g. <30, 30-100, 100-500, >500); (ii) when a population is less than 10, provide a space to record the actual number.

The recording of 'nil' areas, i.e. area explored but no banksias found, could also be beneficial. Such recording is open to error, e.g. when an area is declared "banksia-free" after a less than thorough survey. This is more likely to be a problem with plants having small, inconspicuous flowers than with a genus such as *Banksia* with mainly large, showy flowers.

Flower Code

Apparent confusion over the exact meaning of these codes prompted early amendments to be made. Volunteers were notified of the following new codes in the first newsletter:

- B = majority in bud;
- F = majority in full flower;
- A = mainly old flowers, still with some colour;
- C = flowers absent, old fruiting cones present;
- N = neither flowers nor fruiting cones present.

This somewhat complicated set of codes was designed to show: i) the stage of flowering relative to location and (ii) the presence of immature plants which had not yet flowered. If this was then linked to fire history it should be possible to find out the number of years after fire before flowering recommences.

Unfortunately, the amended codes still proved misleading. To some recorders category C was not applicable to those banksias which retain their old flowers on the fruiting cone. A better description of this category would have been "Flowering finished, fruiting cones present".

If the species under observation had any flowers present volunteers were meant to choose between codes B, F or A. If there were no flowers, either code C or N should be used. At least one recorder (who completed more than 600 sheets!) had a different interpretation, using instead the code that was most applicable to the majority of plants. Therefore, if 30% of plants were flowering and 70% were not, code C was chosen.

Problems such as these were difficult to detect, and in fact were not detected until some way into the project. It is not known how many volunteers misinterpreted either flower codes or instructions. Clearly the coding was too complex. A simple classification into flowering/not flowering would have been better.

New Shoot Growth

This generally worked well. Initially there was a slight problem in definition, current season, semi-mature growth being recorded as 'New Shoot Growth'. It was re-emphasised via the newsletters that 'New Shoot Growth' should only apply to plants showing active vegetative growth usually indicated by soft, hairy leaves of a different colour.

Average Height

Volunteers were instructed to record the most commonly occurring height of the banksia being recorded. In the case of two very definite and distinct height classes for the same type of banksia, they were to use two lines for that species, recording the two heights one under the other.

With species that show a great range of heights, e.g. *B. grandis*, *B. integrifolia*, it was often difficult to select the most appropriate figure. Also, by recording the most commonly occurring height it was likely that both the largest and smallest banksias would be omitted. A better approach may be to have a range of height classes (e.g. <30 cm; 30 cm-2 m; 2-5 m; 5-10 m) with the possibility of any number of these classes being attributed to any banksia recording. Records of seedlings, indicating the regeneration of populations either with or without fire, could be incorporated in this system. In the present study, the omission of regeneration data in the absence of fire was commented on by several volunteers.

Pollinators

Volunteers were requested to look closely at any animals visiting banksia inflorescences and to record only those that were in contact with the tip of the style and which could therefore be effecting pollination. Even with this provision, the pollinators listed for each banksia in Chapter 5 should be regarded as "possible pollinators" only. Confirmed pollination would require more detailed study. With *B. pulchella*, for example, ants were listed as the main pollinator observed. However, it is generally not known whether ants can be effective pollinators especially in a species such as *pulchella* where the style places the pollen presenter at some distance from the nectaries at the base of the flower. It is likely that ants visiting the nectaries may by-pass the pollen presenter.

The lists of possible pollinators should not be regarded as comprehensive. Certain groups of animals (e.g. mammals), may be important pollinators, but due to their mainly nocturnal activities, were under-recorded.

From returned questionnaires, the two main problems expressed by volunteers were: (i) uncertainty whether the presence of an animal meant that it was actually pollinating or not; (ii) the fact that pollinators were rarely seen, implying to many recorders that they should spend more time looking for them.

If this item is included in future projects, it would seem that volunteers need more comprehensive instructions to reduce the above mentioned uncertainties.

Summary of Points to Consider for Future Atlasses

- More extensive field testing of record sheets needed (to reduce/eliminate problem areas) before final printing.
- Computer codes should be associated with whatever they represent and therefore easily remembered.
- Investigate ways of recording large areas with same species (other than filling in repeat record sheets).
- Do not ask for local authority to be recorded unless a map can be provided showing local authority boundaries.
- Reduce large number of altitude and locality resolution codes.
- Include a space on record sheet to indicate specimen has been collected.
- Divide record sheet into core data (essential) and optional data (non-essential).
- Define what is meant by "Conservation Area" in each State. Include space on record sheet to record either name or number.
- Provision to record either lat./long. or grid reference, whichever is the easier.
- Produce leaflet explaining how to record: (i) altitude; (ii) lat./long.; (iii) grid references. The number of erroneous recordings (at least 50%) justifies greater attention in this area.
- Avoid complex coding. In *The Banksia Atlas*, flower codes were too complex. It would have been better to have a simple yes/no question.
- Consider recording 'nil' areas, i.e. area explored but no species found.
- Consider separate, specialist sheet for recording fire response.
- Consider separate, specialist sheet for recording variability within species, especially when the number of species in an area is restricted, e.g. *B. marginata* in Tasmania.

3. ACHIEVEMENTS AND FINDINGS

This chapter summarises achievements and findings of the Atlas contributors. Full results for each species are presented in Chapter 5. We have not attempted detailed statistical analyses of the maps and data set, preferring to leave these interesting avenues of research to other workers. Rather, we aim here to highlight major conclusions that can be derived from the wealth of data gathered by the contributors.

Between July 1984 and August 1986, 12 991 record sheets were received from 421 contributors. This is estimated to represent at least 7 000 recording hours. The specified number of contributors is without doubt an understatement. Many registered names represented families or groups. In other cases, observers passed on their information to registered volunteers without themselves being registered. The total number of contributors was probably well over 1 000.

Table 2 illustrates the number of people sent recording kits, the number of active contributors, and the number of record sheets from each State and Territory.

The number of record sheets sent in by individual contributors varied from 1 to over 1 000. Table 3 shows the number of sheets per contributor. In each case the number in brackets is the percentage of the total number of contributors.

Of all contributors, 49% sent in up to 10 record sheets each and 15% sent in more than 50. About 9 000 sheets were collectively submitted by approximately 60 contributors. Without denying the value of the many small contributions, it is apparent that the bulk of the work was carried out by a few very keen recorders.

The most frequently reported banksia was *B. marginata* (on 3 775 record sheets), followed by *B. grandis* (1 952), *B. integrifolia* (1 736), *B. serrata* (1 472), *B. spinulosa* (1 386), *B. attenuata* (1 280) and *B. integrifolia* var. *integrifolia* (1 147). The least frequently reported were the two new species *B. epica* (1) and *B. oligantha* (1), followed by *B. cuneata* (4), *B. sphaerocarpa* var. *dolichostyla* (5) and *B. conferta* var. *conferta* (6). The number of records received for each species is shown in the first column of Table 4.

The highest number of species recorded from a single record locality (500 m x 500 m) was 9. This was from the south coast of W.A. between Mt Manypeaks and Cape Riche.

Table 2. Volunteer statistics

State	Number sent recording kits	Number of contributors and (% of number of people sent kits)	Number of record sheets
NSW	200	84 (42%)	2 046
VIC	73	40 (55%)	2 628
TAS	31	18 (58%)	625
SA	36	19 (53%)	1 324
WA	491	185 (38%)	5 143
QLD	94	47 (50%)	729
NT	14	2 (14%)	9
ACT	52	24 (46%)	487

MAJOR ACHIEVEMENTS

New Taxonomic Discoveries

During the course of The Banksia Atlas two new species were identified in Western Australia (B. epica and B. oligantha). A new variety of B. spinulosa was also identified in northern New South Wales (var. neoanglica). These new taxa are described by George (1988). For details of the location, habit and habitat of these new taxa, see Chapter 5 and the colour plates.

B. epica grows alongside B. media on the western coast of the Great Australian Bight. Its name commemorates two epic journeys by individuals who have sighted the plants in their remote location. In 1841, the explorer Edward John Eyre sighted banksias, possibly including this species, towards the end of his journey along the Bight. Over a century later (1973), the botanist E. Charles Nelson made the first herbarium collection of banksias here but this species was not recognised as new. In 1985 two recorders for The Banksia Atlas (John and Lalage Falconer) observed that there were in fact 3 species of banksia in the area. They first collected leaves and old flowers of the new species but were unable to collect fresh flowers as it was the wrong season.

On the basis of these specimens, it did indeed seem that a new species had been found. The following year John Falconer returned to collect flowers and fruits. He drove about 2 000 km from Warburton to Point Culver and back, mostly on unformed tracks over a long weekend, a second truly epic journey.

B. epica differs from B. media (Fig. 9) in its generally shorter leaves, larger flowers and distinctive fruiting cone with persistent old flower parts which are curled and point upwards. With B. media the old flowers are straight and point downwards.

The name B. oligantha, the second species identified during the Atlas, is derived from the Greek oligos (few) and anthos (flower). This species (Fig. 10) has generally fewer flowers per inflorescence (less than 30) than any other species of Banksia. In comparison, B. grandis may have up to 6 000. The closely related B. cuneata generally has more than 50. The species was discovered in 1984 by

Table 3. Number and percentage of contributors grouped according to the number of Sight Record Sheets each completed.

No. of record sheets	No. of contributors (%)	No. of record sheets	No. of contributors (%)
1	15 (4)	46-50	3 (1)
2-5	118 (28)	51-60	10 (2)
6-10	73 (17)	61-70	5 (1)
11-15	42 (10)	71-100	21 (5)
16-20	27 (6)	101-150	12 (3)
21-25	29 (7)	151-200	7 (2)
26-30	20 (5)	201-30	0
31-35	16 (4)	301-400	3 (1)
36-40	7 (2)	401-500	1 (0)
41-45	9 (2)	501-1000	2 (1)
		1001+	1 (0)

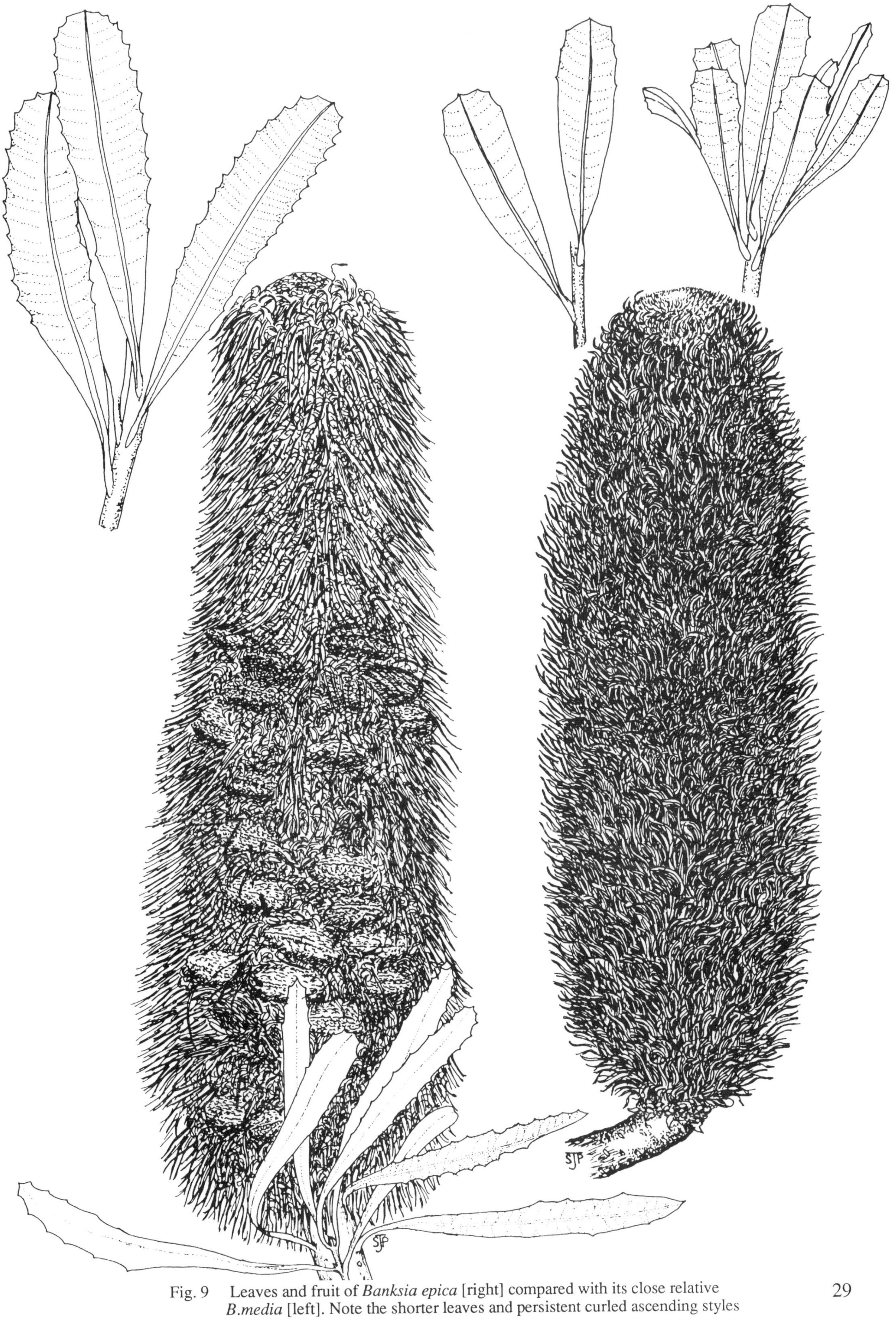

Fig. 9 Leaves and fruit of *Banksia epica* [right] compared with its close relative *B.media* [left]. Note the shorter leaves and persistent curled ascending styles on the fruit of *B.epica*. Drawn by Susan Patrick.

Ken Wallace, Wheatbelt Regional Manager, Department of Conservation and Land Management, during the first few months of the Atlas. It is known from only one location near Wagin in the western central wheatbelt of Western Australia. The species differs from *B. cuneata* in its fewer flowers, its thicker glossy leaves (resembling those of *B. ilicifolia* but smaller) and in its yellow flower colour.

B. spinulosa var. *neoanglica* is so-called because of its occurrence predominantly in that part of New South Wales referred to as the New England Tableland. Plants now identified as var. *neoanglica* had been previously classified as either var. *cunninghamii* or var. *collina*. However, observations by Atlas recorders revealed distinct differences which led to the naming of this new variety. Its lignotuberous habit, which distinguishes it from var. *cunninghamii*, is illustrated in the colour plates.

In addition to these new taxa named by George (1988), other new varieties and subspecies of *Banksia* have been identified but the results of this research were not known in time for inclusion in the Atlas. Conran and Clifford (1987) recognised *B. oblongifolia* var. *minor*, a tall variant with long leaves occurring on upland sandstone outcrops, compared with *B. oblongifolia* var. *oblongifolia* which is low growing with shorter leaves and inhabits near-coastal swamp margins. George (1988) recognised two varieties of *B. leptophylla*, one with small flowers produced in winter (var. *melletica*) and the other with large flowers produced in summer (var. *leptophylla*).

Hopper (1988) named *B. seminuda* subsp. *remanens*, which has shorter leaves, a more shrubby habit and more densely floriferous canopy than subsp. *seminuda*. Subsp. *remanens* has a coastal distribution favouring sandy slopes below granite outcrops, whereas subsp. *seminuda* is usually confined to watercourses. Hopper also named *B. occidentalis* subsp. *formosa*, an attractive low-growing short-leaved race apparently confined to two sits on the south coast of Western Australia.

Increased Knowledge of Current Distributions

Previously, the distribution of banksias was known as scattered point locations based on herbarium collections (Fig. 11b). For almost all species, *The Banksia Atlas* produced an excellent degree of coverage and significantly increased our knowledge of their current distributions (Fig. 11c). *B. attenuata* is a good example with the species now shown to be well distributed throughout the south west. There are many other examples and the following notes serve to highlight a few.

B. baxteri is another example where herbarium collections provided a less than adequate picture. Atlas records show the species to have an almost continuous distribution along the south coast between the Stirling Range and Munglinup Beach.

With *B. elderiana*, herbarium collections had indicated two disjunct populations, one to the west and the other to the east of Kalgoorlie. The eastern population was known from only two locations. Atlas records have increased this to 17 known locations and the species appears common throughout an area of approximately 150 km x 150 km.

A total of only 7 collections of *B. lindleyana* in the Western Australian Herbarium had suggested that this species was probably uncommon (George 1981). Moreover, most records were

Fig. 10 Flowering sprigs and leaves of *Banksia oligantha* [top] and its close relative *B.cuneata* [bottom]. Note the few- flowered inflorescence of *B. oligantha*. Drawn by Susan Patrick

centred on the lower Murchison River, the only exceptions being outliers at Binnu and towards Hamelin Pool. Atlas records included at least 23 locations for this species and also showed the northern and southern populations to be linked.

With both B. caleyi and B. baueri, herbarium collections had suggested two disjunct populations, one inland and another along the south-west coast. With both these species, Atlas records have shown their distributions to be more or less continuous.

Similarly with B. sphaerocarpa var. sphaerocarpa, herbarium collections had indicated populations near the south coast from Mt Lindesay eastwards separated from populations towards the west coast by a gap of some 140 km. Atlas records have linked the two areas.

With rare banksias, any new finds are significant even if they fall within the existing known range of a species. They increase our knowledge of the species' current distributions and their conservation status. New records of this kind were provided for B. aculeata, B. brownii, B. chamaephyton, B. cuneata, B. meisneri var. ascendens, B. plagiocarpa and B. verticillata. These are all considered in more detail under the section "Increased Knowledge of Conservation Status".

Range Extensions

When banksia distributions resulting from Atlas records were compared with herbarium collections, many range extensions became apparent. There were at least 40 banksia species or varieties where significant new locations extended their previously known ranges. These are listed below:

aemula
attenuata
audax
baueri
blechnifolia
burdettii
chamaephyton
caleyi
conferta var. penicillata
dryandroides
elderiana
ericifolia var. ericifolia
gardneri var. gardneri
gardneri var. hiemalis
goodii
grandis
ilicifolia
integrifolia var. integrifolia
laevigata subsp. fuscolutea
lemanniana
lindleyana
littoralis
lullfitzii
marginata
meisneri var. meisneri
nutans var. cernuella
nutans var. nutans
oblongifolia
paludosa
petiolaris
praemorsa
prionotes
pulchella
robur
scabrella
speciosa
sphaerocarpa var. caesia
sphaerocarpa var. sphaerocarpa
spinulosa var. spinulosa
victoriae
violacea

Considering Western Australia firstly, the range of B. prionotes was extended almost 160 km to the south-east by new records from Holland Rocks and Lake Lockhart.

B. speciosa was confirmed at Point Culver. This location appears to be an outlying population, 150 km eastwards of its previously known range. B. ilicifolia was previously thought to occur no further east than Albany. It is now recorded 90 km eastwards near the Cordinup River.

In the eastern part of its distribution Atlas records showed B. violacea to occur within 20 km of Peak Charles. This was almost 90 km north of its nearest previously recorded location. The range of B. blechnifolia was extended some 80 km westwards by the recording of an outlying population near Borden. To the north-east of Esperance new records extended its range approximately 50 km eastwards.

In the south-east of its range, *B. sphaerocarpa* var. *sphaerocarpa* was recorded beyond Jerramungup, an extension of some 100 km. In the north of its range, it was found to extend both closer to the coast (near Lancelin) and further inland (at Watheroo) than was previously known.

B. sphaerocarpa var. *caesia* was recorded north of Mt Holland, a range extension of 120 km eastwards, and near the Stirling Range National Park, an extension of 100 km southwards.

B. meisneri var. *meisneri* was found at Tone Bridge, 70 km south-west of previous occurrences. Both *B. lemanniana* and *B. caleyi* had similar 70 km range extensions westwards, the former occurring near the Pallinup River estuary, the latter near Broomehill.

A location of *B. dryandroides* close to the northern boundary of Stirling Range National Park was a 40 km range extension and provided the first recorded occurrence of this banksia from this national park. In the south-west of the park there were also first recordings of *B. meisneri* var. *meisneri* and *B. caleyi*.

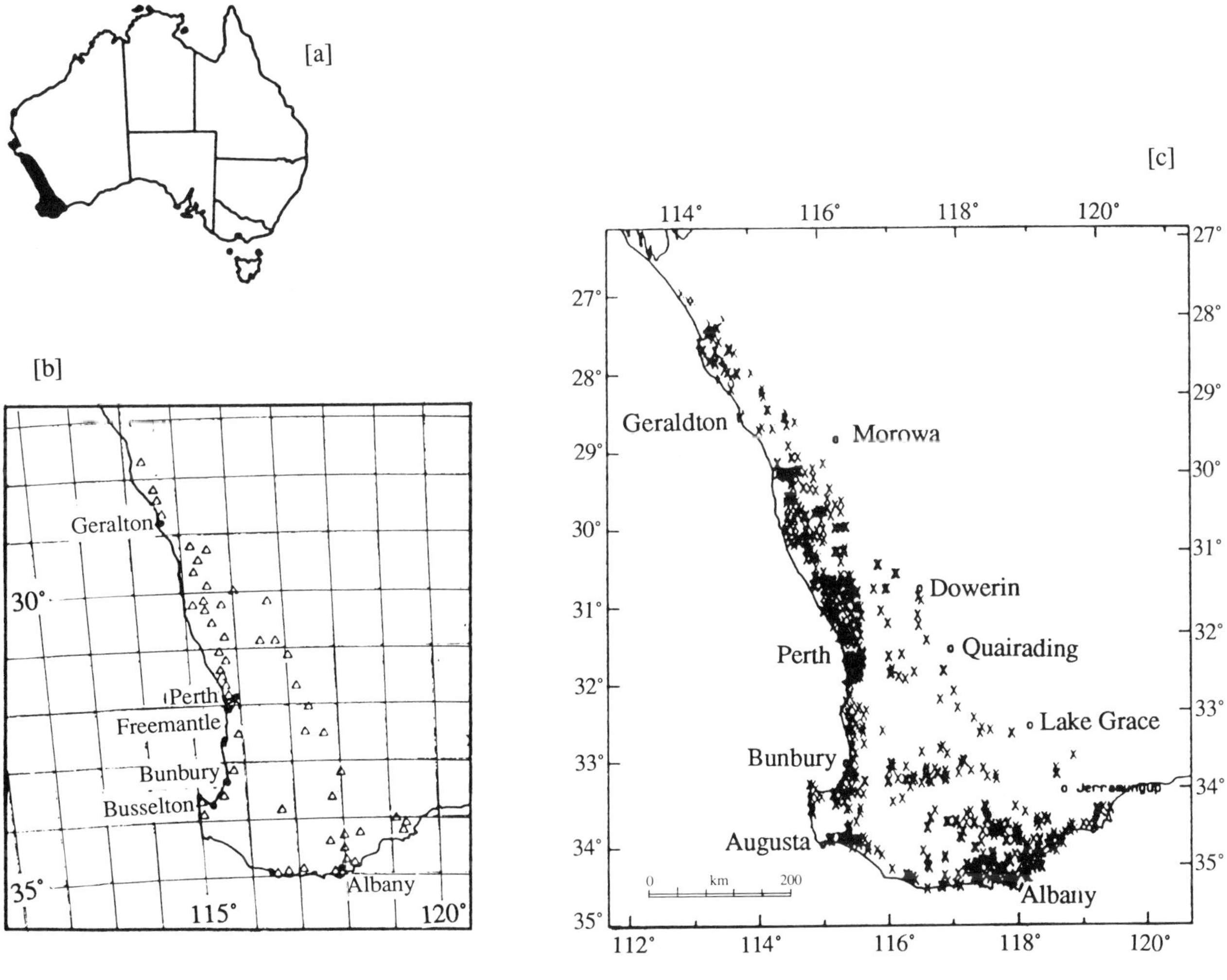

Fig. 11 Maps of the distribution of *Banksia attenuata* showing improvements in knowledge from data in [a] Holliday and Watton's [1975] Field Guide; [b] A.S.George's [1981] taxonomic revision and mapping of herbarium collections; [c] sight records reported by Banksia Atlas contributors.

Range extensions of approximately 50 km westwards were recorded for both B. pulchella and B. petiolaris, the former occurred at Twin Bay in the Fitzgerald River National Park, the latter close to the Rabbit Proof Fence on the Ravensthorpe to Esperance road.

In the central south of Western Australia B. laevigata subsp. fuscolutea was recorded near the headwaters of the Young River, 50 km south of its previously known distribution. B. elderiana was recorded at the old Davyhurst Mine and at Bungalbin Hill, both about 90 km north of previous locations. B. audax was found near Lake Grace and Lake Magenta, range extensions southwards of approximately 80 km. A recording at Wallaroo (west of Kalgoorlie) was a northwards extension of about 50 km.

In the Kalbarri area there were 60 km range extensions northwards to Zuytdorp National Park for B. victoriae and B. attenuata. Also B. lindleyana was recorded at Wandana Nature Reserve, 65 km beyond its previously known range.

Even a relatively common and well-known species such as B. grandis had its known range extended eastwards to Badgebup (45 km) and Dongolocking Nature Reserve (55 km).

Range extensions of some of the rare species were especially important. Such species often have a very restricted range and what appears to be a small extension can be highly significant. B. scabrella was previously thought to be restricted to a small area east and south-east of Walkaway. Its total range was only 40 km long. Atlas records showed a second and disjunct population in the Mt Adams area, 85 km south of the previous locations. The species is in fact far more common at Mt Adams than at Walkaway. Its total range was increased to an area 90 km across.

B. goodii is a declared rare species with only a few known populations. It occurs over a small area just north of Albany.

Prior to the Atlas all known locations were east of the Albany Highway. Atlas records confirmed five new sites all to the west of the highway, a range extension of some 30 km. The width of the total range of this species has increased from 20 to 50 km.

B. burdettii is represented by only 5 collections in the Western Australian Herbarium. It appeared to have a restricted distribution of 120 km and was considered potentially endangered. Atlas records showed it to occur 60 km north and 15 km south of the previously known range, thus increasing its total range to an area almost 200 km long.

The range of the rare species, B. chamaephyton has also been extended southwards approximately 25 km.

Along the south coast, B. praemorsa was recorded 15 km westwards (at Torbay) and 30 km eastwards (on the coast opposite Hauloff Rock), thus increasing the width of its total range from 55 km to 90 km.

There were also possible range extensions amongst some of the banksias which have varieties. However, some of these varieties were at times difficult to identify. Therefore, apparent range extensions may either be real extensions or be due to incorrect identification. Either may suggest the need for further taxonomic studies. Such dilemmas are considered under the section 'Taxonomic Problem Areas Highlighted'.

In eastern Australia there were fewer range extensions than in the West. The most significant was the recording of the rare *B. conferta* var. *penicillata* at a new location in New South Wales 100 km south of its previously known range. *B. aemula* generally occurs near coastal areas. *Atlas* records confirmed interesting inland localities from Agnes Banks (50 km inland) and from south west of Bundaberg (30 km inland). Both *B. aemula* and *B. oblongifolia* were recorded north of Bundaberg towards Agnes Waters. For both species this was a range extension of some 65 km. Both *B. aemula* and *B. robur* were recorded on Fraser Island. These were new recordings for both these species. In western Victoria *B. marginata* had a range extension of some 80 km northwards to Wyperfeld National Park.

Extensions were also recorded for varieties of *B. integrifolia*, *B. spinulosa* and *B. ericifolia*. However, comments made previously for western banksias regarding the problem of apparent range extension versus incorrect identification applies to eastern banksias also. Nevertheless, in the following cases identification does seem correct and range extensions are therefore implied. *B. integrifolia* var. *integrifolia* is generally a coastal species but in southern Queensland were two new *Atlas* records up to 270 km inland. Illustrations of leaves indicate they have been correctly identified, but full specimens have not been seen by the authors. Complete specimens are, however, in the Queensland Herbarium. The occurrence of *B. integrifolia* var. *integrifolia* on the western side of Port Phillip Bay, Victoria, also was recorded.

Other possible range extensions of *B. integrifolia* var. *integrifolia*, *B. ericifolia* var. *ericifolia* and *B. spinulosa* var. *spinulosa* are not supported by herbarium collections. They are considered in more detail under the section 'Taxonomic Problem Areas Highlighted'.

Increased Knowledge of Conservation Status of Certain Species

The conservation status of a banksia is a measure of both its rarity and security. In its assessment, a number of factors must be taken into account. The number of recorded populations is an obvious starting point. Other factors are population size, percentage of populations in conservation reserves and percentage restricted to road verges. For example, there may be a species with few records but each record consisting of a large number of plants and all plants being located in conservation reserves. This hypothetical species would be more secure than another with the same number of records but with only a few plants in each population, nearly all plants occurring on road verges and scarcely any in conservation reserves.

Table 4 illustrates the factors used to assess conservation status of banksias.

Another factor to consider in Western Australia relates to the spread of dieback disease, and the susceptible nature of most banksias to this disease (Fig. 12). Susceptible species occurring where the disease is most prevalent may be at greater risk than less susceptible species in other areas.

The commercial harvesting of some banksia species' flowers and fruits for the cut flower trade should also be considered. This is especially so for species which are both heavily picked and susceptible to dieback because pickers may contribute to the spread of this disease.

Those species that are killed by fire and regenerate from seed are also more at risk than species which resprout from lignotubers or epicormic buds. Successive

Table 4. Data for assessing conservation status of banksias. Taxa are listed according to the number of Atlas Sight Records made, from the least to the most.

	Number of sight records	% of populations >100 plants	% of populations on road verges	% of populations in conservation areas
B. epica	1	0	0	100
B. oligantha	1	100	0	100*
B. cuneata	4	0	100	0
B. sphaerocarpa var. *dolichostyla*	5	20	20	0
B. conferta var. *conferta*	6	17	0	100
B. lullfitzii	7	29	0	71
B. aculeata	12	33	0	92
B. tricuspis	13	23	100	0
B. laevigata subsp. *laevigata*	14	36	0	64
B. goodii	17	29	41	18
B. lanata	17	53	41	18
B. scabrella	17	59	53	6
B. brownii	18	28	28	50
B. solandri	19	26	0	100
B. plagiocarpa	21	53	0	57
B. verticillata	21	24	0	81
B. meisneri var. *meisneri*	22	23	5	45
B. laricina	23	70	9	65
B. victoriae	23	48	17	48
B. elegans	24	46	0	25
B. conferta var. *penicillata*	26	15	0	31
B. lindleyana	28	18	0	54
B. praemorsa	28	43	0	74
B. gardneri var. *brevidentata*	29	31	0	100
B. meisneri var. *ascendens*	29	38	10	55
B. micrantha	31	3	35	23
B. oreophila	31	29	0	100
B. conferta	32	16	0	44
B. audax	33	27	6	36
B. spinulosa var. *neoanglica*	35	43	3	63
B. benthamiana	36	17	47	17
B. chamaephyton	36	17	17	17
B. hookeriana	36	72	3	11
B. blechnifolia	37	38	49	14
B. incana	38	37	24	32
B. laevigata subsp. *fuscolutea*	41	10	20	15
B. telmatiaea	43	54	23	26
B. canei	48	75	0	19
B. petiolaris	51	65	37	18

Table 4 continued.

	Number of sight records	% of populations >100 plants	% of populations on road verges	% of populations in conservation areas
B. meisneri	52	31	8	50
B. grossa	53	45	42	19
B. saxicola	54	28	0	94
B. laevigata	59	19	15	25
B. dryandroides	65	22	25	40
B. lemanniana	69	35	1	74
B. ashbyi	70	24	3	9
B. ericifolia var. *macrantha*	76	68	5	67
B. occidentalis	80	29	17	41
B. sceptrum	80	39	10	45
B. burdettii	81	30	40	32
B. gardneri var. *hiemalis*	81	25	4	72
B. integrifolia var. *aquilonia*	92	12	5	36
B. candolleana	95	56	21	21
B. sphaerocarpa var. *caesia*	98	32	41	41
B. caleyi	100	16	39	37
B. nutans var. *cernuella*	100	20	33	36
B. pilostylis	102	56	44	14
B. leptophylla	104	52	25	30
B. baueri	132	22	33	46
B. dentata	133	17	0	28
B. pulchella	147	37	24	24
B. violacea	166	22	19	48
B. elderiana	167	34	12	22
B. coccinea	171	39	13	47
B. gardneri var. *gardneri*	180	25	21	44
B. quercifolia	180	34	4	17
B. baxteri	188	40	27	37
B. seminuda	190	26	4	12
B. spinulosa var. *collina*	191	37	2	46
B. nutans var. *nutans*	204	17	14	59
B. speciosa	216	56	29	27
B. robur	261	62	5	34
B. aemula	273	65	3	52
B. sphaerocarpa var. *sphaerocarpa*	283	23	29	35
B. paludosa	291	57	0	54
B. gardneri	329	25	14	55
B. spinulosa var. *cunninghamii*	331	49	6	43
B. media	336	39	18	43
B. repens	346	40	21	39
B. nutans	360	20	21	50

Table 4 continued.

	Number of sight records	% of populations >100 plants	% of populations on road verges	% of populations in conservation areas
B. prionotes	386	43	26	31
B. ericifolia var. ericifolia	423	58	1	60
B. sphaerocarpa	433	24	33	35
B. integrifolia var. compar	437	22	13	27
B. menziesii	442	43	9	21
B. oblongifolia	492	61	1	50
B. ilicifolia	557	24	3	32
B. ericifolia	565	59	1	61
B. littoralis	726	14	11	26
B. spinulosa var. spinulosa	765	58	2	39
B. ornata	839	65	27	35
B. integrifolia var. integrifolia	1147	55	8	50
B. attenuata	1280	44	14	26
B. spinulosa	1386	51	3	42
B. serrata	1472	60	3	51
B. integrifolia	1736	43	9	44
B. grandis	1952	29	10	19
B. marginata	3775	53	29	31

*a population of B. cuneata missed by Atlas recorders is known to be on a nature reserve.

fires at intervals too frequent to allow the seedlings time to reach maturity, flower and set seed, could eliminate some populations.

In Western Australia, the following banksias are currently declared as rare or endangered and are afforded special protection under the Wildlife Conservation Act 1950 - B. brownii, B. cuneata, B. goodii, B. meisneri var. ascendens, B. oligantha, B. sphaerocarpa var. dolichostyla, B. tricuspis B. verticillata. For most of these, Atlas records confirmed their rarity. For B. sphaerocarpa var. dolichostyla and B. tricuspis there were no new records. New populations were recorded for B. brownii (2), B. cuneata (1), B. goodii (5), B. meisneri var. ascendens (a few) and B. oligantha.

B. chamaephyton, previously declared rare, appears to be more common than was previously thought. Although its populations are generally less than 100 plants, and only 17% of recorded populations are in conservation reserves, the number of known populations is such that the species was taken off the list of declared rare plants but continues to be monitored.

Two species have been added to the list recently. The newly discovered B. oligantha is currently known from one location only. Although there are approximately 300 plants, these

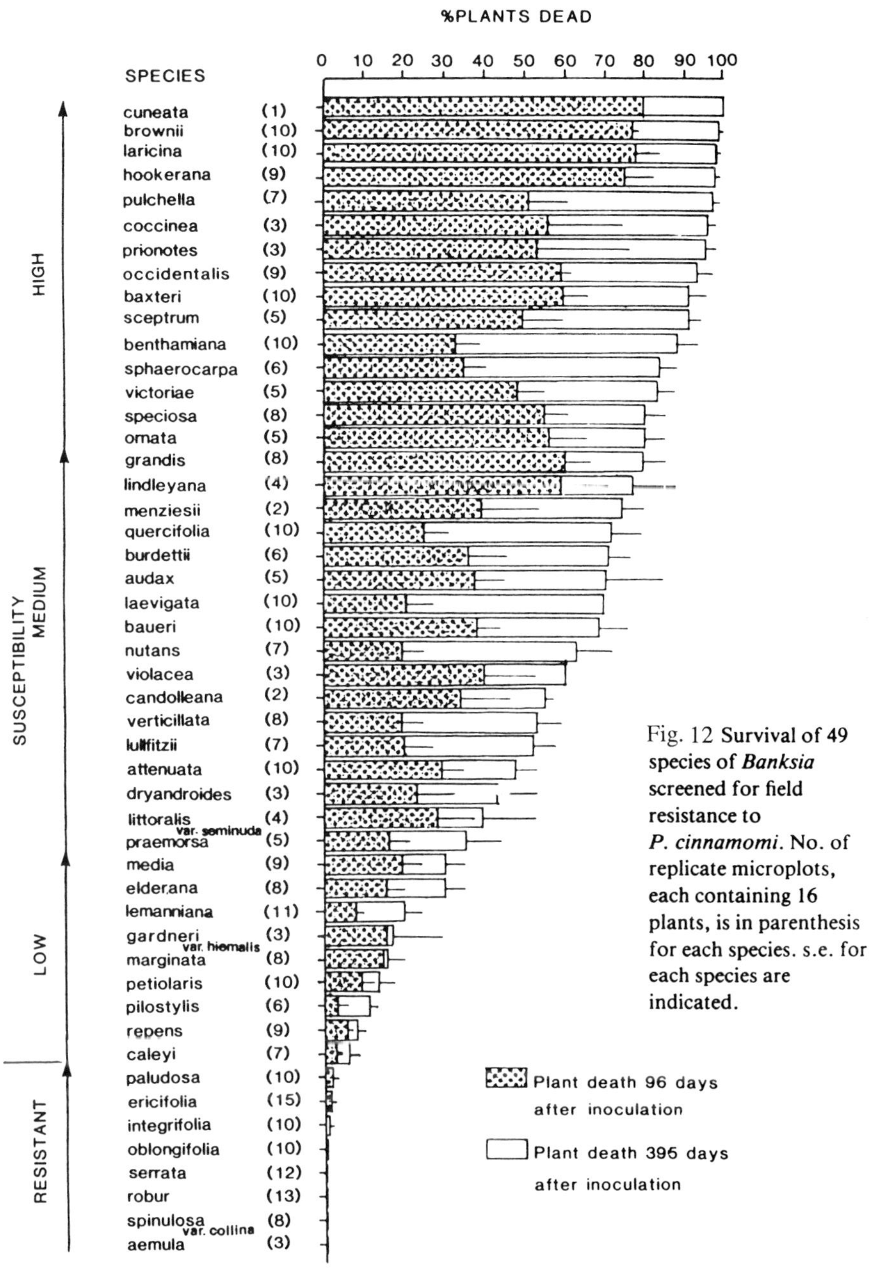

Fig. 12 Survival of 49 *Banksia* species 96 days [dotted bars] and 395 days [open bars] after inoculation with the dieback fungus *Phytophora cinnamomi*. Reprinted with permission from McCredie *et al.* [1985, *Australian Journal of Botany* 33 (6): 629-37].

are an even-aged stand with no seedlings. The species appears to be non-lignotuberous and would therefore be susceptible to frequent fires. B. verticillata is now declared rare flora. It occurs in 10 or 11 scattered populations along the south coast. Atlas records have shown that most of these are relatively small, with possibly only 3 containing more than 100 plants. Although well represented in conservation reserves the species is highly susceptible to dieback disease. It is likely to become increasingly rare.

In Western Australia Atlas records suggest that another six banksias should be closely monitored. These are: B. aculeata, B. benthamiana, B. elegans, B. laevigata subsp. laevigata, B. praemorsa and B. scabrella.

At less risk but still in need of monitoring are B. blechnifolia, B. coccinea, B. baxteri, B. epica, B. meisneri var. meisneri, B. lanata and B. speciosa. Each of the above mentioned species and the various factors effecting its conservation status are summarised in Table 5.

A few banksia species have been shown to be more common than was previously thought and continued monitoring of their populations is probably not required. Banksias in this category are B. burdettii, B. laricina and B. lindleyana.

Amongst eastern banksias only B. plagiocarpa and the two varieties of B. conferta are regarded as rare. The former is known from only three populations, one of which was a new recording during the Atlas. It appears to be susceptible to fire and one of these populations has been burnt during the last few years.

There was a significant new population recorded for B. conferta var. penicillata but no new records for B. conferta var. conferta. Both varieties are well protected in conservation reserves. Too-frequent burning would probably be the main threat to their survival.

Some Taxonomic Problem Areas Highlighted

Most problems were concerned with identifying varieties of B. nutans, B. gardneri, B. sphaerocarpa, B. spinulosa and B. integrifolia. Only one species, B. micrantha consistently gave problems in identification, being readily confused with B. sphaerocarpa var. sphaerocarpa. The latter, in its northern heathland form, is a low (c. 0.5 m) sprawling shrub with short pungent leaves, small flower spikes and flowers sometimes falling well within the specified measurements of B. micrantha. These plants nevertheless were identified as B. sphaerocarpa because they had hairy perianth limbs (not glabrous as in micrantha). Many volunteers, however, found such a small difference difficult to detect.

The two varieties of B. nutans (var. nutans and var. cernuella) were often difficult to separate. They are identified by flower size and characteristics of the fruiting cone, var. cernuella having smaller flowers appearing later, and smaller, less knobbly follicles on the fruiting cone. The distributions of the two varieties are thought to be disjunct with var. cernuella occurring further west. However, there are several Atlas records of plants with large flowers and follicles in the western part of the species' range and of plants with small flowers and follicles in the east. Depending on interpretation, these should be regarded either as range extensions or as reasons for further taxonomic studies of this species.

Table 5. Data for assessing conservation status of banksias at risk, including those in need of monitoring.

TAXON	Number of Records	% of Populations >100 Plants	% of Populations on Road Verges	% of Populations in Conservation Areas	Susceptible to Dieback[1]	Use in Cut-flower Trade[2]	Response to Fire[3]
Rare or vulnerable							
B. brownii	18	28	28	50	high	-	S
B. chamaephyton	36	17	17	17	-	-	R
B. conferta var. *conferta*	6	17	0	100	-	-	S
B. conferta var. *penicillata*	26	15	0	31	high	-	S
B. cuneata	4	0	100	0	-	-	S
B. eipca	1	0	0	100	-	-	-
B. goodii	17	29	41	18	-	-	R
B. meisneri var. *ascendens*	29	38	10	55	high	-	S
B. oligantha	1	100	0	100	-	-	-
B. plagiocarpa	21	53	0	57	medium	-	S & R
B. sphaerocarpa var. *dolichostyla*	5	20	20	0	high	-	R
B. tricuspis	13	23	100	0	-	-	S & R
B. verticillata	21	24	0	81	medium	-	-
In need of regular monitoring							
B. aculeata	12	33	0	92	-	-	S
B. benthamiana	36	17	47	17	high	-	S
B. elegans	24	46	0	25	-	-	R
B. laevigata subsp. *laevigata*	14	36	0	64	medium	low	S
B. praemorsa	28	43	0	74	-	low	S
B. scabrella	17	59	53	6	-	-	S
Occasional monitoring							
B. baxteri	188	40	27	37	high	high	S
B. blechnifolia	37	38	49	14	-	-	S
B. coccinea	171	39	13	47	high	high	S
B. meisneri var. *meisneri*	22	23	5	45	-	-	S
B. speciosa	216	56	29	27	high	medium	S

1. High, medium or low as given in Fig. 1.
2. High (200 000+ stems harvested as reported in Burgman & Hopper 1982), medium (50 000 - 200 000 stems), low (less than 50 000 stems).
3. S = killed by fire and regenerates by seed; R = resprouts from lignotuber and/or epicormic buds, as reported by George (1981, 1984), or recorded by Atlas contributors.

A similar problem occurred with B. gardneri var. gardneri and B. gardneri var. hiemalis, which are identified by flower colour and flowering time. If flowers were absent it was impossible to distinguish these varieties. Apparent range extensions of var. gardneri eastwards and var. hiemalis westwards may be interpreted as either real extensions or wrong identifications.

B. repens was also sometimes confused with B. gardneri. Its juvenile leaves sometimes resemble leaves of B. gardneri (Fig. 13). However, the flowers are generally a different colour, so if these were present, identification could be confirmed. Also the stems of B. gardneri are always on the soil surface, whereas those of B. repens are underground.

B. sphaerocarpa var. caesia and var. sphaerocarpa are identified by the former's larger habit, bluish-green foliage and smaller follicles. However, there were several reports of var. sphaerocarpa with blue-green foliage and this often caused confusion. Generally, the two varieties do not overlap in their distribution. However, Atlas records of var. sphaerocarpa west of Wagin and var. caesia near the Stirling Range National Park do not conform to this pattern. The Wagin records may be wrongly identified but specimens of the Stirling Range records have been confirmed as var. caesia.

The discovery of lignotuberous populations of B. ashbyi and confirmation of the same in B. violacea by Atlas contributors suggests that the formal recognition of varieties may be warranted. This would be consistent with George's (1981) treatment of the B. spinulosa complex.

In eastern Australia the varieties of B. spinulosa caused

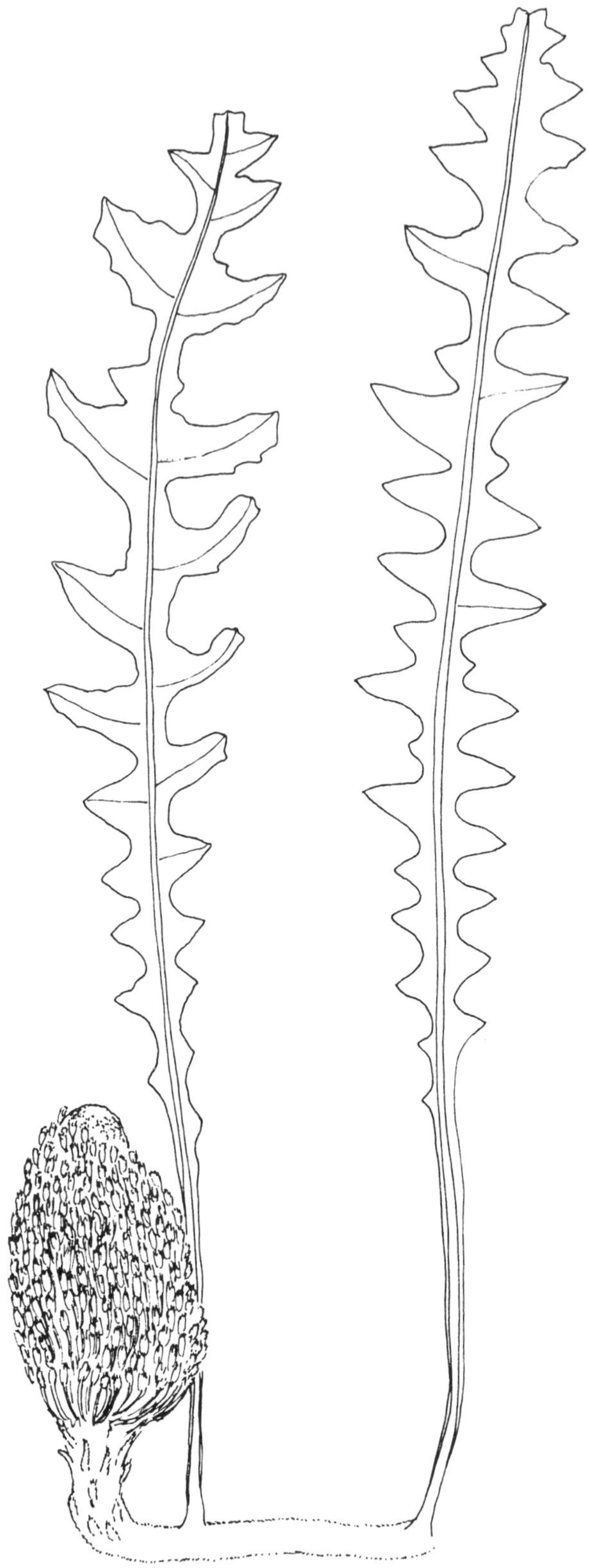

Fig. 13

Figure 13. Banksia repens from Camel Lake Nature Reserve, north of the Stirling Range (collected by John Chilvers). Note the resemblance of the leaves to those of B. gardneri. Drawn by Susan Patrick.

occasional problems particularly in parts of south-eastern Queensland and north of Sydney where the ranges of var. collina and var. spinulosa overlap. The varieties are identified solely by leaf shape, the mature leaves of var. spinulosa being narrow with revolute margins and serrations generally in the upper half only. In contrast, var. collina has wider leaves with generally serrated edges and recurved margins. On the undersides of the leaves the lateral venation is also clearly evident. In the two areas mentioned, examples of intermediate forms were often recorded. Sometimes they were recorded as one variety (var. spinulosa) when herbarium collections had indicated the other variety (var. collina) for the same location. Thus in the area north and north-west of Sydney there are several Atlas records of var. spinulosa, whereas previously this area had been thought to contain only var. collina. As before, this can be interpreted either as a range extension of var. spinulosa or as an indication of a taxonomic problem needing study.

With B. integrifolia there was sometimes confusion between var. compar and var. integrifolia especially where they occurred together, e.g. in coastal parts of south-eastern Queensland and northern New South Wales. For example, a site near Caloundra had typical var. integrifolia on coastal sand dunes, typical var. compar only one kilometre inland, whilst in between were plants showing leaf forms that were intermediate between the two varieties. Thus some leaves had the typical blunt apex of var. integrifolia but were up to 15 cm long, others had the typical pointed apex of var. compar but were only 8-9 cm long. Similar intermediate forms were recorded from areas close to Grafton, Singleton, and Gosford (all New South Wales). North of Fraser island (Queensland), populations which had been identified as var. compar by George (1981) were submitted as var. integrifolia by A. Salkin who stated that the only difference between these plants and those further south was in the leaves being slightly longer. Once again, depending on interpretation, this can be interpreted either as a range extension of var. integrifolia or as an indication of a taxonomic problem needing study.

Hybrids - Newly Recorded and Previously Known

Several presumed hybrids were recorded for The Banksia Atlas. Some were new discoveries. Others had been recorded previously and were re-found by Atlas contributors who often provided new locations. For the sake of completeness, both the new discoveries and the previously known forms are included in Table 6. In most cases, they should be regarded as presumed hybrids only. Confirmation of their status relies on a number of criteria including: (i) the presence of both parent species in the immediate area, (ii) morphological characteristics of the hybrid (e.g. leaf size and shape) being intermediate between both parent types, (iii) a degree of pollen sterility, (iv) use of techniques such as protein electrophoresis to document inheritance of enzymes of both presumed parents. Most of the presumed hybrid banksias have yet to be assessed by such techniques in criteria (iii) and (iv).

The hybrid between B. prionotes and B. lindleyana was the most significant recorded for the Atlas (Fig. 15). The two species involved are not closely related, being classified in separate taxonomic series (George 1981). Only a single presumed hybrid exists in an area north of Kalbarri National Park. The fruit is like that of B.

lindleyana, but leaves are intermediate between the two parent species.

In the Caloundra area, southern Queensland, a single presumed hybrid of *B. oblongifolia* x *B. integrifolia* var. *integrifolia* was found. Leaves are the shape of *integrifolia* except for a few blunt teeth. They are generally white on the undersurface (*integrifolia*) but have a firmer texture and rusty hairs on the midrib (*oblongifolia*). The flowers have a silky aspect more characteristic of *integrifolia* than *oblongifolia*.

Table 6. Hybrid banksias recorded for the Atlas (* = new discovery)

robur x *oblongifolia*
aemula x *serrata*
paludosa x *integrifolia* var. *integrifolia*
**paludosa* x *marginata*
**saxicola* x *marginata*
integrifolia var. *integrifolia* x *marginata*
**integrifolia* var. *integrifolia* x *oblongifolia*
**ericifolia* var. *ericifolia* x *spinulosa* var. *spinulosa*
ericifolia var. *ericifolia* x *spinulosa* var. *cunninghamii*
prionotes x *hookeriana*
**prionotes* x *lindleyana*
**hookeriana* x *attenuata*
hookeriana x *menziesii*

A presumed *B. saxicola* x *B. marginata* hybrid was again a single plant from near the summit of Mt William in The Grampians, Victoria. The leaves were similar to *B. saxicola* but the fruiting cones retained the old flowers and appeared intermediate between the two parent species.

A presumed hybrid of *B. hookeriana* x *B. attenuata* occured in the Mt Adams area of Western Australia. There was a single, large, old plant with more than 1 000 old flower spikes. All had retained their old flowers (cf. *hookeriana*). Flower spikes were much reduced in size and had only c. 120 flowers on each spike (compared to an average of 1 000 per spike for *B. hookeriana*). The flowers were somewhat malformed and never opened properly. The plant was infertile and had not produced any seed. Leaves were intermediate between *B. hookeriana* and *B. attenuata*. Old leaves were retained on the plant as with *B. hookeriana*.

Variation Within Populations

Variants of many kinds were observed by *Atlas* contributors. Some are discussed below under the various fields of information recorded by contributors on Sight Record Sheets, or in the discussions of individual species in Chapter 5. Others of note include the following.

Rare flower colour variants were observed, some of considerable horticultural merit. A few are illustrated in the colour plates. These variants included yellow *B. praemorsa*; yellow, orange and bronze *B. ornata*; cream *B. occidentalis*: yellow and bronze *B. menziesii*; orange *B. coccinea*; red *B. seminuda*; and pink and green *B. candolleana*.

Several variations on typical species characteristics were reported. Populations of *B. media*, *B. serrata* and *B. ericifolia* var. *macrantha* had extra-large flower spikes. Both large-leaved and small-leaved forms of *B. integrifolia* var. *aquilonia* were recorded. An attractive variant of *B. laricina* with wavy margins on the follicles was observed in the south-west corner of Moore River National Park, W.A. (Fig. 15). Near Gibson, W.A., several plants of *B. speciosa* had an upright form with short upright leaves. Attractive individual specimens of *B. marginata* with pendulous foliage were noted.

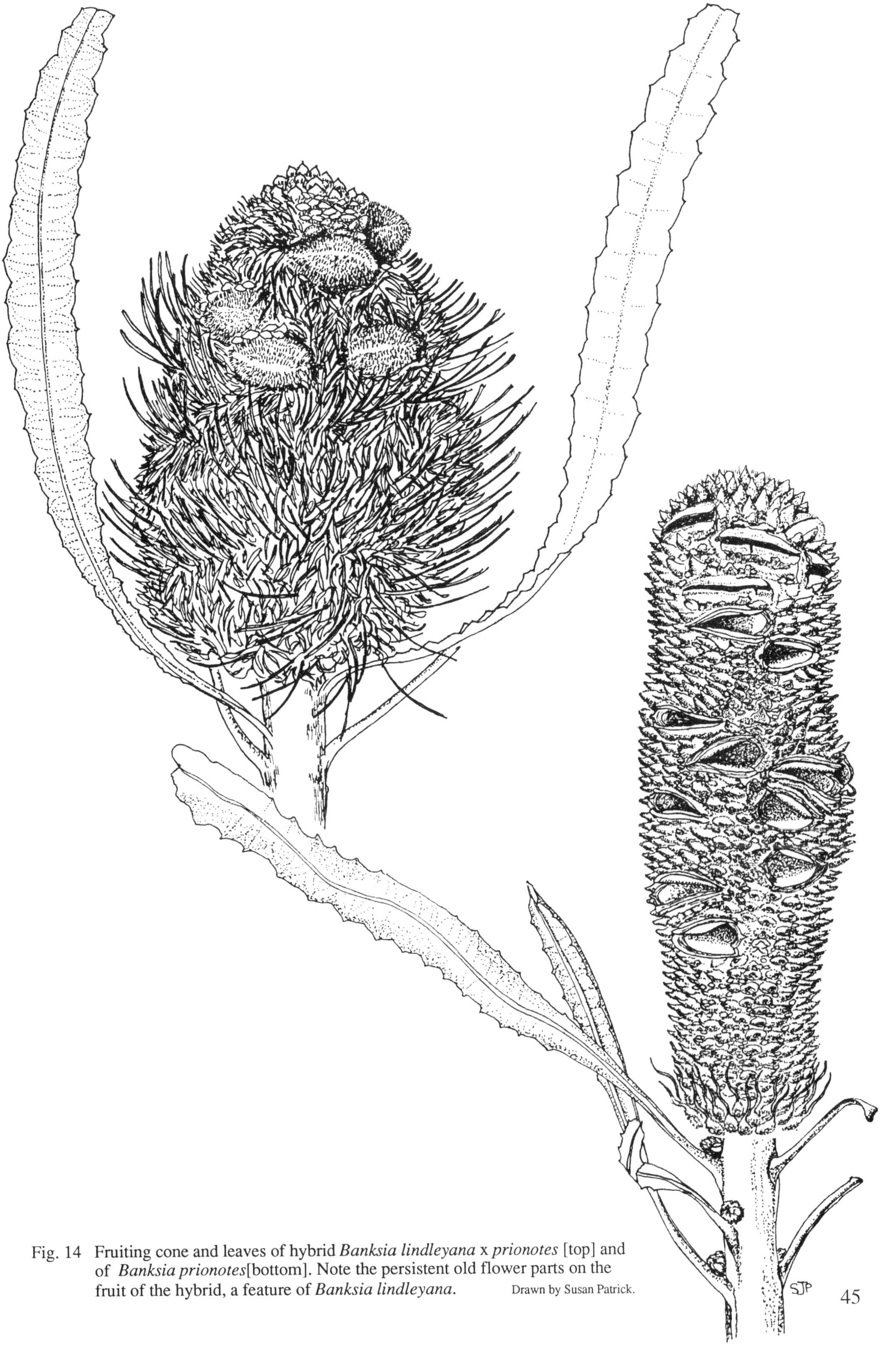

Fig. 14 Fruiting cone and leaves of hybrid *Banksia lindleyana* x *prionotes* [top] and of *Banksia prionotes*[bottom]. Note the persistent old flower parts on the fruit of the hybrid, a feature of *Banksia lindleyana*. Drawn by Susan Patrick.

Tree/Shrub Habit

Slightly less than half (40) of the 98 taxa were always shrubs, while the remainder were recorded as either trees or shrubs. Species and varieties for which 80% or more of their records were as trees included *B. dentata*, *B. grandis*, *B. ilicifolia*, *B. integrifolia* var. *aquilonia*, *B. integrifolia* var. *compar*, *B. littoralis*, *B. menziesii*, *B. prionotes*, *B. seminuda* and *B. serrata*. Coastal conditions sometimes restricted some of these tree banksias to low shrubs less than 0.5 m tall (e.g. *B. serrata* at Green Cape lighthouse, N.S.W., and *B. grandis* at Torbay, W.A.).

Height

The most common height class recorded for banksias was in the range 1-1.9 m (32 taxa), closely followed by 2-3.9 m (30 taxa) and 0-0.9 m (23 taxa). Only 13 taxa were commonly recorded as 4-9.9 m tall. *Banksia seminuda* and *B. integrifolia* each had a significant proportion of their records in the 10-19.9 m range, and a small number greater than 20 m tall.

There was some dispute as to the tallest banksia observed. Potential candidates included *B. integrifolia* var. *compar* growing in coachwood rainforest to 30 m+ with a diameter at breast height of almost 1m in Washpool National Park, N.S.W. *Banksia seminuda* was reported to grow 30-35 m tall along riverbanks in Karri forest in W.A., and *B. marginata* was reputed to attain a similar height in western Tasmania.

Contributors recorded unusually large specimens of a number of shrubby banksias. For example, *B. robur* grew to 5 m as a small single-stemmed tree in Wooroi State Forest near Caloundra in Queensland. In Western Australia the normally low *B. praemorsa* attained heights of 6-7 m in protected dunes at Two Peoples Bay Nature Reserve east of Albany. Variation in height was also evident in the two colour plates of *B. epica.*

New Shoot Growth

Data on new shoot growth were adequate to assess seasonal variation for c. 60 of the 98 banksias. All 60 showed a summer peak with new shoots appearing at lower frequencies in other seasons.

Flowering Season

Greater divergence was documented in the seasonality of flowering than for new shoot growth. Of the 89 taxa with an adequate number of observations, 34 had a maximum number of flowering records in autumn, 24 in summer, 14 in spring, 13 in winter and 3 appeared to flower throughout the year. Most species flowered over several months of the year.

Growth Response After Fire

Thirty-six banksias had 20 or more observations of their response to fire. Of the 36, 10 had the majority of their adult plants killed by fire and regenerated by germination of seed. These fire-sensitive banksias included *B. coccinea*, *B. ericifolia* var. *ericifolia*, *B. nutans*, *B. ornata*, *B. prionotes*, *B. quercifolia*, *B. speciosa* and *B. spinulosa* var. *cunninghamii*.

Of the 26 resprouting species, 15 were trees capable of resprouting from the ground and the trunk (e.g. *B. grandis*, *B. integrifolia*, *B. menziesii*, *B. serrata*). The remaining 11 were prostrate species (e.g. *B. gardneri*, *B. repens*) or erect shrubs (e.g. *B. elderiana*, *B. paludosa*, *B. spinulosa* var. *collina*) that resprouted only from the lignotuber.

Data from records on most banksias were too few to draw conclusions on their growth response after fire but are given for most taxa by George (1981, 1984) on the basis of his own observations, on herbarium label data and on unpublished observations of others (George pers. comm.).

Flowering Response After Fire

Data were generally insufficient for reliable conclusions to be drawn. There were also problems in recording technique due to the variation and patchiness of observed fires and the complex responses shown by banksias in the face of such variation (see relevant sections in Chapters 2 and 5). For these reasons, this section has not been summarised. However, for the sake of completeness, the data for each species are provided in Chapter 5.

Possible Pollinators

Data on possible pollinators ranged from no observations to 920 for B. integrifolia. Animal activity at flowers was usually not observed, however, as more than 75% of pollinator records were negative observations. The study of Banksia pollination clearly requires special techniques if it is to be adequately documented.

Birds featured prominently as possible pollinators for well-recorded banksias, with species such as the ubiquitous New Holland Honeyeater being commonly observed. This is no doubt a reflection of the conspicuousness of honeyeaters. Bees, wasps and ants were also observed frequently. Their role as pollinators needs careful research, however, because their body size and behaviour are usually inappropriate to regularly place pollen on stigmas.

Mammals were rarely reported at flowers, which is understandable given their nocturnal feeding behaviour. Nevertheless, recent research suggests that honey possums, Antechinus spp., gliders and bats may be important pollinators of banksias in some circumstances (Whelan & Goldingay 1986; Goldingay et al. 1987 and references therein).

Coastal Banksias

Five banksias were reported to grow within 2 km of the coast for

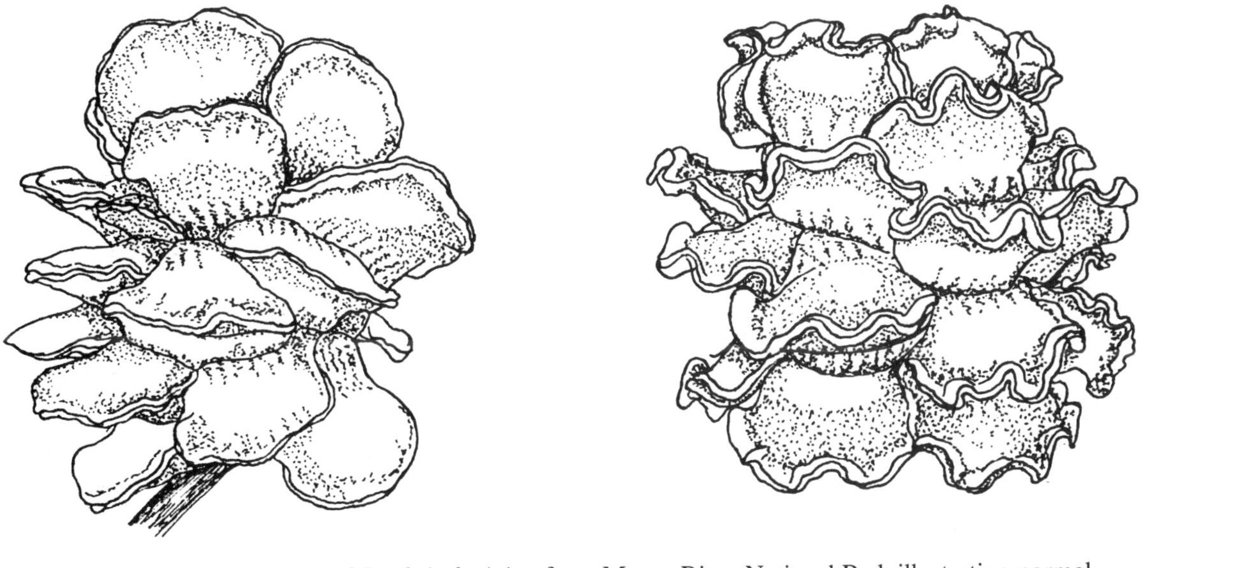

Fig. 15 Fruiting cones of *Banksia laricina* from Moore River National Park illustrating normal follicles [left] and a variant with wavy margins [right]. Drawn by Susan Patrick

the majority of their Atlas records - *B. epica*, *B. ericifolia* var. *macrantha*, *B. integrifolia* var. *integrifolia*, *B. praemorsa* and *B. verticillata*. A number of others had a significant percentage of their populations near the coast but also extended further inland (e.g. *B. aemula*, *B. occidentalis*, *B. serrata*).

Altitude

Most banksias (51 of 98 taxa) were most commonly recorded at altitudes between 100 and 249 m, a reflection of the species richness of banksias on the subdued sandplains of Western Australia. Two (*B. integrifolia* var. *compar* and *B. spinulosa* var. *neoanglica*) occurred most frequently in uplands more than 1 000 m above sea level, and another nine were usually in mountains between 500 and 999 m (*B. canei*, *B. conferta*, *B. conferta* var. *penicillata*, *B. integrifolia* var. *aquilonia*, *B. paludosa*, *B. plagiocarpa*, *B. saxicola* and *B. spinulosa* in eastern Australia, *B. solandri* in the Stirling Range of Western Australia).

At the other extreme, 13 banksias were found mainly in low-lying often near-coastal areas less than 20 m above sea level (*B. aemula*, *B. ericifolia* var. *macrantha*, *B. ilicifolia*, *B. integrifolia* var. *integrifolia*, *B. littoralis*, *B. meisneri* var. *ascendens*, *B. oblongifolia*, *B. occidentalis*, *B. praemorsa*, *B. robur* and *B. serrata*).

Soil Type

Sand was the preferred soil type of 83 of the 98 banksias, with three more favouring sand over laterite. Rocky soils, sometimes with laterite, and principally in montane areas, were occupied by 12 taxa. *B. seminuda* was exceptional in its preference for rich loams along watercourses in the forests of Western Australia.

Vegetation

Banksias were usually recorded growing in heathlands less than 2 m tall (46 taxa) or in woodlands dominated by low to tall trees whose canopies shade up to 30% of the ground (33 taxa). Two were exceptional in growing in forests (*B. seminuda* and *B. integrifolia* var. *aquilonia*). Indeed, a significant number of records of the latter were made in rainforest.

Large shrub and mallee communities were favoured by a small number of banksias. Grasslands and cleared farmland were rarely reported as the habitat.

Landform

Flat or gradually sloping terrains were favoured by most banksias (52 and 37 taxa respectively). Steep montane slopes were preferred by *B. gardneri* var. *brevidentata*, *B. integrifolia* var. *aquilonia*, while *B. praemorsa* occured mainly on steep coastal slopes.

Banksias growing on hill or mountain tops included *B. conferta* var. *conferta*, *B. oreophila*, *B. plagiocarpa* and *B. sphaerocarpa* var. *dolichostyla*.

At the other extreme, banksias such as *B. littoralis*, *B. robur*, *B. telmatiaea*, *B. quercifolia* and *B. occidentalis* favoured seasonally waterlogged swamps. A few other taxa often occured on the drier raised margins of such swamps, e.g. *B. aemula*, *B. oblongifolia*, *B. meisneri* var. *ascendens*, *B. paludosa*, *B. ericifolia* var. *macrantha* and *B. dentata*.

Two unusual landforms were river banks, the preferred habitat of *B. seminuda*, and large coastal granite outcrops, occupied by *B. verticillata*.

4. CONCLUSIONS

It is important to note that The Banksia Atlas was a project based on observations of field workers, rather than on experimental scientific research. As such, one of its values is in the many questions raised about the genus Banksia. Each of these questions must await experimental research work before it can be answered convincingly.

In the short time available to prepare the Atlas for publication, we have barely been able to scratch the surface in analyzing the wealth of data now on computer. There are many correlations between geographical distribution, habitat and phenology as yet unexplored. Patterns of geographical variation, natural hybridisation and confusing taxonomic relationships await the skilled attention of botanists interested in population genetics and evolution. Many fascinating problems requiring research become apparent on reading the data summarised in Chapter 3 and presented in the main Atlas below. It is clear that much remains to be learnt about the biology and ecology of banksias.

To facilitate future work on the Atlas data base, we have placed a copy on computer tape and lodged it with the Bureau of Flora and Fauna in Canberra. Interested researchers are invited to obtain a copy from that institution or from the Western Australian Department of Conservation and Land Management if they wish to pursue further analyses of the data.

Being the first of its kind in Australia, The Banksia Atlas has highlighted a number of points worth bearing in mind should similar projects be undertaken in the future. These are summarised at the end of Chapters 1 and 2. Suffice it to say here that we are convinced that every effort should be made to ensure that cooperative projects of this kind are continued. It is pleasing to note that an atlas of rare and poorly known eucalypts will follow on from The Banksia Atlas in Western Australia. Perhaps other States or the Commonwealth might similarly seize the initiative in view of the demonstrable success of The Banksia Atlas.

Apart from the valuable contributions to scientific knowledge made by The Banksia Atlas, it was also hoped that taking part in the project would stimulate interest and enjoyment of the rich Australian native flora and fauna. Towards the end of the project, numerous letters were received stressing both how much volunteers had enjoyed taking part, and how much they had learnt. The following comments are a representative sample of those received:

> "I'm really sorry the Banksia Atlas has ended and regret, of course, that we didn't get cracking a bit earlier. I hope... that the whole exercise has been worthwhile - it certainly was for our little crew down in the Deep South."

> "I found the whole exercise most stimulating and rewarding... I would strongly encourage you to repeat the exercise with some of the more common and widespread plants."

> "Taking part in the project has transformed my whole appreciation of Australian flora and given me a new perspective on this continent."

> "Thank you for the opportunity to participate... I thoroughly enjoyed the whole system. It would be a shame if

the expertise and technology could not be employed in a future program... I look forward to future developments."

"Through the project I have made a whole new group of people aware of banksias. I certainly enjoy the new knowledge which I and my family have gained."

Many also commented on the effect that their activities had had on other people.

"I think the idea of volunteers undertaking survey work is a great way of both getting the job done and also creating a public awareness of our bush."

"The project has been most enjoyable. One thing I've noticed is the interest shown by local people when asked questions about localities, fire histories, etc."

"We are finding that our enthusiasm for banksias is rubbing off on to people who have previously had little thought for this genus. Perhaps this will be one of the most important benefits generated by the Banksia Atlas project."

"Reading about the Cooma farmer who was against banksias reminds me of one we met. He too, 'killed Banksias wherever we find them, they're a pesky nuisance'. However in spite of this he was kind enough to send a man with us to show us where we would find some. I later wrote to thank him and to tell him the name of the type we found, also a few other names and I asked him wherever possible to please spare the trees. About 3 months later I received a delightful letter in answer thanking me for 'opening his eyes' and to say he had no idea there were so many types nor that they had such a variety of names, shapes and colours. He promised to leave them alone where possible and invited us to call again when we were near."

This spin-off effect amongst the wider community should not be under-estimated. Conservation of native vegetation is not only a matter of protection through reservation, erection of fences and bans on clearing. An equally important element is to have a well-informed, interested and enthusiastic public. Projects like The Banksia Atlas greatly assist in this process.

LAMENT OF A BANKSIA OBSERVER (FAILED!)

Oh I love to look at Banksias
(It's heaven in the field)
I love to count their little heads
And calculate their yield
It's fun to look for little bugs
Which go from flower to flower
It's fun to see what shrubs grow round
and name the trees that tower.
Discussions too I do enjoy
I get a lot of pleasure
Out of talking atlas talk
With observers at their leisure
But when at home the record sheet
Lies open on the table
To find the time to fill it in
I find that I'm unable.

Glenys Trigwell

SPINULOSA VARIETIES

I know a spinulosa
From the rarer paludosa
And from marginata, aemula, and all.
I know it from dentata,
And from robur and serrata,
But varieties can send me up the wall.

It could be spinulosa,
Perhaps collina's closer,
Or maybe it is cunninghamii.
To add to the confusion
There's now a new intrusion,
The one they want to call New
Englandi.

I lose all my momentum
When that leafy indumentum
Could be tomentose, pubescent,
or hirsute.
And if I look more fully,
Is it villous or just woolly,
Or maybe there's a better name
to suit.

And the edges of the leaves
Are a factor that deceives
When changes in one tree can be
so great.
A further complication
Is if every variation
Has a leaf which is entire to
serrate.

Perhaps it's not dentated,
But rather more serrated,
Apiculate, retuse, or
mucronate,
And this one is a beaut,
Recurved or revolute,
Acute, and just a bit
emarginate.

I'll leave those varied edges
To study shapes like wedges,
Or sickle-like or truncate
turbinate.
Now are the leaves mature,
Or just a little newer,
Like juveniles with quite a
different trait?

With botanists so clever,
I wonder if they'll ever
Produce a tree that grows its
printed name.
An inbuilt appellation
Would cause a big sensation,
But botany would never be the
same.

Well spinulosa chasing
Is brain and body bracing,
Despite the fact it sends me up
the wall.
I won't be broken hearted
If another "Atlas" started;
Perhaps I'll even be the first
on call.

Eric Williamson

5. INTERPRETING THE BANKSIA MAPS AND ASSOCIATED DATA

The rest of the book presents all data submitted on the Sight Record Sheets for each banksia. The sequence of banksias is alphabetical. The botanical name of each banksia is followed by the botanist who named it with the date of publication in brackets. Common names are also provided.

For each species, the reader is requested to first consult the total number of records and their seasonal distribution. The accuracy and sample size of any subsequent data entry may then be judged with regard to these figures.

The total number of records for any species does not always tally with the number of records illustrated on a monthly basis. With B. burdettii as an extreme example, there is a total of 81 records, but the monthly figures add up to only 35. This discrepancy is a result of the month not always being recorded on sight records.

Percentages have generally been used to summarise results. There are just two exceptions where actual numbers have been used instead. With Specific Pollinators there are generally very few records and percentages could convey an incorrect picture of results. With Population Size, the use of actual numbers illustrates more vividly the relative abundance of each species than could percentages.

In most cases, the total number of records for each species is used to calculate subsequent percentages. Thus for B. attenuata most results are expressed as a percentage of 1 280 records. Sections that were not filled in on the record sheets were classified as "Unspecified" and are included in the results.

The "Response to Fire" and "Pollinator" sections are treated differently. For any species the number of recordings within these sections is indicated and results are expressed as percentages of these figures rather than of the species as a whole.

Whenever percentages have been used figures have generally been rounded off to the nearest whole number. However, this process could eliminate some potentially interesting records which, due to their low frequency (0.5%), should theoretically be rounded off to zero. The infrequent recording of mammals as a type of pollinator is a good example. Such records are difficult to obtain and in this case any record may significantly increase our knowledge of a species' pollination ecology. Small percentage figures are therefore included and have been rounded off to the first decimal point.

Interpreting Fire Response

There were several problems experienced in the accurate recording of fire response (see relevant section in Chapter 2). The "patchiness" of fires meant that within a generally burnt locality there could be small areas that remained unburnt. Accurate recording required the use of two record sheets - one for the burnt and another for the unburnt area. Understandably, there was reluctance to repeat record sheets in this way. Sometimes volunteers combined the information on a single record sheet. It had been intended that the "Minimum number of months to flowering" refer only to burnt plants. It would illustrate the minimum time required after fire for a species to become reproductively active again and therefore capable of producing seed. However, when non-burnt plants were included on the same record sheet, and were not detected through the normal checking process, they could give an incorrect impression of a short time between fire and flowering.

With some tall species, e.g. B. attenuata, fire does not always reach the canopy and it is possible that burnt plants could be flowering within 1-2 months of a fire. With smaller species, especially if they are killed by fire and regenerate from seed, a small time interval is unlikely and most probably refers to unburnt plants. Since it was not possible to separate this information, the reader is cautioned to interpret these data carefully.

Interpreting Pollinators

The list of pollinators for any banksia species should be regarded as "possible pollinators" only (see Chapter 2). The lists are also not comprehensive. Certain groups of animals (e.g. mammals) are likely to be under-recorded due to their mainly nocturnal activities.

Due to the design of the record sheet, the pollinator class frequencies do not necessarily tally with the specific pollinator frequencies. For example, a single recording of "bird" in pollinator class could be followed by a list of several specific bird species, thus creating an apparent imbalance between the two categories. This is particularly the case with birds. Species of other potential pollinators, e.g. ants, were not known and were therefore confined to pollinator class.

Distribution Maps

The base map data were provided by the Western Australian Department of Lands and Surveys. The projection used is an Alber Equal Area on Clarke's 1858 spheroid, with standard parallels of 17°30' and 31°30', and a central meridian of 130°E. Computer projection routines are Hutchinson's (1980), modified by N.Hall.

The distribution maps show all recorded occurrences of banksias during the period July 1984 - August 1986. The presence of a banksia at a location is indicated by a cross. The precision of these locations varies as already explained in Chapter 2 under 'Locality Resolution Code'.

Almost all sight records are accurate to the nearest 1' (1.8 km). Most are accurate to the nearest 30" (900 m) and many to the nearest 10" (300 m). Tasmania was the most accurately recorded State with almost all records specified and checked to the nearest 10" (300 m).

No attempt has been made to illustrate the abundance of a species at each record locality. A cross may represent anything from a single plant to more than 1 000 plants within a record locality. Population data had been collected on the record sheets and are included in each species' report. A visual illustration of these data was not considered justifiable since in most areas there had been no attempt made at a thorough survey and an illustration of abundance based on incomplete data would be misleading. A heavy concentration of crosses in any area therefore indicates an abundance of recording and not necessarily an abundance of plants.

The following example illustrates this point. Banksia marginata appears on maps in Chapter 5 to be very common and widespread in parts of South Australia. However, this impression of abundance is misleading since the area is largely cleared and most populations exist as road verge remnants only. In this case, the diligence of a single recorder has resulted in the systematic recording of many small populations in adjoining record localities.

In other cases, an abundance of crosses is consistent with an abundance of plants, e.g. B. grandis along the coastal plain and Darling Scarp of Western Australia, and B. serrata and B. integrifolia along coastal parts of eastern Victoria.

On all maps, the reverse situation also applies. Blank areas without any crosses may mean an absence of sight records rather than an absence of banksias. The recording of 'nil' areas would have helped to clarify this point.

Generally, the distribution maps have been drawn to the largest scale possible. Place names, and in some cases national park boundaries, are provided, to help the reader locate actual distributions. In Western Australia, there are eight banksias currently gazetted as declared rare or endangered flora. For these nine species only small scale maps are provided in order to retain the confidentiality of their exact locations.

Herbarium collections which generally pre-date the Atlas are indicated on the small inset map within each large distribution map. For Western Australian banksias data were collected from all specimens housed in the Western Australian Herbarium up until December 1984. For eastern banksias, records were compiled from all herbaria (both national and international) where relevant collections were known to be lodged up until 1980.

Herbarium records are thus largely historical records dating back up to 217 years. The original collectors often did not specify latitude and longitude and these have been subsequently added as precisely as the information on locality allowed. Sometimes the latter was very vague and inaccuracies of the order of 20 or even 50 kilometres are met. Therefore, in terms of latitude and longitude, maps based on herbarium collections generally are less accurate than Atlas maps.

In almost all cases the Atlas maps show many more crosses than maps based on herbarium collections. This reflects the two different methods of data collection employed and should not be taken to mean an increase in abundance of any species. Where crosses now exist where there were none before, this obviously reflects an increase in knowledge of the banksias' distribution and not a spread of the species into new territory.

Finally, it must be stressed that the results are presented only as summaries of raw data from the record sheets. There are many subsequent avenues that could be explored using the existing Atlas data base, e.g. the relationship between plant height and location or between plant height and altitude; the effect of location or altitude on flowering time or new shoot growth; the relationship between fire intensity and banksia response; the banksia assemblages of different areas; the most important factors limiting species' distributions. Such topics will doubtless be the subject of future studies. Given publication time-contraints, they were beyond the scope of this book.

REFERENCES

Baird, A.M. (1977). Regeneration after fire in Kings Park, Perth, Western Australia. J. Roy. Soc. W. Australia 60 : 1-22.

Barker, M.J., & Lamont, B.B. (1986). The conservation status and reproductive biology of Banksia burdettii. Unpubl. report to World Wildlife Fund Australia.

Barrett, G.J. (1985). The reproductive biology and conservation of two rare Banksia species. M.App.Sci. thesis, W. Austral. Inst. Techn., Perth.

Blakers, M., Davies, S.J., & Reilly, P.N. (1984). The Atlas of Australian Birds. Melbourne University Press, Melbourne

Burgman, M.A., & Hopper, S.D. (1982). The Western Australian Wildflower Industry 1980-81. Dept. Fish. Wildl. W. Austral. Rep No. 53.

Busby, J.R. (1984). Bioclimate Prediction System Users' Manual. Bureau of Flora & Fauna, Canberra.

Conran, J.G., & Clifford, H.T. (1987). Variation in Banksia oblongifolia Cav.: Proteaceae. Brunonia, in press.

George, A.S. (1981). The genus Banksia L.f. (Proteaceae). Nuytsia 3: 239-474.

George, A.S. (1984). The Banksia Book. Kangaroo Press, Sydney.

George, A.S. (1988). New taxa and notes on Banksia L.f. (Proteaceae). Nuytsia 6(2), in press.

Goldingay, R., Carthew, S., & Whelan, R. (1987). Transfer of Banksia spinulosa pollen by mammals : implications for pollination. Austral. J. Ecol., (in press).

Holliday, I., & Watton, G. (1975). A Field Guide to Banksias. Rigby, Adelaide.

Hopper, S.D. (1980). Bird and mammal pollen vectors in banksia communities at Cheyne Beach, Western Australia. Austral. J. Bot. 28: 61-75.

Hopper, S.D. (1981). Honey Possums. Swans 11: 21-23.

Hopper, S.D. (1988). New subspecies of Banksia seminuda and B. occidentalis (Proteaceae) from the south coast of Western Australia. Nuytsia, (in press).

Hutchinson, M.F. (1980). MAPROJ, a computer map projection system. CSIRO Div. Land Use Res. Tech. Pap. No. 39.

McCredie, T.A., Dixon, K.W., & Sivasithamparam, K. (1985). Variability in the resistance of Banksia L.f. species to Phytophthora cinnamomi Rands. Austral. J. Bot. 33: 629-37.

Newbey, K.R., & Hnatiuk, R.J. (1985). Vegetation and flora. In The Biological Survey of the Eastern Goldfields of Western Australia. Part 3. Jackson-Kalgoorlie Study Area. Rec. W. Austral. Mus. Suppl. No. 23.

Nix, H.A. (1986). A biogeographic analysis of Australian elapid snakes. Australian Flora & Fauna Series 7: 4-15.

Perring, F., & Walters, S. (1962). Atlas of the British Flora. Botanical Society of the British Isles, London.

Van Leewen, S.J. (1985). An investigation of the biology and conservation status of Banksia tricuspis, a gazetted rare species. Thesis, Grad. Dip. Nat. Resources, W. Australia. Institute of Technology, Perth.

Whelan, R., & Goldingay, R. (1986). Do pollinators influence seed set in Banksia paludosa Sm. (sic) and Banksia spinulosa R.Br.(sic) Austral. J. Ecol. 11: 181-6.

1

2

3

4

5

1 *Banksia oligantha* near Wagin in W.A. (P.C. Taylor)

2 A mature plant of *Banksia epica* (J. Falconer)

3 *Banksia epica* growing as a postrate shrub on limestone cliffs at the head of the Great Australian Bight, Western Australia (G.J. Keighery)

4 Flowers of *Banksia oligantha* (P.C. Taylor)

5 *Banksia spinulosa* var. *neoanglica* resprouting from its lignotuber following fire (A. Taylor)

6

7

9

10

11

12
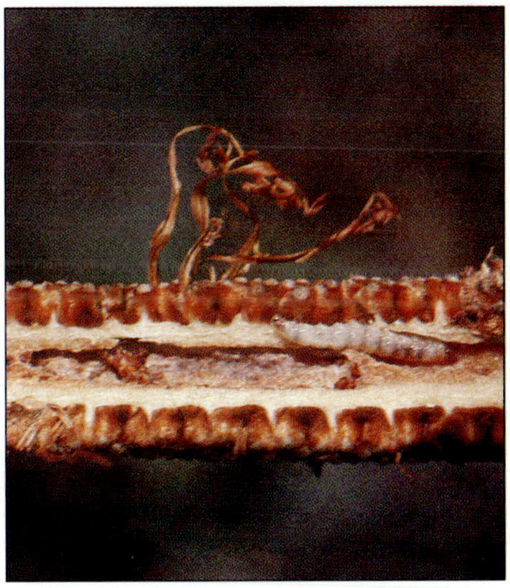

6 *Banksia marginata* growing as a 10m tall tree in eastern Tas. (S.D. Hopper)

7 Possibly a new variety of *Banksia ashbyi* growing as a low lignotuberous shrub on Quobba Station (G. Schmidt)

8 Excavated lignotuber of *Banksia ashbyi* shrub at Cape Range (G. Schmidt)

9 Honey possum on *Banksia coccinea*. Pollination by mammals was rarely recorded by contributors but may be important for some species (S.D. Hopper)

10 *Banksia marginata*, the most commonly recorded Australian species, growing as a 1.5m tall shrub in Deua National Park, N.S.W. (S.D. Hopper)

11 *Banksia conferta* var. *conferta* — a seedling that has germinated after fire (A. Taylor)

12 Predation of the floral axis of *Banksia tricuspis* by a moth larva. Such predation was commonly observed by contributors. The insects themselves are a food source for cockatoos in W.A. (S.D.Hopper)

13

14

15

16

17

18

19

13 *Atlas* contributors and Frenchman Peak, Cape Le Grand National Park, W.A. (P.C. Taylor)

14 *Atlas* contributors using the Instruction Booklet at Cannington, W.A. (P.C. Taylor)

15 *Banksia speciosa* killed by dieback fungus on the south coast of W.A. (T. Hill)

16 *Banksia verticillata,* a species rarer than previously thought, near Bald Island, W.A. (S.D. Hopper)

17 *Banksia scabrella* inconspicuous on a road verge north-east of Dongara, W.A. Road verges are important for the conservation of several banksias in agricultural regions (S.D. Hopper)

18 *Banksia brownii* on Bluff Knoll in the Stirling Range, W.A. A rare species threatened by dieback infection (S.D. Hopper)

19 *Banksia robur* (left), *B. oblongifolia* (right) and a natural hybrid (centre) from Caloundra, Qld (A. Taylor)

20 

21 

22 

23 

24 

25 

26 

20 Normal colour-form of *Banksia candolleana* (G. Velterop)

21 Rare pink colour-form of *Banksia candolleana* (G. Velterop)

22 Normal colour-form of *Banksia ornata* (A.S. George)

23 Rare brown colour-form of *Banksia ornata* (S.D. Hopper)

24 Normal colour-form of *Banksia menziesii* (S.D. Hopper)

25 Rare yellow colour-form of *Banksia menziesii* (S.D. Hopper)

26 Rare bronze colour-form of *Banksia menziesii* (G. Velterop)

BANKSIA ACULEATA A.S.George (1981)

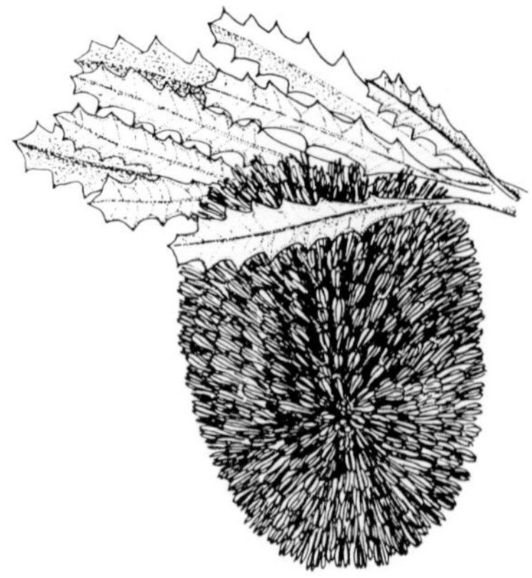

12 RECORDS: Jan (0) Feb (2) Mar (0) Apr (0) May (1) Jun (6) Jul (0) Aug (0) Sep (1) Oct (2) Nov (0) Dec (0).

Population Size: 1-10 (4) 10-100 (3) >100 (4) Unspecified (1).

Conservation Status: Restricted to road verge (0%) Not (100%) Unspecified (0%). In conservation reserve (92%) Not (8%) Unspecified (0%).

Tree/Shrub: Tree form (0%) Shrub form (100%) Unspecified (0%).

Height (metres)

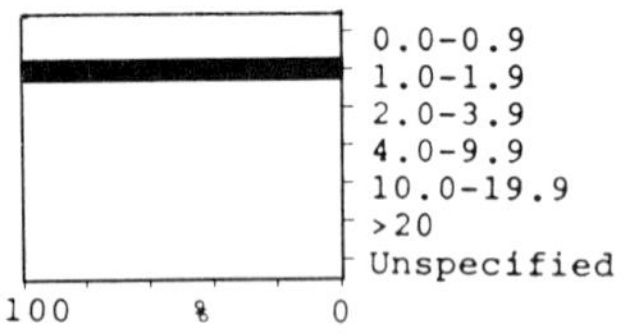

New Shoot Growth

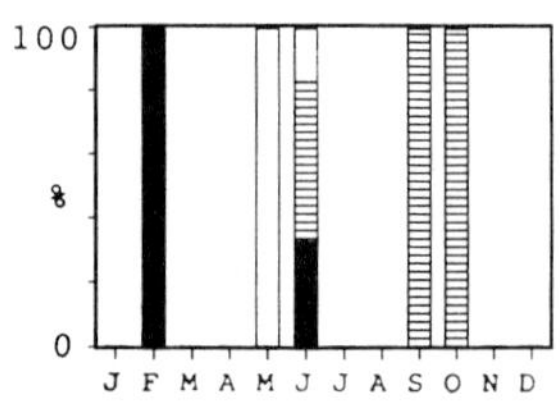

Yes
No
Unspecified

HABITAT

Within 2 km of coast: Yes (0%) No (100%) Unspecified (0%).

Altitude (metres)

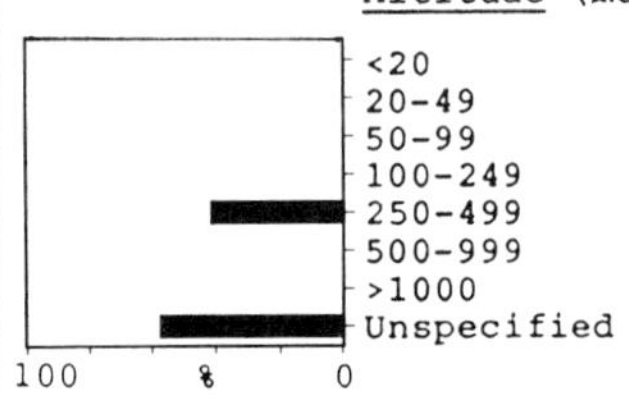

Vegetation

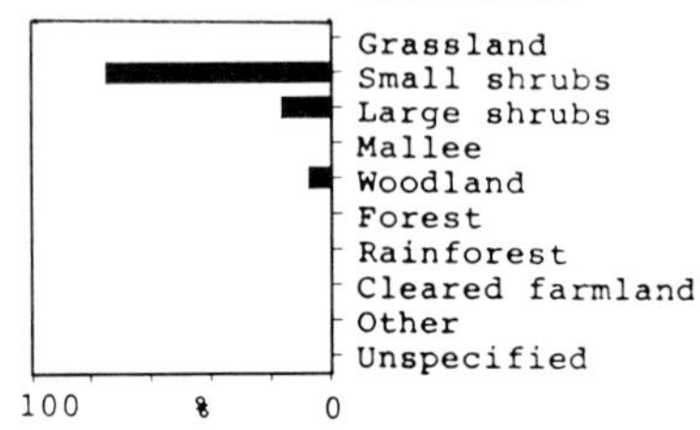

Soil Type

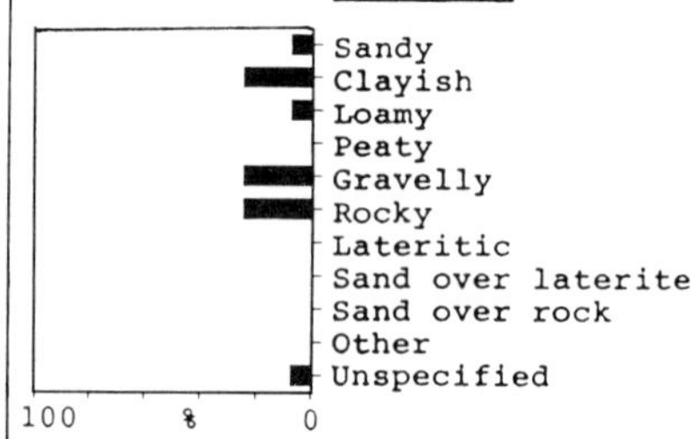

Landform

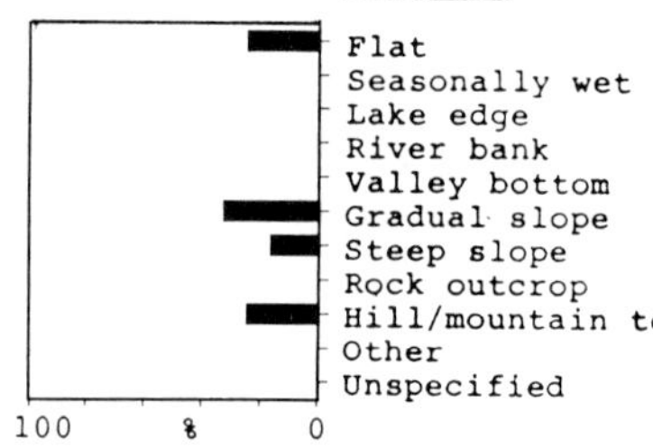

Response to Fire: 9 records

Growth response: Killed, new seedlings (78%) unspecified (22%).

Flowering response: Number of records where flowering had occurred after fire (6). Median reported time to flowering (25-36 months); minimum reported time to flowering (<1 month). Number of records where flowering had not occurred after fire (3). Median reported time to no flowering (13-24 months); maximum reported time to no flowering (25-36 months).

Flowering

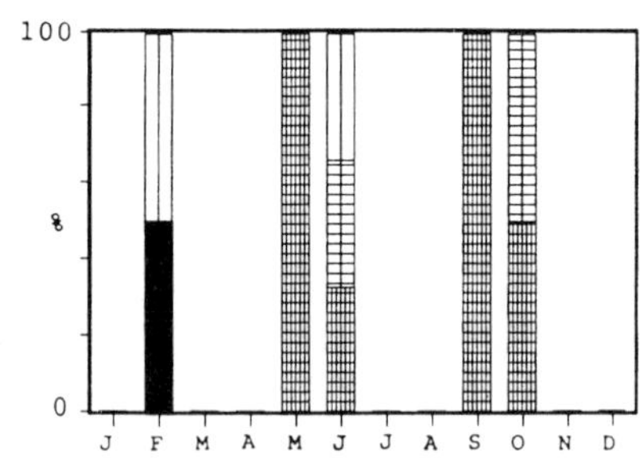

Majority of flowers fully open
Majority of flowers in bud
Majority of flowers recently finished
Flowers finished, fruiting cones present
Neither flowers nor fruiting cones present
Unspecified

Possible pollinators

Pollinator type 3 records: no pollinator observed (100%).

Specific pollinator 0 records.

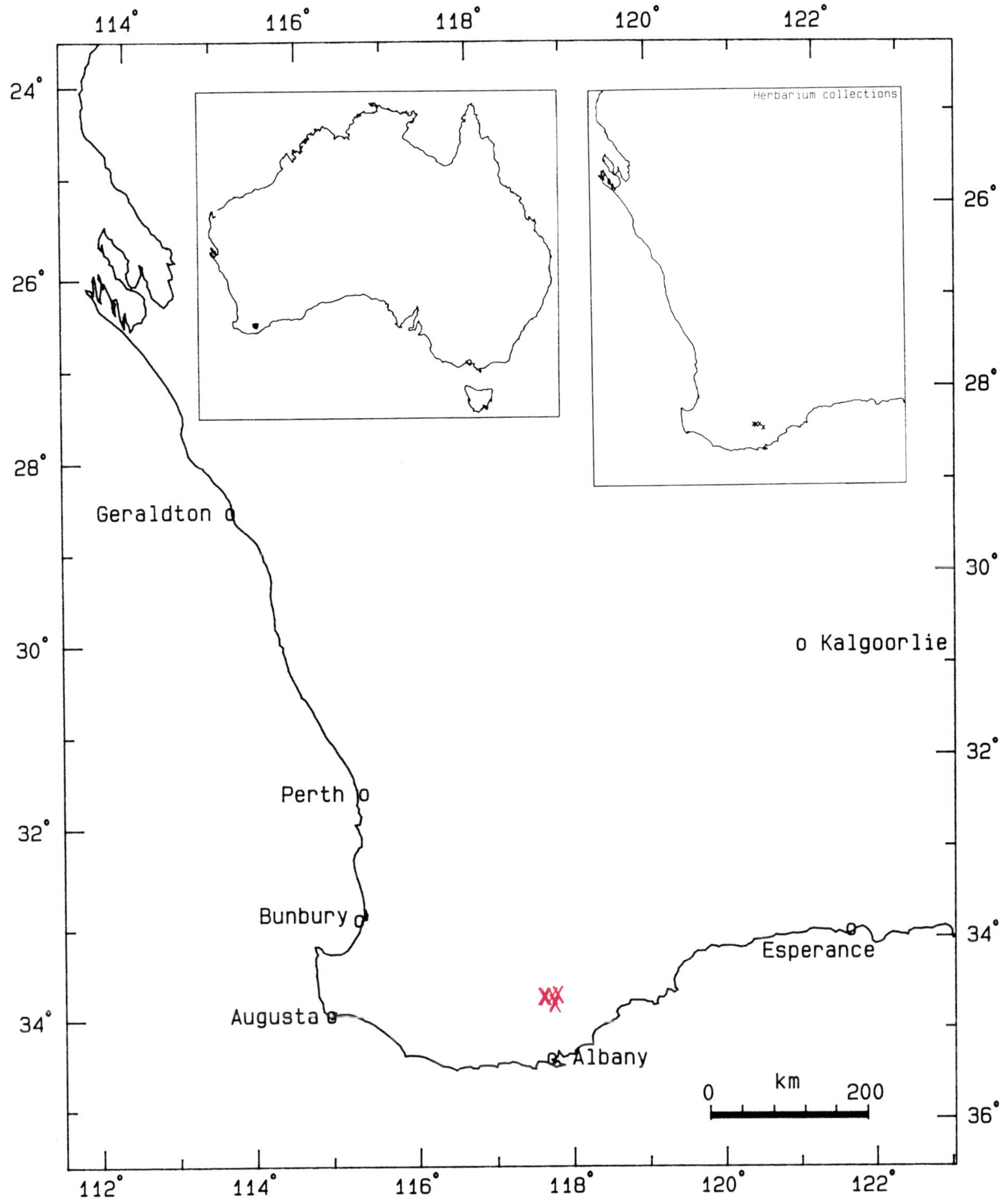

DISCUSSION

Atlas records have confirmed that B. aculeata is restricted to the Stirling Range area. A herbarium collection of G.J. Keighery's in 1983 from the western boundary of the national park was not recorded by Atlas contributors. However contributors did report at least 5 new locations of this species. This is a rare banksia with few sight records and most of its populations consisting of less than 100 plants. It occurs mainly within the Stirling Range National Park and, therefore, should be well protected. However, its susceptibility to dieback disease is unknown, and further research on its response to fire is needed for appropriate fire management.

B. aculeata matures to a shrub of 1-2 m in height. It grows in shrublands on the slopes and tops of foothills of the Stirling Range at an altitude of 250-500 m, usually on skeletal rocky soils that are gravelly and clayish. Flowering was recorded once (in February). No pollinators were observed. The species is killed by fire and regenerates from seed. It may take 3-4 years following fire for seedlings to mature to flowering.

BANKSIA AEMULA R. Brown (1810)

Wallum Banksia

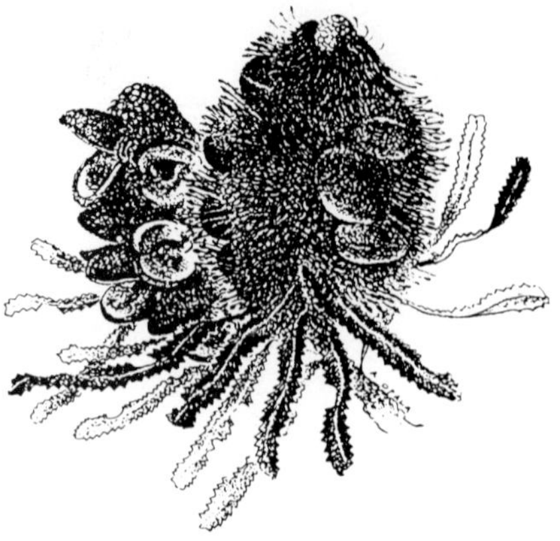

273 RECORDS: Jan (33) Feb (5) Mar (23) Apr (20) May (33) Jun (12) Jul (30) Aug (47) Sep (21) Oct (25) Nov (15) Dec (8).

Population Size: 1-10 (20) 10-100 (70) >100 (177) Unspecified (6)

Conservation Status: Restricted to road verge (3%) Not (95%) Unspecified (2%). In conservation reserve (52%) Not (32%) Unspecified (15%).

Tree/Shrub: Tree form (60%) Shrub form (38%) Unspecified (3%).

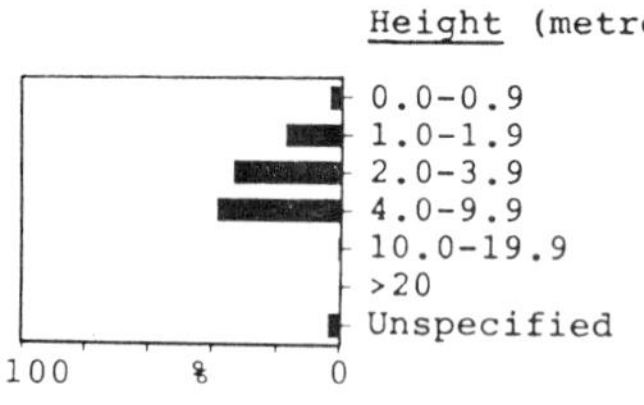

HABITAT

Within 2 km of coast: Yes (69%) No (30%) Unspecified (1%).

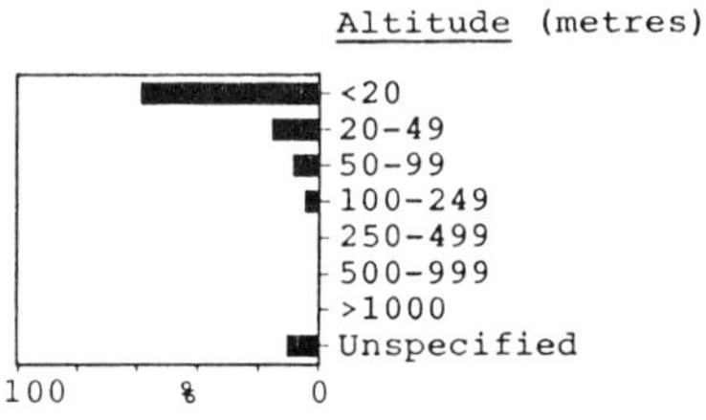

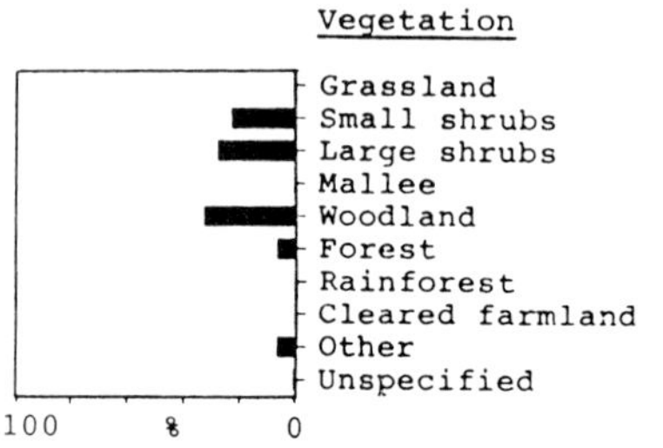

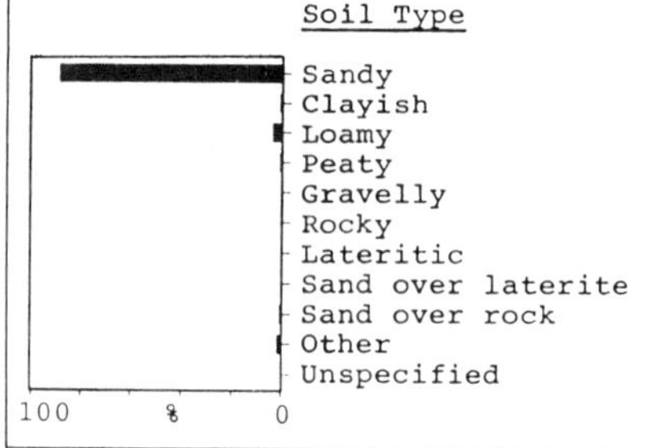

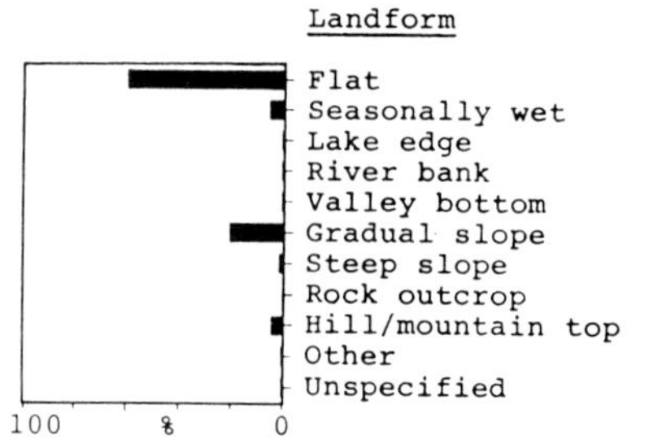

New Shoot Growth

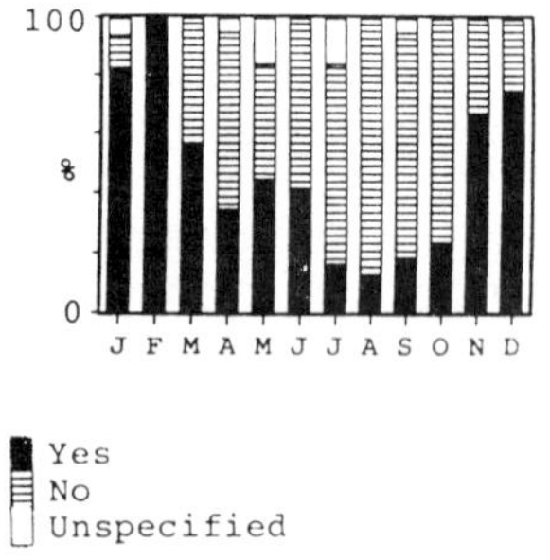

Response to Fire: 51 records

Growth response: Ground resprout (18%) ground resprout and trunk resprout (20%) killed, new seedlings and trunk resprout (2%) trunk resprout (39%) unspecified (22%).

Flowering response: Number of records where flowering had occurred after fire (30). Median reported time to flowering (1-12 months); minimum reported time to flowering (<1 month). Number of records where flowering had not occurred after fire (7). Median reported time to no flowering (1-12 months); maximum reported time to no flowering (25-36 months).

Flowering

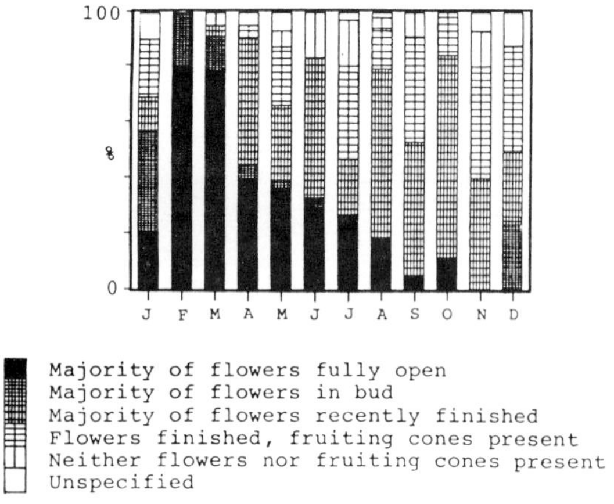

Possible pollinators

Pollinator type 109 records: bird (29%) butterflies, moths (1%) bees, wasps, ants (18%) no pollinator observed (51%).

Specific pollinator 50 records: honeyeater (10) New Holland Honeyeater (2) Brown Honeyeater (4) White-cheeked Honeyeater (6) Tawny-crowned Honeyeater (1) Noisy Miner (1) wattlebird (2) Little Wattlebird (2) friarbird (2) Noisy Friarbird (1) lorikeets and allies (2) Rainbow Lorikeet (1) flycatchers and allies (1) ant (3) bee (9) European honey bee (3).

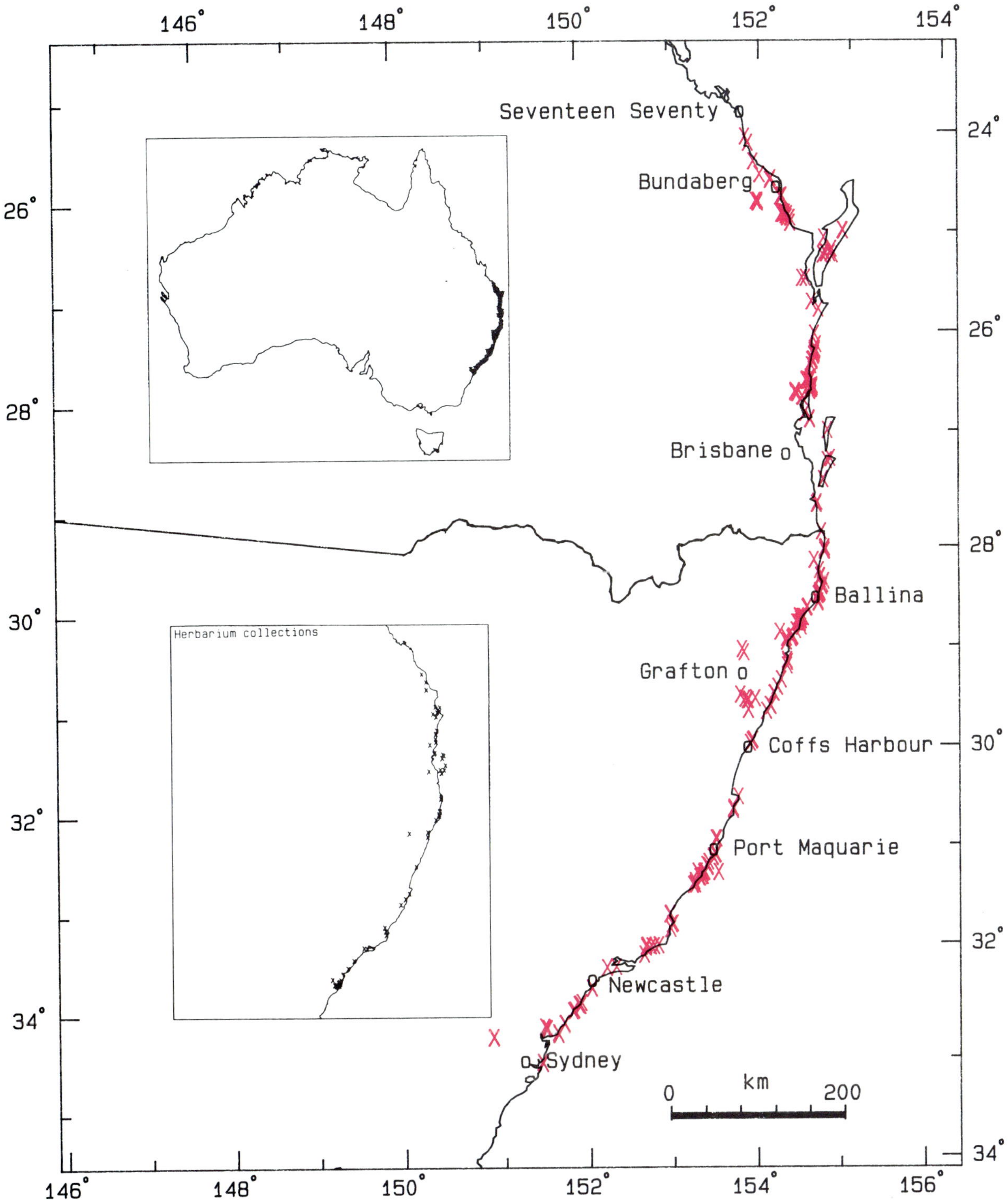

DISCUSSION

B. aemula has a coastal distribution from Sydney to ca. 70 km north of Bundaberg. It is also present on Moreton Island, North Stradbroke Island and Fraser Island. Previously, the species was known to occur only as far north as Bundaberg. Also, it had not been recorded from Fraser Island. B. aemula rarely occurs more than a few kilometres inland. Notable exceptions are at Agnes Banks Nature Reserve west of Sydney (ca. 50 km inland) and both north and south of Grafton near Coaldale and Glenreagh respectively. Another inland location is about 30 km south-west of Bundaberg. Of these, only the Coaldale population had been previously collected (by W. Blakely and D. Shiress in 1922). An early collection (no date or collector available) from Paramatta, west of Sydney was not recorded by Atlas contributors. The species is often found in association with its close relative B. serrata and presumed hybrids have been recorded. The possibility of observers confusing B. aemula with B. serrata meant that specimens were always requested for verification in the case of new locations.

In line with its coastal distribution, 75% of recorded populations occur at altitudes of less than 50 m. B. aemula may either be a tree up to 10 m in height or a large bushy shrub. Deep sandy soils are strongly preferred - either coastal dunes or sandy flats which are sometimes seasonally wet. Surrounding vegetation may be woodland or shrubland. In Queensland, B. aemula is often associated with B. robur and B. oblongifolia in the distinctive Wallum communities. A network of low sandy rises and intervening swamps create a patchwork of vegetation types with B. robur in the wetter areas and B. aemula on the slightly higher, drier sites. Banksia aemula flowers from January to October though mainly in February and March. The conspicuous flowers are attractive to many different birds including honeyeaters, wattlebirds, friarbirds and lorikeets. New shoot growth is mainly in spring and summer. Following fire the species may resprout from its lignotuber or from epicormic buds. The opening of the seed follicles is also stimulated by fire and resultant seed germination is usually prolific.

BANKSIA ASHBYI E.G. Baker (1934)

Ashby's Banksia

70 RECORDS: Jan (3) Feb (3) Mar (0) Apr (4) May (12) Jun (8) Jul (2) Aug (10) Sep (21) Oct (1) Nov (3) Dec (0).

Population Size: 1-10 (24) 10-100 (26) >100 (17) Unspecified (3).

Conservation Status: Restricted to road verge (3%) Not (94%) Unspecified (3%). In conservation reserve (9%) Not (81%) Unspecified (10%).

Tree/Shrub: Tree form (46%) Shrub form (51%) Unspecified (3%).

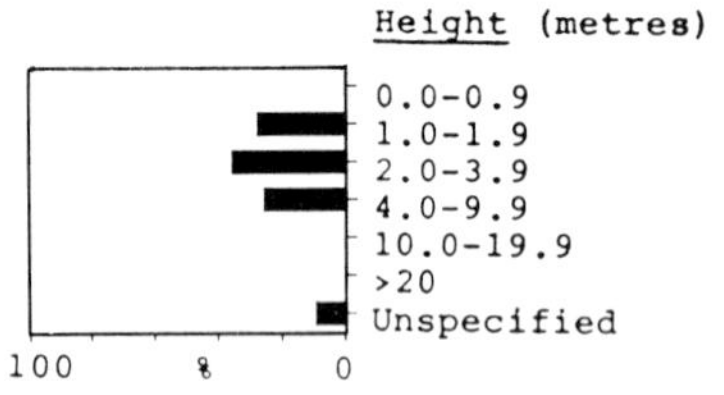

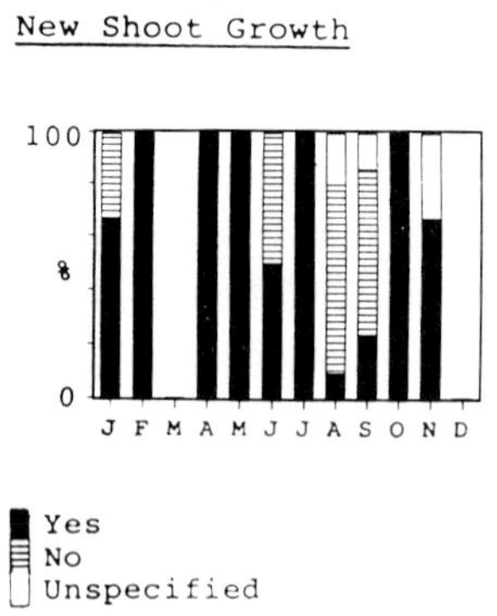

HABITAT

Within 2 km of coast: Yes (30%) No (70%) Unspecified (0%).

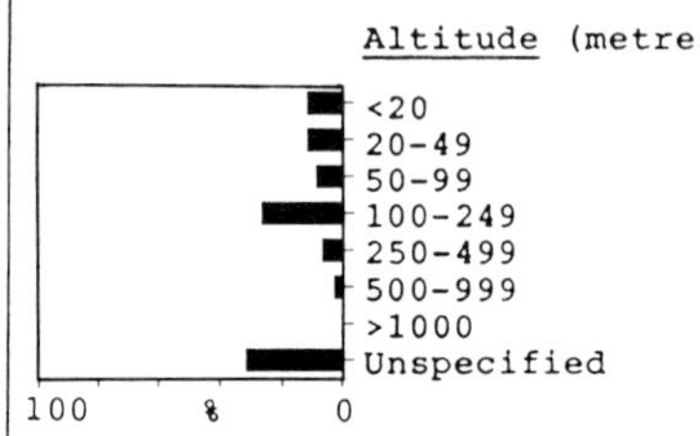

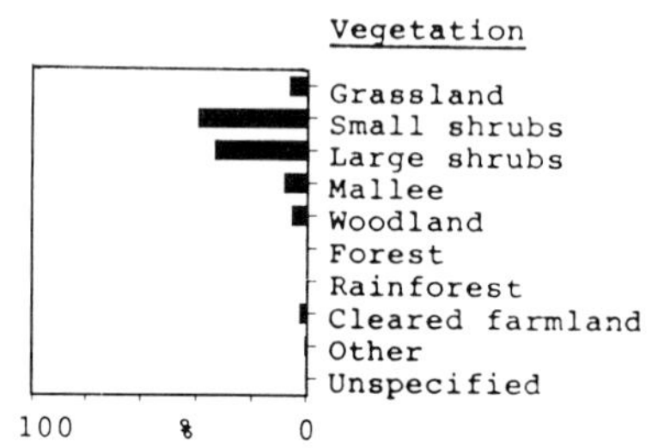

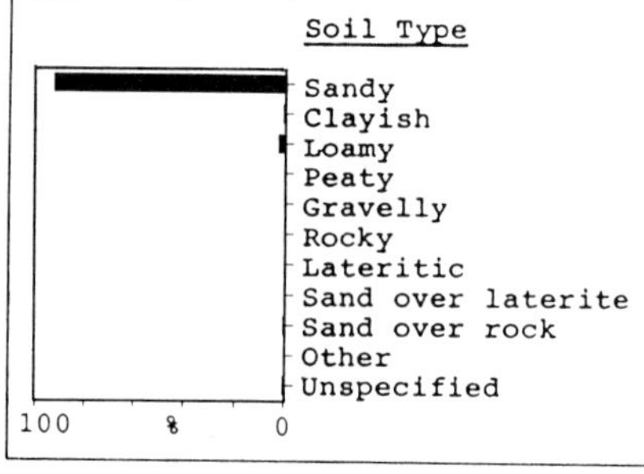

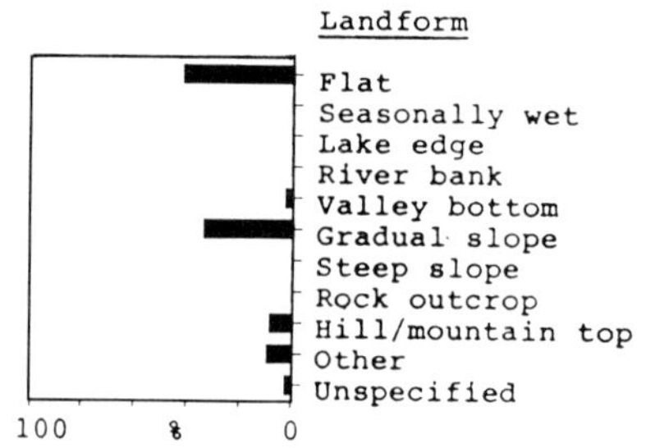

Response to Fire: 8 records

Growth response: Ground resprout (12%) killed, new seedlings (37%) killed, new seedlings and trunk resprout (12%) unspecified (37%).

Flowering response: Number of records where flowering had occurred after fire (5). Median reported time to flowering (13-24 months); minimum reported time to flowering (<1 month). Number of records where flowering had not occurred after fire (2). Median reported time to no flowering (13-24 months); maximum reported time to no flowering (25-36 months).

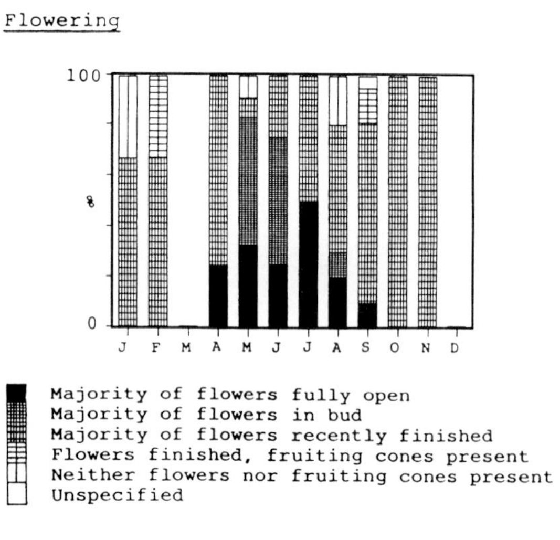

Possible pollinators

Pollinator type 17 records: bird (18%) butterflies, moths (6%) bees, wasps, ants (24%) no pollinator observed (53%).

Specific pollinator 7 records: honeyeater (3) Singing Honeyeater (1) ant (3).

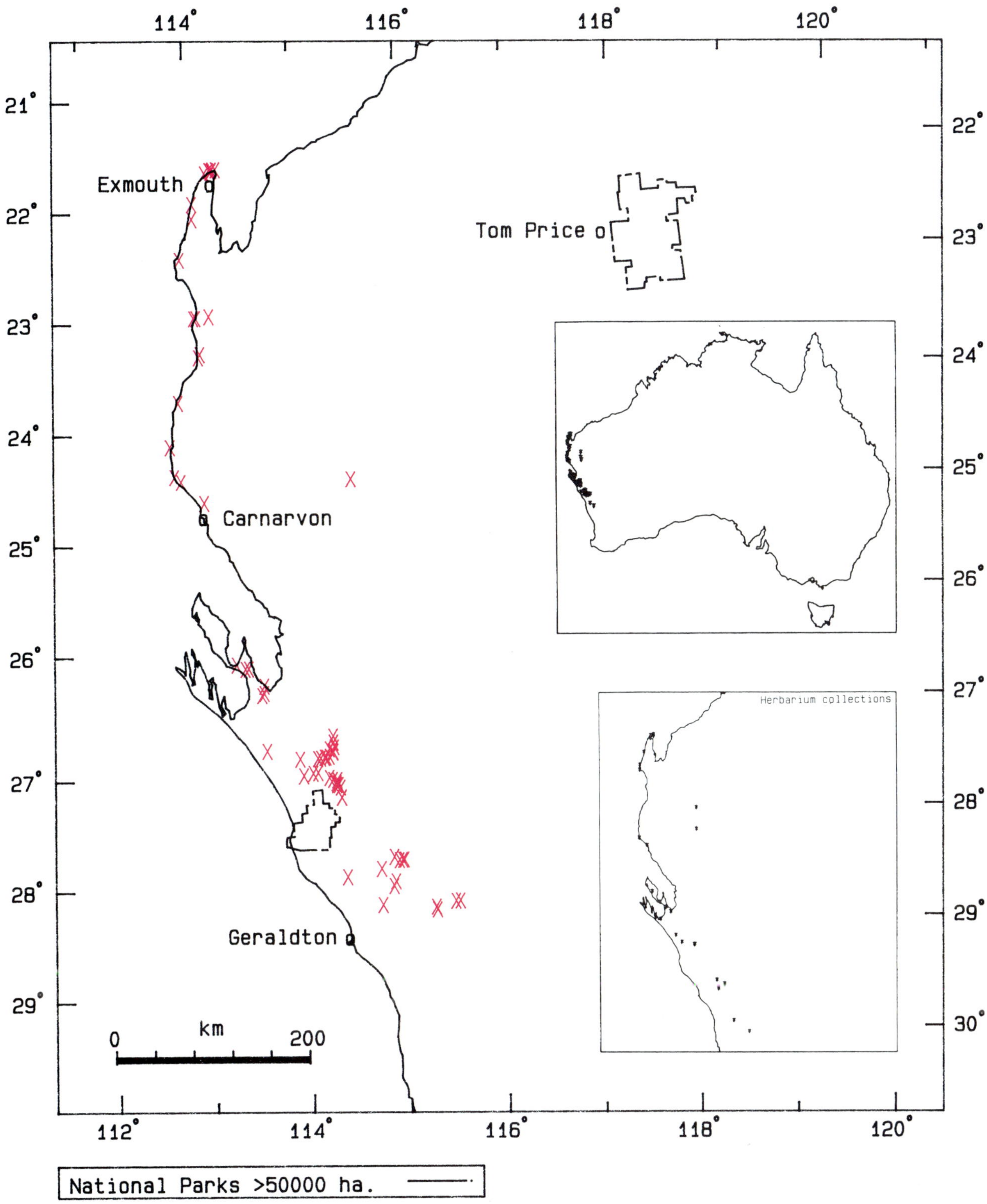

DISCUSSION

Atlas data confirmed herbarium records for B. ashbyi in showing that it extends from Geraldton to Exmouth, with an inland outlier in the Kennedy Range. Collections of W.E. Blackall and C.A. Gardner, both in 1940 and from the Mingenew and Coorow areas, were not supported by sight records. Contributors did extend the known distribution further east from Geraldton (to Nungulya homestead, 20 km E of Mullewa) and established that previous records from North West Cape and Quobba Station were connected by populations along the intervening coastline. B. ashbyi is reasonably secure occurring in such reserves as Cape Range National Park, Cooloomia Nature Reserve and Wandana Nature Reserve.

Northern populations occur as shrubs in heath and spinifex grassland on near-coastal red sand dunes. In this area, Atlas records suggest that the species is lignotuberous. At the Kennedy Range, the single recorded population grows on a red sandy hilltop with small shrubs and spinifex. Populations from around Shark Bay also occupy red sand dune country. Further south the species grows away from the coast on undulating well-drained yellow sandplains. Its stature increases to a small tree. From Shark Bay southwards the species is non-lignotuberous, regenerating from seed after fire. This presence or absence of a lignotuber amongst distinct populations within a species is a relatively rare phenomenon. It may warrant the division of the species into two varieties. Flowering was observed mainly in winter with birds the most frequently observed potential pollinators. New shoot growth was reported throughout the year, but chiefly in autumn ahd had declined to low levels in late winter - early spring.

BANKSIA ATTENUATA R. Brown (1810)

Slender Banksia , Candlestick Banksia, Coast Banksia

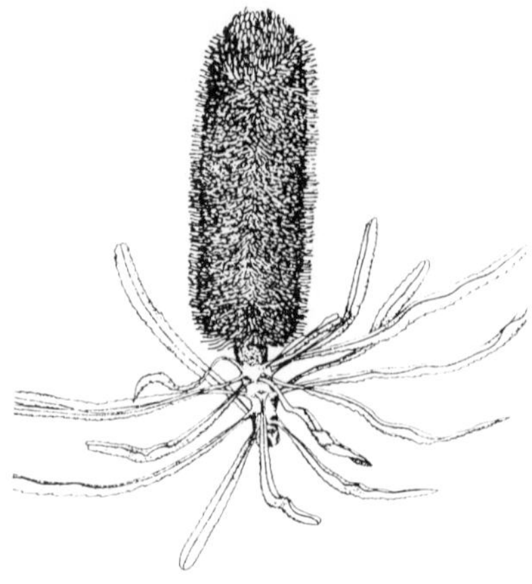

1280 RECORDS: Jan (120) Feb (42) Mar (160) Apr (142) May (106) Jun (76) Jul (71) Aug (131) Sep (182) Oct (90) Nov (73) Dec (80).

Population Size: 1-10 (181) 10-100 (524) >100 (559) Unspecified (16).

Conservation Status: Restricted to road verge (14%) Not (83%) Unspecified (4%). In conservation reserve (26%) Not (69%) Unspecified (5%).

Tree/Shrub: Tree form (76%) Shrub form (23%) Unspecified (1%).

Height (metres)

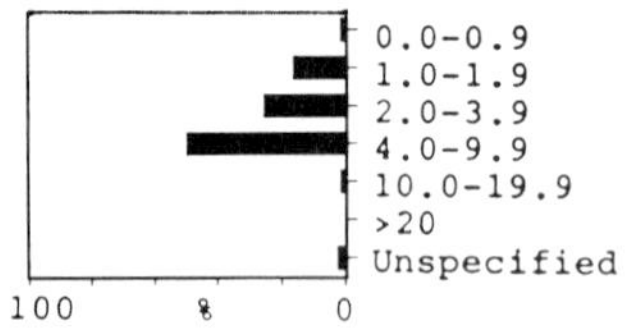

New Shoot Growth

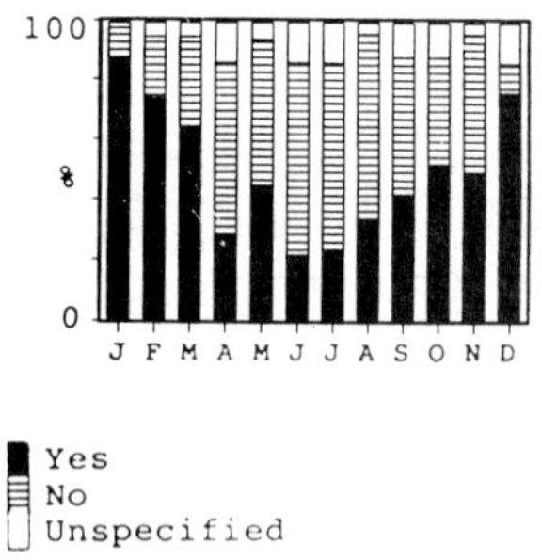

HABITAT

Within 2 km of coast: Yes (8%) No (92%) Unspecified (1%).

Altitude (metres)

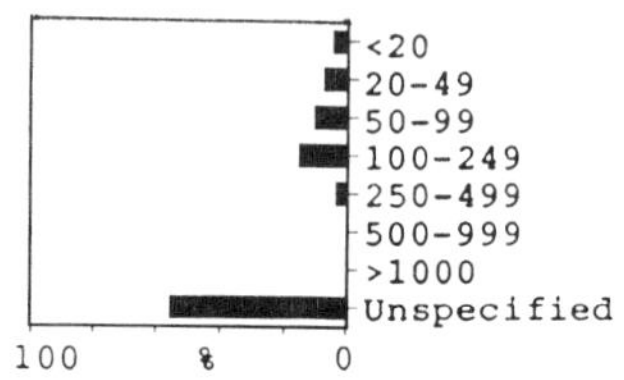

Vegetation

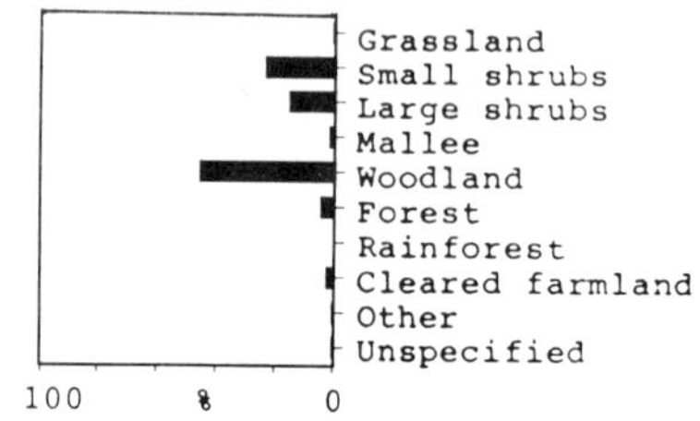

Soil Type

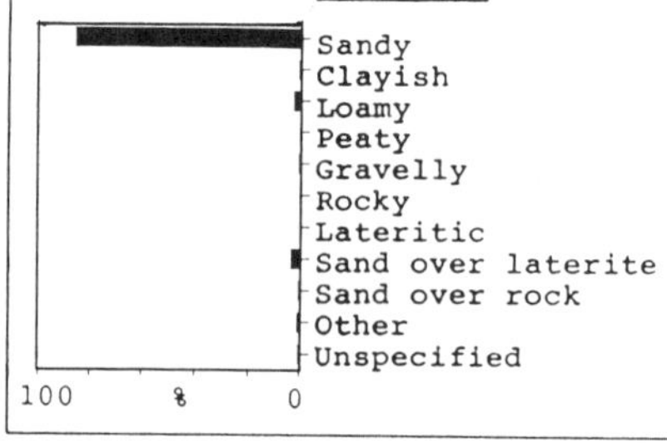

Landform

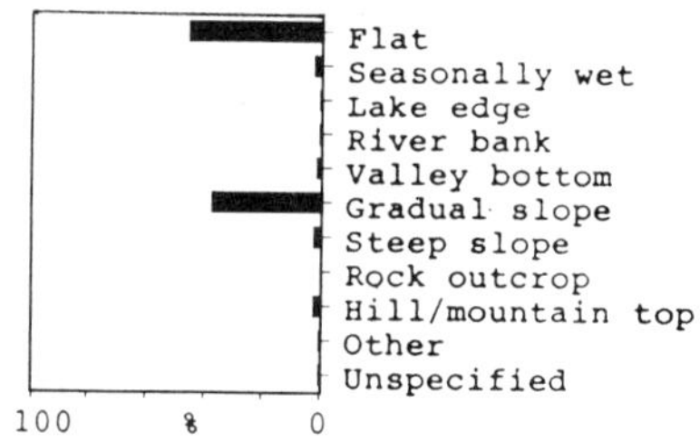

Response to Fire: 191 records

Growth response: Ground resprout (18%); ground resprout and killed, new seedlings (5%); ground resprout and trunk resprout (14%); killed, new seedlings (2%); killed, new seedlings and trunk resprout (2%); trunk resprout (27%); unspecified (34%).

Flowering response: Number of records where flowering had occurred after fire (134). Median reported time to flowering (13-24 months); minimum reported time to flowering (<1 month).
Number of records where flowering had not occurred after fire (20). Median reported time to no flowering (13-24 months); maximum reported time to no flowering (61-72 months).

Flowering

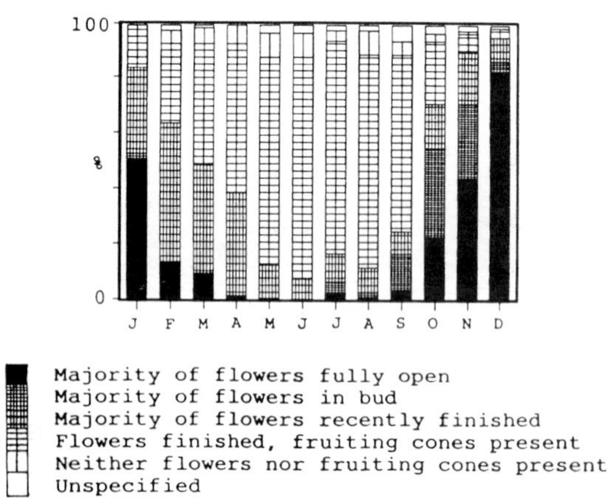

Possible pollinators

Pollinator type 436 records: bird (6%) butterflies, moths (1%) bees, wasps, ants (11%) flies (1%) no pollinator observed (82%).

Specific pollinator 66 records: honeyeater (5) New Holland Honeyeater (1) Brown Honeyeater (1) Singing Honeyeater (1) Miner (1) wattlebird (3) Western Spinebill (2) Port Lincoln Parrot (1) Red-tailed Black Cockatoo (1) ant (9) bee (29) European honey bee (1) blowfly (1) dragonfly (1).

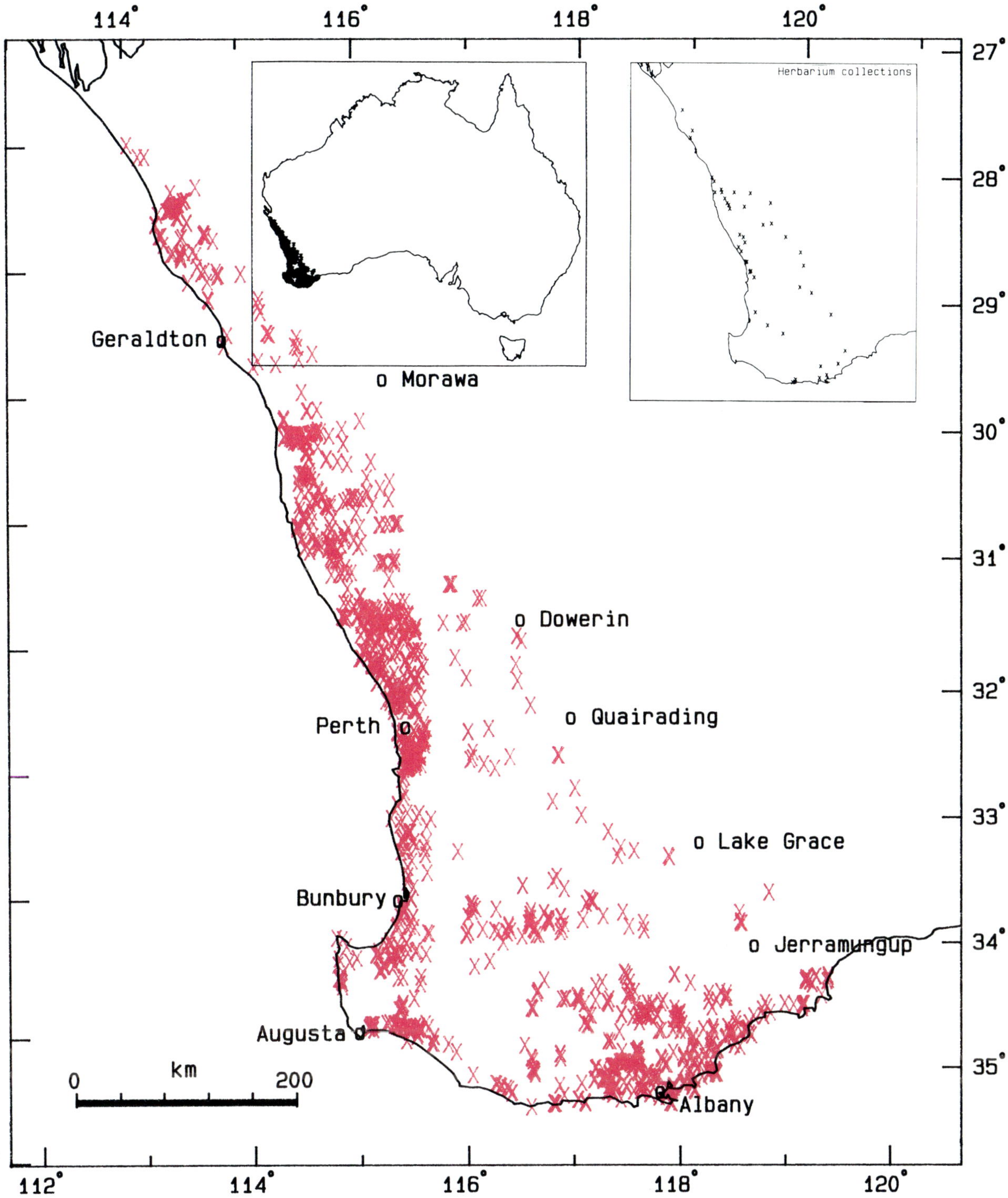

DISCUSSION

The 1 280 sight records of B. attenuata show it to be a common component of woodlands and shrublands on sandplain from the Kalbarri region south to Augusta, Albany and the western end of Fitzgerald River National Park. Heavy soils constitute a significant barrier in its distribution. For example, it is clearly confined to the deep sands of the western coastal plain and does not occur east of the Darling Scarp except for small isolated pockets of sand scattered throughout the Jarrah forest and wheatbelt. Only in the Collie - Boyup Brook - Kojonup area east of Bunbury is there sufficient sandplain development to allow a more continuous distribution of B. attenuata. This belt effectively divides the northern Jarrah forest from the southern Jarrah forest. Interesting outliers include observations near Lake Magenta (N of Jerramungup), and in Zuytdorp National Park south of Shark Bay. Atlas records show the species to be common on the Scott River sandplains and south-eastwards along the south coast, areas not represented in the herbarium collections mapped.

Populations from Badgingarra northwards consist of shrubs 1-2 m tall. To the south, stature increases to a small tree of 4-10 m. B. attenuata flowers and produces new shoots mainly in late spring and summer. It resprouts well from the lignotuber and from epicormic buds after fire, and was observed to then flower 1-2 years later on average. Honeyeaters may be important pollinators.

BANKSIA AUDAX C.Gardner (1928)

33 RECORDS: Jan (3) Feb (0) Mar (0) Apr (7) May (6) Jun (0) Jul (2) Aug (5) Sep (6) Oct (2) Nov (0) Dec (0).

Population Size: 1-10 (5) 10-100 (17) >100 (9) Unspecified (2).

Conservation Status: Restricted to road verge (6%) Not (94%) Unspecified (0%). In conservation reserve (36%) Not (55%) Unspecified (9%).

Tree/Shrub: Tree form (0%) Shrub form (94%) Unspecified (6%).

Height (metres)

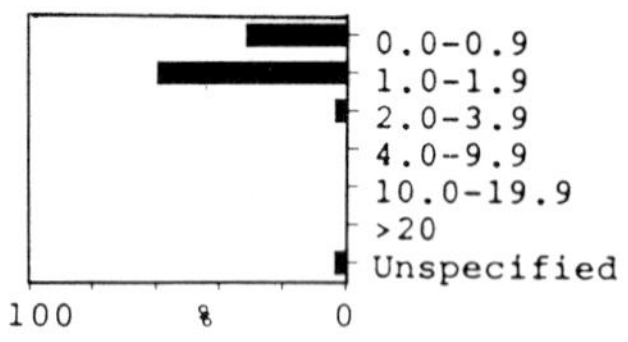

New Shoot Growth

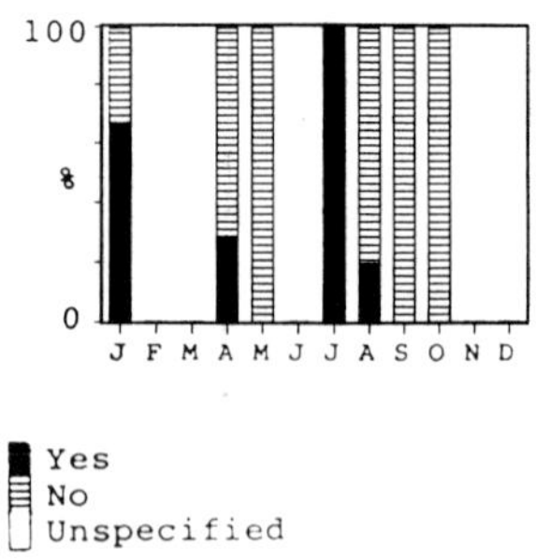

HABITAT

Within 2 km of coast: Yes (0%) No (100%) Unspecified (0%).

Altitude (metres)

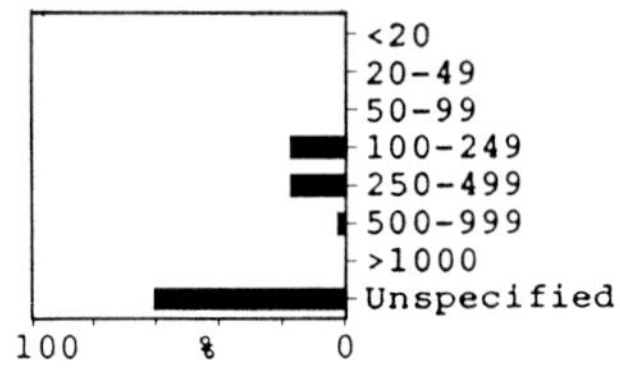

Vegetation

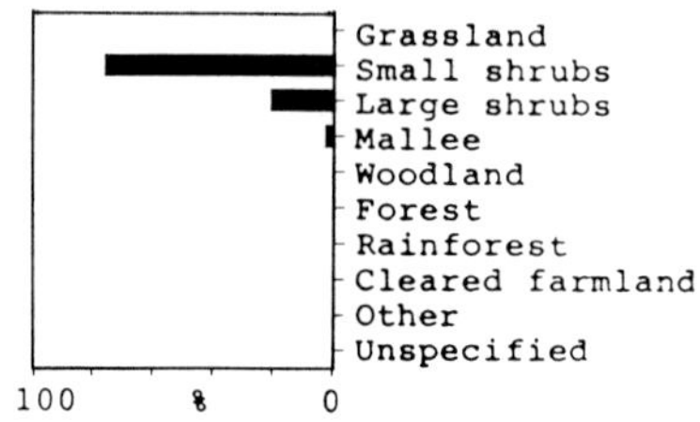

Soil Type

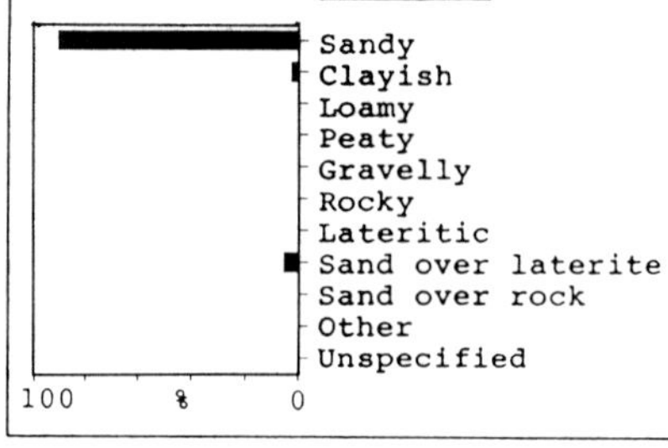

Landform

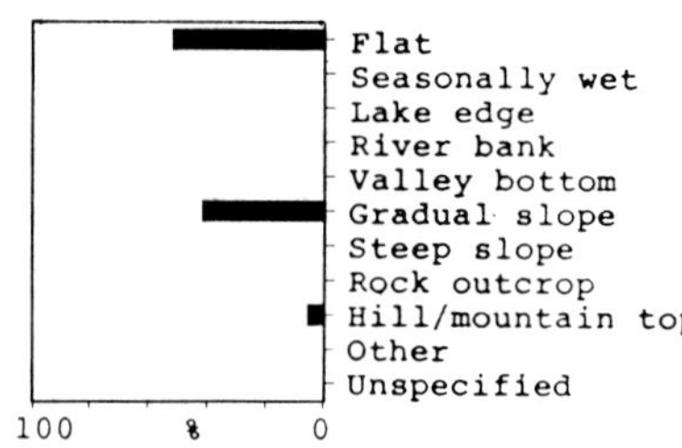

Response to Fire: 4 records

Growth response: Ground resprout (50%) unspecified (50%).

Flowering response: Number of records where flowering had occurred after fire (4). Median reported time to flowering (25-36 months); minimum reported time to flowering (13-24 months). Number of records where flowering had not occurred after fire (0).

Flowering

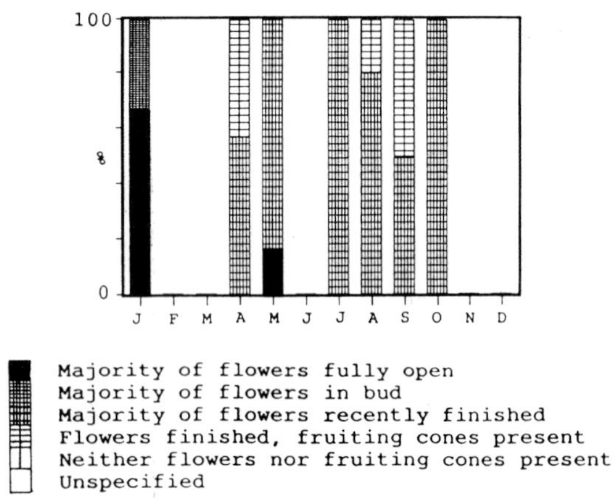

Possible pollinators

Pollinator type 1 record: bees, wasps, ants (100%).

Specific pollinator 0 records.

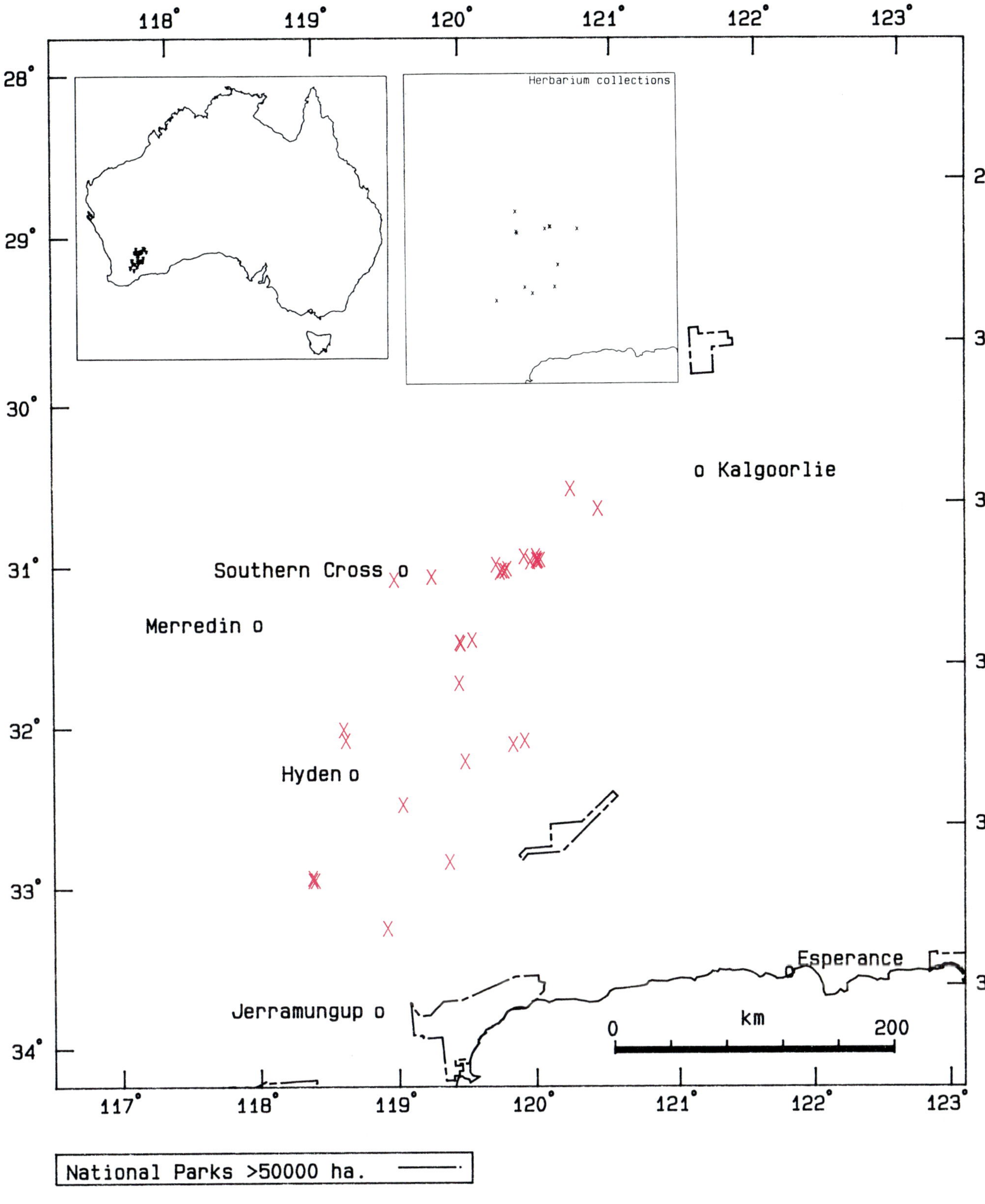

DISCUSSION

B. audax was recorded in scattered populations from the goldfields region west of Kalgoorlie to the Great Southern wheatbelt near Lake Magenta. Herbarium collections from Koolyanobbing and south-west of Coolgardie were not recorded by Atlas contributors, but significant range extensions included populations near Lake Grace, Lake Magenta and Wallaroo to the west of Kalgoorlie. The species is probably under-recorded because access is difficult in much of its habitat. It is also a relatively inconspicuous shrub which flowers in summer at a time when field work is unpleasant due to extreme temperatures. Most populations were of less than 100 plants and about a third were recorded from conservation reserves (notably Boorabbin National Park). Much of the range of the species is in uncleared land to the east of the wheatbelt. Its conservation status therefore seems relatively secure.

The species grows as a shrub usually 1-2 m tall in flat to gently undulating sandplain. Typically it grows in shrubland. It appears to regenerate from a lignotuber after fire. Further data on pollinators and new shoot growth are needed for reliable conclusions to be drawn.

BANKSIA BAUERI R.Brown (1830)

Woolly Banksia, Woolly-spiked Banksia, Possum Banksia

132 RECORDS: Jan (11) Feb (6) Mar (12) Apr (9) May (7) Jun (6) Jul (2) Aug (13) Sep (24) Oct (14) Nov (14) Dec (12).

Population Size: 1-10 (24) 10-100 (68) >100 (29) Unspecified (11).

Conservation Status: Restricted to road verge (33%) Not (58%) Unspecified (9%). In conservation reserve (46%) Not (50%) Unspecified (4%).

Tree/Shrub: Tree form (2%) Shrub form (96%) Unspecified (2%).

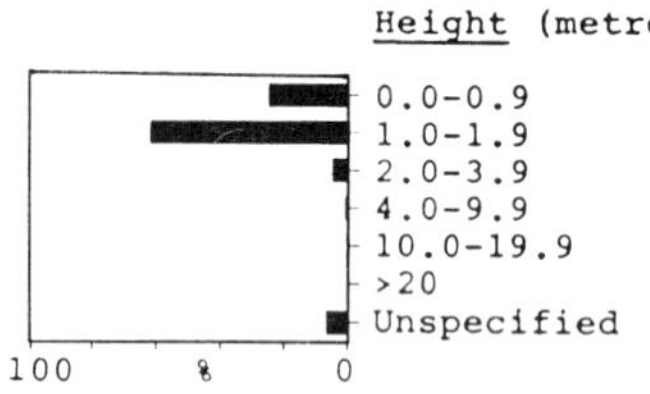

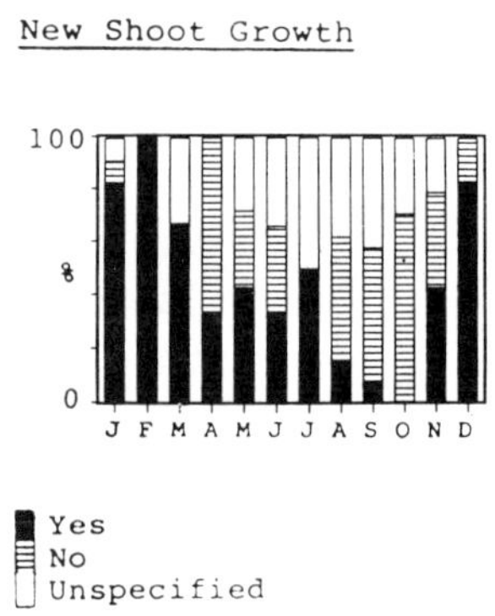

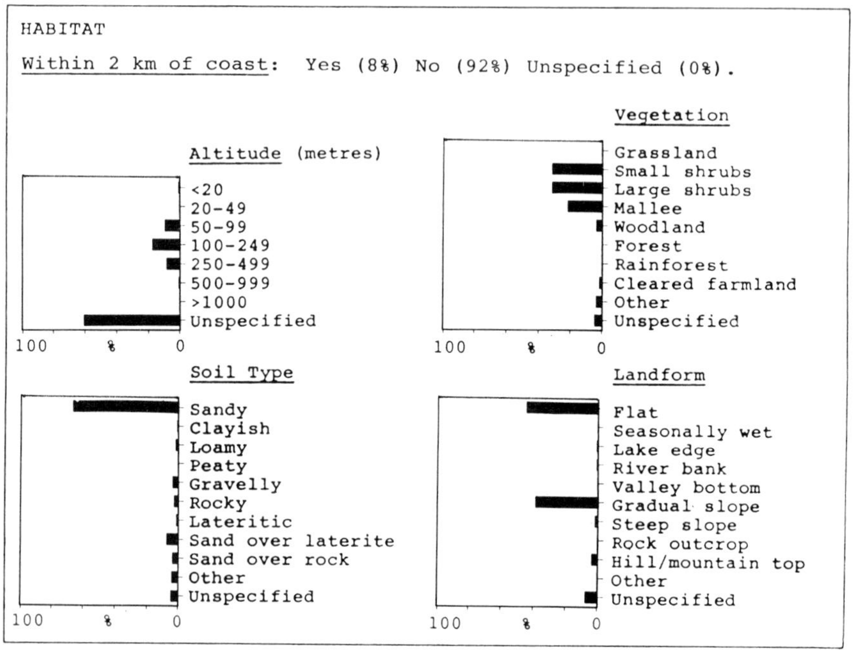

Response to Fire: 3 records

Growth response: Killed, new seedlings (33%) killed, new seedlings and trunk resprout (33%) unspecified (33%).

Flowering response: Number of records where flowering had occurred after fire (2). Median reported time to flowering (13-24 months); minimum reported time to flowering (1-12 months). Number of records where flowering had not occurred after fire (0).

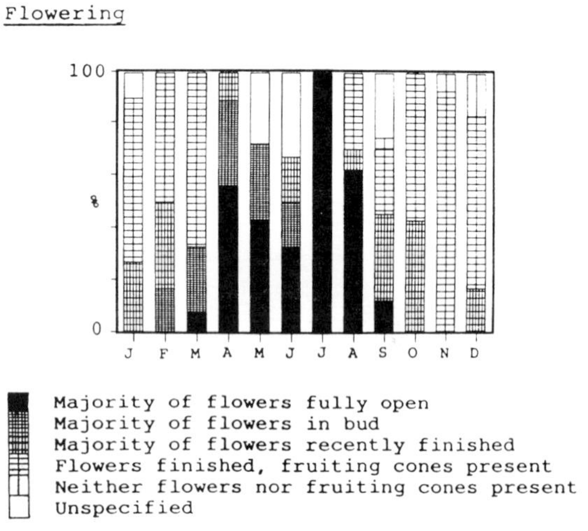

Possible pollinators

Pollinator type 21 records: bees, wasps, ants (5%) no pollinator observed (95%).

Specific pollinator 1 record: ant (1).

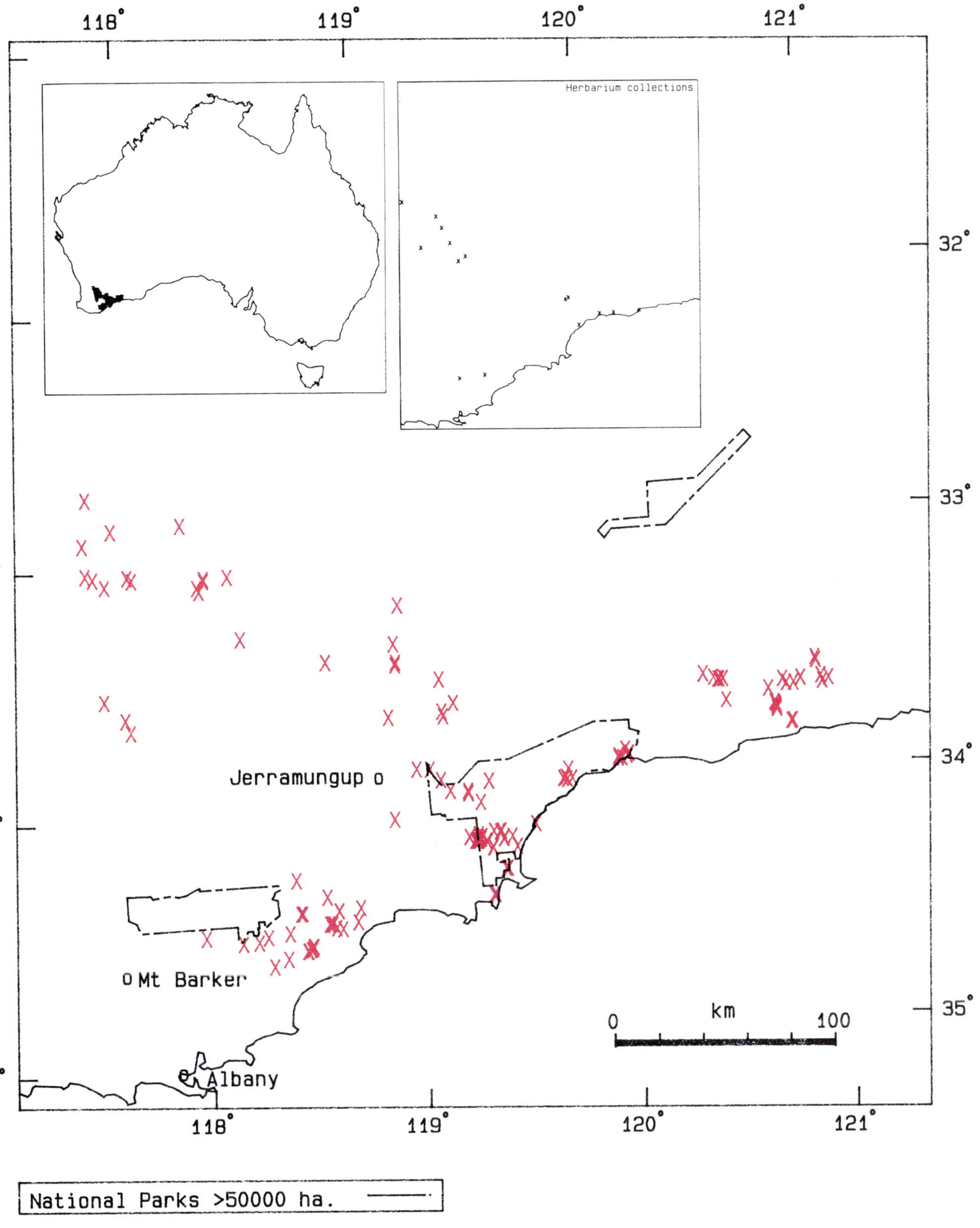

DISCUSSION

Herbarium records had indicated three disjunct areas of occurrence for B. baueri, one along the south coast between Bremer Bay and Jerdacuttup, one on the South Stirling sandplains and the other inland between Kweda and Tarin Rock. The Atlas provides many new sightings in the area between Fitzgerald, Pingrup, Tarin Rock and Lake Lockhart thus linking the previously recorded inland and south coast occurrences. A new outlier is reported to the east of Woodanilling and Katanning and the easternmost occurrence of B. baueri is now known to be beyond Munglinup. Most populations are between 10-100 plants with almost 50% in conservation reserves.

Nearly all plants grow as shrubs, generally between 1-2 m tall though sometimes less than a metre. Flat or gently sloping land with sandy soil is the preferred habitat. Surrounding vegetation is of large or small shrubs and sometimes mallee. Flowering occurs in autumn and winter, with both the usual blue-grey and occasional golden or rusty coloured variants being recorded. New shoot growth occurs in summer. Observed pollinator activity was very low. This supports work by R.D. Wooller and B.J. Collins (pers. comm.) who have established that B. baueri is an important food source for the secretive Honey-possums and not for the more conspicuous honey eating birds. The species is killed by fire and regenerates from seed.

BANKSIA BAXTERI R.Brown (1830)

Baxter's Banksia, Bird's Nest Banksia

188 RECORDS: Jan (20) Feb (20) Mar (16) Apr (12) May (9) Jun (12) Jul (8) Aug (14) Sep (26) Oct (24) Nov (15) Dec (11).

Population Size: 1-10 (22) 10-100 (86) >100 (76) Unspecified (4).

Conservation Status: Restricted to road verge (27%) Not (67%) Unspecified (6%). In conservation reserve (37%) Not (57%) Unspecified (5%).

Tree/Shrub: Tree form (29%) Shrub form (69%) Unspecified (2%).

Height (metres)

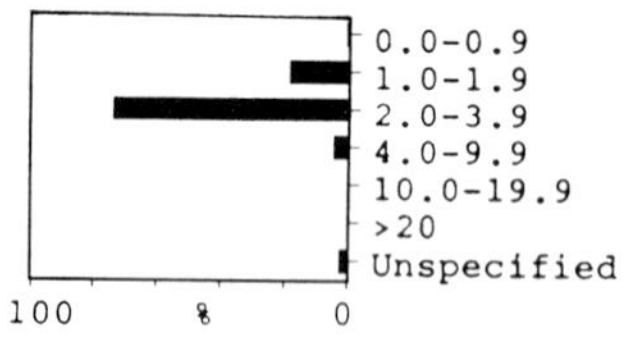

New Shoot Growth

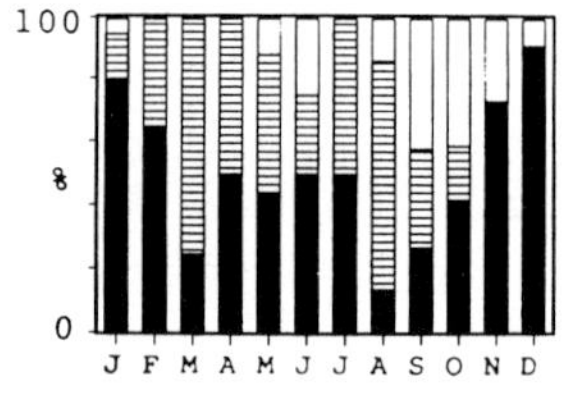

Yes
No
Unspecified

HABITAT

Within 2 km of coast: Yes (15%) No (84%) Unspecified (1%).

Altitude (metres)

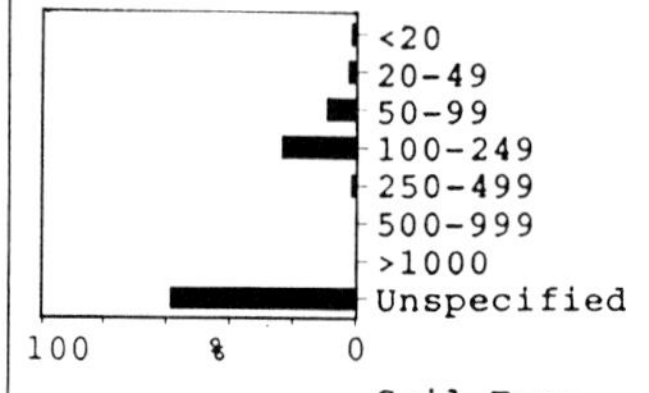

Vegetation

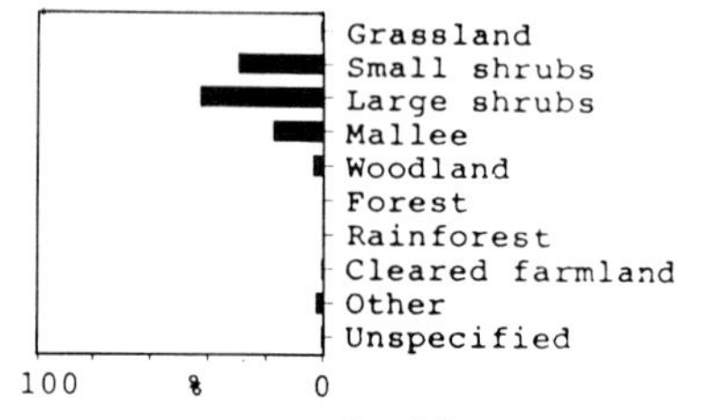

Soil Type

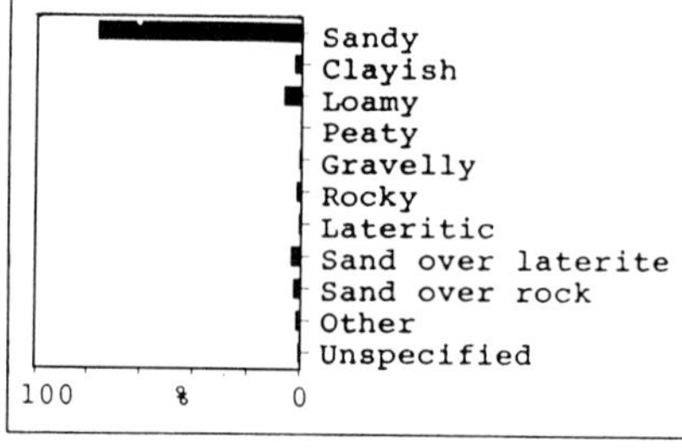

Landform

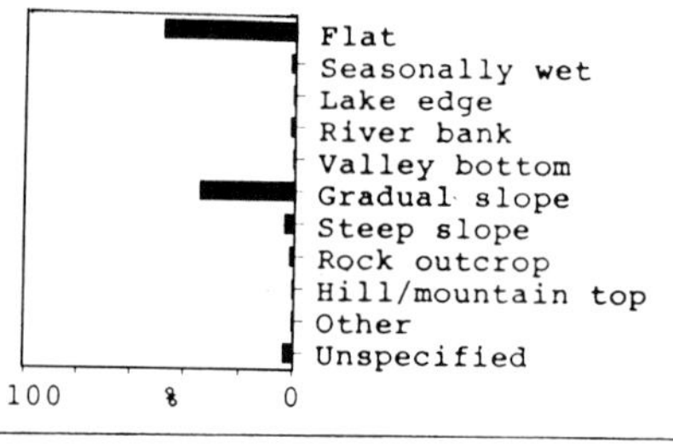

Response to Fire: 24 records

Growth response: Ground resprout and killed, new seedlings (4%) killed, new seedlings (62%) unspecified (33%).

Flowering response: Number of records where flowering had occurred after fire (12). Median reported time to flowering (13-24 months); minimum reported time to flowering (1-12 months). Number of records where flowering had not occurred after fire (5). Median reported time to no flowering (37-48 months); maximum reported time to no flowering (61-72 months).

Flowering

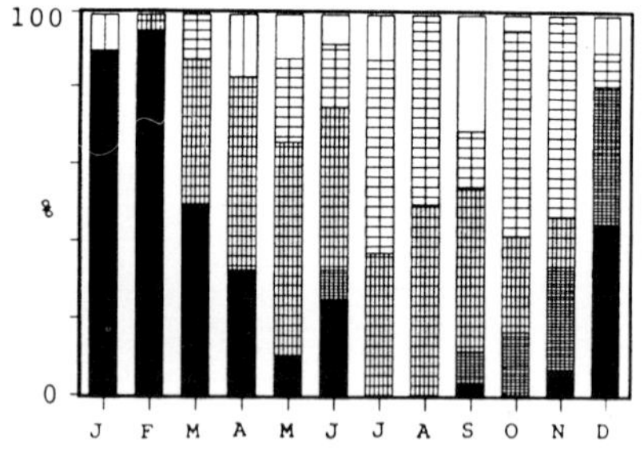

Majority of flowers fully open
Majority of flowers in bud
Majority of flowers recently finished
Flowers finished, fruiting cones present
Neither flowers nor fruiting cones present
Unspecified

Possible pollinators

Pollinator type 61 records: bird (23%) bees, wasps, ants (7%) no pollinator observed (70%).

Specific pollinator 14 records: honeyeater (2) New Holland Honeyeater (5) White-cheeked Honeyeater (1) Little Wattlebird (2) Western Spinebill (1) Red-tailed Black Cockatoo, (2) ant (1).

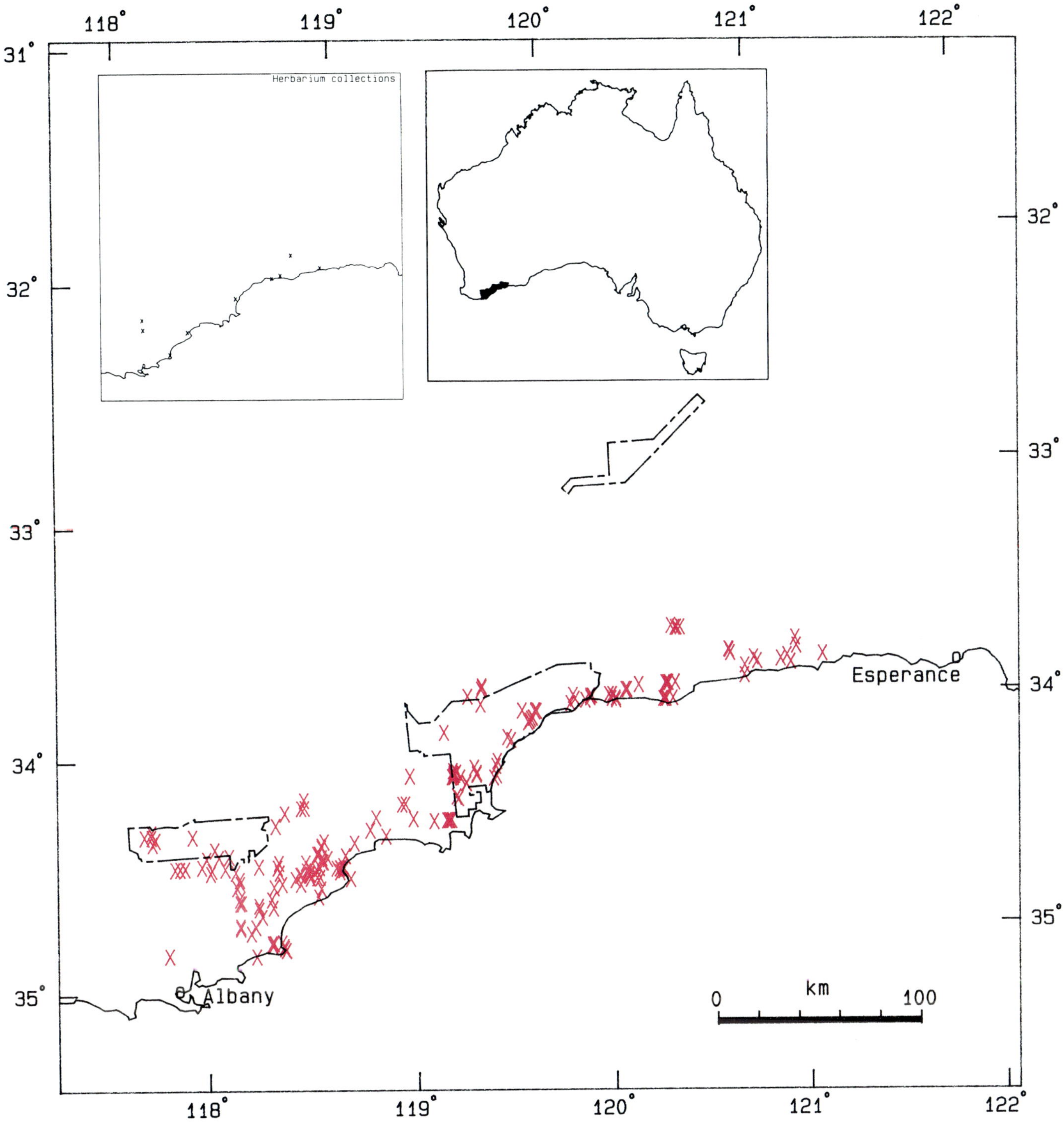

National Parks >50000 ha.

DISCUSSION

The few herbarium records for B. baxteri indicated a scattered distribution along the south coast between the Stirling Range and Munglinup Beach. Atlas records have confirmed this distribution and have shown it to be fairly continuous. An interesting outlier occurs at Millbrook Nature Reserve to the north of Albany. Other range extensions are along the northern boundary of Fitzgerald River National Park, and to the north-east of Stirling Range National Park. Many populations are of more than 100 plants and 37% are in conservation reserves. However, a significant proportion (27%) is confined to road verges, the species is highly susceptible to dieback disease and it is heavily used in the cut wildflower trade (Burgman & Hopper 1982). Its conservation status therefore needs monitoring.

Plants usually grow as shrubs 2-4 m in height. The preferred landform is flat or gently sloping and surrounding vegetation is mainly shrubland or less often mallee. Sandy soils are strongly preferred. This is a summer flowering species which reaches its peak in January and February. Some flowers were recorded as late as June, a few as early as September. Birds were frequently seen visiting flowers (see also Hopper, 1980). New shoot growth occurs in summer. B. baxteri appears to be killed by fire and to take at least 3-4 years for seedlings to mature and flower.

BANKSIA BENTHAMIANA

C.Gardner (1964)

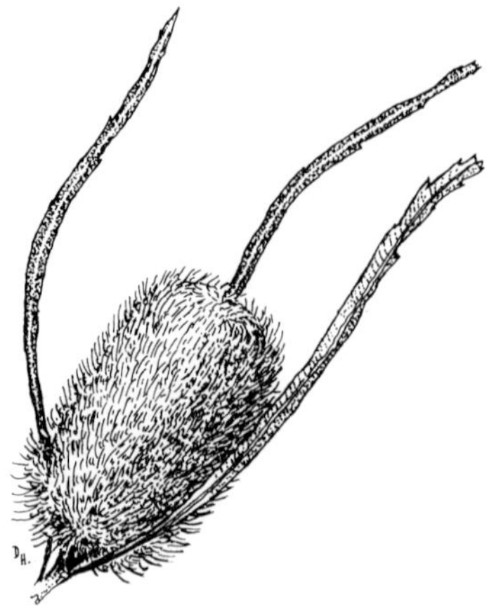

36 RECORDS: Jan (0) Feb (1) Mar (0) Apr (0) May (0) Jun (8) Jul (1) Aug (11) Sep (5) Oct (6) Nov (1) Dec (2).

Population Size: 1-10 (10) 10-100 (18) >100 (6) Unspecified (2).

Conservation Status: Restricted to road verge (47%) Not (47%) Unspecified (6%). In conservation reserve (17%) Not (75%) Unspecified (8%).

Tree/Shrub: Tree form (36%) Shrub form (58%) Unspecified (6%).

Height (metres)

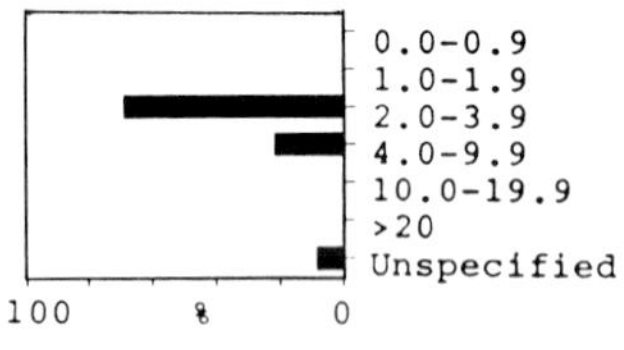

New Shoot Growth

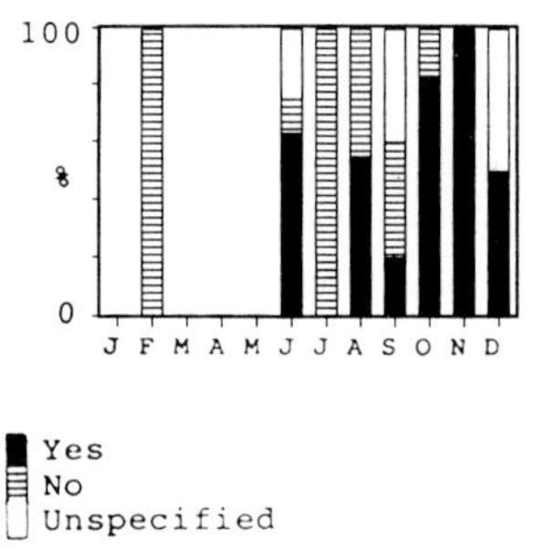

HABITAT

Within 2 km of coast: Yes (0%) No (100%) Unspecified (0%).

Altitude (metres)

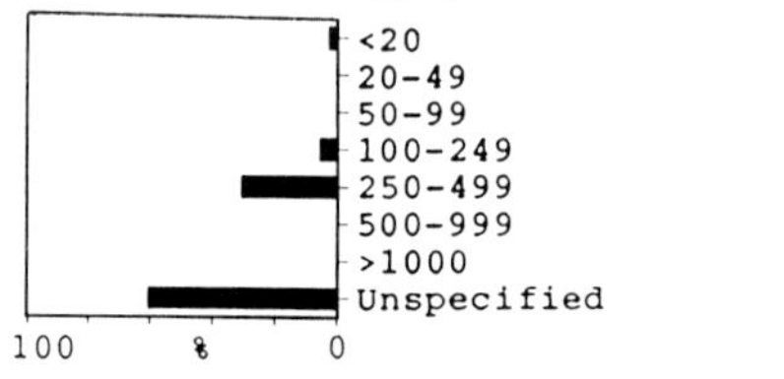

Vegetation

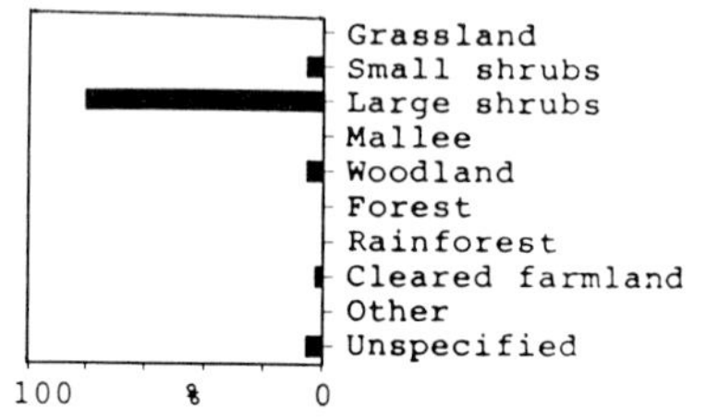

Soil Type

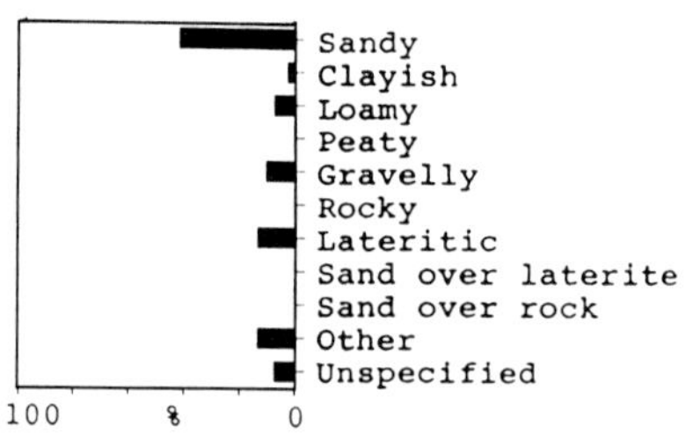

Landform

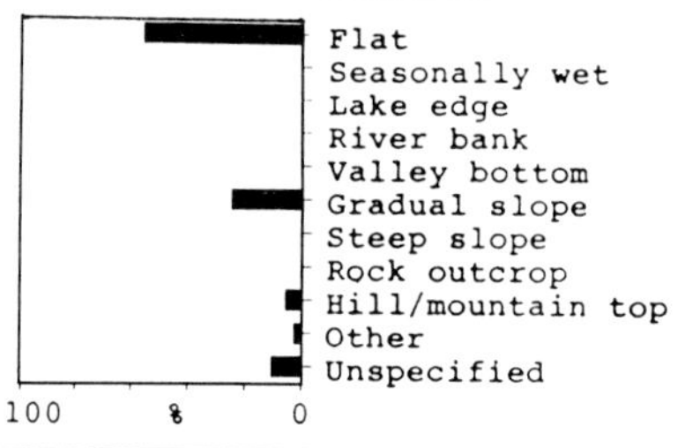

Response to Fire: 2 records

Growth response: Killed, new seedlings (100%).

Flowering response: Number of records where flowering had occurred after fire (1). Minimum reported time to flowering (85-96 months).
Number of records where flowering had not occurred after fire (0).

Flowering

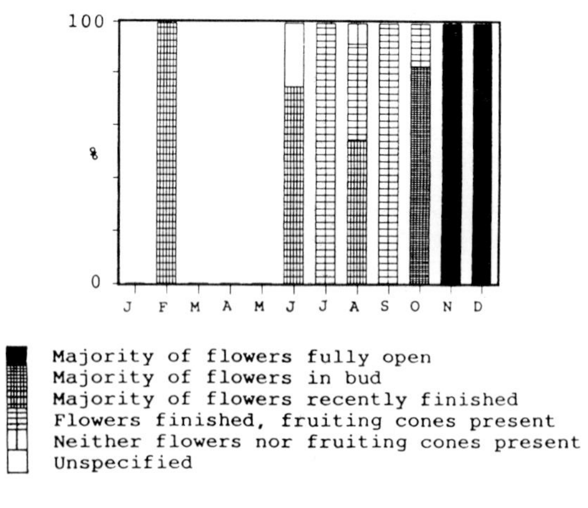

Possible pollinators

Pollinator type 6 records: butterflies, moths (17%) no pollinator observed (83%).

Specific pollinator 0 records.

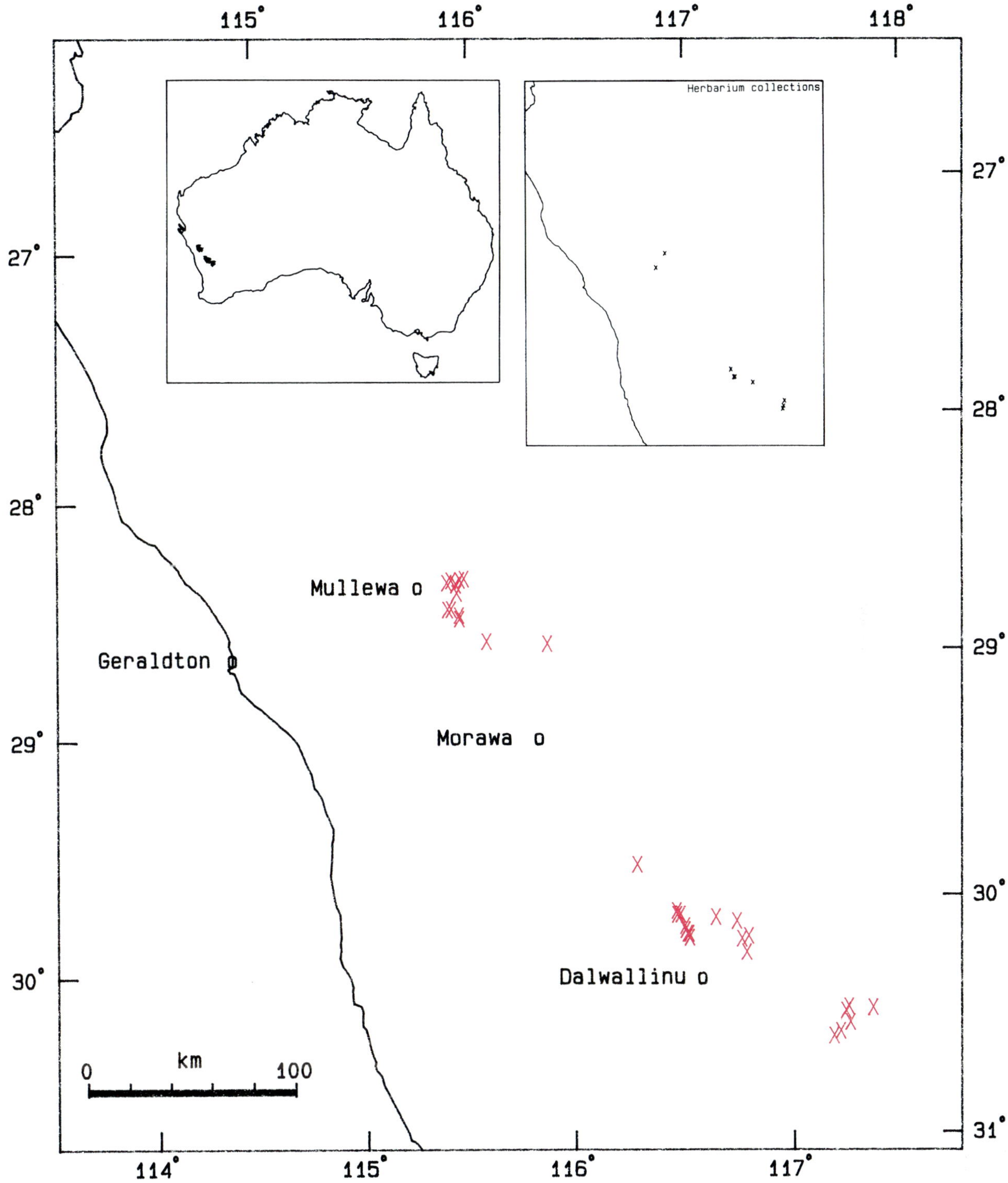

DISCUSSION

Atlas data have confirmed the previously known distribution for B. benthamiana of two disjunct populations in the northern wheatbelt, one to the east and south-east of Mullewa, the other to the north and south-east of Dalwallinu. Range extensions were recorded from the intervening Tardun and Latham areas. The species' status in the wild needs monitoring because most reported populations are of less than 100 plants, with almost 50% restricted to road verges.

B. benthamiana grows as either a shrub or tree, usually 2-4 m tall. It occurs on flat or gently sloping land among large shrubs in sand or lateritic gravelly sand. Reported flowering data are few but summer appears to be the main season. There is a similar lack of data for the season of new shoot growth and for growth and reproductive responses to fire. However, George's (1981) observations of a non-lignotuberous habit were supported by two records of adult plants killed and regenerating only from seed following fire.

BANKSIA BLECHNIFOLIA F.Mueller (1864)

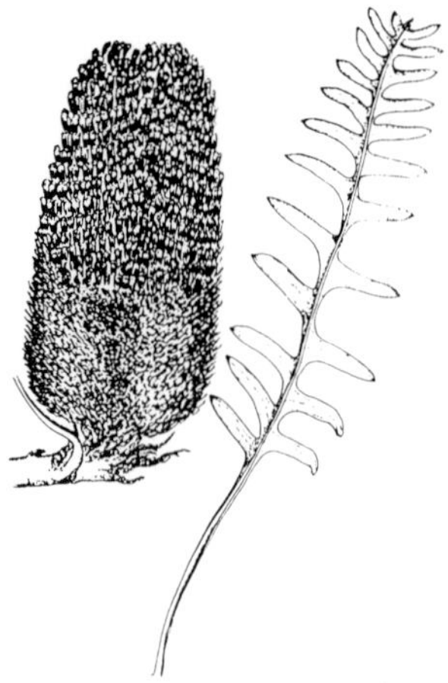

37 RECORDS: Jan (1) Feb (6) Mar (1) Apr (2) May (0) Jun (6) Jul (3) Aug (2) Sep (6) Oct (4) Nov (4) Dec (1).

Population Size: 1-10 (9) 10-100 (13) >100 (14) Unspecified (1).

Conservation Status: Restricted to road verge (49%) Not (40%) Unspecified (11%). In conservation reserve (14%) Not (86%) Unspecified (0%).

Tree/Shrub: Tree form (0%) Shrub form (100%) Unspecified (0%).

Height (metres)

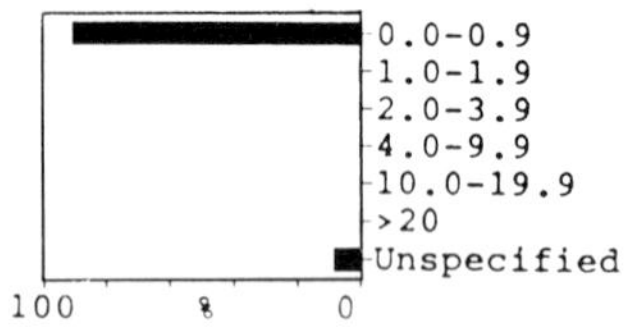

New Shoot Growth

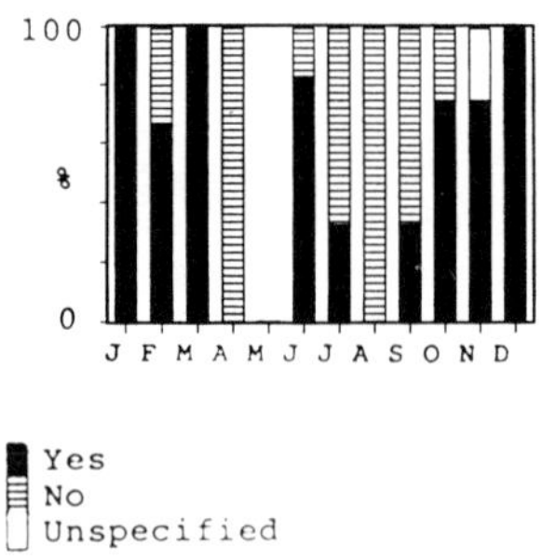

HABITAT

Within 2 km of coast: Yes (0%) No (100%) Unspecified (0%).

Altitude (metres)

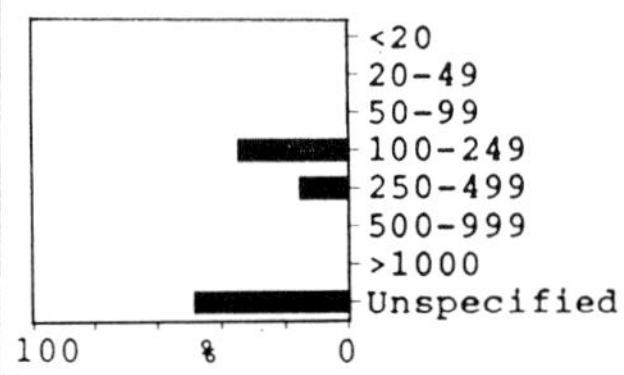

Vegetation

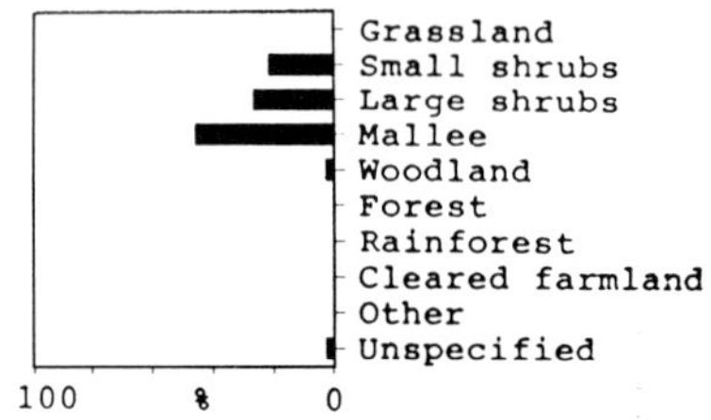

Soil Type

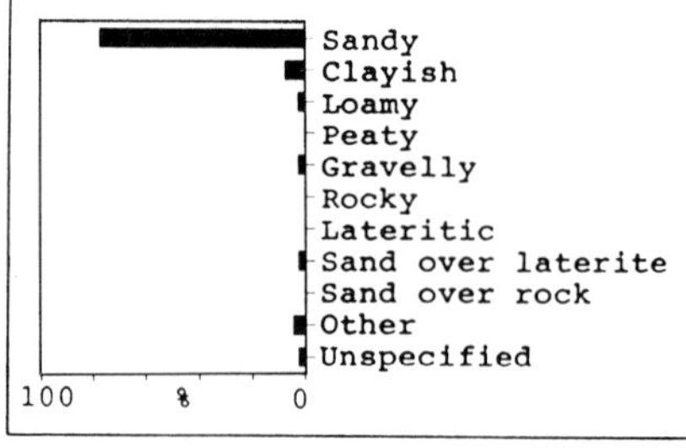

Landform

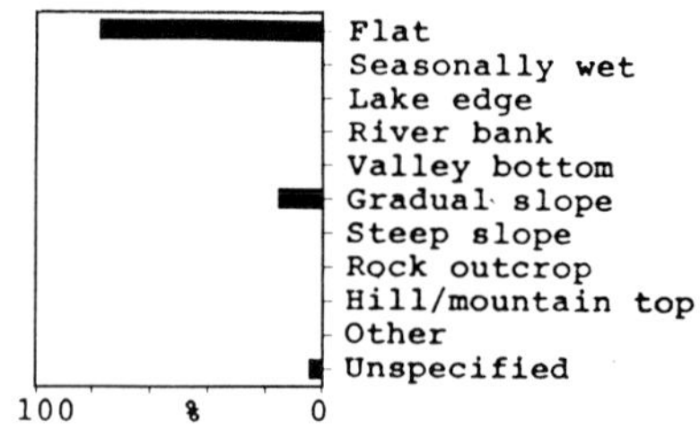

Response to Fire: 2 records

Growth response: Killed, no seedlings and killed, new seedlings (50%) unspecified (50%).

Flowering response: Number of records where flowering had occurred after fire (1). Minimum reported time to flowering (85-96 months).
Number of records where flowering had not occurred after fire (1). Maximum reported time to no flowering (13-24 months).

Flowering

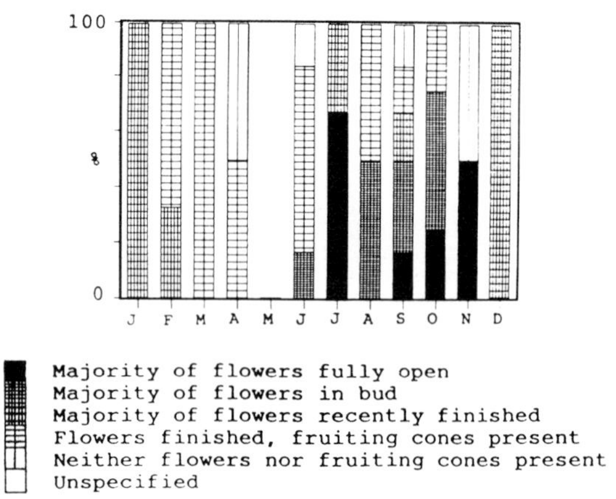

Possible pollinators

Pollinator type 12 records: bees, wasps, ants (17%) flies (17%) no pollinator observed (66%).

Specific pollinator 3 records: ant (2) blowfly (1)

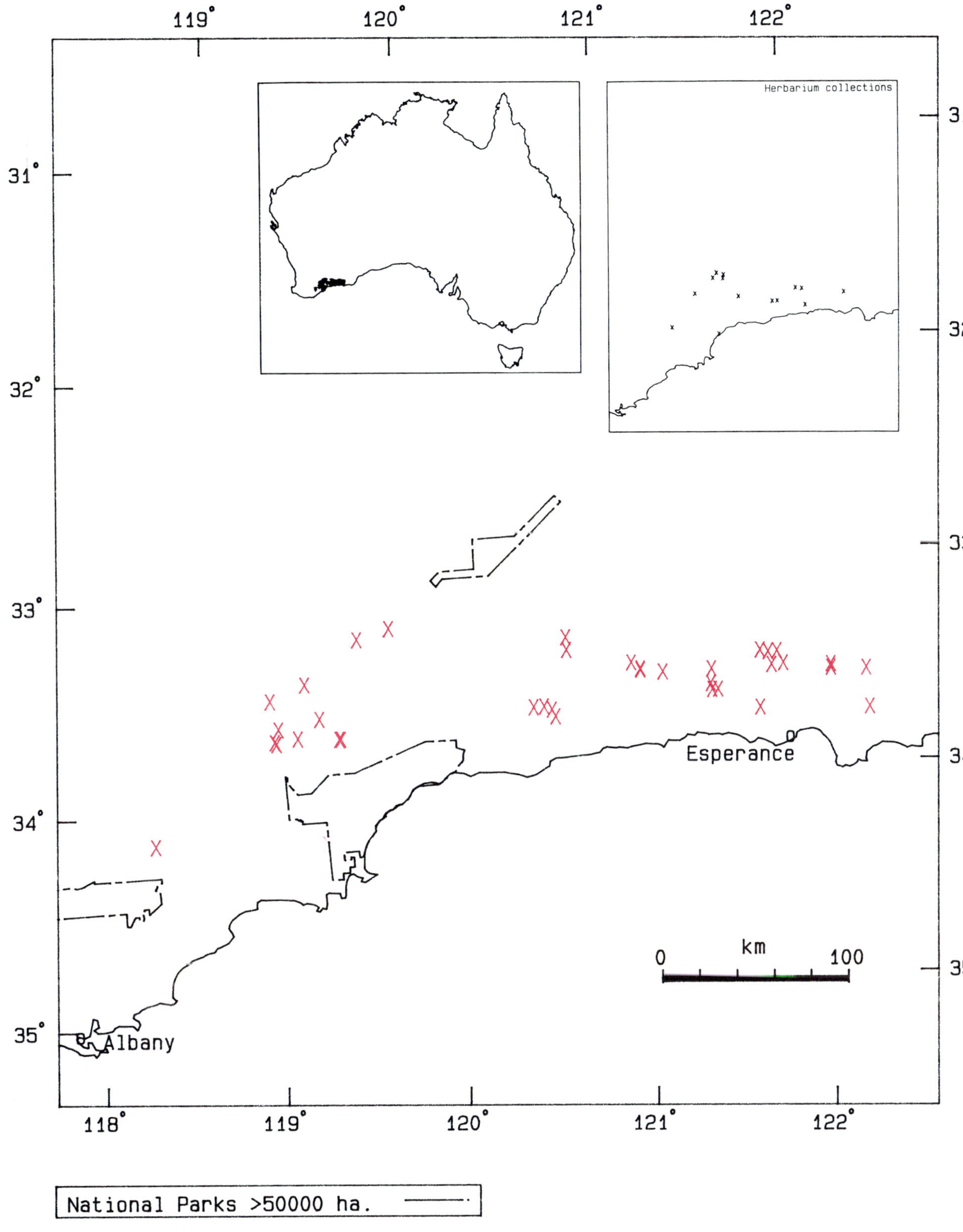

DISCUSSION

The previously known distribution of B. blechnifolia between Jerramungup and Gibson has been confirmed by Atlas contributors. However, a 1965 collection by C.A. Gardner and W.E. Blackall from near Middle Mt Barren in the Fitzgerald River National Park was not recorded. An Atlas record of an outlying population near Borden is a range extension of ca. 80 km westwards. Other new records are from the western boundary of Fitzgerald River National Park, and three recordings from north-east of Esperance. Recent years have seen the area in which B. blechnifolia grows being increasingly cleared for agriculture. With almost 50% of recorded populations on road verges and only 14% in conservation reserves, it is potentially vulnerable.

The species grows as a prostrate shrub with flower spikes and fruiting cones at ground level. It occurs on flat or gently undulating sandplain in heath or mallee-heath. Flowering is mainly in spring although it may begin as early as June. New shoot growth was recorded in many months though largely from late spring to early autumn. A single post-fire record suggests that B. blechnifolia is killed by fire and regenerates from seed. This is supported by observations of George (1981).

BANKSIA BROWNII Baxter ex R.Brown (1830)

Brown's Banksia, Feather-leaved Banksia

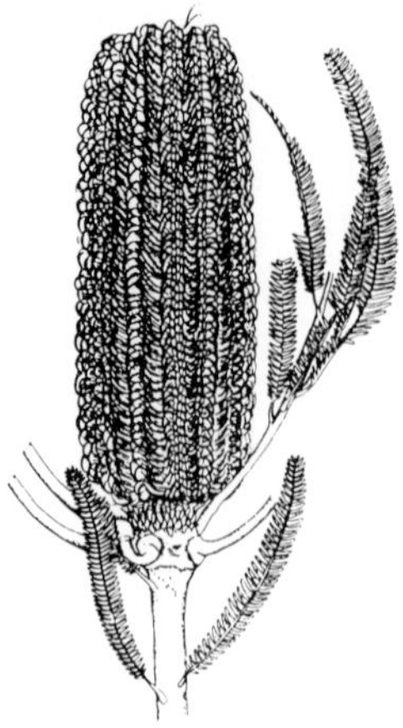

18 RECORDS: Jan (3) Feb (0) Mar (0) Apr (1) May (3) Jun (0) Jul (1) Aug (2) Sep (1) Oct (1) Nov (6) Dec (0).

Population Size: 1-10 (10) 10-100 (2) >100 (5) Unspecified (1).

Conservation Status: Restricted to road verge (28%) Not (72%) Unspecified (0%). In conservation reserve (50%) Not (39%) Unspecified (11%).

Tree/Shrub: Tree form (56%) Shrub form (39%) Unspecified (5%).

Height (metres)

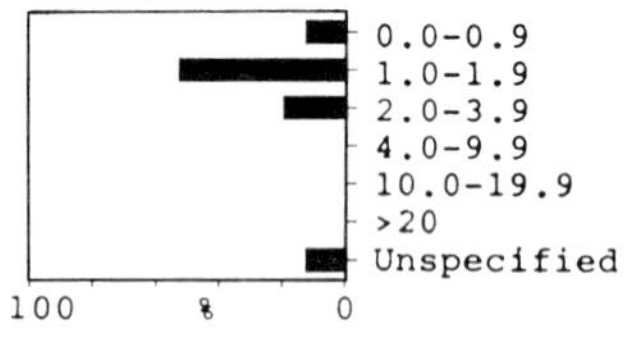

New Shoot Growth

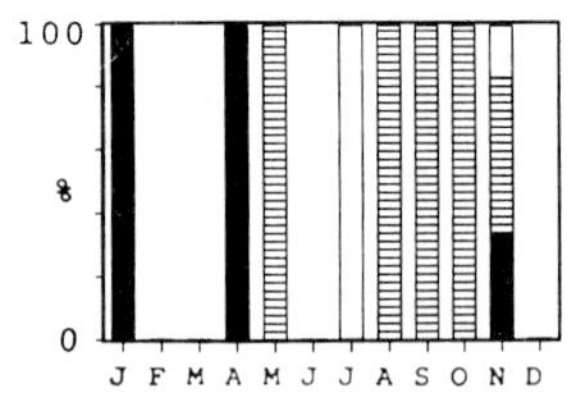

Yes
No
Unspecified

HABITAT

Within 2 km of coast: Yes (0%) No (100%) Unspecified (0%).

Altitude (metres)

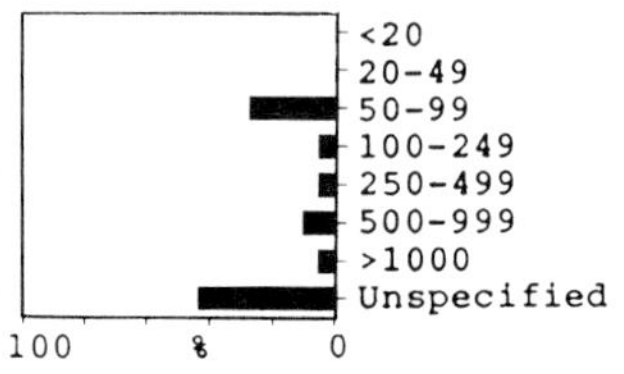

Vegetation

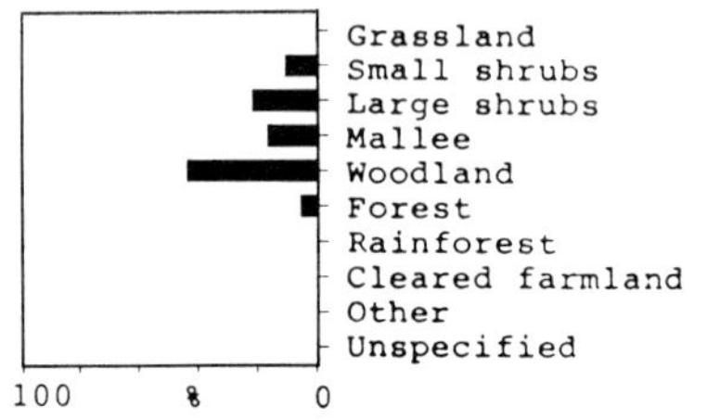

Soil Type

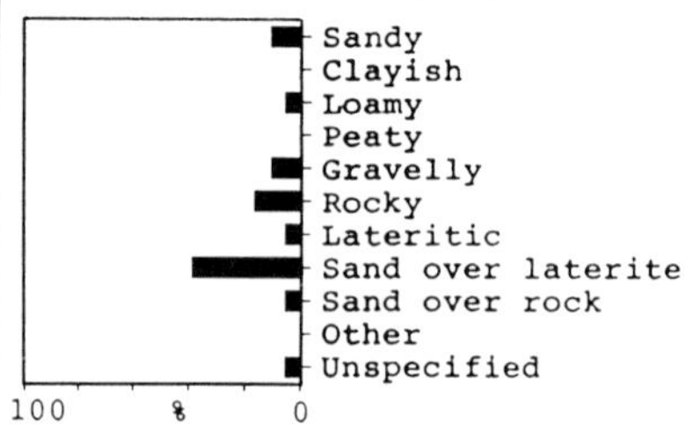

Landform

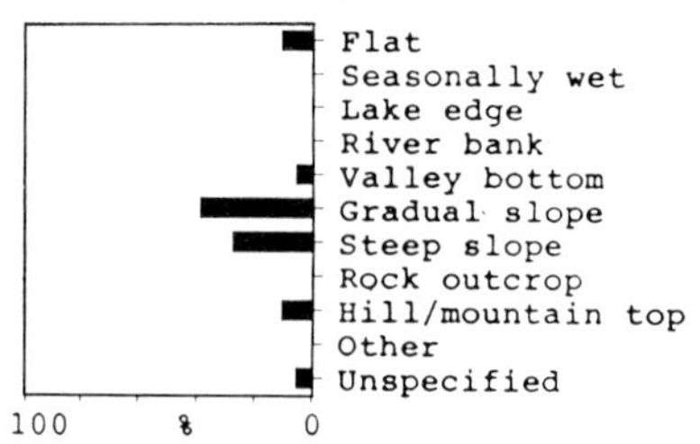

Response to Fire: 2 records

Growth response: Killed, new seedlings (50%) unspecified (50%).

Flowering response: Number of records where flowering had occurred after fire (2). Minimum reported time to flowering (61-72 months).
Number of records where flowering had not occurred after fire (0).

Flowering

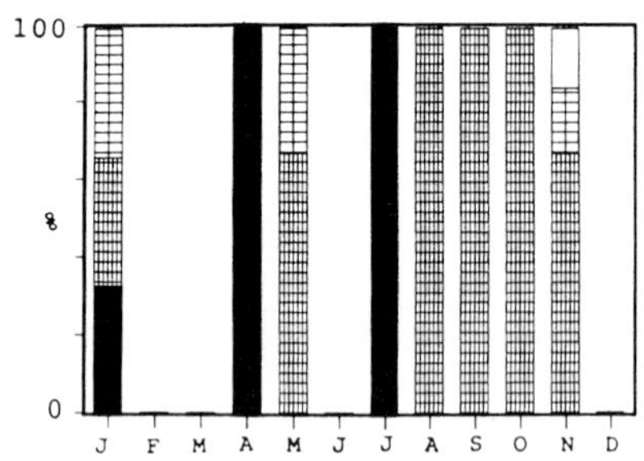

Majority of flowers fully open
Majority of flowers in bud
Majority of flowers recently finished
Flowers finished, fruiting cones present
Neither flowers nor fruiting cones present
Unspecified

Possible pollinators

Pollinator type 2 records: no pollinator observed (100%).

Specific pollinator 0 records.

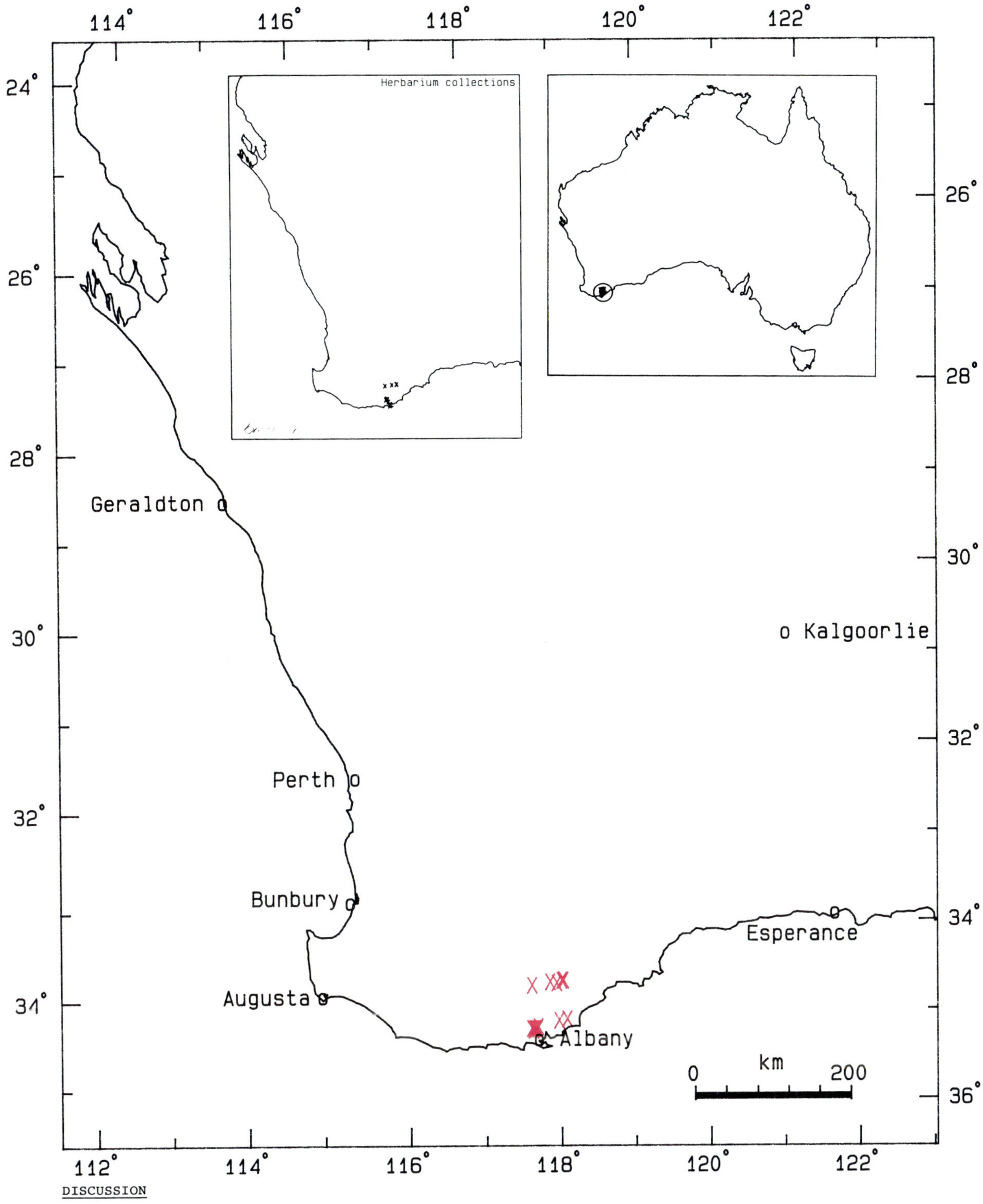

DISCUSSION

B. brownii was previously recorded in the Stirling Range and to the north of Albany. Atlas records have confirmed and extended this distribution. New populations have been recorded from two sites north-east of Albany. This is a rare banksia with few sight records reported and the majority of its populations being of less than ten plants. Although 50% of the populations are in conservation reserves and a few are reasonably large, the species is threatened by dieback disease, to which it is highly susceptible.

B. brownii grows as either a tree or shrub generally between 1-2 m though sometimes up to 4 m. In the Stirling Range the species grows on mountain slopes and tops at altitudes between 500 and 1 100 m. Here, the habitat is heath and open mallee on rocky skeletal soils. Southern populations occur in heath and woodland on gradual slopes in gravelly lateritic sands. Flowering appears to be in winter and new shoot growth in summer. Limited data suggest that the species is killed by fire and regenerates solely through seedling germination.

BANKSIA BURDETTII E.G.Baker (1934)

Burdett's Banksia

81 RECORDS: Jan (2) Feb (4) Mar (1) Apr (2) May (0) Jun (1) Jul (1) Aug (6) Sep (12) Oct (0) Nov (2) Dec (4).

Population Size: 1-10 (10) 10-100 (12) >100 (24) Unspecified (35).

Conservation Status: Restricted to road verge (40%) Not (59%) Unspecified (1%). In conservation reserve (32%) Not (67%) Unspecified (1%).

Tree/Shrub: Tree form (17%) Shrub form (26%) Unspecified (57%).

Height (metres)

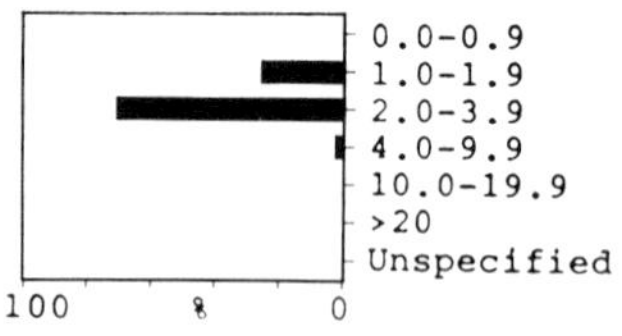

New Shoot Growth

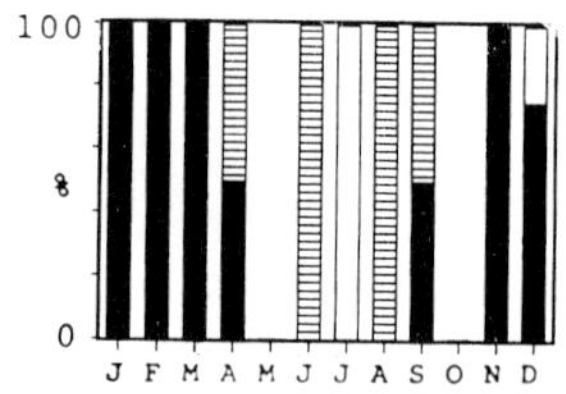

Yes
No
Unspecified

HABITAT

Within 2 km of coast: Yes (0%) No (100%) Unspecified (0%).

Altitude (metres)

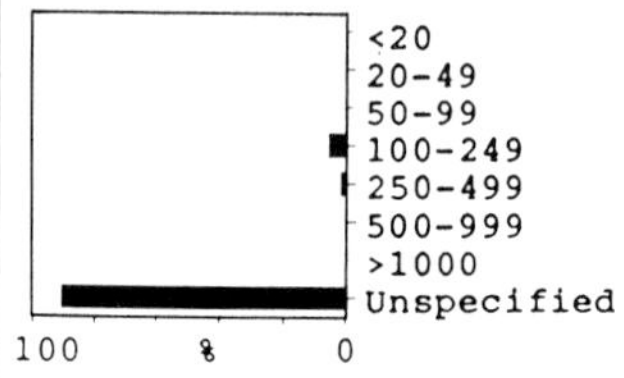

Vegetation

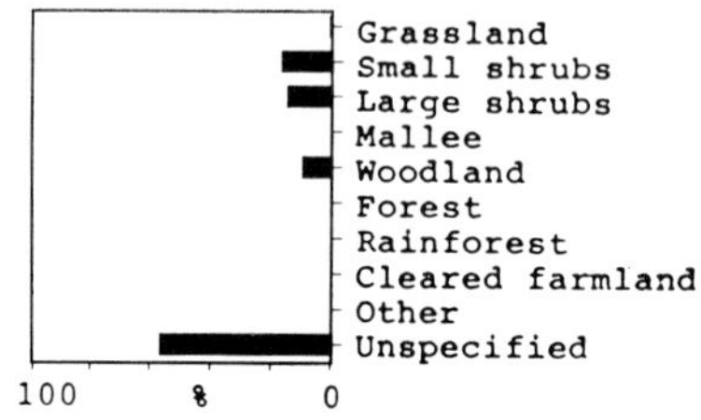

Soil Type

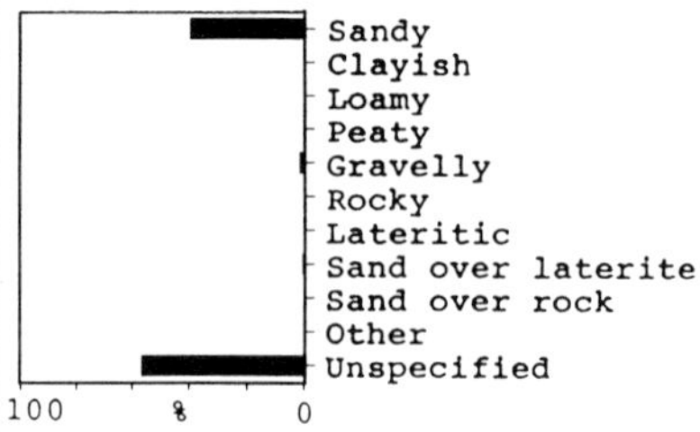

Landform

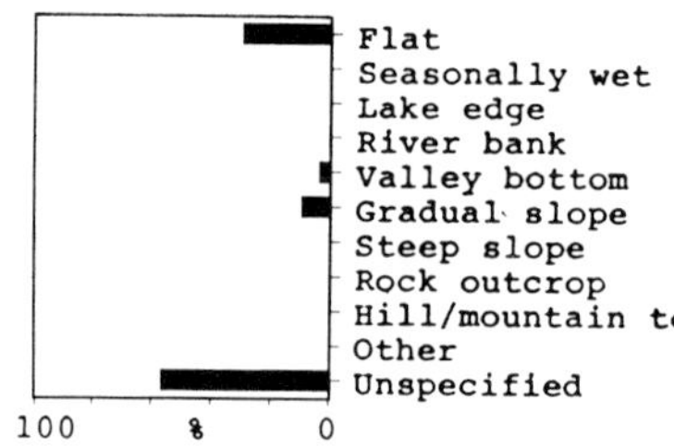

Response to Fire: 1 record

Growth response: Unspecified (100%).

Flowering response: Number of records where flowering had occurred after fire (1). Minimum reported time to flowering (49-60 months).
Number of records where flowering had not occurred after fire (0).

Flowering

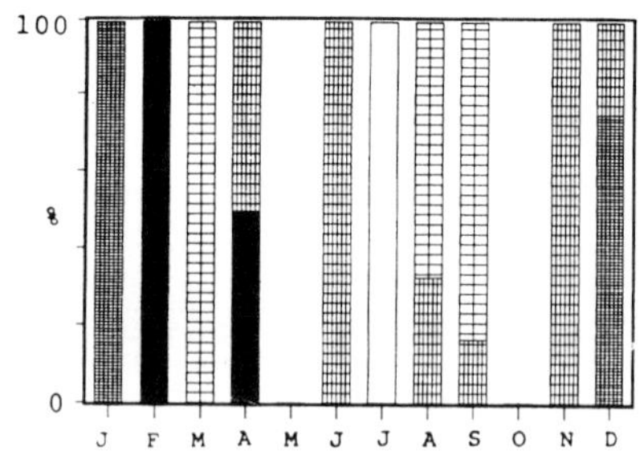

Majority of flowers fully open
Majority of flowers in bud
Majority of flowers recently finished
Flowers finished, fruiting cones present
Neither flowers nor fruiting cones present
Unspecified

Possible pollinators

Pollinator type 2 records: bird (100%).

Specific pollinator 1 record: Brown Honeyeater (1).

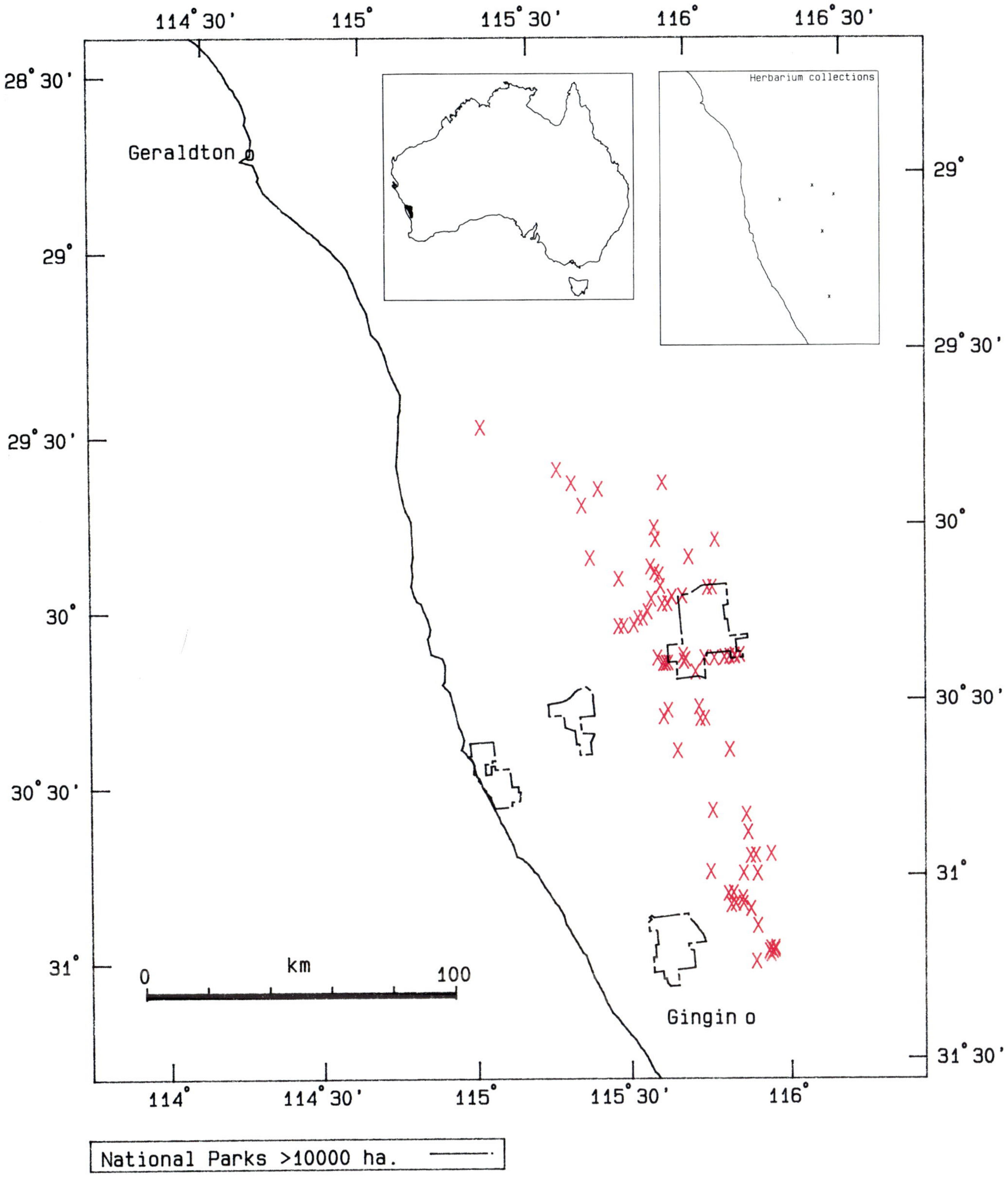

DISCUSSION

Western Australian Herbarium collections had indicated an intermittent and restricted distribution for *B. burdettii* between Eneabba and Mogumber. Atlas records have confirmed this general distribution whilst extending the range *ca.* 40 km further north and 15 km further south. Atlas records have also shown the species to be more common than was previously thought. Almost one-third of populations recorded were in conservation reserves (chiefly Watheroo National Park, but also Alexander Morrison and Tathra National Parks). Most other populations occur on road verges bordering agricultural land. A relatively large percentage of populations have over 100 plants. Indeed, Barker and Lamont (1986) found that 20% of the populations they studied had over 1 000 plants with the largest stand estimated at 10 000 plants. Although the species is used in the cut wildflower trade, and it is also susceptible to dieback disease, its present status in the wild does not seem under threat.

B. burdettii is restricted to white or yellow sandplain country where it is a component of shrubland, occasionally occurring in woodland. It grows as a large shrub or small tree usually up to 4 m in height. Flowering is in summer and early autumn and new shoot growth from November through to March. The two observations of bird-pollinators by Atlas contributors support the study by Barker and Lamont (1986), where the following birds were all found to be carrying pollen of *B. burdettii* : Little Wattlebird, Red Wattlebird, Brown Honeyeater, White-cheeked Honeyeater, Spiny-cheeked Honeyeater. In addition, a Honey-possum (*Tarsipes rostratus*) was observed feeding on *B. burdettii* flower spikes, and trapped specimens carried significant pollen loads. There is a lack of Atlas data for the species' response to fire but detailed studies by Barker and Lamont (1986) show it to be highly sensitive, being killed by fire and regenerating from seed. An average of four years was needed for resultant seedlings to reach flowering stage.

BANKSIA CALEYI R.Brown (1830)

Cayley's Banksia

100 RECORDS: Jan (9) Feb (1) Mar (3) Apr (5) May (5) Jun (3) Jul (9) Aug (9) Sep (15) Oct (18) Nov (13) Dec (7).

Population Size: 1-10 (31) 10-100 (43) >100 (16) Unspecified (10).

Conservation Status: Restricted to road verge (39%) Not (50%) Unspecified (11%). In conservation reserve (37%) Not (59%) Unspecified (4%).

Tree/Shrub: Tree form (2%) Shrub form (95%) Unspecified (3%).

Height (metres)

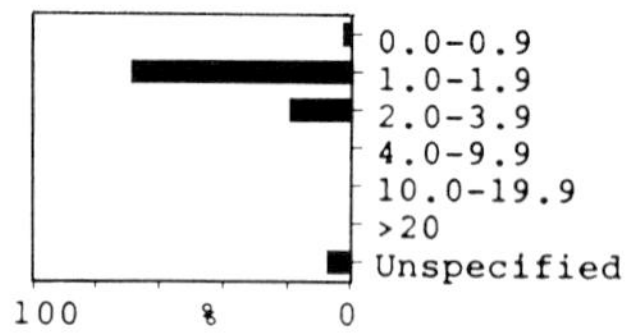

New Shoot Growth

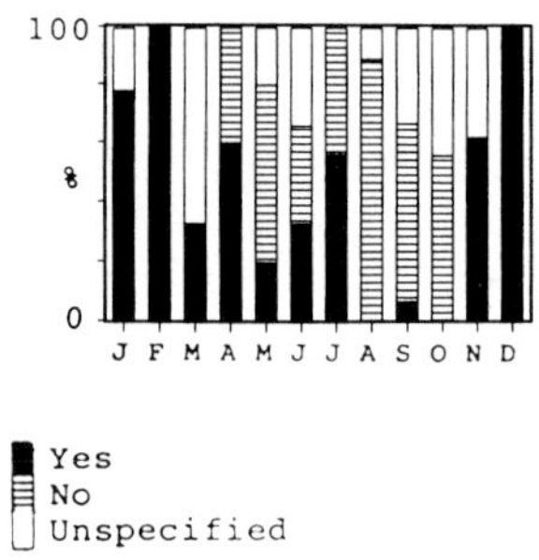

HABITAT

Within 2 km of coast: Yes (3%) No (97%) Unspecified (0%).

Altitude (metres)

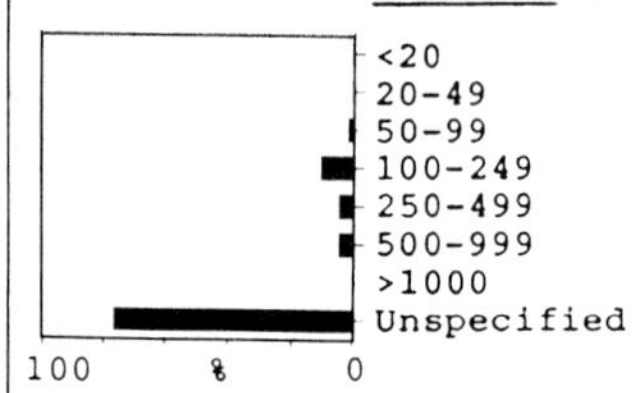

Vegetation

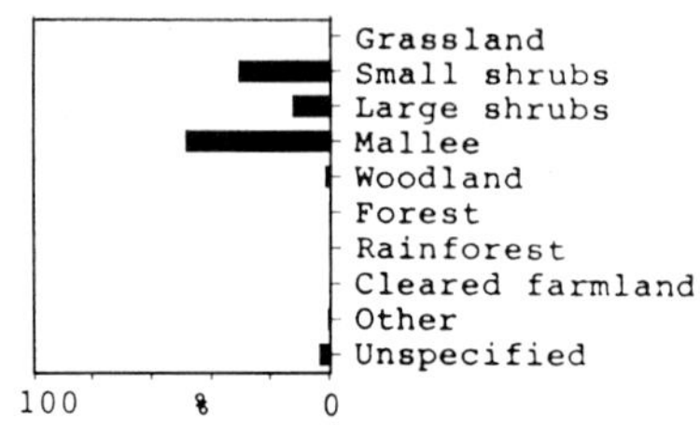

Soil Type

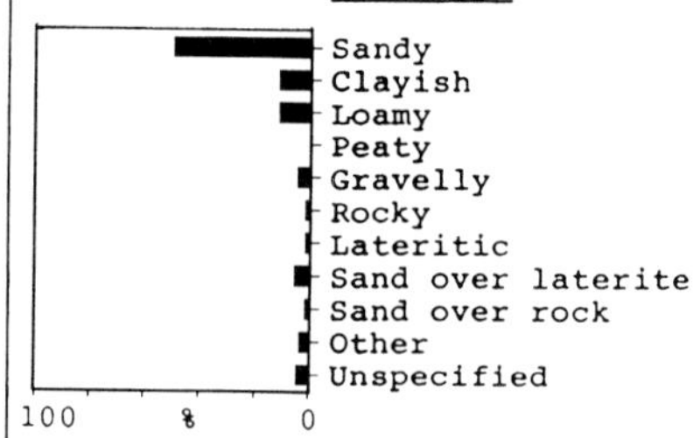

Landform

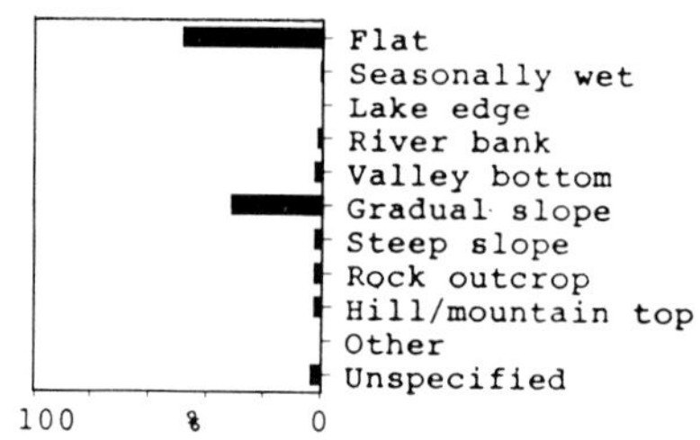

Response to Fire: 3 records

Growth response: Killed, new seedlings (33%) killed, no seedlings (33%) unspecified (33%).

Flowering response: Number of records where flowering had occurred after fire (1). Minimum reported time to flowering (25-36 months).
Number of records where flowering had not occurred after fire (1). Maximum repoted time to no flowering (25-36 months).

Flowering

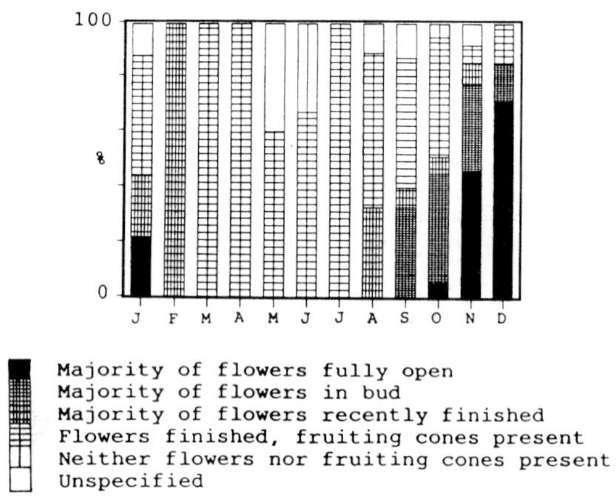

Possible pollinators

Pollinator type 29 records: bees, wasps, ants (3%) no pollinator observed (97%).

Specific pollinator 1 record: ant (1).

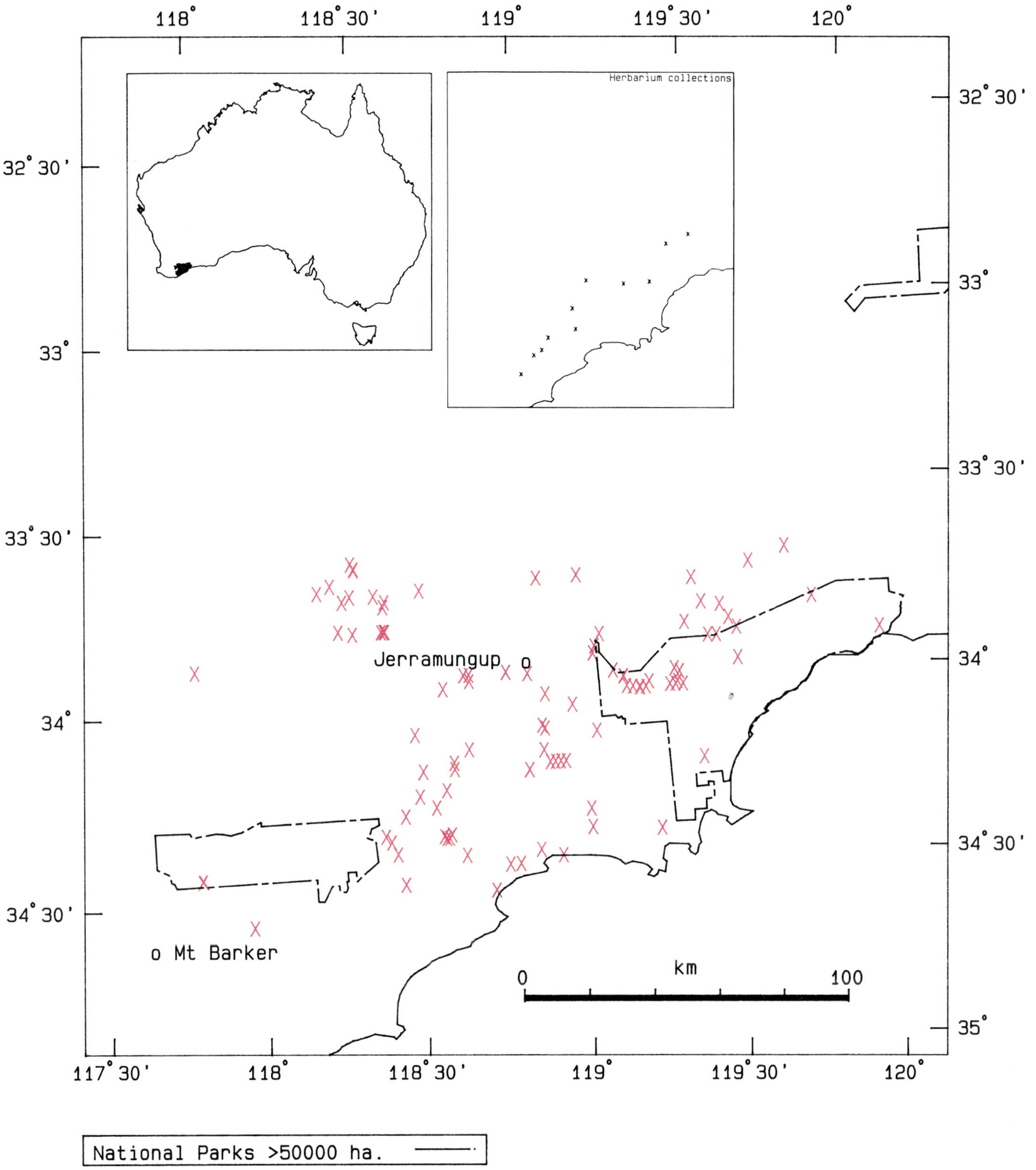

DISCUSSION

Western Australian Herbarium collections indicated a near-coastal distribution for B. caleyi between South Stirling and the West River. George (1981) also cited a 1968 collection of A.M. Ashby's from an outlying population south of Pingrup. Sight records have confirmed this distribution but also suggest a more continuous pattern between the southern populations and those near Pingrup. There are also new recordings from several near-coastal locations such as Cheyne Bay to Beaufort Inlet, near Bremer Bay, West and East Mt Barren. (The last record is not supported by a voucher specimen.) Outlier populations near Broomehill and in the western Stirling Range National Park are significant range extensions. A collection made by A.S. George in 1978 from South Stirling was not recorded by Atlas contributors. However, a new population to the west of Chester Pass Rd was reported. The species is relatively secure with 37% of its populations in conservation reserves, notably Fitzgerald River National Park. It is also one of the least susceptible banksias to dieback disease.

B. caleyi grows as a shrub usually 1-2 m, but sometimes 2-4 m in height. It occurs on flat or gently undulating land amongst mallee eucalypts or small shrubs. Soils are of sand, sandy-loam or sandy-clay. To the north-east of Cape Riche the species occurs in near-coastal communities of similar habitat and structure to inland types. Flowering is from September to January and a new shoot growth occurs predominantly in summer. The species appears to be killed by fire and to require at least three years for seedlings to reach maturity and flower.

BANKSIA CANDOLLEANA Meissner (1855)

Propeller Banksia

95 RECORDS: Jan (2) Feb (5) Mar (4) Apr (8) May (19) Jun (0) Jul (9) Aug (13) Sep (19) Oct (4) Nov (1) Dec (11).

Population Size: 1-10 (5) 10-100 (34) >100 (53) Unspecified (3).

Conservation Status: Restricted to road verge (21%) Not (78%) Unspecified (1%). In conservation reserve (21%) Not (77%) Unspecified (2%).

Tree/Shrub: Tree form (0%) Shrub form (99%) Unspecified (1%).

Height (metres)

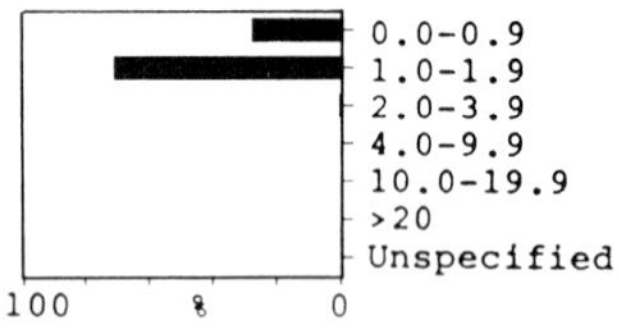

New Shoot Growth

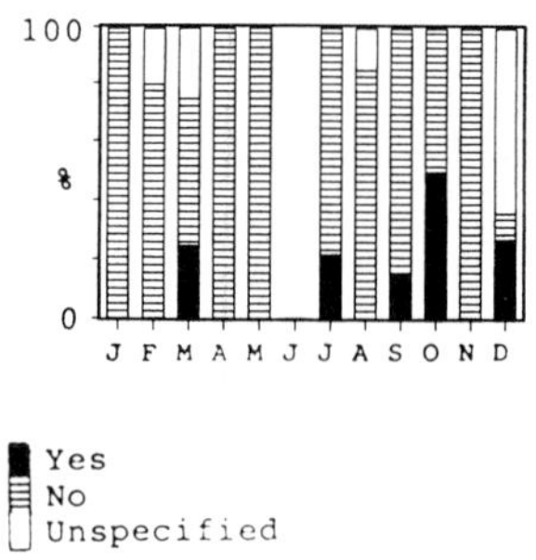

HABITAT

Within 2 km of coast: Yes (0%) No (99%) Unspecified (1%).

Altitude (metres)

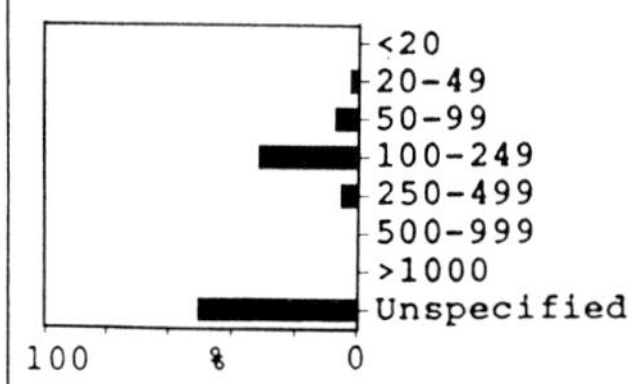

Vegetation

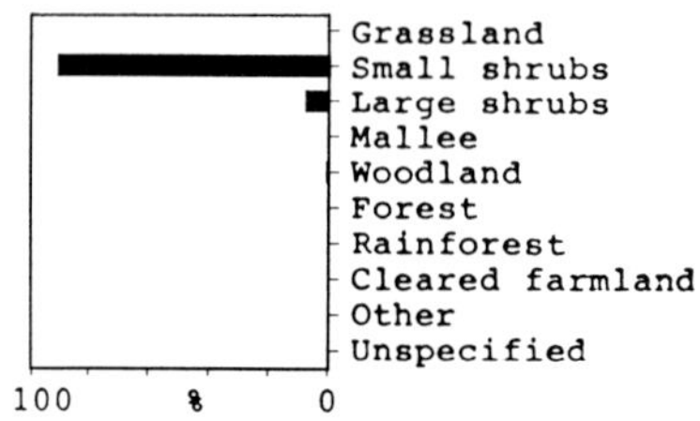

Soil Type

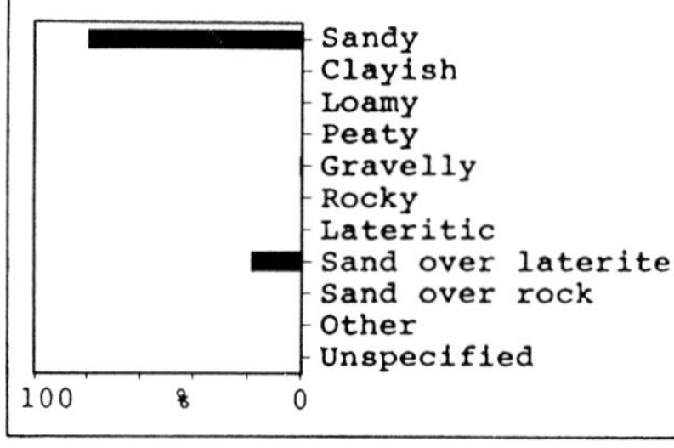

Landform

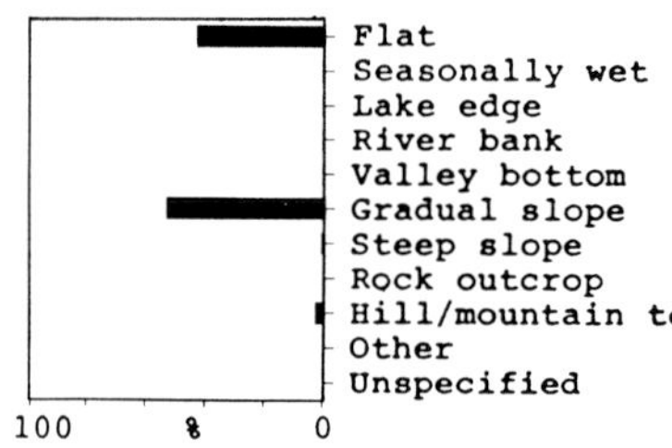

Response to Fire: 15 records

Growth response: Ground resprout (60%) ground resprout and killed, new seedlings (7%) unspecified (33%).

Flowering response: Number of records where flowering had occurred after fire (10). Median reported time to flowering (13-24 months); minimum reported time to flowering (13-24 months). Number of records where flowering had not occurred after fire (3). Median reported time to no flowering (13-24 months); maximum reported time to no flowering (13-24 months).

Flowering

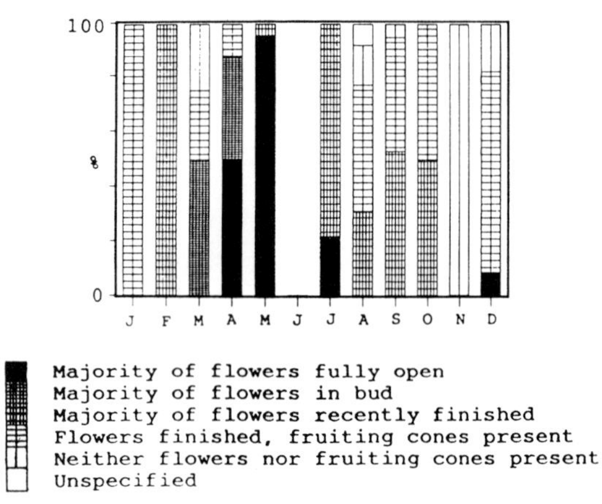

Possible pollinators

Pollinator type 26 records: bees, wasps, ants (12%) no pollinator observed (88%).

Specific pollinator 3 records: ant (2) European honey bee (1).

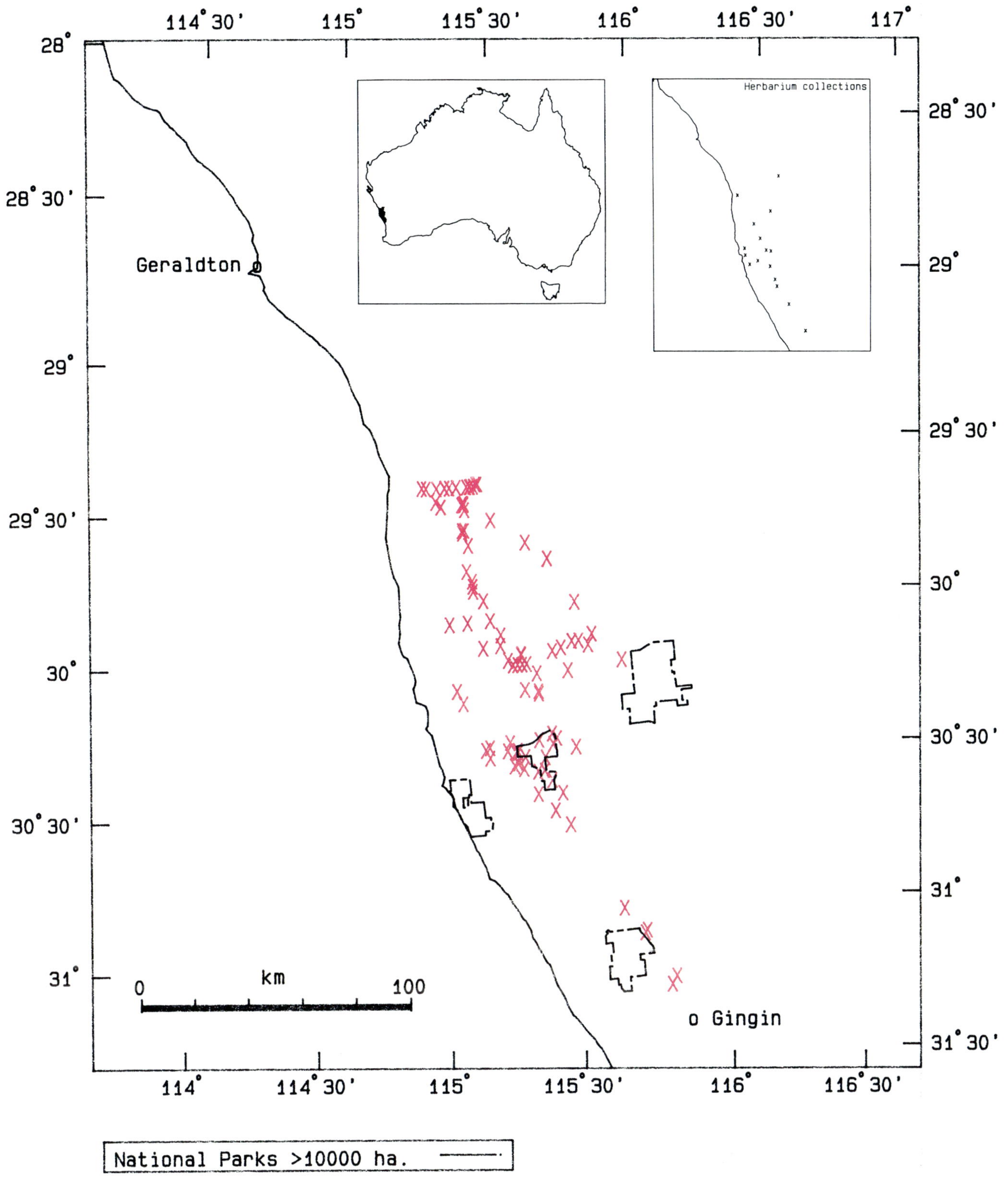

DISCUSSION

Western Australian Herbarium records of B. candolleana indicated a distribution between Arrowsmith and Gingin, generally with 50 km of the coast. Atlas contributors confirmed this general distribution and showed it to extend further inland than was previously known e.g. almost to the north-west corner of Watheroo National Park. A 1962 collection of C.A. Gardner's from north-west of Three Springs was not recorded. Most populations are of more than 100 plants and 21% occur in conservation reserves, notably Badgingarra and Alexander Morrison National Parks.

B. candolleana is a shrub up to 1.5 m in height. It grows on flat or gently undulating sandplain sometimes in sand overlying laterite. Generally, surrounding vegetation is of low heath. Flowering is from March to July. There is also a single record of flowering in December. There is an ill-defined period of new shoot growth, though spring-early summer appears most likely. Following fire, the species resprouts from its lignotuber and is flowering again within one-two years.

BANKSIA CANEI J.H.Willis (1967)

Mountain Banksia

48 RECORDS: Jan (20) Feb (5) Mar (10) Apr (1) May (7) Jun (0) Jul (0) Aug (1) Sep (2) Oct (0) Nov (0) Dec (2).

Population Size: 1-10 (4) 10-100 (7) >100 (36) Unspecified (1).

Conservation Status: Restricted to road verge (0%) Not (96%) Unspecified (4%). In conservation reserve (19%) Not (79%) Unspecified (2%).

Tree/Shrub: Tree form (6%) Shrub form (92%) Unspecified (2%).

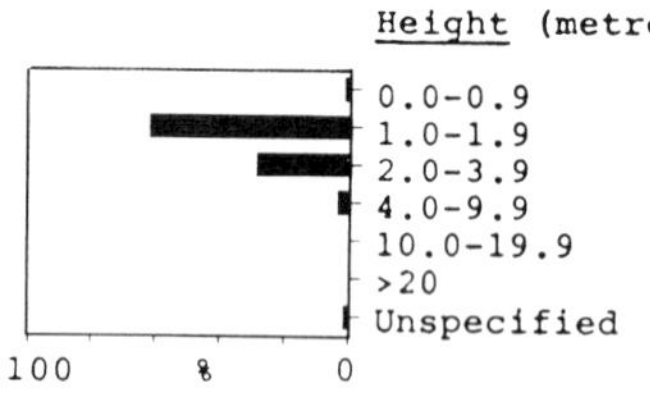

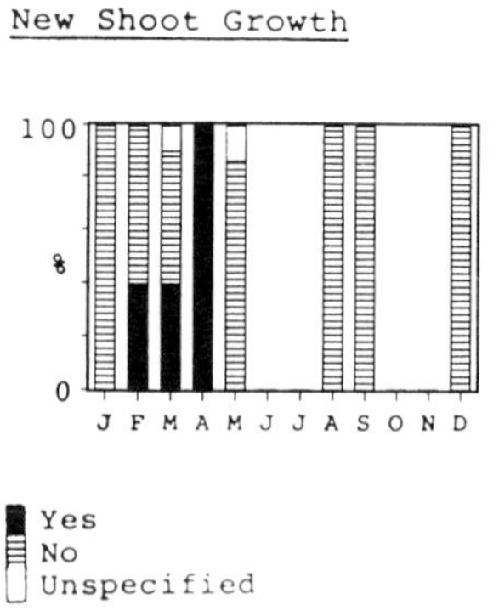

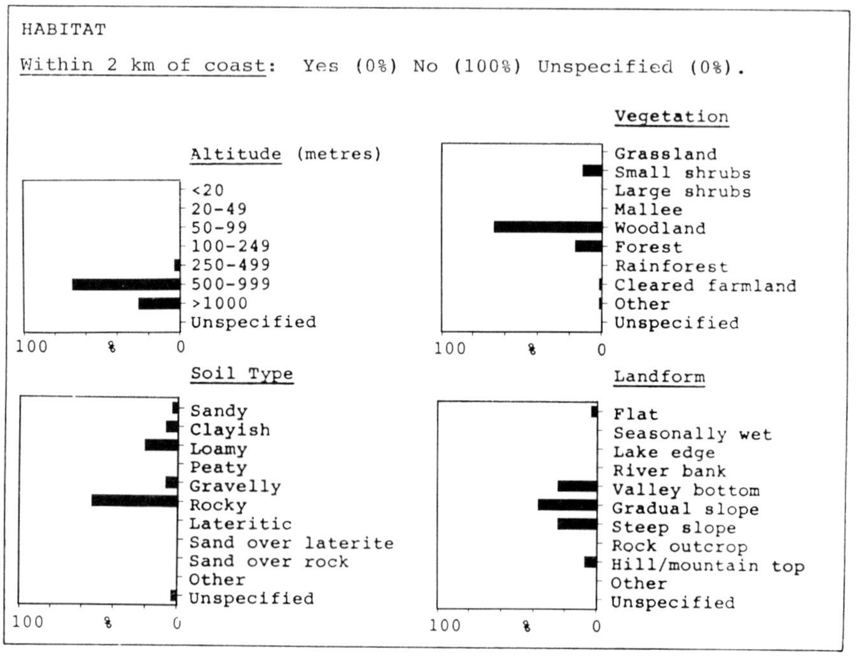

Response to Fire: 3 records

Growth response: Unspecified (100%).

Flowering response: Number of records where flowering had occurred after fire (2). Minimum reported time to flowering (<1 month).
Number of records where flowering had not occurred after fire (0).

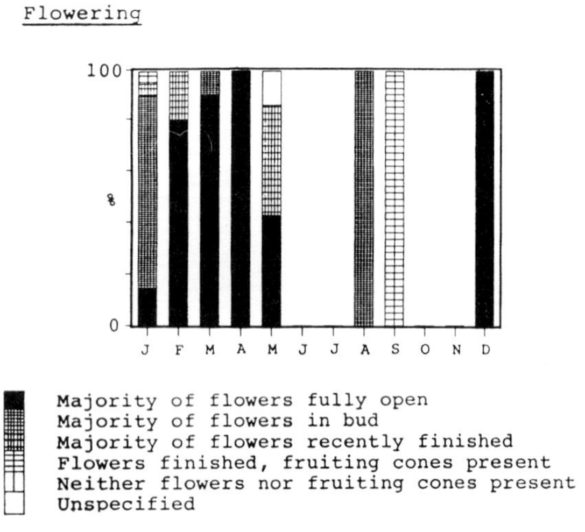

Possible pollinators

Pollinator type 42 records: bird (5%) bees, wasps, ants (7%) no pollinator observed (88%)

Specific pollinator 4 records: honeyeater (1) Yellow-tufted Honeyeater (1) bee (2)

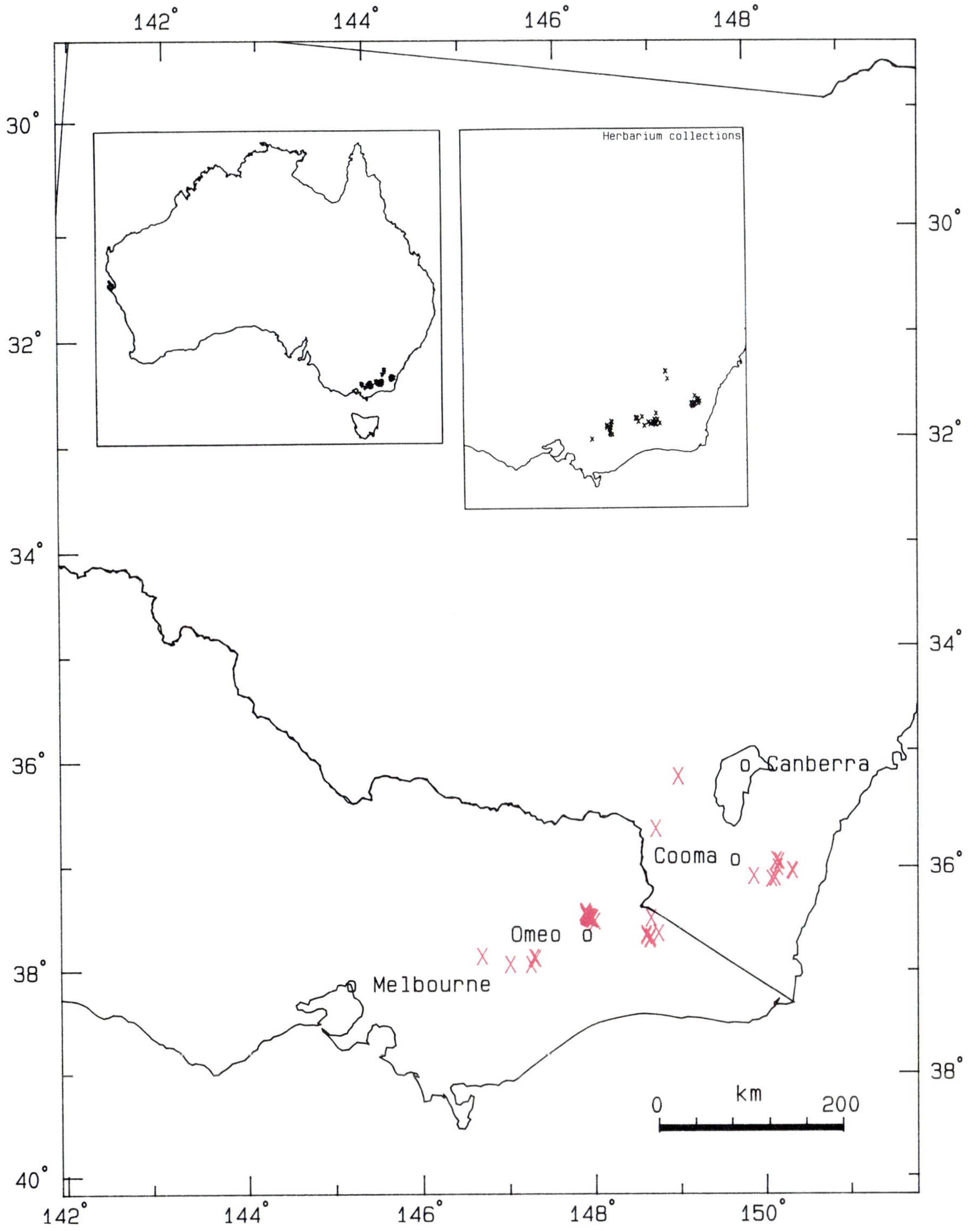

DISCUSSION

Herbarium records indicated that B. canei had a scattered distribution in subalpine areas of the Great Dividing Range, with the bulk of its populations located in four disjunct areas; (i) the Snowy Range (Victoria) between the Maroka and Barkly Rivers, as far north as Mt Howitt; (ii) Omeo eastwards to Wulgulmerang and the Little River (Victoria); (iii) the Kybean Range and Tuross River (New South Wales); (iv) Talbingo in the Bogong Mountains west of Canberra. There is also an outlier in the Bogong Mountains west of the A.C.T. Atlas contributors have confirmed this distribution and located a new outlier west of Mt Kosciusko. An interesting range extension of the Tuross River population was located at Yowrie where the plants occur as isolated shrubs on otherwise cleared farmland at an altitude of only 250 m. At all other known sites, B. canei grows at altitudes in excess of 500 m, often over 1 000 m.

B. canei is usually a shrub up to 4 m in height. It grows in rocky, sometimes loamy soils generally derived from granite. It is found on mountain slopes and in adjacent valleys usually in eucalypt woodland or forest. Flowers were recorded from December to May with a peak from February to April. The latter months are also the season for new shoot growth.

BANKSIA CHAMAEPHYTON A.S.George (1981)

Fishbone Banksia

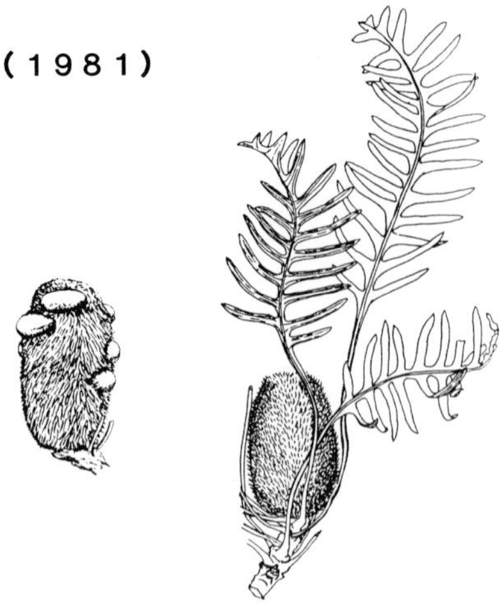

36 RECORDS: Jan (1) Feb (1) Mar (4) Apr (9) May (4) Jun (1) Jul (0) Aug (4) Sep (8) Oct (1) Nov (2) Dec (1).

Population Size: 1-10 (8) 10-100 (21) >100 (6) Unspecified (1).

Conservation Status: Restricted to road verge (17%) Not (83%) Unspecified (0%). In conservation reserve (17%) Not (81%) Unspecified (3%).

Tree/Shrub: Tree form (0%) Shrub form (97%) Unspecified (3%).

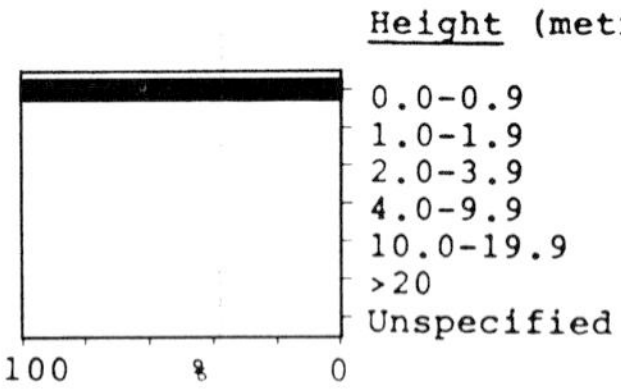

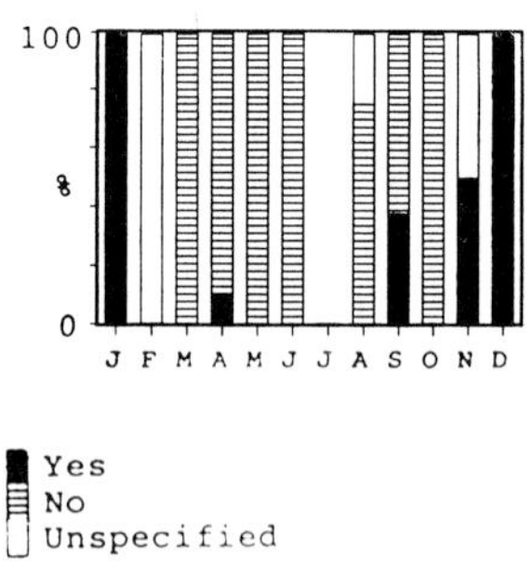

HABITAT

Within 2 km of coast: Yes (0%) No (97%) Unspecified (3%).

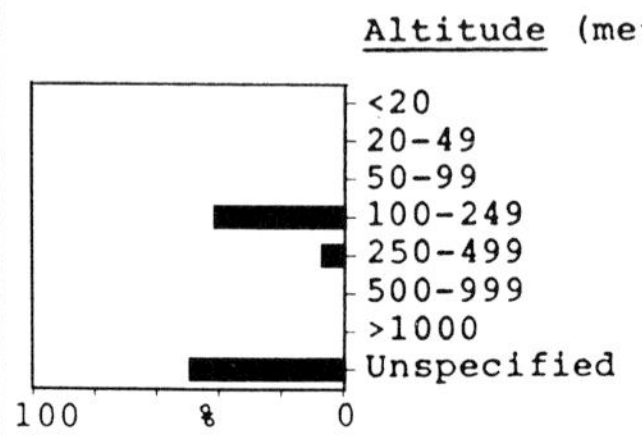

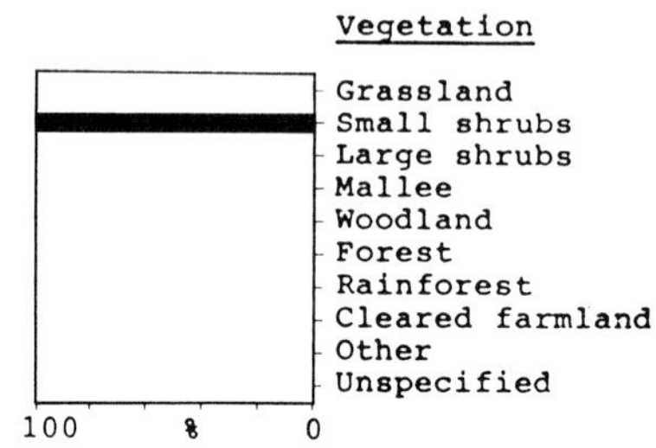

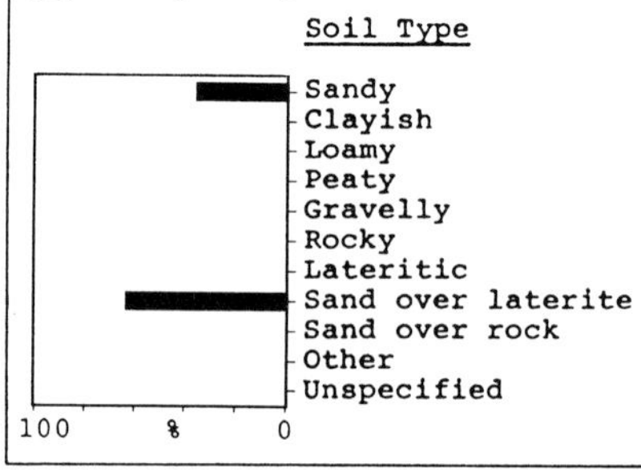

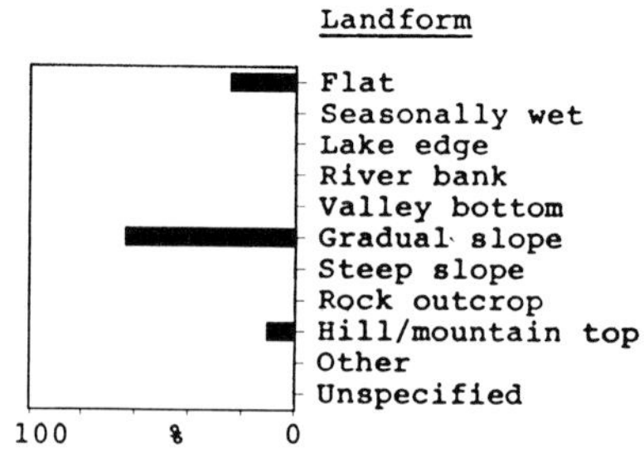

Response to Fire: 0 records

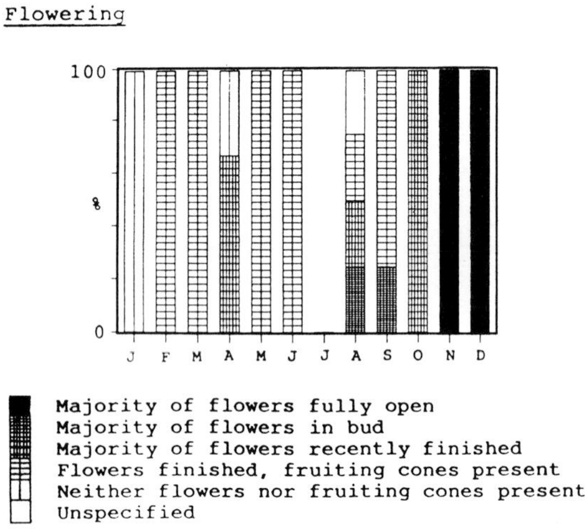

Possible pollinators

Pollinator type 13 records: no pollinator observed (100%).

Specific pollinator 0 records.

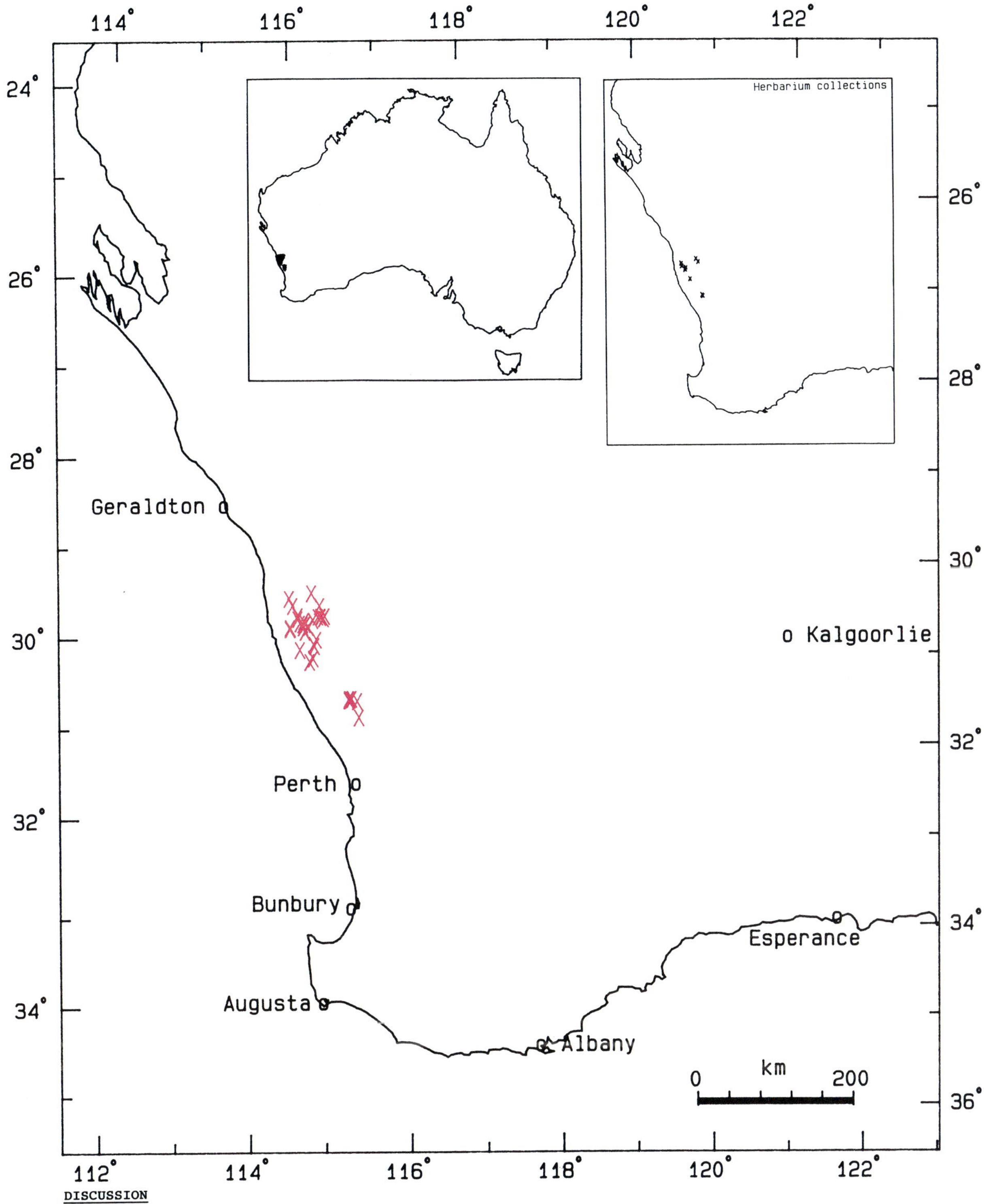

DISCUSSION

B. chamaephyton occurs in two disjunct populations. The northern population extends from Eneabba to the southern tip of Badgingarra National Park and inland to Alexander Morrison National Park. Localities of two herbarium collections from further inland (Chapman 1970 and George 1971) were not refound. The southern population is from the Mogumber area. This is separated by ca. 70 km from the populations further north. The southern-most record is a new location which extends the range of this species by about 25 km. The 36 Atlas records suggest that this species is more common than was previously thought. However, most populations are relatively small (less than 100 plants) and only 17% are in national parks and nature reserves. The southern group is particularly at risk since at least one of the sites at which B. chamaephyton occurs is currently being considered for agricultural clearance. For all these reasons, the species needs to be monitored closely to ensure it remains secure in the wild.

B. chamaephyton is the only low-growing prostrate banksia from the heaths north of Perth. It grows in clumps connected by underground stems, which make it difficult to assess population size. Barrett (1985) suggested that clumps less than 50 cm apart should be regarded as one individual. The species occurs in sand over laterite or sand amongst low heath, often co-existing with B. incana. Landform is usually gently undulating. Flowering is in November and December. No pollinators were observed, but research on other prostrate species suggests that small mammals may be implicated (Hopper 1981, Collins, pers. comm.). Although there were no observations of fire response by Atlas contributors, A. Taylor has noted B. chamaephyton resprouting strongly from its lignotuber following removal of its above ground shoots for firebreak clearance. Similar observations were made by Barrett (1985). It is likely that plants would respond to fire in a similar fashion.

BANKSIA COCCINEA R.Brown (1810)

Scarlet Banksia, Waratah Banksia, Albany Banksia

171 RECORDS: Jan (10) Feb (8) Mar (13) Apr (11) May (6) Jun (15) Jul (15) Aug (21) Sep (34) Oct (19) Nov (10) Dec (8).

Population Size: 1-10 (27) 10-100 (75) >100 (67) Unspecified (2).

Conservation Status: Restricted to road verge (13%) Not (82%) Unspecified (5%). In conservation reserve (47%) (Not (46%) Unspecified (7%).

Tree/Shrub: Tree form (33%) Shrub form (65%) Unspecified (2%).

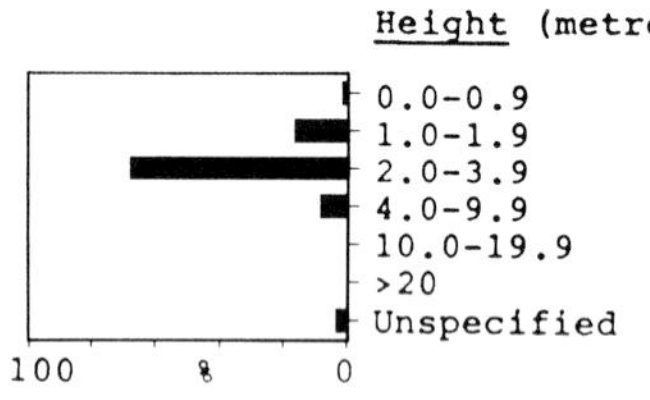

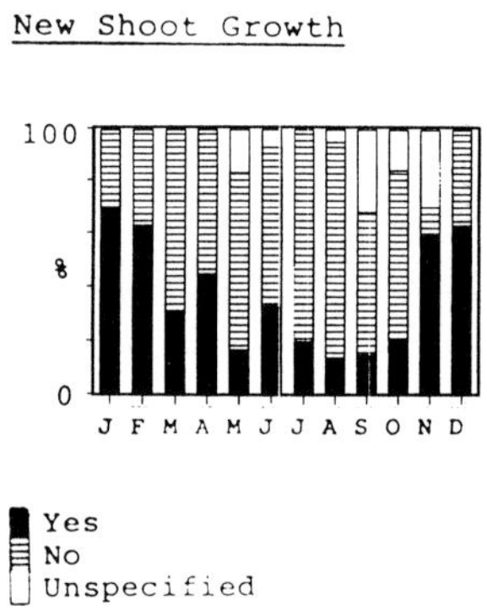

HABITAT

Within 2 km of coast: Yes (25%) No (74%) Unspecified (1%).

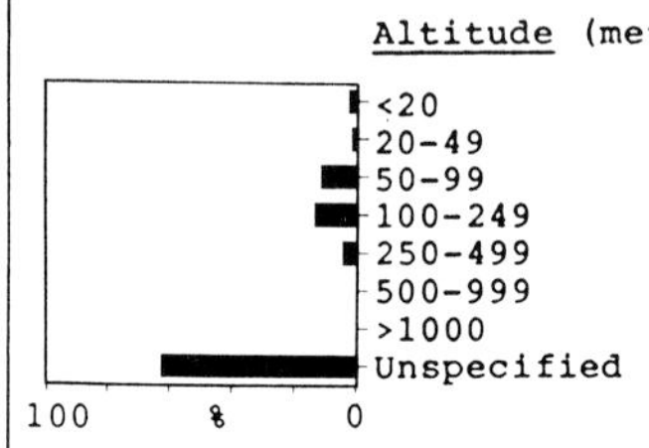

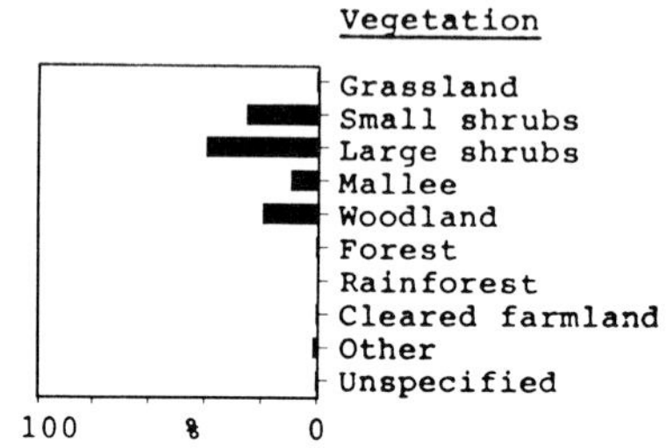

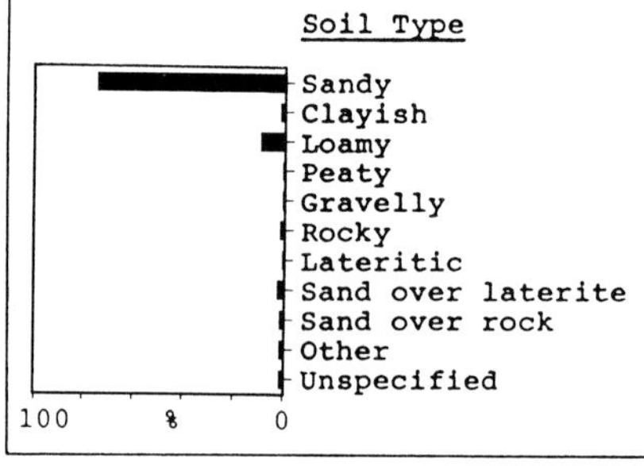

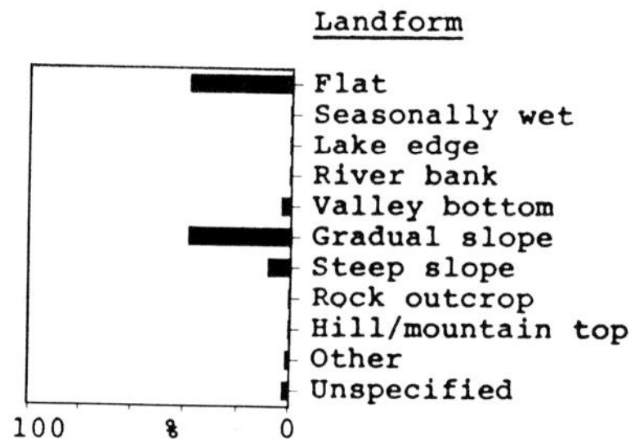

Response to Fire: 29 records

Growth response: Ground resprout and killed, new seedlings (3%) killed, new seedlings (59%) unspecified (38%).

Flowering response: Number of records where flowering had occurred after fire (15). Median reported time to flowering (13-24 months); minimum reported time to flowering (<1 month). Number of records where flowering had not occurred after fire (8). Median reported time to no flowering (24-25 months); maximum reported time to no flowering (49-60 months).

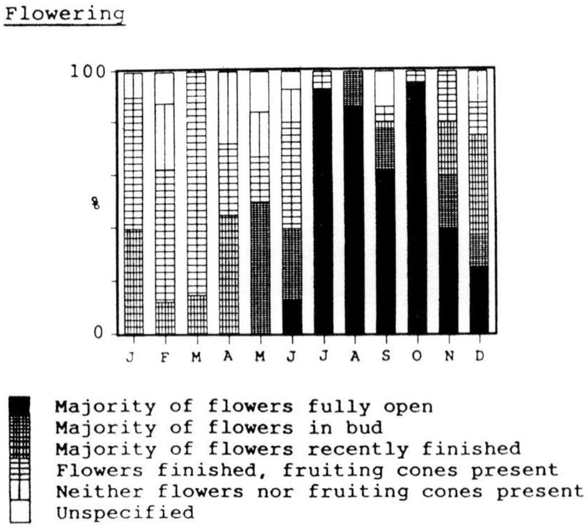

Possible pollinators

Pollinator type 43 records: bird (21%) bees, wasps, ants (9%) no pollinator observed (70%).

Specific pollinator 13 records: New Holland Honeyeater (7) Red Wattlebird (1) Western Spinebill (1) rosella (1) ant (3)

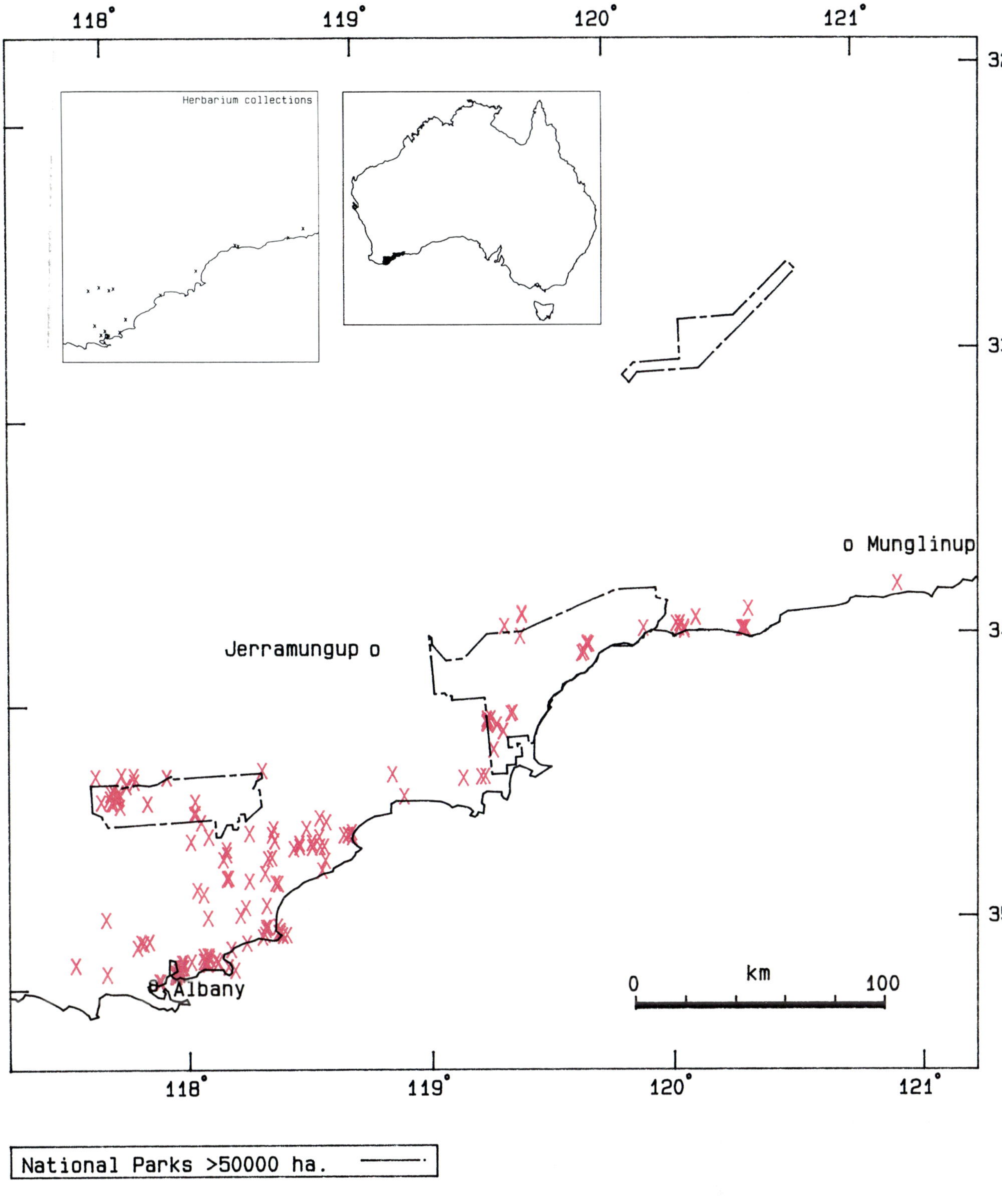

DISCUSSION

Atlas records have confirmed a near-coastal distribution with the most inland records coming from the Stirling Range. In the southern part of its distribution, B. coccinea's range has been extended to the Hay River, north-east of Denmark. The records on the northern boundary of Fitzgerald River National Park also represent a range extension. The eastern limit of the species is at Stokes National Park south-east of Munglinup. Although many of the recorded populations are of more than 100 plants and almost 50% are in conservation reserves, the species needs close monitoring since it is highly susceptible to dieback disease and many of its populations are decreasing at an alarming rate. B. coccinea is also heavily used in the cut wildflower trade (Burgman and Hopper 1982).

The species grows as a shrub or small tree usually 2-4 m in height, rarely larger. White or grey sandy soils are preferred with surrounding vegetation being tall shrubland, heath or mallee-heath. In some western populations it grows in Eucalyptus woodland with B. attenuata and B. ilicifolia. Landforms are generally flat or undulating. An unusual occurrence is at Ellen Peak in the Stirling Range where B. coccinea grows on a steep rocky slope. Flowering is from May to December with a peak from July to October. The conspicuous flowers are attractive to birds with the New Holland Honeyeater being the most frequently recorded. Orange-flowering variants of B. coccinea were occasionally recorded. New shoot growth peaks in summer. The species is killed by fire and regenerates from seed. Seedlings may take up to 5 years to reach flowering stage.

BANKSIA CONFERTA A.S.George (1981)

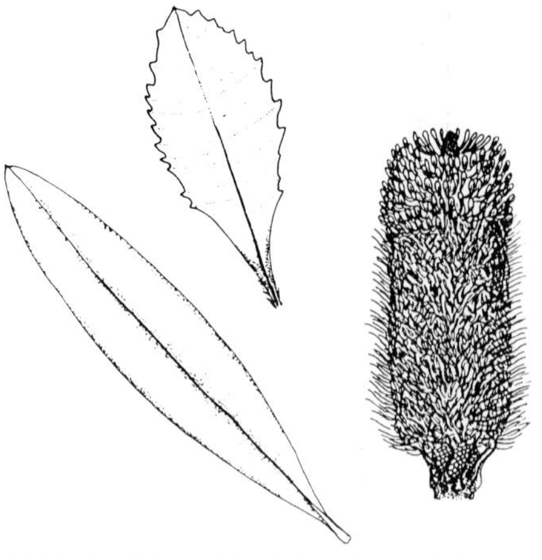

32 RECORDS: Jan (1) Feb (2) Mar (7) Apr (6) May (4) Jun (3) Jul (2) Aug (1) Sep (2) Oct (0) Nov (0) Dec (3).

Population Size: 1-10 (10) 10-100 (17) >100 (5) Unspecified (0).

Conservation Status: Restricted to road verge (0%) Not (97%) Unspecified (3%). In conservation reserve (44%) Not (56%) Unspecified (0%).

Tree/Shrub: Tree form (9%) Shrub form (91%) Unspecified (0%).

Height (metres)

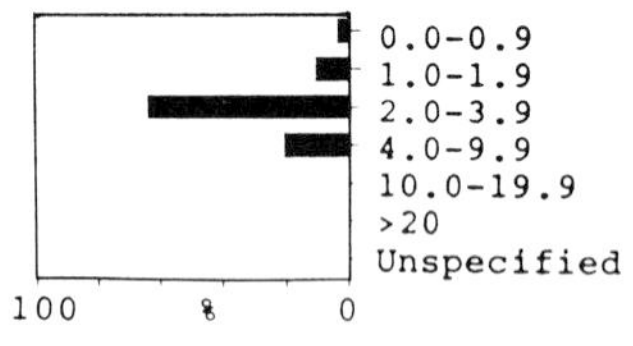

New Shoot Growth

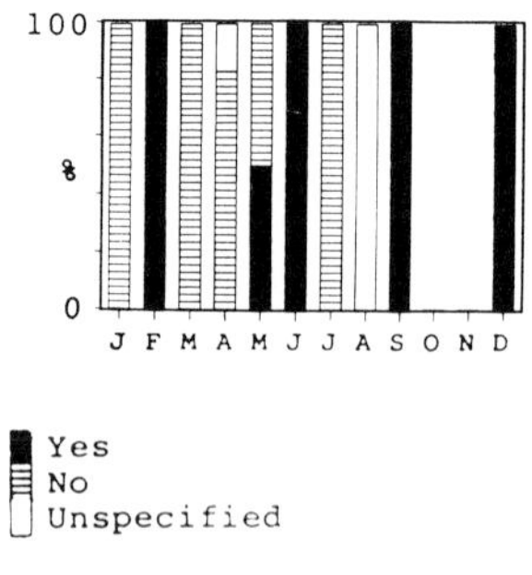

HABITAT

Within 2 km of coast: Yes (0%) No (100%) Unspecified (0%).

Altitude (metres)

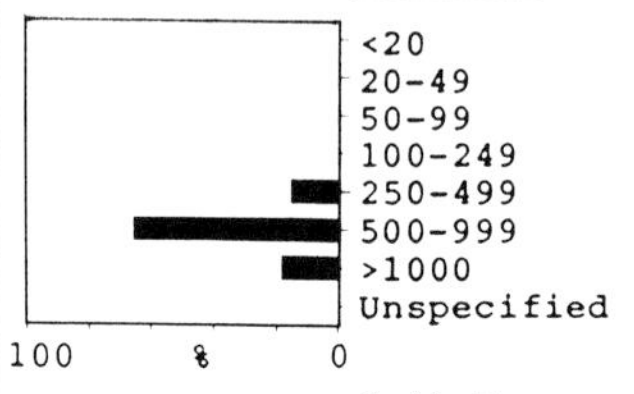

Vegetation

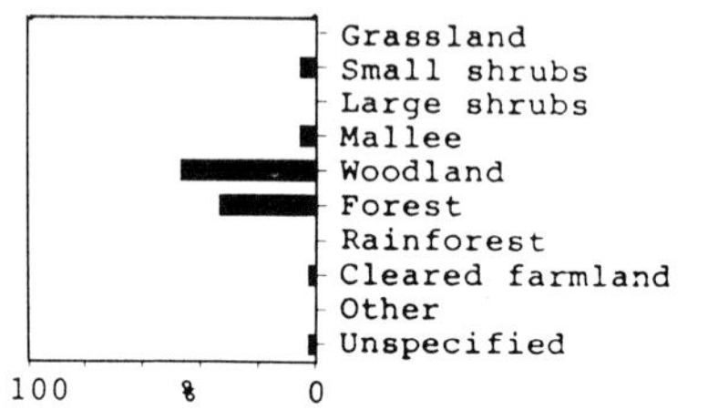

Soil Type

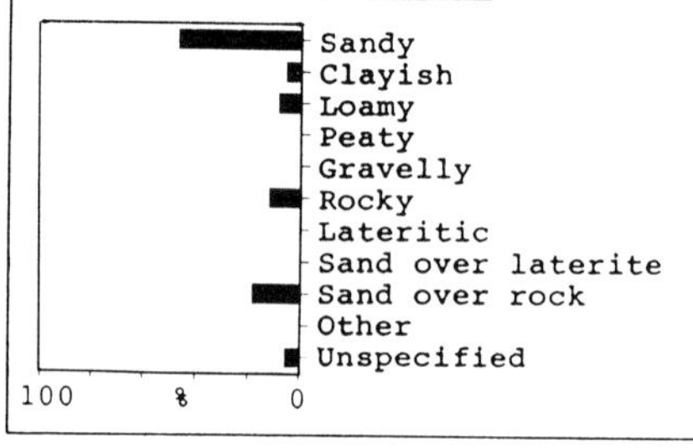

Landform

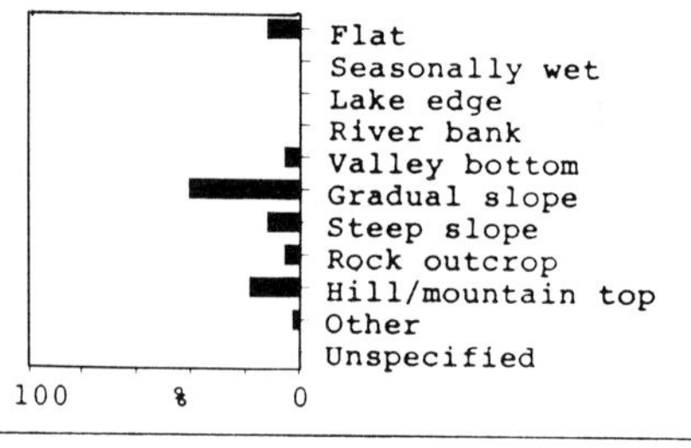

Response to Fire: 6 records

Growth response: Killed, new seedlings (67%) unspecified (33%).

Flowering response: Number of records where flowering had occurred after fire (4). Median reported time to flowering (60-72 months); minimum reported time to flowering (49-60 months). Number of records where flowering had not occurred after fire (1). Maximum reported time to no flowering (37-48 months).

Flowering

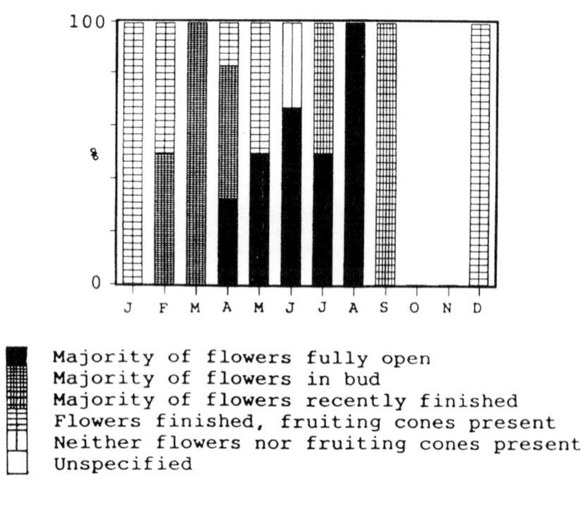

Possible pollinators

Pollinator type 22 records: bird (5%) bees, wasps, ants (14%) flies (5%) no pollinator observed (77%)

Specific pollinator 2 records: ant (1) bee (1)

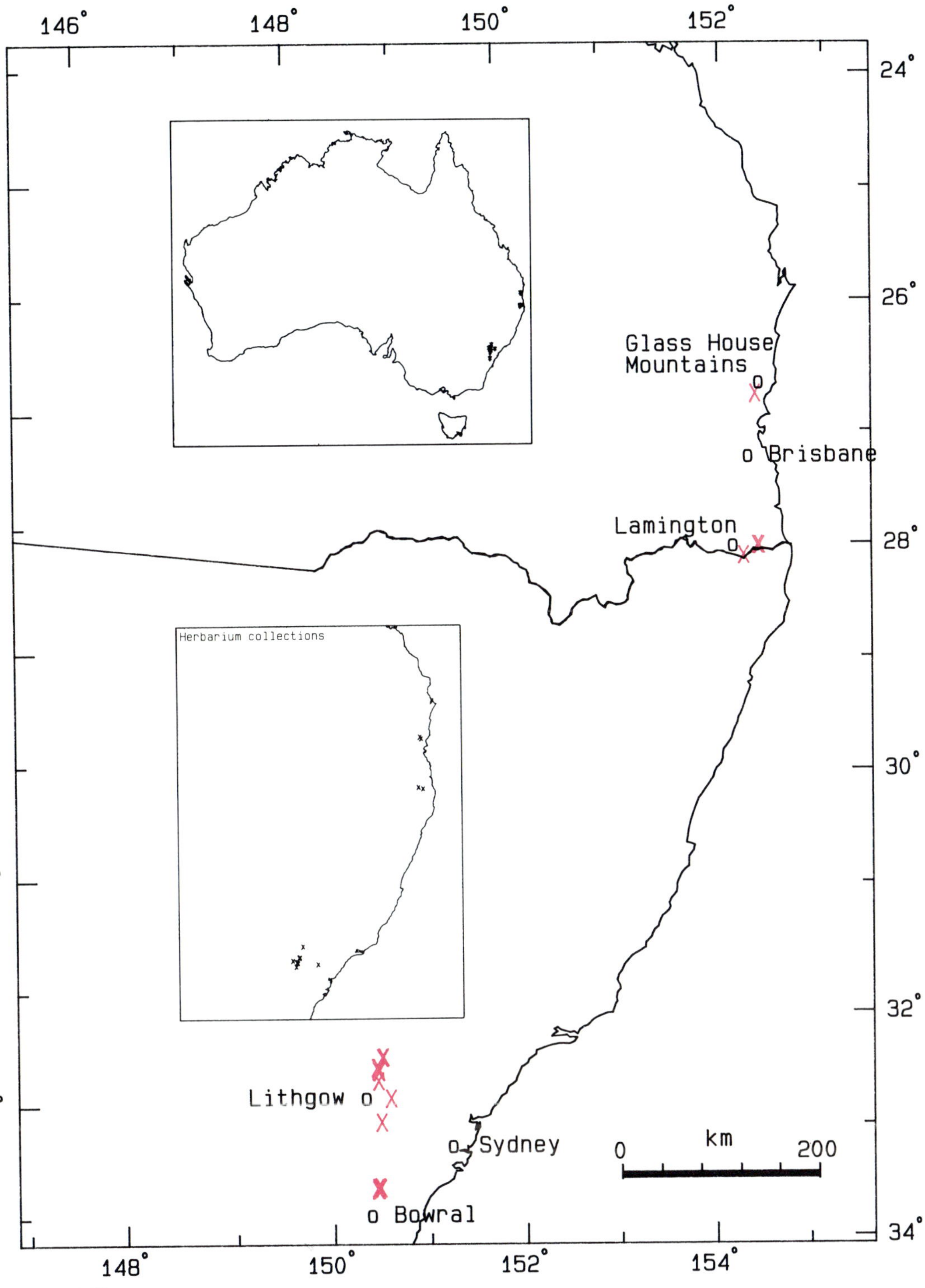

DISCUSSION

B. conferta is a rare montane shrub or small tree usually found in woodland or forest. It occurs as two varieties in two disjunct areas. Populations in Queensland (var. conferta) are found near the New South Wales border on the Lamington Plateau, Mt Barney, and further north in the Glass House Mountains. New South Wales populations (var. penicillata) occur in the Blue Mountains north-west of Sydney and south near Bowral.

George (1981) stated that one of the differences between var. conferta and var. penicillata is in the shape of mature leaves with conferta having entire leaf margins compared to the serrated margins of penicillata. This requires further investigation because a population of var. penicillata in the Blue Mountains included 2 m high plants with both serrated and entire leaves. The population of B. conferta var. conferta on Mt Barney has the majority of its leaves serrated. Most plants represented seedling growth resulting from fire. However, even mature (unburnt) specimens were noticeable for their serrated leaves.

BANKSIA CONFERTA A.S.George (1981) var. CONFERTA

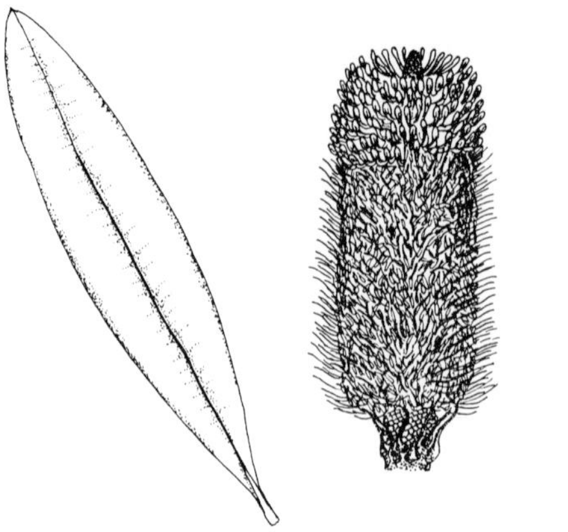

6 RECORDS: Jan (0) Feb (0) Mar (0) Apr (1) May (2) Jun (0) Jul (0) Aug (1) Sep (0) Oct (0) Nov (0) Dec (1).

Population Size: 1-10 (1) 10-100 (4) >100 (1) Unspecified (0).

Conservation Status: Restricted to road verge (0%) Not (83%) Unspecified (17%). In conservation reserve (100%) Not (0%) Unspecified (0%).

Tree/Shrub: Tree form (17%) Shrub form (83%) Unspecified (0%).

Height (metres)

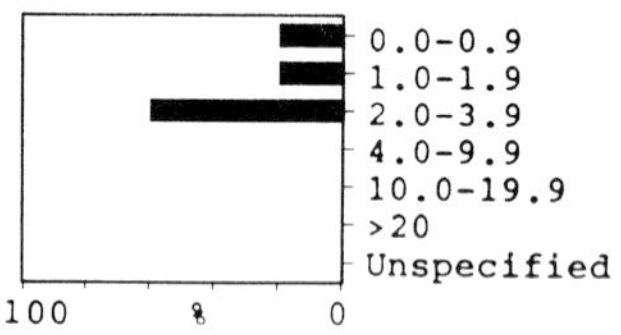

New Shoot Growth

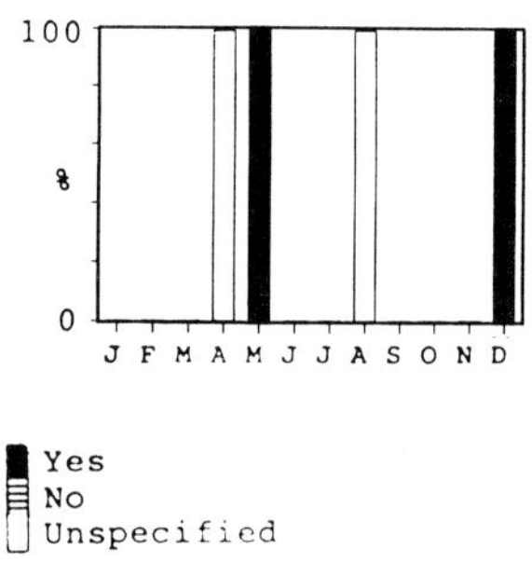

HABITAT

Within 2 km of coast: Yes (0%) No (100%) Unspecified (0%).

Altitude (metres)

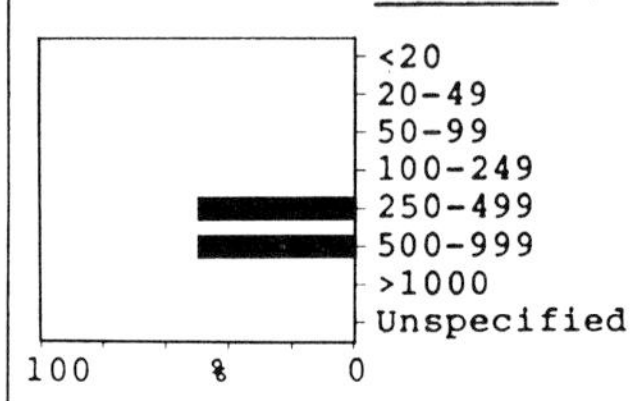

Vegetation

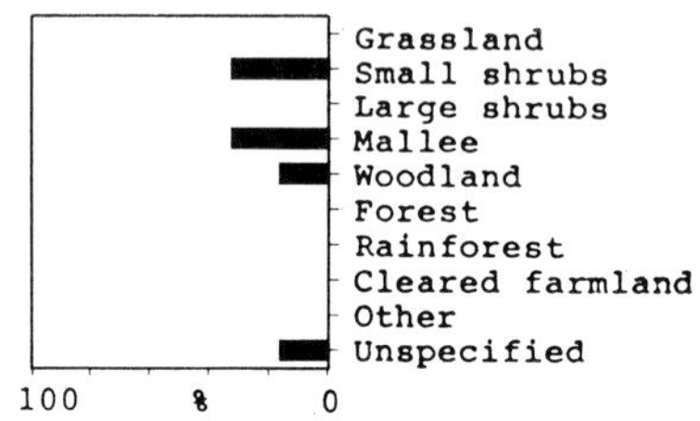

Soil Type

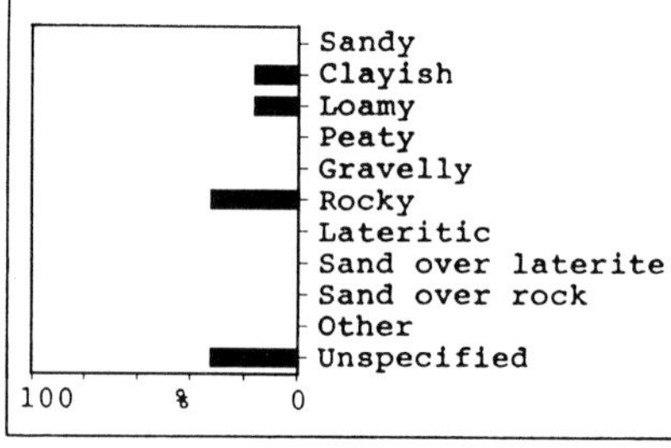

Landform

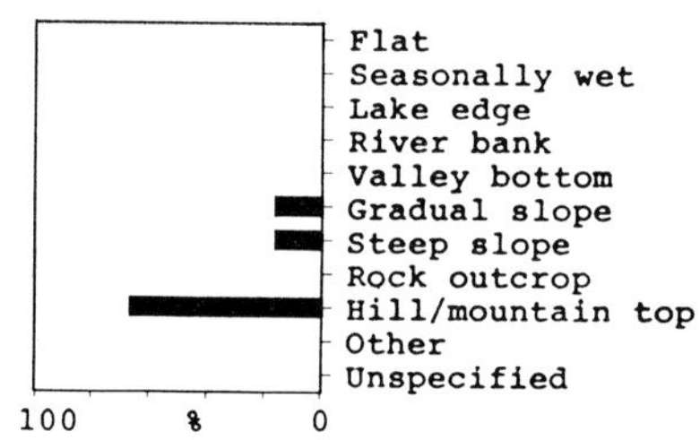

Response to Fire: 5 records

Growth response: Killed, new seedlings (60%) unspecified (40%).

Flowering response: Number of records where flowering had occurred after fire (4). Median reported time to flowering (61-72 months); minimum reported time to flowering (49-60 months). Number of records where flowering had not occurred after fire (0).

Flowering

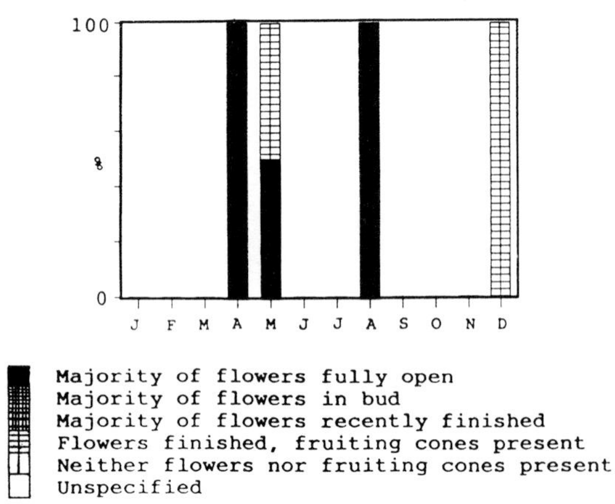

Possible pollinators

Pollinator type 2 records: bird (50%) bees, wasps, ants (50%)

Specific pollinator 0 records:

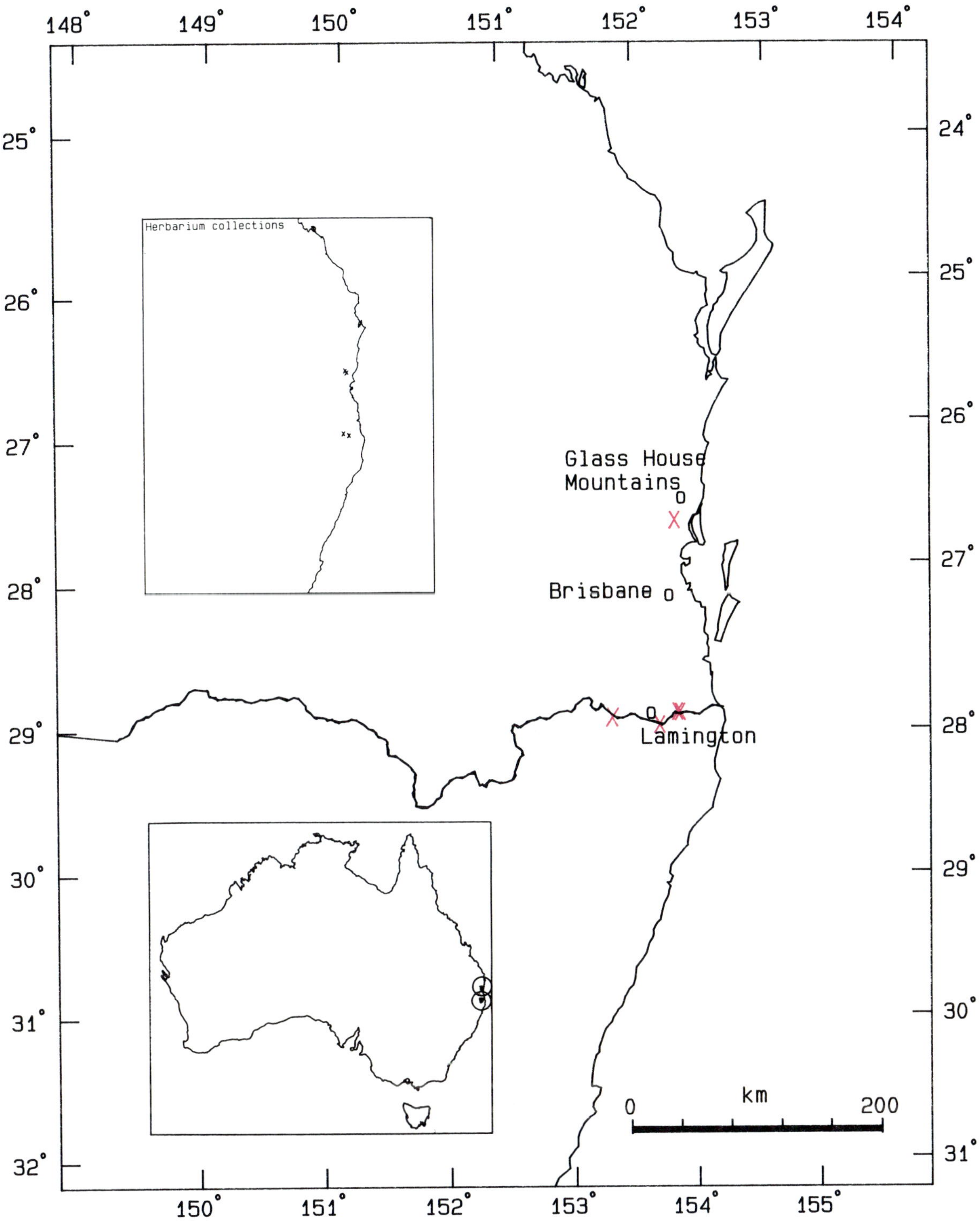

DISCUSSION

B. conferta var. conferta is one of the rarest banksias of the eastern States occurring only in the Glass House Mountains, on the Lamington Plateau and on Mt Barney. Its conservation status needs close monitoring especially since it appears to be killed by fire and to regenerate from seed. Repeated fires at too short intervals could eliminate one of these populations.

The variety grows as a shrub or small tree up to 4 m in height. It occurs on steep rocky slopes and hilltops up to 1 000 m above sea level. Surrounding vegetation is generally shrubland and scattered mallee eucalypts.

Limited records suggest a flowering period from autumn to late winter. There are insufficient data to determine the season of new shoot growth. George (1984) suggested that B. conferta var. conferta was probably fire tolerant, sprouting from epicormic buds on the trunk following fire. It now seems certain that this is not the case since Atlas contributors investigated two sites where recent fires had occurred, and in both cases the mature plants had been killed but there was abundant new seedling regrowth. In one case the fire was known to have occurred almost 5 years previously. Some of the young B. conferta plants had flowered for the first time in the previous (their fourth) year.

BANKSIA CONFERTA A.S.George var. PENICILLATA A.S.George (1981)

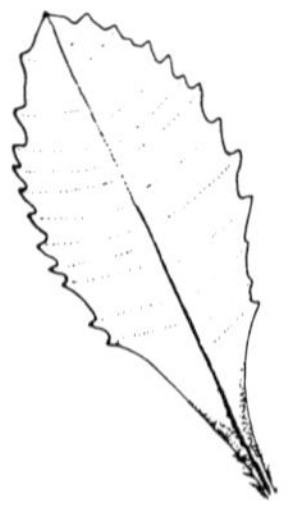

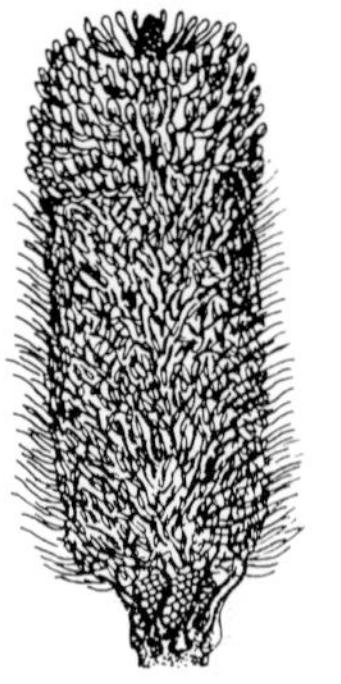

26 RECORDS: Jan (1) Feb (2) Mar (7) Apr (5) May (2) Jun (3) Jul (2) Aug (0) Sep (2) Oct (0) Nov (0) Dec (2).

Population Size: 1-10 (9) 10-100 (13) >100 (4) Unspecified (0).

Conservation Status: Restricted to road verge (0%) Not (100%) Unspecified (0%). In conservation reserve (31%) Not (69%) Unspecified (0%).

Tree/Shrub: Tree form (8%) Shrub form (92%) Unspecified (0%).

Height (metres)

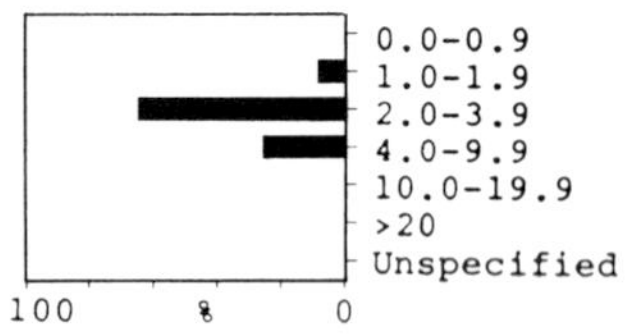

New Shoot Growth

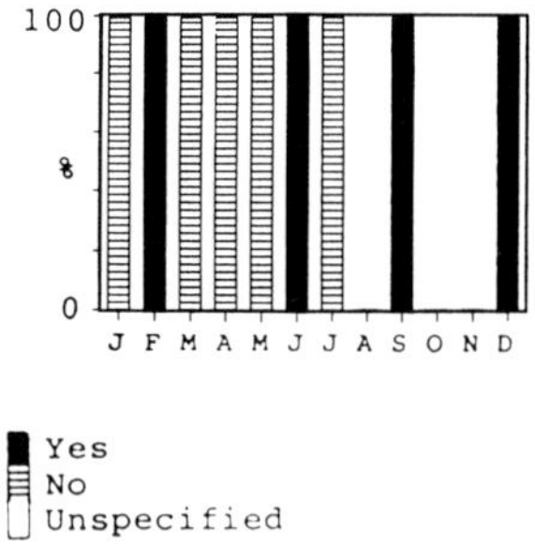

HABITAT

Within 2 km of coast: Yes (0%) No (100%) Unspecified (0%).

Altitude (metres)

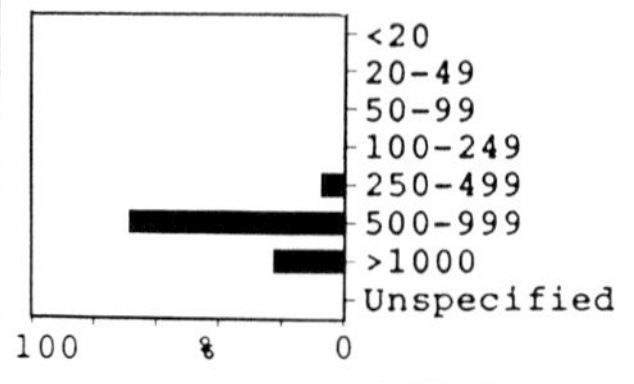

Vegetation

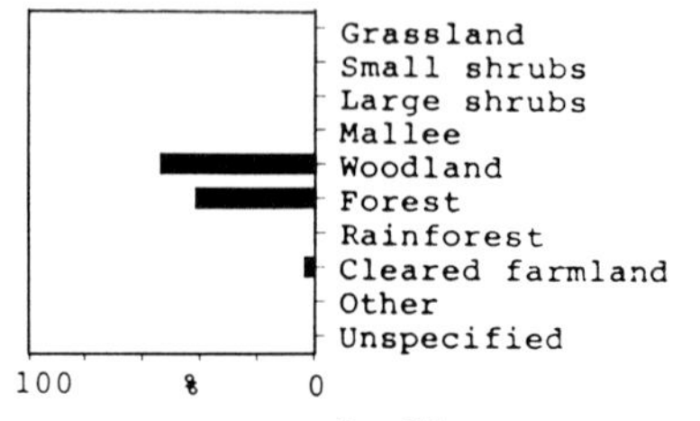

Soil Type

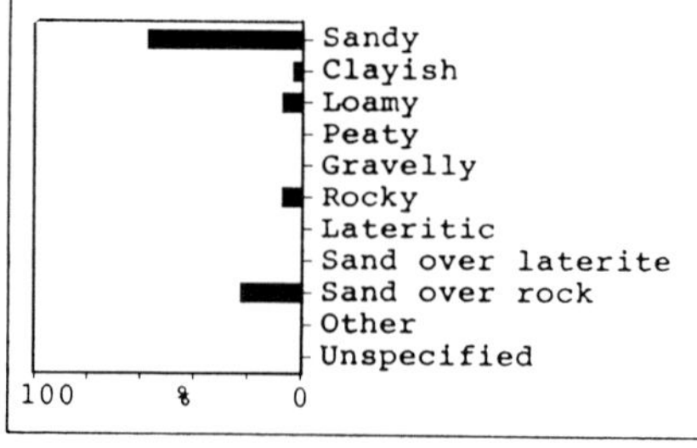

Landform

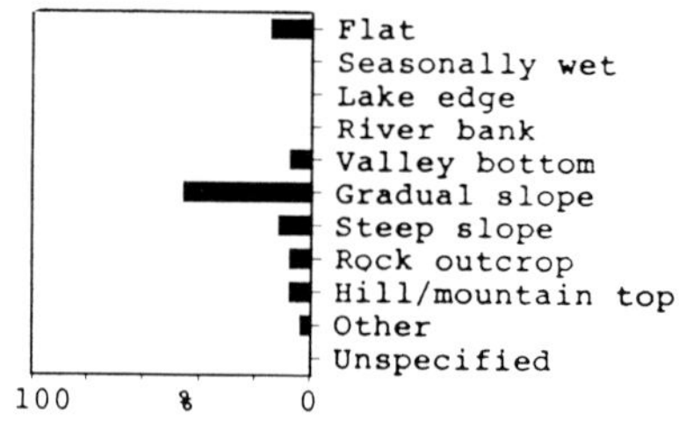

Response to Fire: 1 record

Growth response: Killed, new seedlings (100%).

Flowering response: Number of records where flowering had occurred after fire (0). Number of records where flowering had not occurred after fire (1). Maximum reported time to no flowering (37-48 months).

Flowering

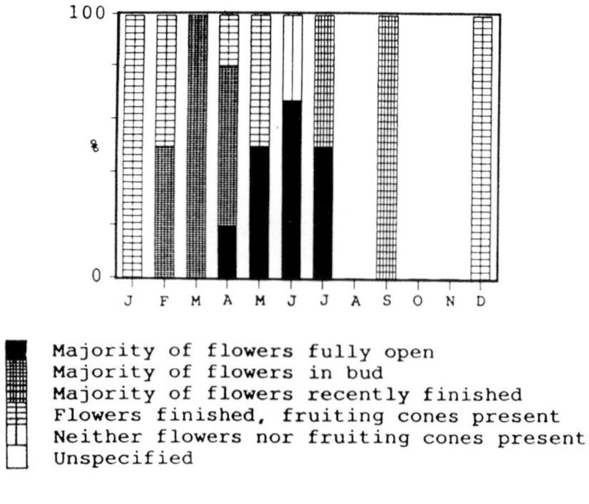

Possible pollinators

Pollinator type 20 records: bees, wasps, ants (10%) flies (5%) no pollinator observed (85%).

Specific pollinator 2 records: ant (1) bee (1).

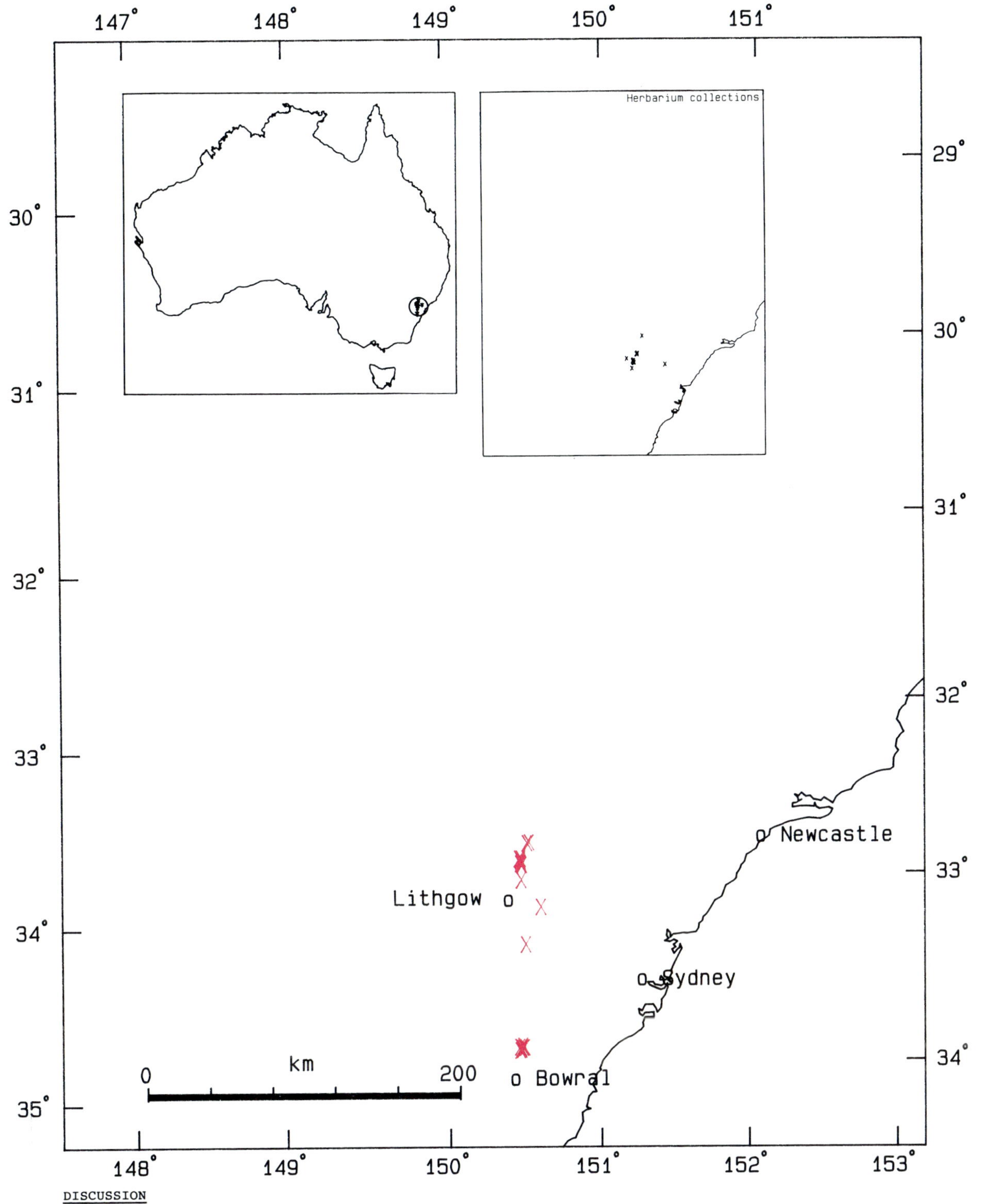

DISCUSSION

Herbarium collections indicated that B. conferta var. penicillata was restricted to a few locations in the Blue Mountains north-west of Sydney. Sight records have confirmed its occurrence in the Newnes Plateau area but did not record it at the site of R. Coveny's 1977 collection on the Boorai Ridge, north-west of St Albans. However, the known range has been extended by about 100 km further south by the discovery of an outlying population north of Bowral This outlier is atypical, and may be closer morphologically to B. paludosa (George pers. comm.). A single isolated plant near Katoomba, west of Sydney, probably represents part of a larger population which has since been cleared for urban development. The conservation status of this banksia needs monitoring since although 31% of populations are located in conservation reserves (notably Wollemi National Park), the bulk of the plants are located in just two main areas (Newnes Plateau and north of Bowral). Moreover, plants appear to be killed by fire and to take at least four years for seedlings to mature and reach flowering stage. Frequent fires at either of these two sites could easily eliminate a large percentage of the plants.

The variety usually grows as a shrub 2-4 m in height. It occurs on upland plateaus, rocky slopes and hilltops at altitudes generally between 500 and 1 000 metres, in woodland or forest. Soils are of sand derived from underlying sandstone becoming shallow in more exposed locations. Flowering is from autumn through winter.

BANKSIA CUNEATA A.S. George (1981)

Matchstick Banksia , Quairading Banksia

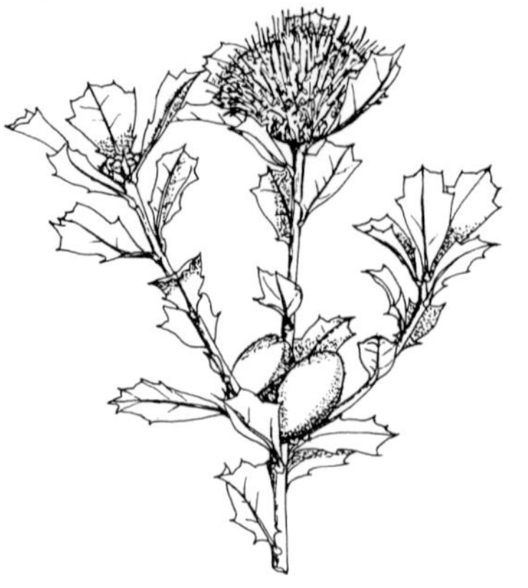

4 RECORDS: Jan (0) Feb (0) Mar (0) Apr (0) May (0) Jun (2) Jul (0) Aug (0) Sep (0) Oct (0) Nov (1) Dec (1).

Population Size: 1-10 (1) 10-100 (3) >100 (0) Unspecified (0).

Conservation Status: Restricted to road verge (100%) Not (0%) Unspecified (0%). In conservation reserve (0%) Not (75%) Unspecified (25%).

Tree/Shrub: Tree form (50%) Shrub form (50%) Unspecified (0%).

Height (metres)

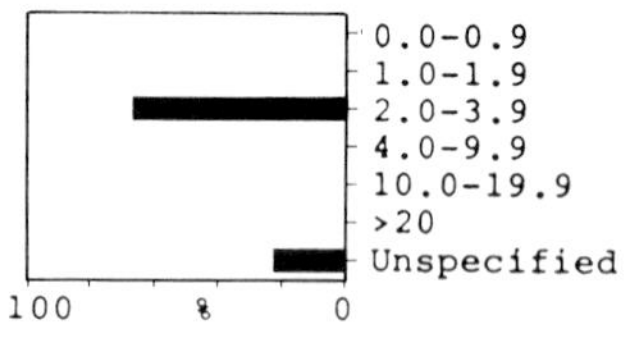

New Shoot Growth

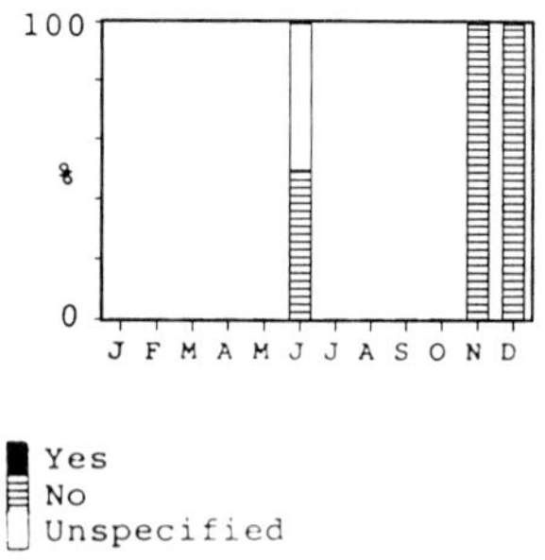

HABITAT

Within 2 km of coast: Yes (0%) No (100%) Unspecified (0%).

Altitude (metres)

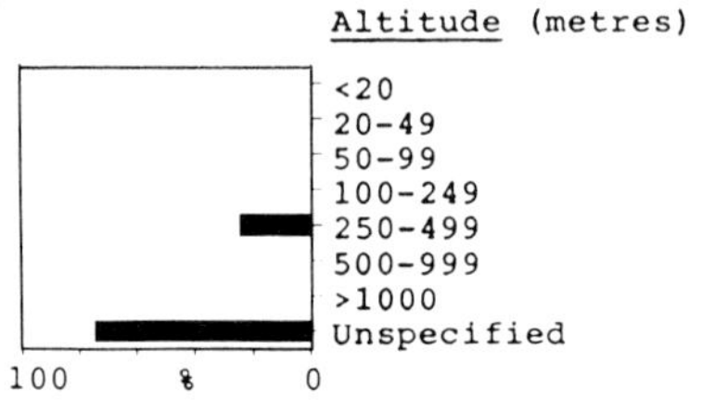

Vegetation

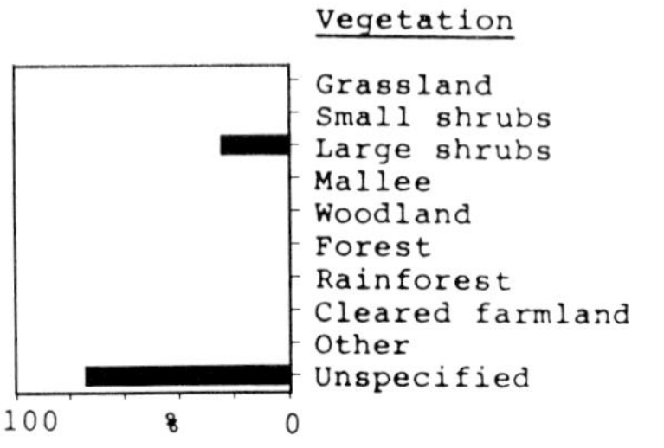

Soil Type

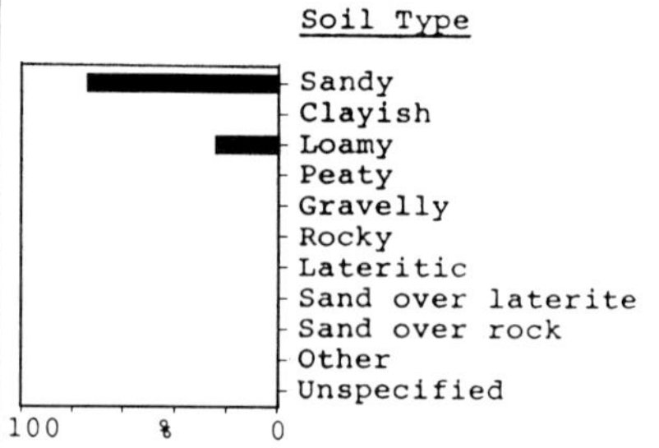

Landform

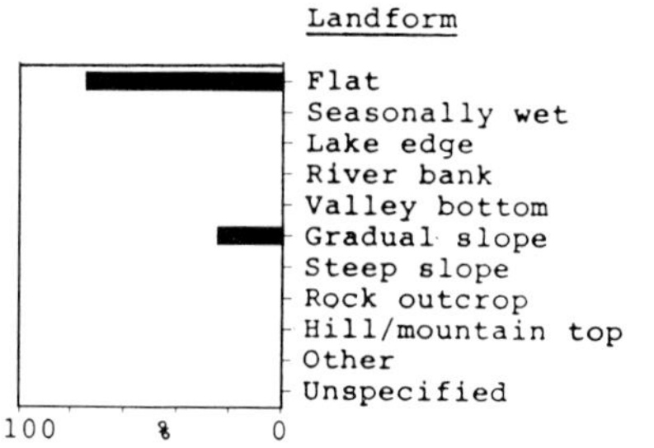

Response to Fire: 0 records

Flowering

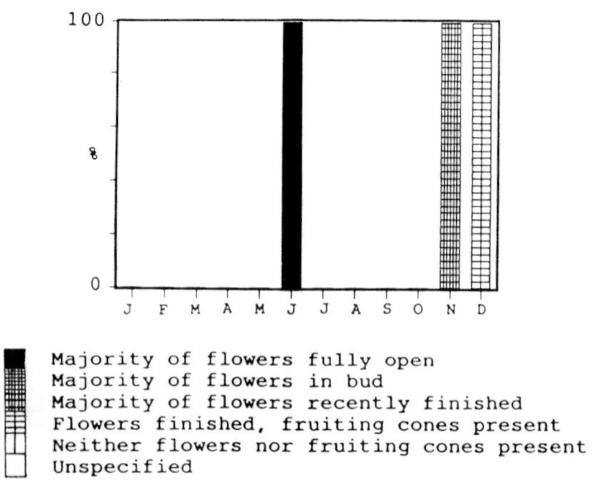

Possible pollinators

Pollinator type 0 records:

Specific pollinator 0 records:

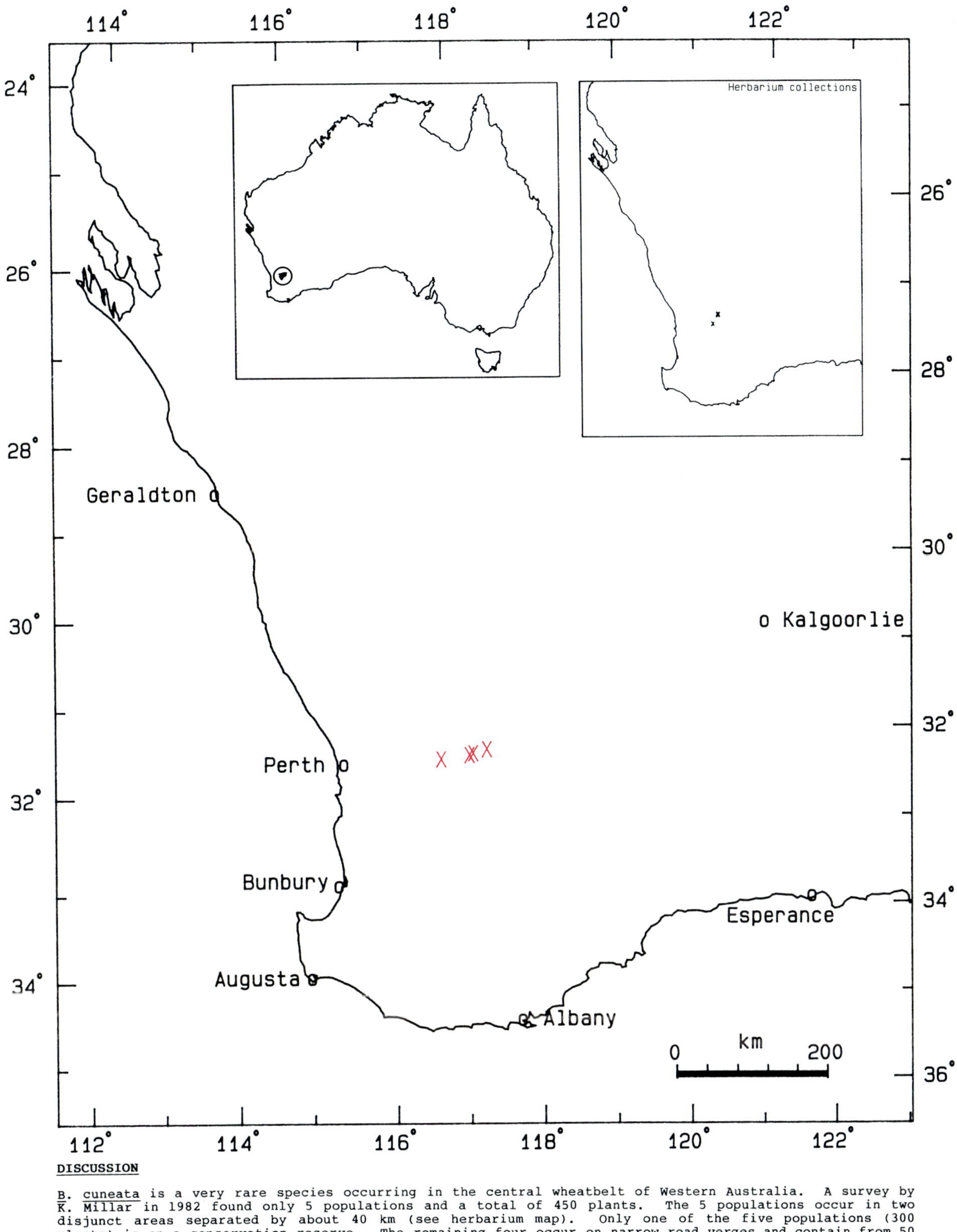

DISCUSSION

B. cuneata is a very rare species occurring in the central wheatbelt of Western Australia. A survey by K. Millar in 1982 found only 5 populations and a total of 450 plants. The 5 populations occur in two disjunct areas separated by about 40 km (see herbarium map). Only one of the five populations (300 plants) is on a conservation reserve. The remaining four occur on narrow road verges and contain from 50 to 70 individuals. Atlas contributors recorded only one of these populations, but also reported two new sites both east and west of the main population. One of these consists of ca. 12 plants. Details of the other are not known. Both require urgent investigation and confirmation. B. cuneata is unlikely to be found between the two areas where it is known to occur since most of the land is subject to inundation and contains unsuitable habitats. The remainder is largely cleared for agriculture.

Recent investigations have shown at least one of the road verge populations to be senescing with little or no seedling recruitment. The road verge is also weed infested. B. cuneata is highly susceptible to dieback disease. It is a species which is clearly at risk and needs urgent attention in order to ensure its long term conservation.

B. cuneata occurs as a large shrub or small tree up to 4 m in height. It grows in yellow or yellow-brown sand as a component of low woodland or tall shrubland, being associated with species such as B. prionotes and Xylomelum angustifolium. The species flowers in spring. It is not know at what season new shoot growth occurs. B. cuneata appears to be non-lignotuberous (George, 1981) and to rely on seed germination for post fire survival. It would therefore be vulnerable to fires at intervals too short to allow regeneration from seed and subsequent seed production. However, the species may need fire at infrequent intervals as its follicles appear to remain closed until burnt.

BANKSIA DENTATA Linnaeus f. (1782)

Tropical Banksia

133 RECORDS: Jan (0) Feb (0) Mar (0) Apr (5) May (26) Jun (52) Jul (32) Aug (6) Sep (10) Oct (0) Nov (0) Dec (2).

Population Size: 1-10 (22) 10-100 (83) >100 (22) Unspecified (6).

Conservation Status: Restricted to road verge (0%) Not (90%) Unspecified (10%). In conservation reserve (28%) Not (64%) Unspecified (8%).

Tree/Shrub: Tree form (92%) Shrub form (5%) Unspecified (3%).

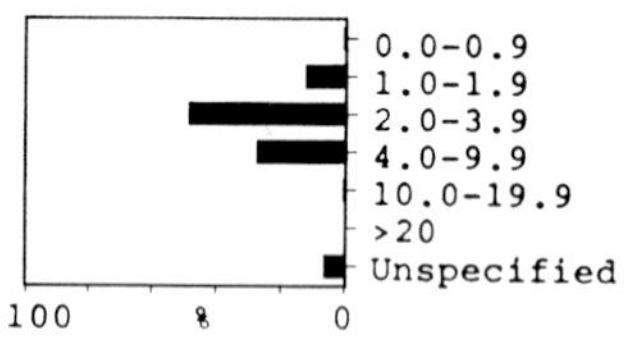

New Shoot Growth

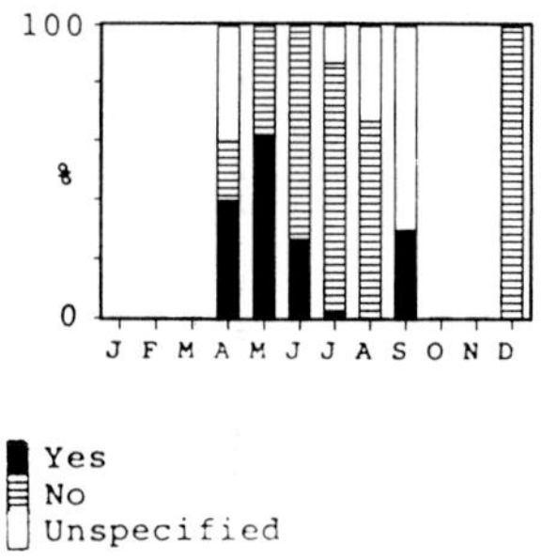

HABITAT

Within 2 km of coast: Yes (7%) No (92%) Unspecified (1%).

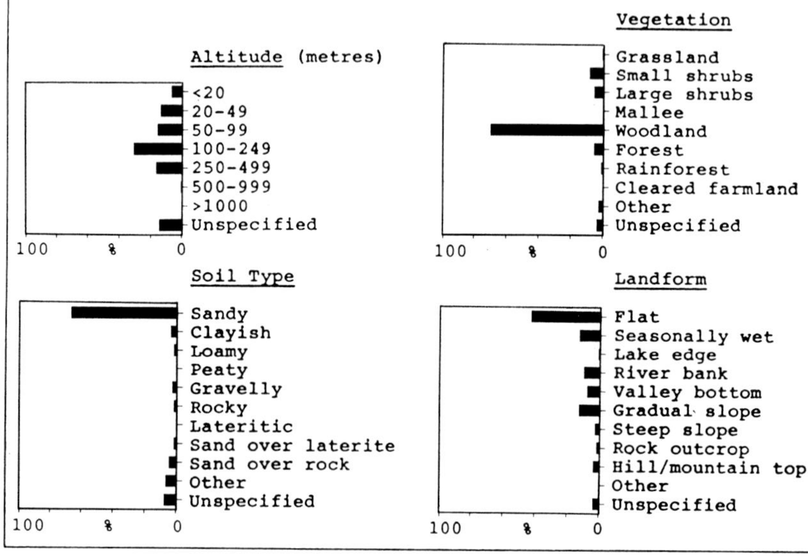

Response to Fire: 36 records

Growth response: Ground resprout (8%) killed, new seedlings (8%) trunk resprout (56%) killed, no seedlings and trunk resprout (6) unspecified (22%).

Flowering response: Number of records where flowering had occurred after fire (23). Median reported time to flowering (1-12 months); minimum reported time to flowering (<1 month). Number of records where flowering had not occurred after fire (6). Median reported time to no flowering (1-12 months); maximum reported time to no flowering (13-24 months).

Flowering

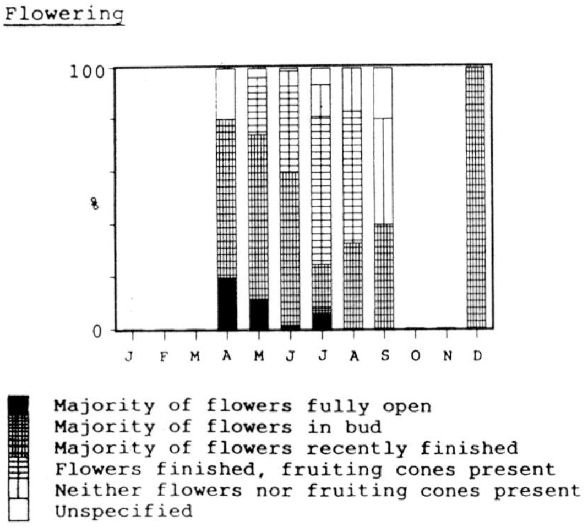

Possible pollinators

Pollinator type 44 records: bird (2%) no pollinator observed (98%).

Specific pollinator 1 record: honeyeater (1).

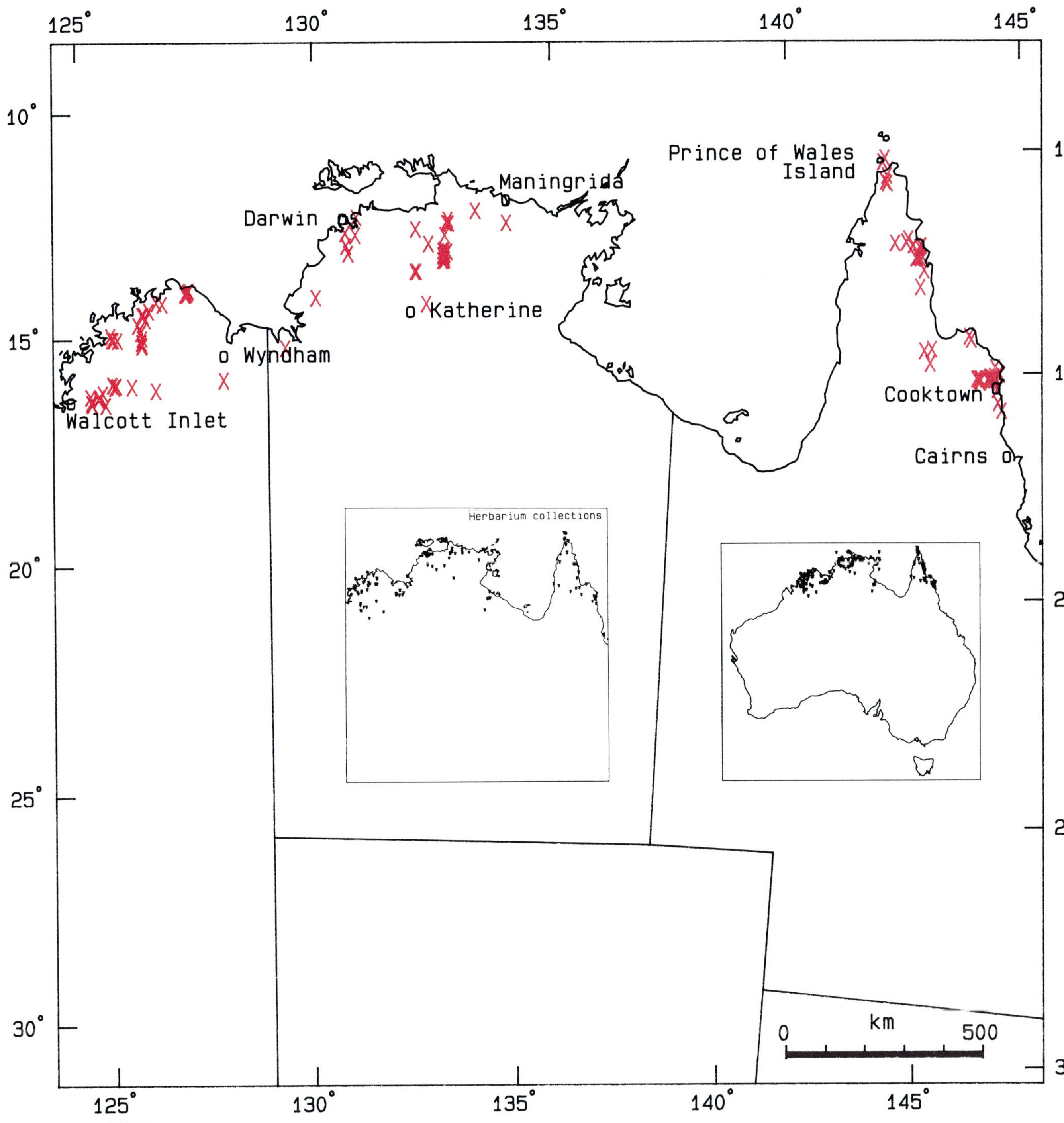

DISCUSSION

Banksia dentata is the only tropical species of banksia, occurring in northern parts of Western Australia, the Northern Territory and Queensland. It also extends into New Guinea. In Western Australia, a herbarium collection of W. Fitzgerald in 1905 indicated that the species may occur almost as far south as the King Leopold Ranges, though Atlas contributors did not record it beyond the Synnot Range. From east of Wyndham there are also several herbarium collections which are not represented in Atlas records. In the Northern Territory, the southernmost record is from Katherine Gorge. In the west, there is a new Atlas record from the vicinity of Moyle River. There is an apparent gap in the species' distribution across the bottom of the Gulf of Carpentaria. However, this gap is probably not as large as Atlas data suggest since herbarium collections indicate populations in Arnhem Land and west of Borroloola (N.T.) and in Western Cape York Peninsula (Queensland). In the Northern Territory, there are also herbarium collections from Groote Eylandt and the Wessel Islands. In Queensland the species occurs from Prince of Wales island south to the Bloomfield River. Since much of the range of B. dentata is remote and difficult of access, it is likely that the species is more widespread than is indicated by present knowledge, based on both herbarium collections and Atlas records. The distribution maps should therefore be regarded as guidelines only.

B. dentata is typically a small tree up to 4 m in height though sometimes larger. The species usually occurs along seasonally-wet flats and river courses in sandy soils. A few populations have also been recorded from steep rocky slopes and hilltops up to 500 m above sea level. Surrounding vegetation is usually open savannah-type woodland, though shrubland is occasionally found. There are few Atlas records of flowering, though George (1981) stated the flowering period to be November to June. This is the tropical wet season when field work is difficult due to problems with access and uncomfortable working conditions. Further data are also needed to define the main period of new shoot growth and to ascertain which are the principal pollinators of B. dentata.

BANKSIA DRYANDROIDES

Baxter ex Sweet (1828)

Dryandra-leaved Banksia

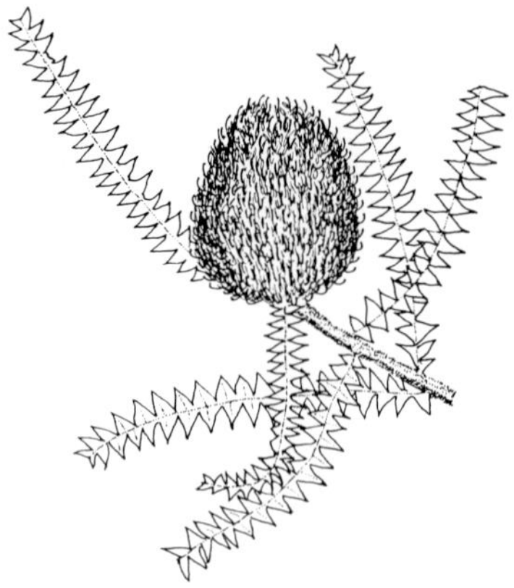

65 RECORDS: Jan (2) Feb (4) Mar (0) Apr (7) May (3) Jun (2) Jul (3) Aug (14) Sep (13) Oct (6) Nov (8) Dec (2).

Population Size: 1-10 (17) 10-100 (33) >100 (14) Unspecified (1).

Conservation Status: Restricted to road verge (25%) Not (65%) Unspecified (11%). In conservation reserve (40%) Not (49%) Unspecified (11%).

Tree/Shrub: Tree form (0%) Shrub form (100%) Unspecified (0%).

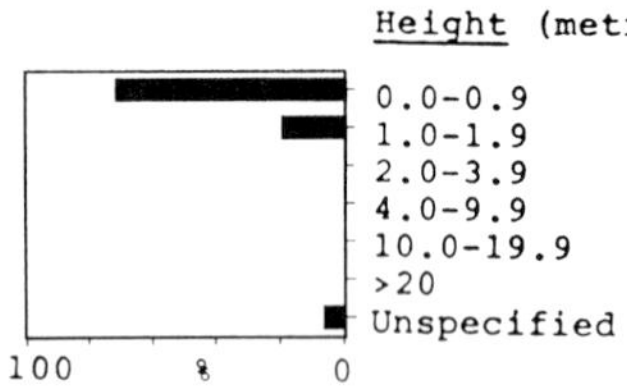

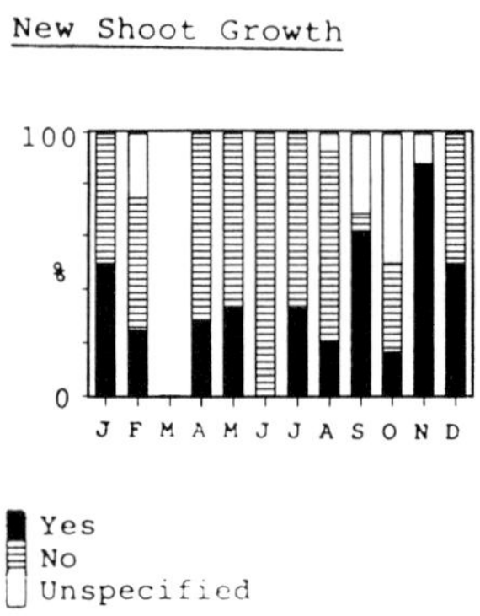

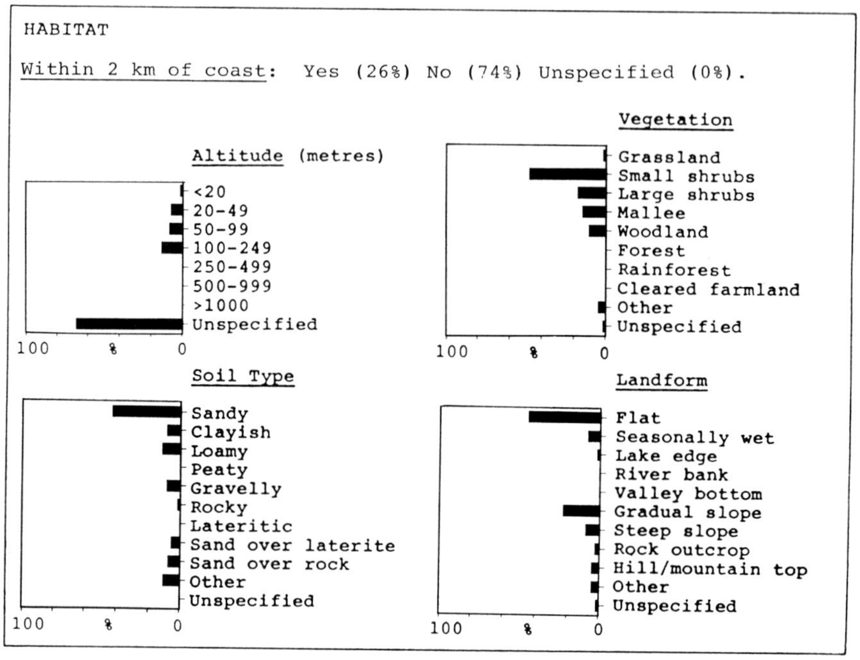

Response to Fire: 7 records

Growth response: Ground resprout (14%) killed, new seedlings (43%) unspecified (43%).

Flowering response: Number of records where flowering had occurred after fire (3). Median reported time to flowering (25-36 months); minimum reported time to flowering (13-24 months). Number of records where flowering had not occurred after fire (3). Median reported time to no flowering (37-48 months); maximum reported time to no flowering (37-48 months).

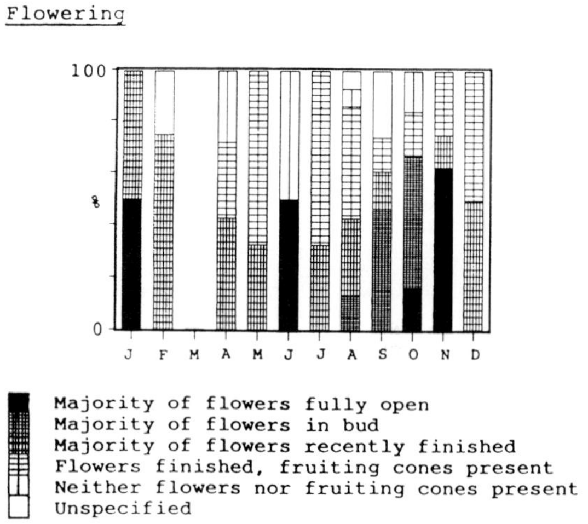

Possible pollinators

Pollinator type 17 records: no pollinator observed (100%).

Specific pollinator 0 records.

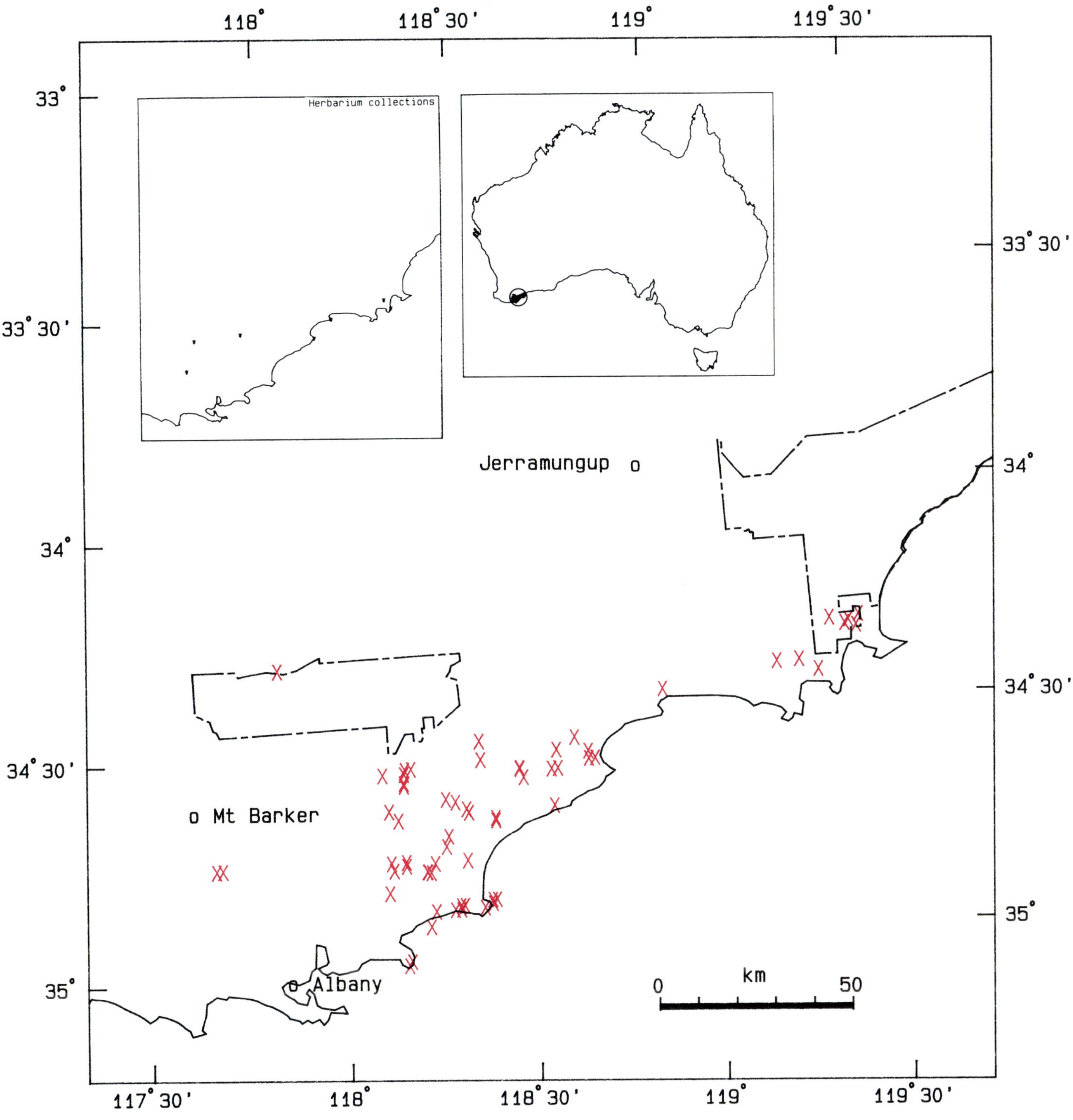

DISCUSSION

Atlas records confirmed a distribution for B. dryandroides between Bremer Bay and Narrikup. The latter apparently is an outlier separated by ca. 40 km from the main range further east. The bulk of the populations are from Two Peoples Bay northwards almost to the Stirling Range National Park and eastwards to Cheyne Bay. There is a single isolated record from Beaufort Inlet and several records from both west and north of Bremer Bay. The latter, occurring in the Fitzgerald River National Park are new recordings. Another new recording from the northern boundary of Stirling Range National Park extends the range of this species by almost 40 km. A 1909 collection of T. Maiden's from north-east of Narrikup was not recorded. B. dryandroides has a restricted range but it is locally common. It is well represented on conservation reserves and does not appear to be under any immediate threat.

The species is typically a small shrub of less than 1 m in height growing in shrubland, coastal heath, in woodland or amongst mallee eucalypts. It occurs predominantly on sandy flats and occasionally on heavier clays, loams and gravels. The main flowering period is from spring through to early summer, though flowers have also been recorded in June and old flowers still with some colour, from every month. New shoot growth appears to occur throughout much of the year with a possible peak in spring. Pollinators were not observed. B. dryandroides is generally killed by fire and regenerates from seed, although a single Atlas record of lignotuberous plants needs to be investigated. Following fire, plants may need at least 4 years to reach maturity and flower.

BANKSIA ELDERIANA F.Muell. & Tate (1893)

Swordfish Banksia, Palm Banksia

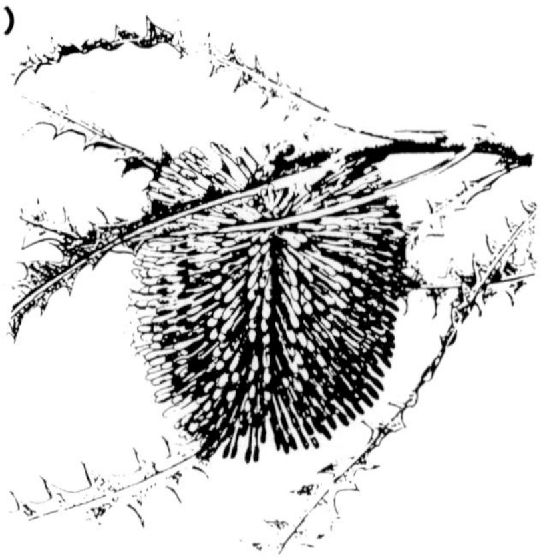

167 RECORDS: Jan (6) Feb (8) Mar (4) Apr (41) May (16) Jun (3) Jul (4) Aug (22) Sep (16) Oct (14) Nov (17) Dec (13).

Population Size: 1-10 (25) 10-100 (84) >100 (57) Unspecified (1).

Conservation Status: Restricted to road verge (12%) Not (85%) Unspecified (3%). In conservation reserve (22%) Not (71%) Unspecified (7%).

Tree/Shrub: Tree form (1%) Shrub form (99%) Unspecified (1%).

Height (metres)

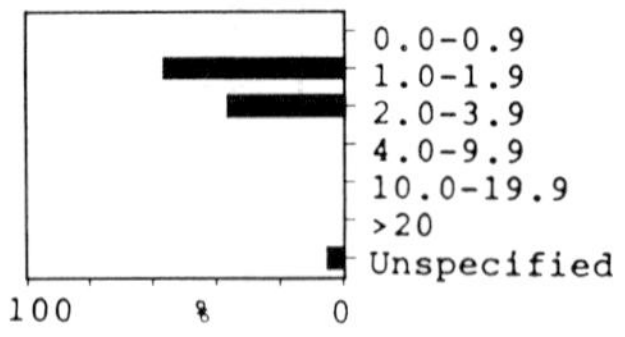

New Shoot Growth

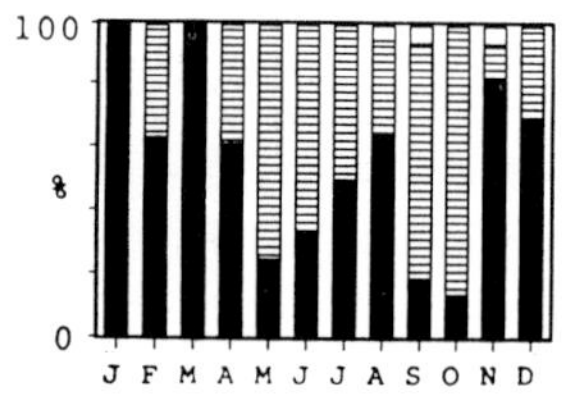

Yes
No
Unspecified

HABITAT

Within 2 km of coast: Yes (0%) No (99%) Unspecified (1%).

Altitude (metres)

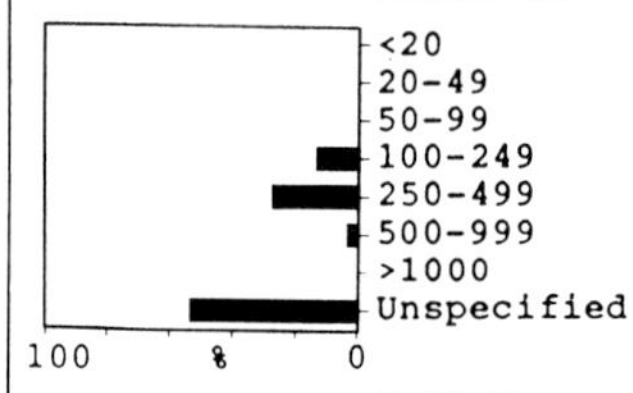

Vegetation

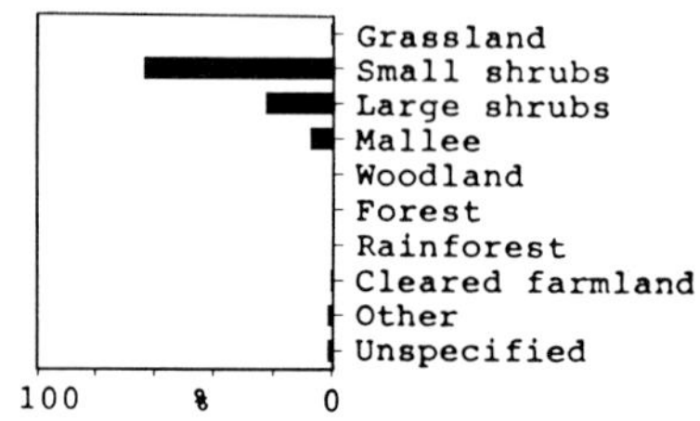

Soil Type

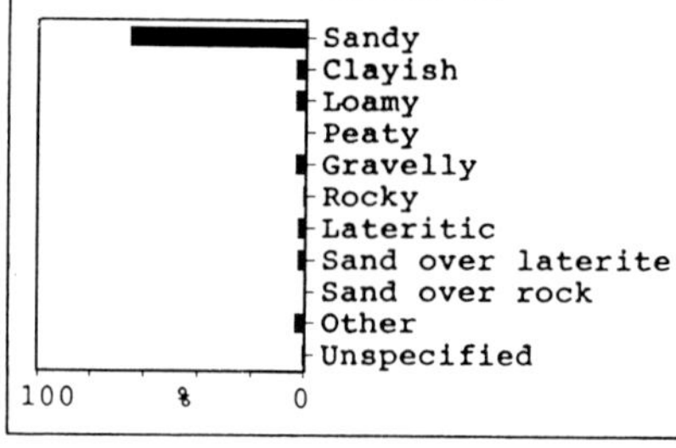

Landform

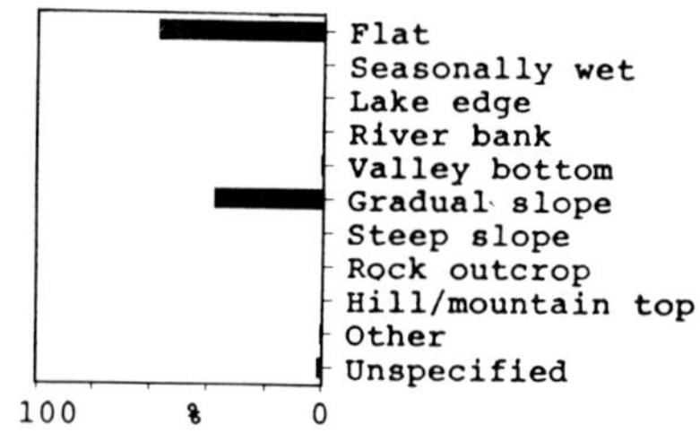

Response to Fire: 30 records

Growth response: Ground resprout (73%) ground resprout and trunk resprout (3%) killed, no seedlings (3%) unspecified (20%).

Flowering response: Number of records where flowering had occurred after fire (18). Median reported time to flowering (25-36 months); minimum reported time to flowering (1-12 months). Number of records where flowering had not occurred after fire (8). Median reported time to no flowering (1-12 months); maximum reported time to no flowering (25-36 months).

Flowering

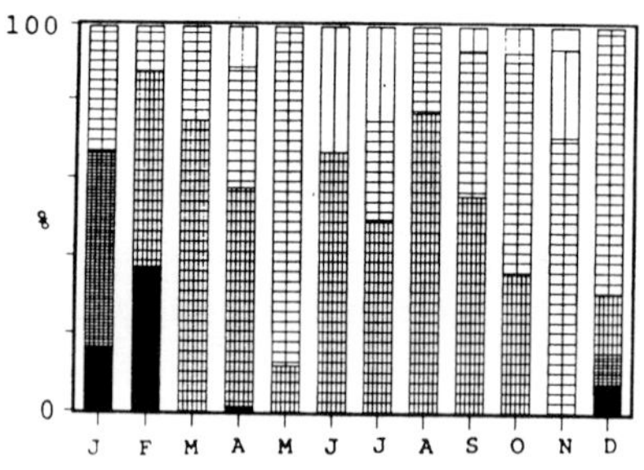

Majority of flowers fully open
Majority of flowers in bud
Majority of flowers recently finished
Flowers finished, fruiting cones present
Neither flowers nor fruiting cones present
Unspecified

Possible pollinators

Pollinator type 27 records: bees, wasps, ants (11%) flys (4%) no pollinator observed (85%).

Specific pollinator 2 records: ant (2).

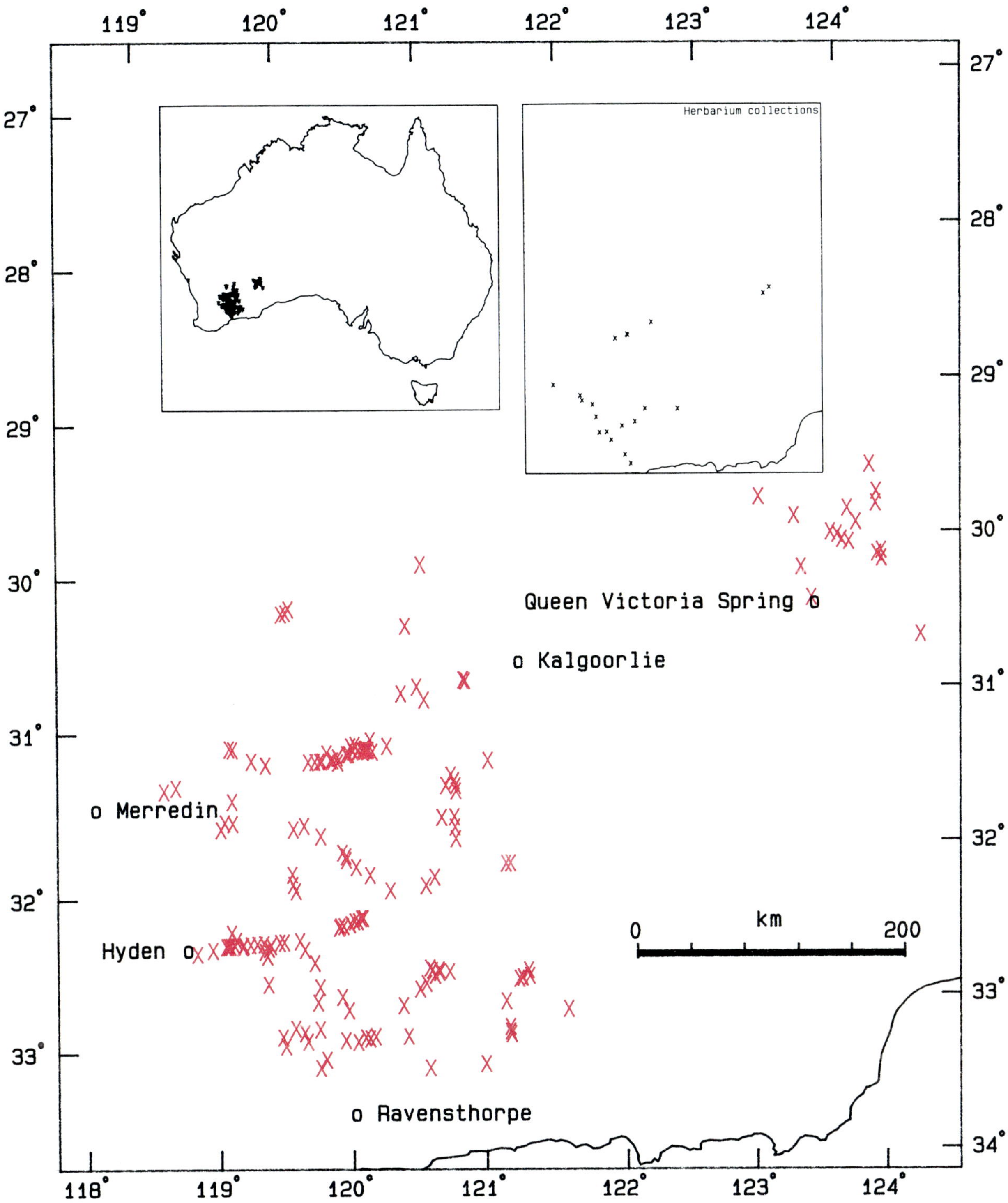

DISCUSSION

B. elderiana occurs in two disjunct areas to the west and east of Kalgoorlie. The former extends from Salmon Gums through the central south region as far as Hyden and from the abandoned Davyhurst mine north-west of Kalgoorlie to Bungalbin Hill and as far west as Bodallin. Herbarium collections from Narembeen and Ravensthorpe areas were not recorded but range extensions from Atlas contributors included the Salmon Gums, Bodallin, Davyhurst and Bungalbin Hill populations. East of Kalgoorlie there are several new recordings in the Great Victoria Desert from Lake Minigwal in the north-west approximately 170 km south-eastwards and 100 km eastwards. On the basis of searches undertaken for the Atlas, it is unlikely that B. elderiana will be found in the area between the western and eastern populations. The species mainly occurs on deep yellow sands which are largely absent from this area where an old drainage system (whose remnant salt lakes still exist) has produced mainly heavier soils.

Although most recorded populations were of less than 100 plants and only about one-fifth were recorded from conservation reserves, most of the range of the species is in uncleared land to the east of the wheatbelt. Its conservation status therefore seems relatively secure.

The species is typically a shrub up to 4 m in height. Western populations grow in low or tall shrubland whilst those to the east occur amongst mallee eucalypts and spinifex grassland. Flowering is from December to February and new shoot growth mainly from late spring to autumn. Following fire, the species resprouts from its lignotuber and up to three years may be required for the regrowth to reach maturity and flower.

BANKSIA ELEGANS Meissner (1855)

Elegant Banksia

24 RECORDS: Jan (1) Feb (0) Mar (3) Apr (1) May (3) Jun (1) Jul (0) Aug (0) Sep (6) Oct (0) Nov (0) Dec (7).

Population Size: 1-10 (5) 10-100 (7) >100 (11) Unspecified (1).

Conservation Status: Restricted to road verge (0%) Not (96%) Unspecified (4%). In conservation reserve (25%) Not (75%) Unspecified (0%).

Tree/Shrub: Tree form (17%) Shrub form (83%) Unspecified (0%).

Height (metres)

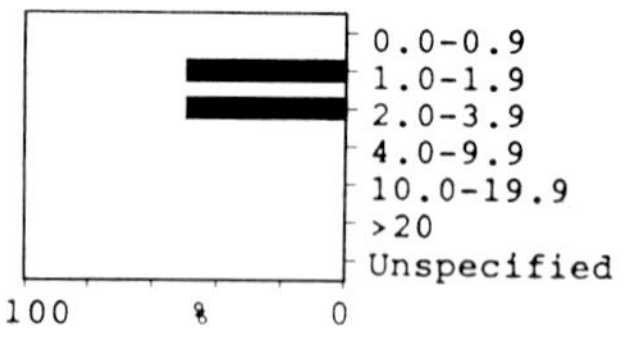

New Shoot Growth

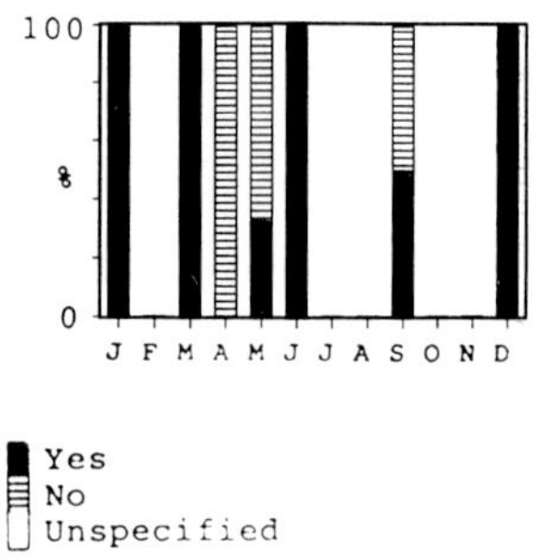

HABITAT

Within 2 km of coast: Yes (0%) No (100%) Unspecified (0%).

Altitude (metres)

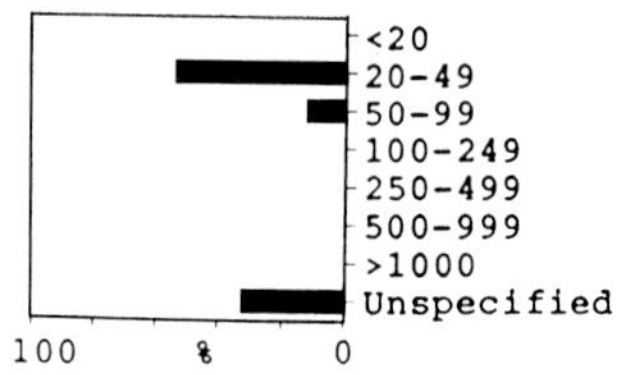

Vegetation

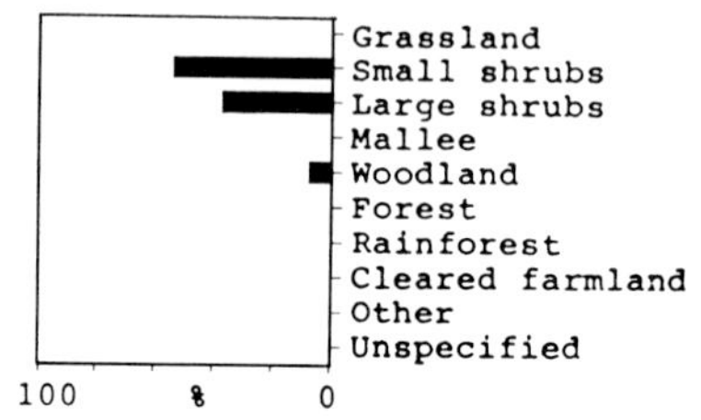

Soil Type

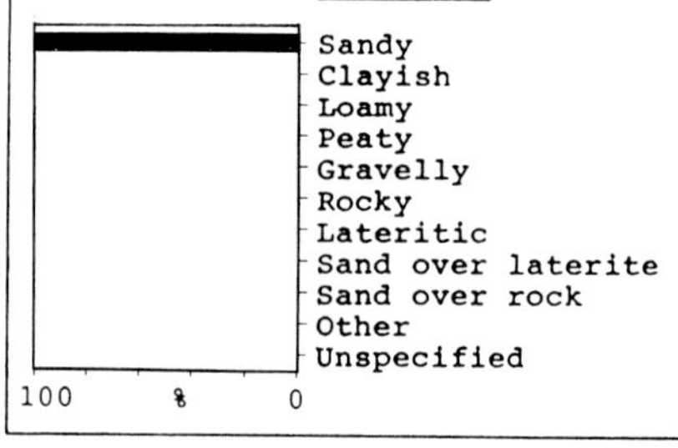

Landform

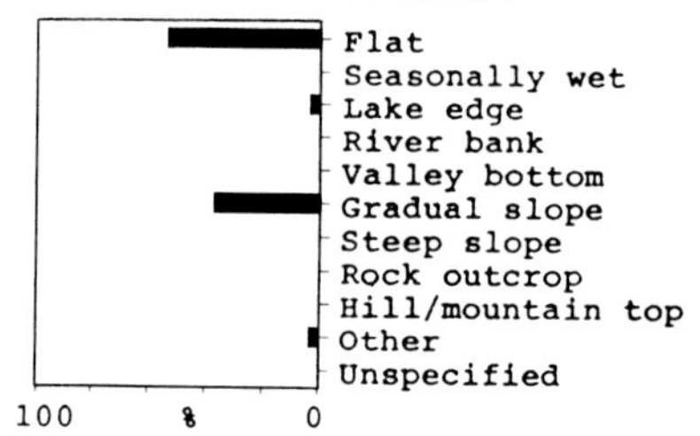

Response to Fire: 15 records

Growth response: Ground resprout (47%) ground resprout and trunk resprout (27%) trunk resprout (27%).

Flowering response: Number of records where flowering had occurred after fire (5). Median reported time to flowering (13-24 months); minimum reported time to flowering (1-12 months).
Number of records where flowering had not occurred after fire (8). Median reported time to no flowering (13-24 months); maximum reported time to no flowering (25-36 months).

Flowering

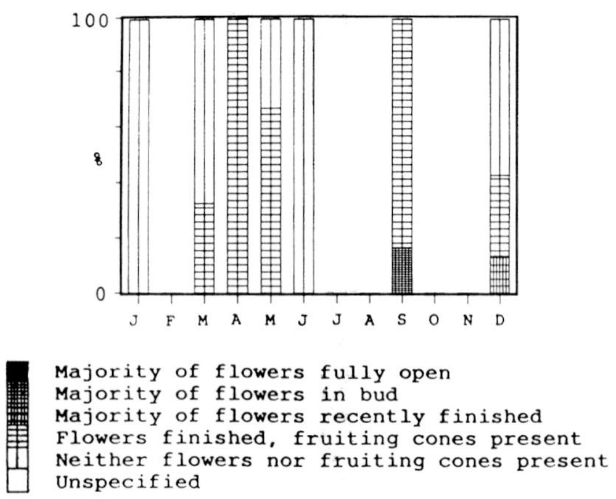

Possible pollinators

Pollinator type 6 records: no pollinator observed (100%).

Specific pollinator 0 records.

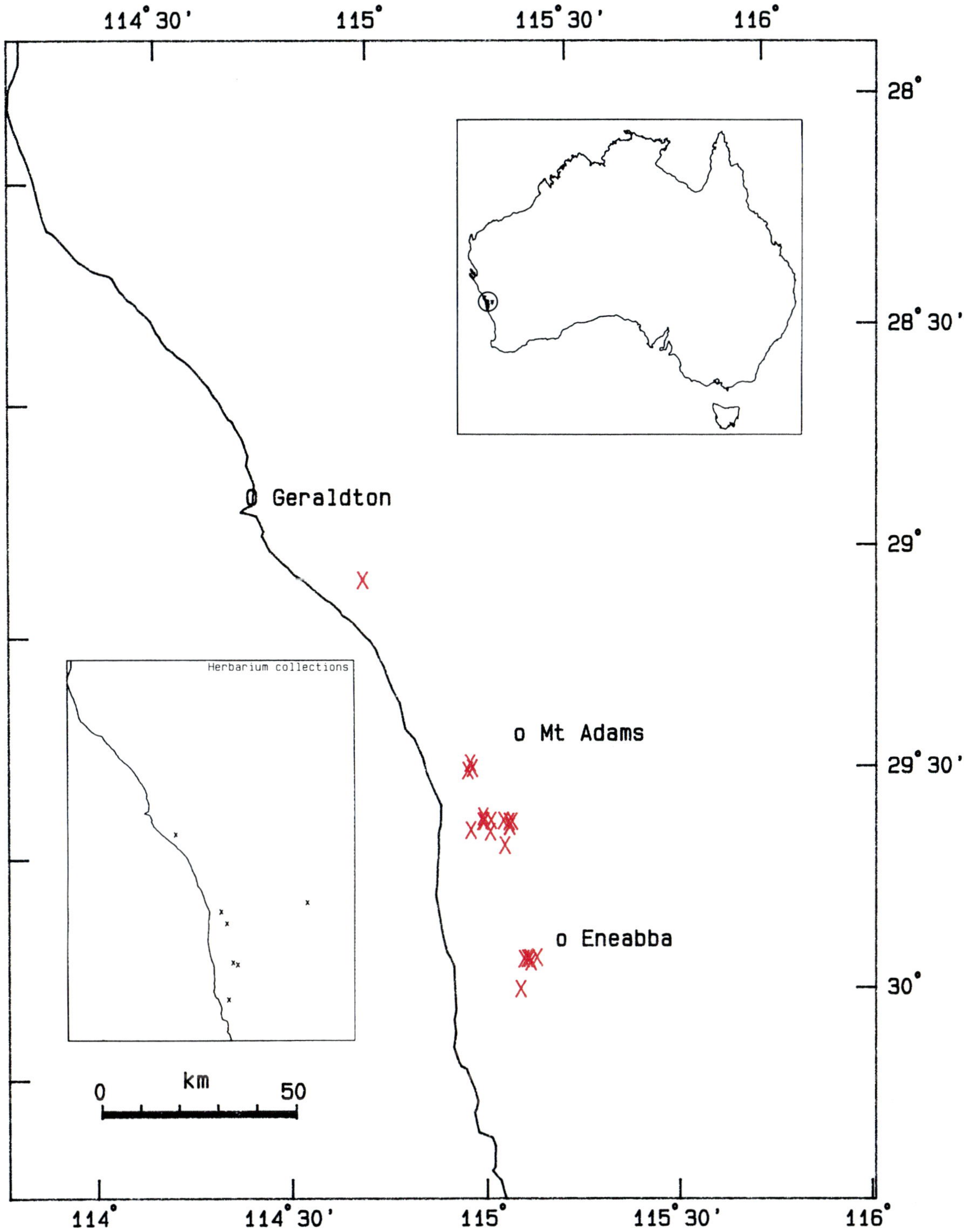

DISCUSSION

Herbarium collections had indicated a restricted distribution for B. elegans between Walkaway and Mt Peron. Atlas records have confirmed this general distribution though neither C. Gardner's 1946 collection from near Three Springs, nor K. Newbey's 1966 collection from near Mt Peron were recorded. The latter is, however, still known to the present (G. Barrett, 1985). The northernmost population at Walkaway was recorded for the Atlas in late 1984 but has since been cleared for agriculture. This population had been reported by George (1981) as having larger inflorescences than all other populations. The total range of the species is therefore now reduced to 65 km. The conservation status of the species needs monitoring since although 25% of populations were recorded on conservation reserves, these generally consist of less than 100 plants each. The largest populations (e.g. Mt Adams area and east of Arrowsmith) are on private land or vacant Crown land. Moreover, the species' unusual reproductive characteristics with few follicles produced, suggests that sexual reproduction is unusally limited.

B. elegans grows as a shrub, rarely a small tree between 1-4 m in height. It occurs on flat or gently undulating sandplain at altitudes of less than 50 m, often near winter-wet depressions or lakes. It grows in shrubland, occasionally in low woodland and is often associated with species such as B. attenuata, B. menziesii, B. prionotes (G. Barrett, 1985). Atlas contributors did not observe flowering plants nor pollinators though George (1981) recorded the flowering period as October and November. Barrett (1985) found birds to be important pollinators especially the White-cheeked Honeyeater. Limited Atlas records suggest that new shoot growth occurs mainly in summer. Following fire, the species resprouts from its lignotuber and from epicormic buds. Barrett (1985) also found sucker regrowth to be greatly stimulated by fire. B. elegans is unique among western banksias in its suckering habit. The only other Banksia species with this habit are some variants of the eastern species B. marginata and B. integrifolia.

BANKSIA EPICA A.S. George (1987)

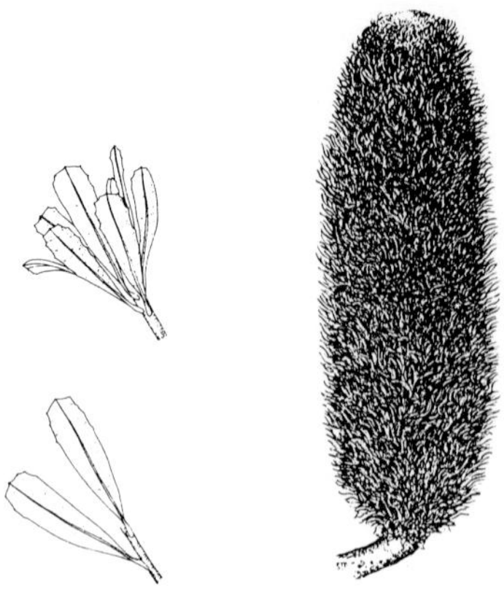

1 RECORD: Jan (0) Feb (0) Mar (0) Apr (0) May (1) Jun (0) Jul (0) Aug (0) Sep (0) Oct (0) Nov (0) Dec (0).

Population Size: 1-10 (0) 10-100 (1) >100 (0) Unspecified (0).

Conservation Status: Restricted to road verge (0%) Not (100%) Unspecified (0%). In conservation reserve (100%) Not (0%) Unspecified (0%).

Tree/Shrub: Tree form (100%) Shrub form (0%) Unspecified (0%).

Height (metres)

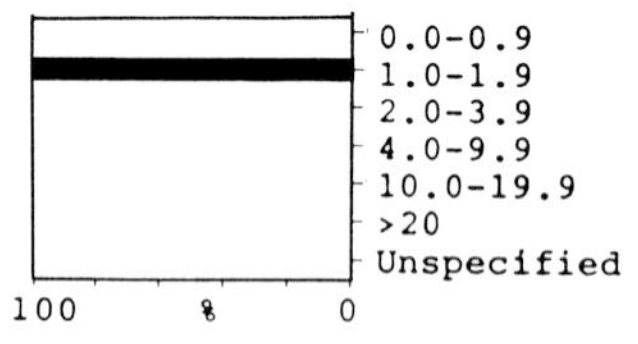

New Shoot Growth

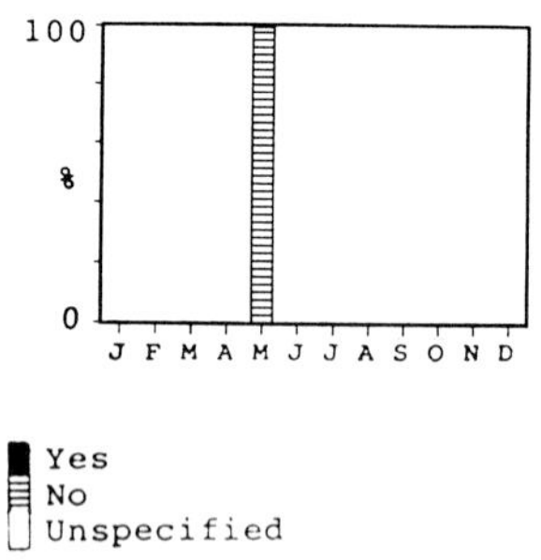

HABITAT

Within 2 km of coast: Yes (100%) No (0%) Unspecified (0%).

Altitude (metres)

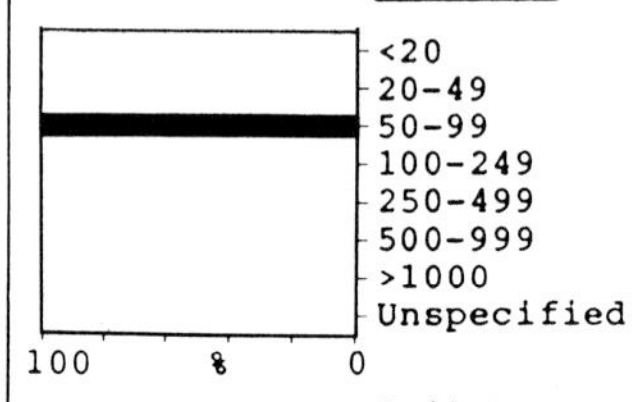

Vegetation

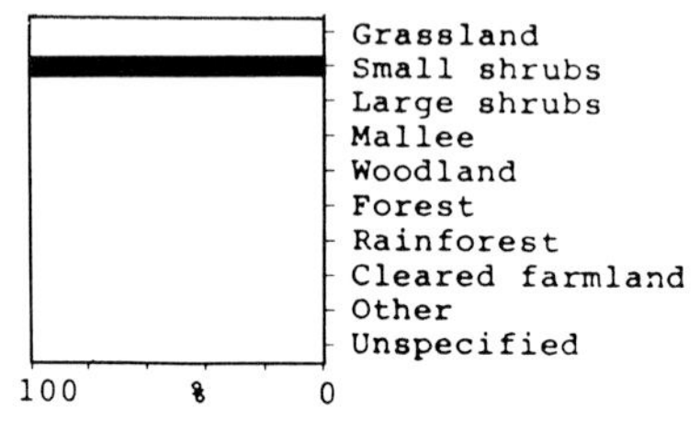

Soil Type

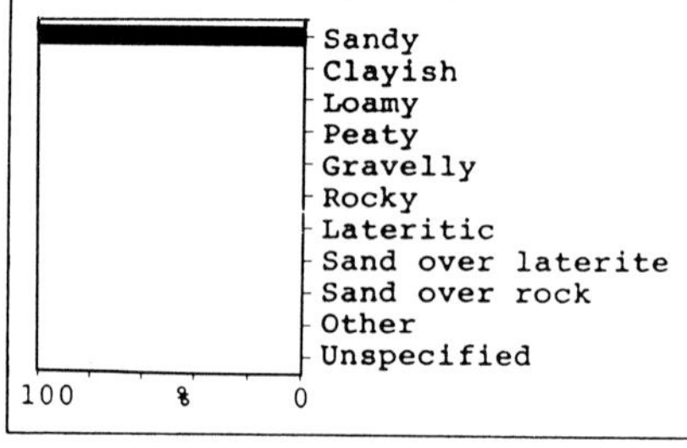

Landform

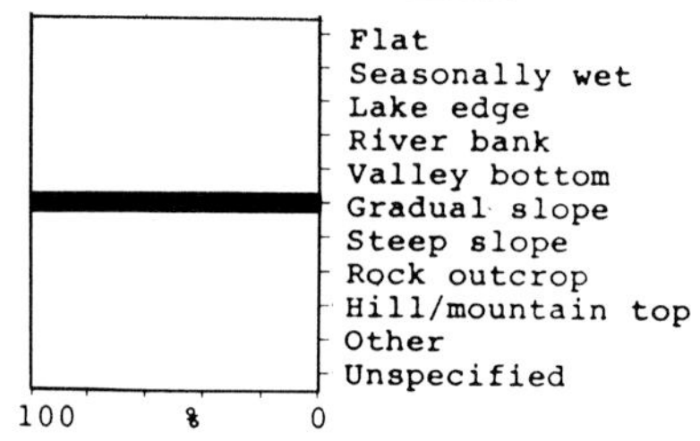

Response to Fire: 0 records

Flowering

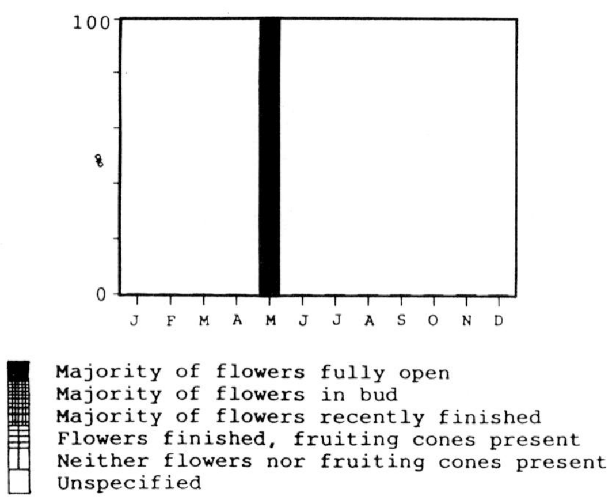

Possible pollinators

Pollinator type 1 record: bird (100%)

Specific pollinator 2 records: New Holland Honeyeater (1) thornbill (1)

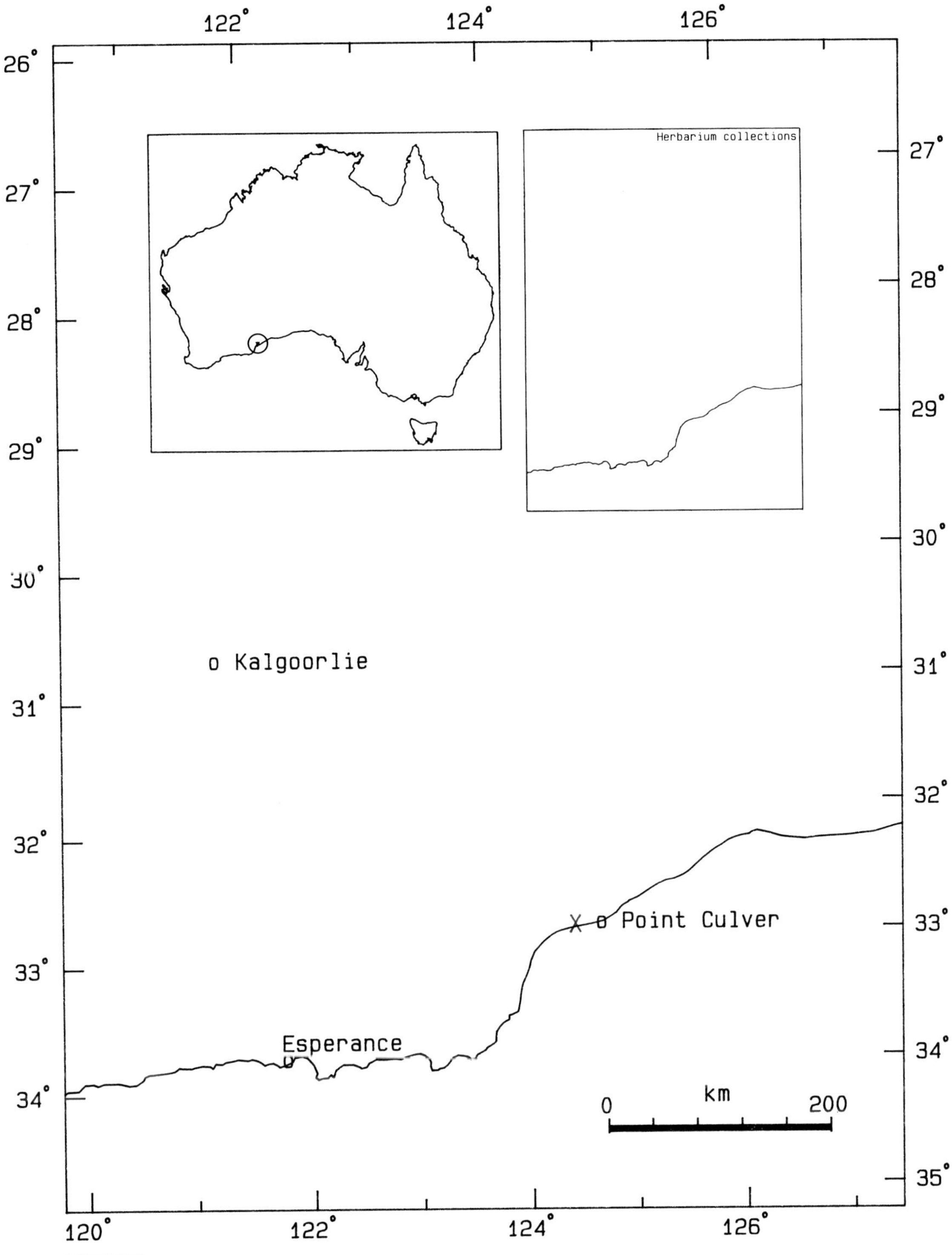

DISCUSSION

B. epica is a new species that was unnamed prior to the Banksia Atlas but first collected in 1973. It has been recorded from only two locations, Toolinna Cove and 30 km west of Point Culver, both along the western coastline of the Great Australian Bight. Only the Point Culver location was recorded for the Atlas. The species is named and described by George (1988). Illustrations of its flowers, overall habit, leaves and fruiting cone appear elsewhere in this text. The following notes present a brief summary of the Point Culver population.

The species occurs as a much-branched and spreading bushy shrub up to 2 m in height. Leaves are like shortened forms of B. media, 1.5-5 cm long, 6-15 mm wide. Flower spikes are of similar size and shape to B. media but flowers are larger and are of a more creamy colour with the apex of the pollen presenter being purple. The fruiting cone is distinctive with persistent old flower parts which are curled and point upwards. On B. media the old flowers are straight and point downwards. The entire flowering season is not known though flowers have been observed in April and May.

At the Point Culver location at least 100 plants were observed. They grow on top of a limestone escarpment in sandy/limestone soil in close association with B. media and B. speciosa. Surrounding vegetation is low heath. The species appears to be non-lignotuberous. Birds such as the New Holland Honeyeater and a type of thornbill were observed feeding on the flower spikes. The period of new shoot growth is not known.

Although only two populations are currently known, the area in which this species occurs has been inadequately surveyed and other populations may exist. Due to its remote location, the species is unlikely to be at risk.

BANKSIA ERICIFOLIA Linnaeus f. (1781)

Heath-leaved Banksia

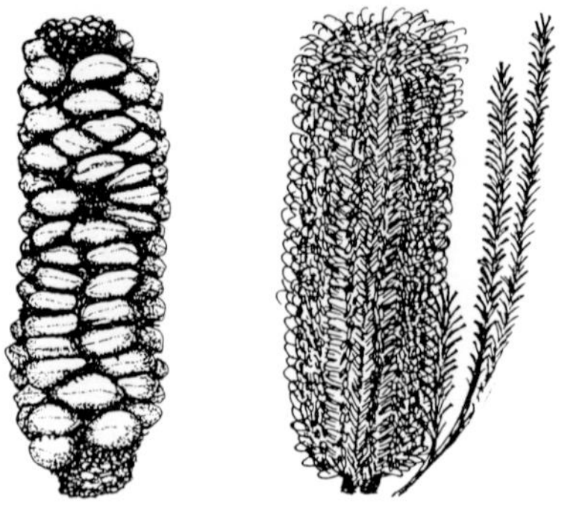

565 RECORDS: Jan (66) Feb (38) Mar (79) Apr (64) May (69) Jun (70) Jul (19) Aug (47) Sep (33) Oct (23) Nov (28) Dec (29).

Population Size: 1-10 (88) 10-100 (136) >100 (334) Unspecified (7).

Conservation Status: Restricted to road verge (1%) Not (98%) Unspecified (1%). In conservation reserve (61%) Not (34%) Unspecified (5%).

Tree/Shrub: Tree form (21%) Shrub form (75%) Unspecified (4%).

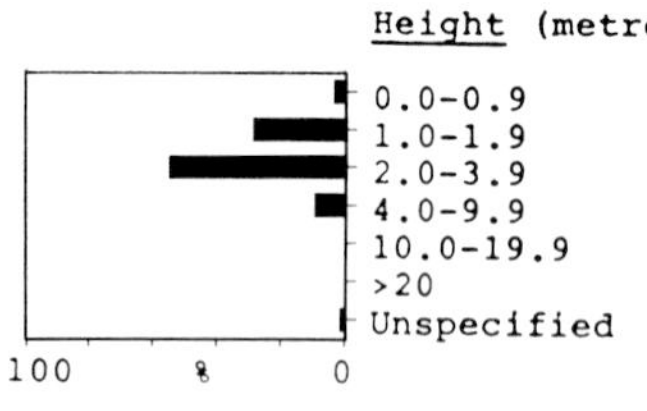

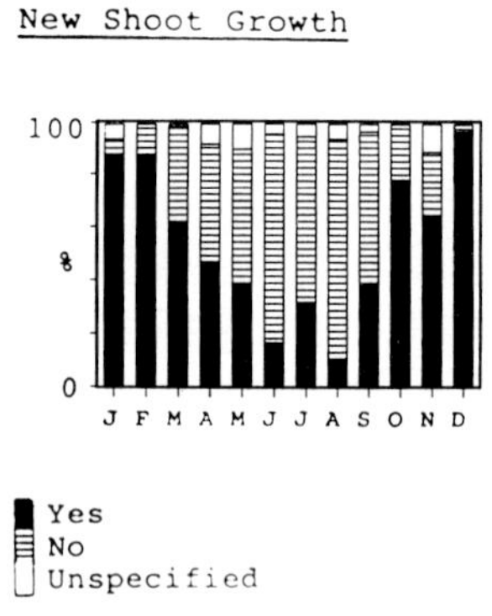

HABITAT

Within 2 km of coast: Yes (31%) No (68%) Unspecified (1%).

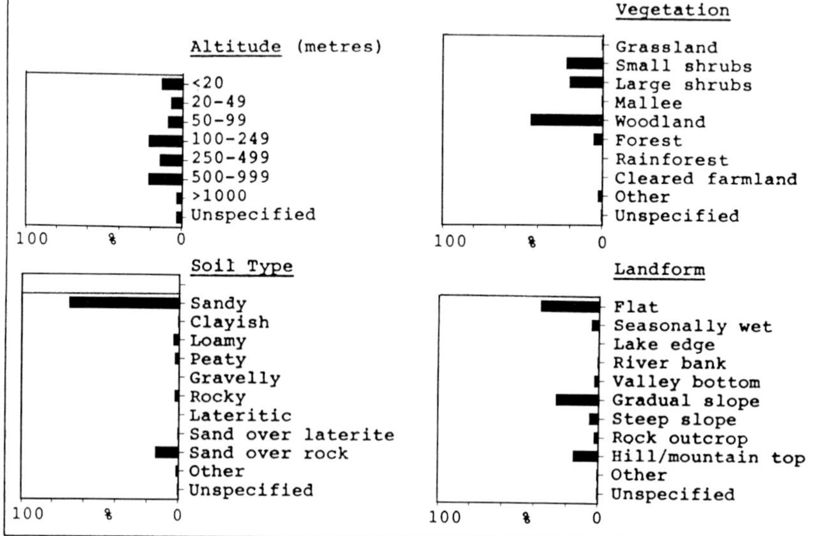

Response to Fire: 97 records

Growth response: Ground resprout (5%) killed, new seedlings (60%) trunk resprout (2%) killed, no seedlings (3%) killed, no seedlings and killed, new seedlings (1%) unspecified (29%).

Flowering response: Number of records where flowering had occurred after fire (32). Median reported time to flowering (13-24 months); minimum reported time to flowering (<1 month). Number of records where flowering had not occurred after fire (32). Median reported time to no flowering (13-24 months); maximum reported time to no flowering (49-60 months).

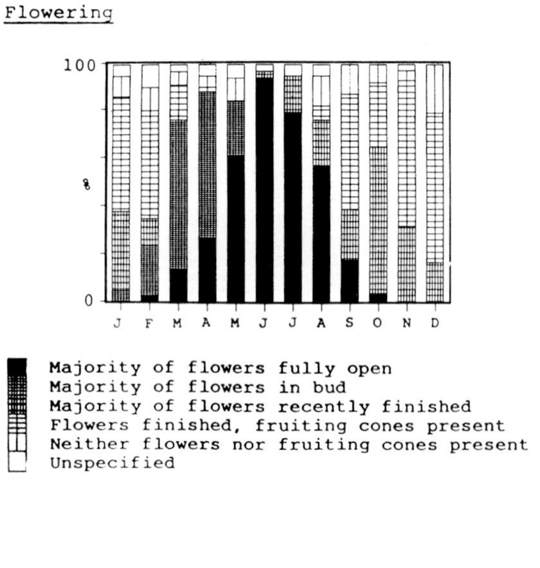

Possible pollinators

Pollinator type 372 records: bird (15%) butterflies, moths (1%) bees, wasps, ants (8%) flies (0.3%) mammals (1%) no pollinator observed (75%) .

Specific pollinator 116 records: honeyeater (12) New Holland Honeyeater (17), Brown Honeyeater (1) White-cheeked Honeyeater (11), White-eared Honeyeater (3) White-plumed Honeyeater (1) Yellow-faced Honeyeater (1) Crescent Honeyeater (3) Tawny-crowned Honeyeater (1) Rainbow Lorikeet (1) Noisy Miner (2) wattlebird (3) Little Wattlebird (10) Red Wattlebird (10) Eastern Spinebill (11) Black-faced Cuckoo-shrike (1) friarbird (1) bee (20) European honey bee (5) possum (2).

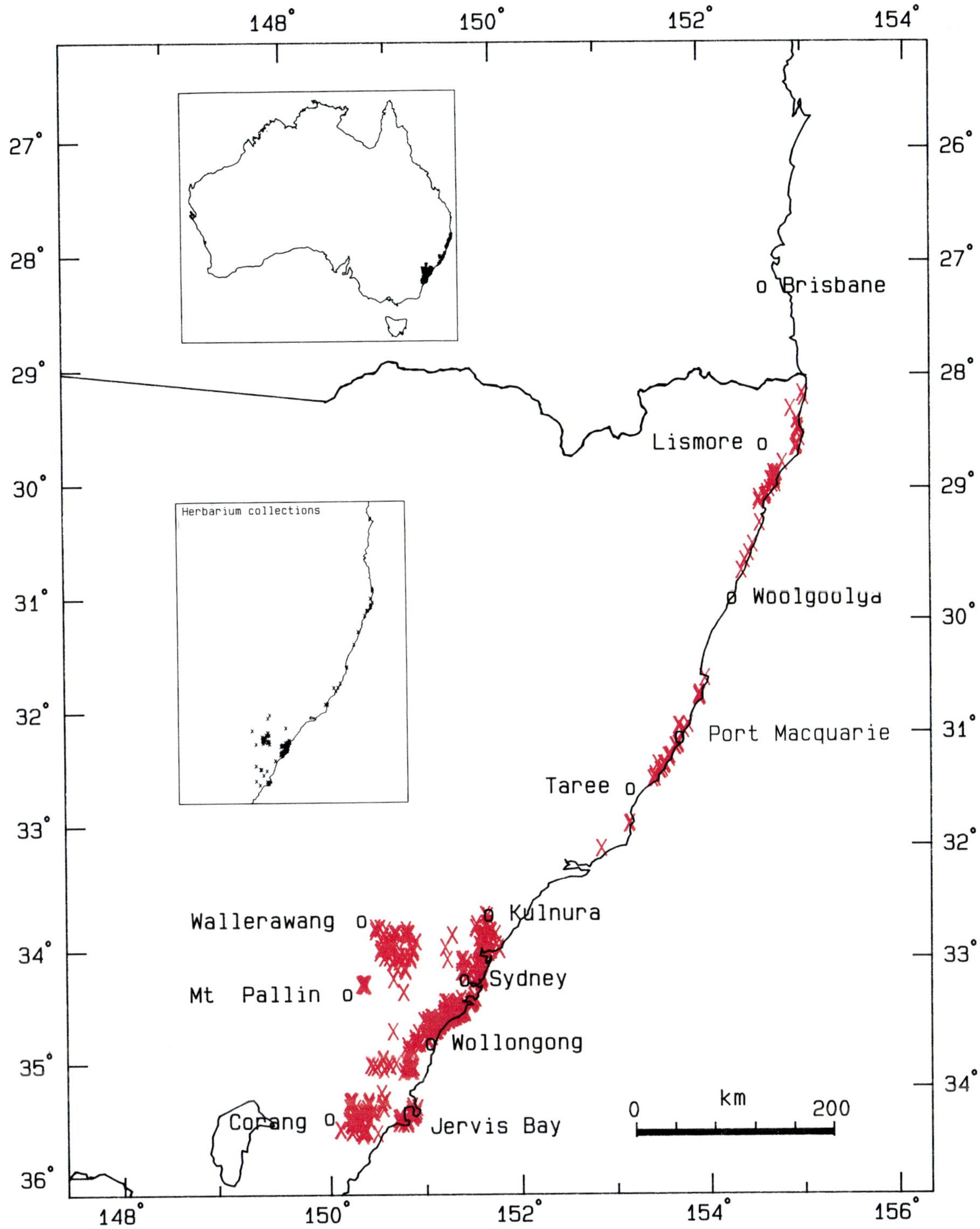

DISCUSSION

B. ericifolia is a large shrub or small tree generally up to 4 m in height, occasionally larger. The species contains two varieties, differentiated mainly by seedling leaves and flower length. The 565 records of B. ericifolia included 499 where a varietal name was specified as well as 66 records where only the species name was noted. Of the latter, all occur within the general distribution of either var. ericifolia or var. macrantha. There were few, if any, problems caused by differentiating between the two varieties. Those records where a varietal name was omitted were most likely caused by observers forgetting to identify at the varietal level rather than being unable to do so.

The distribution map of B. ericifolia shows clearly that the main ranges of the two varieties are geographically disjunct with over 150 km between them. There is, however, an interesting record of B. ericifolia var. ericifolia from Myall Lakes, which, if correctly identified would extend the range of this variety considerably and bring it to within close proximity of that of var. macrantha. If it is in fact var. macrantha it would be a southern range extension for this variety.

Flowering time of B. ericifolia is predominantly in winter with flowers recorded as early as February and as late as October. Conversely, new shoot growth is mainly in summer. Sandy soils are strongly preferred. Other aspects of habitat differ for the two varieties, and are referred to in the next few pages.

BANKSIA ERICIFOLIA Linnaeus f. (1781) var. ERICIFOLIA

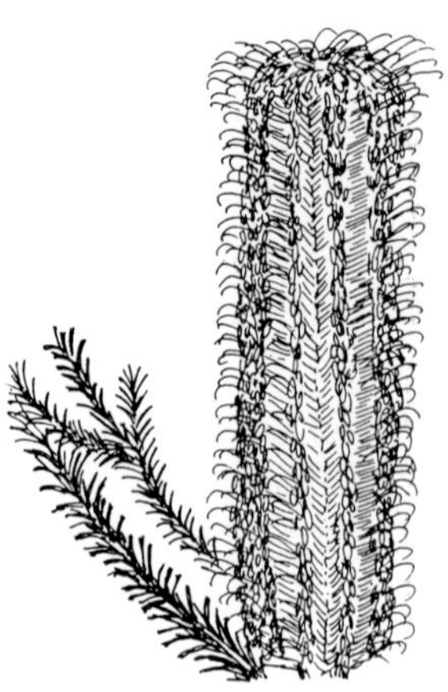

423 RECORDS: Jan (42) Feb (38) Mar (69) Apr (45) May (55) Jun (64) Jul (10) Aug (28) Sep (24) Oct (11) Nov (19) Dec (18).

Population Size: 1-10 (68) 10-100 (106) >100 (246) Unspecified (3).

Conservation Status: Restricted to road verge (1%) Not (98%) Unspecified (1%). In conservation reserve (60%) Not (37%) Unspecified (3%).

Tree/Shrub: Tree form (9%) Shrub form (90%) Unspecified (1%).

Height (metres)

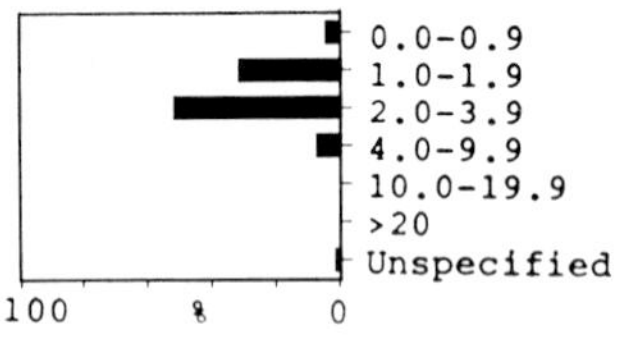

New Shoot Growth

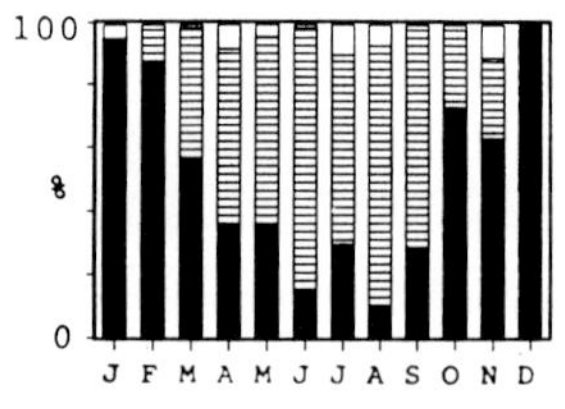

Yes
No
Unspecified

HABITAT

Within 2 km of coast: Yes (22%) No (78%) Unspecified (0%).

Altitude (metres)

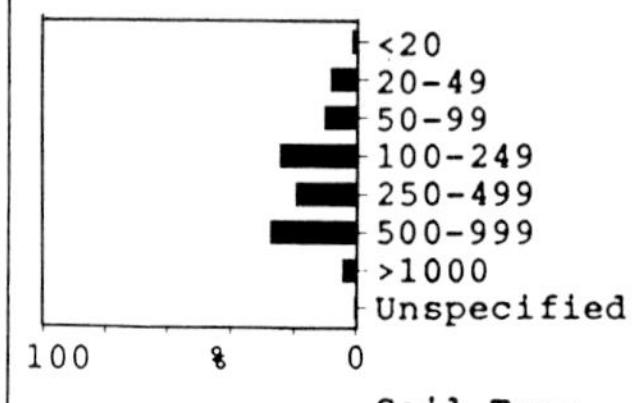

Vegetation

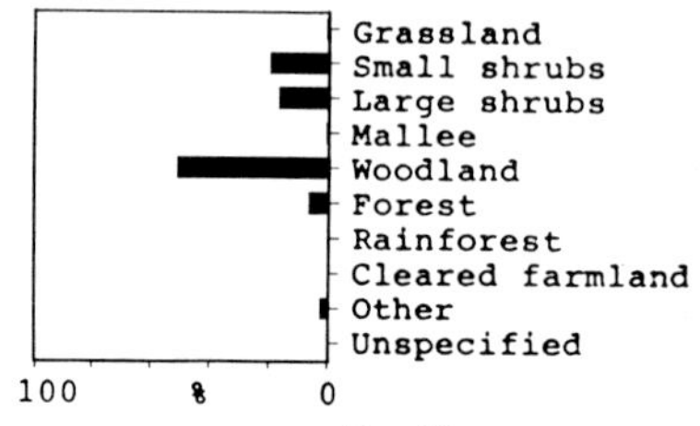

Soil Type

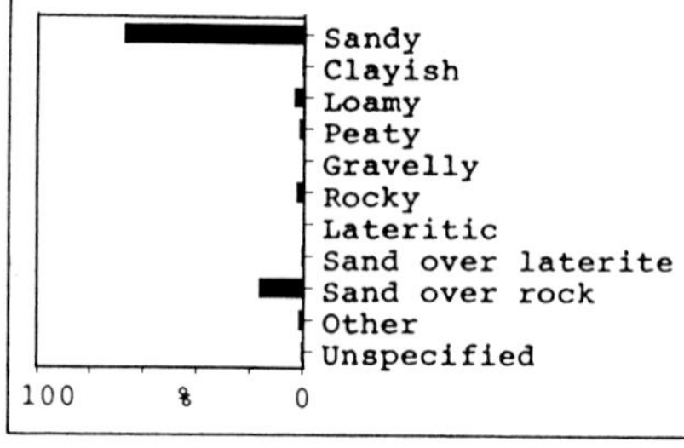

Landform

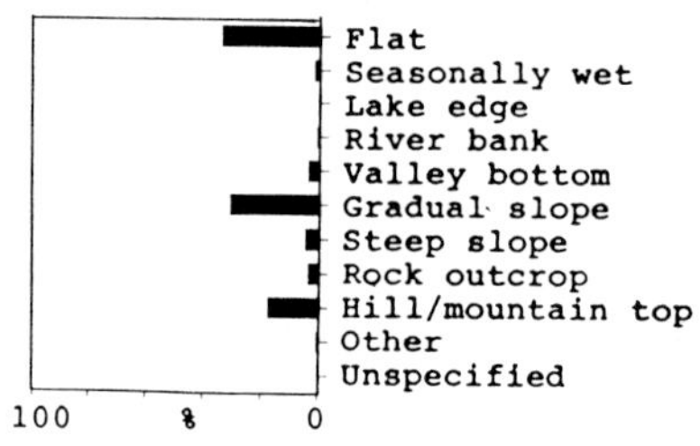

Response to Fire: 69 records

Growth response: Ground resprout (4%) killed, new seedlings (58%) killed, no seedlings (3%) killed, no seedlings and killed, new seedlings (1%) unspecified (33%).

Flowering response: Number of records where flowering had occurred after fire (24). Median reported time to flowering (13-24 months); minimum reported time to flowering (<1 month). Number of records where flowering had not occurred after fire (19). Median reported time to no flowering (13-24 months); maximum reported time to no flowering (37-48 months).

Flowering

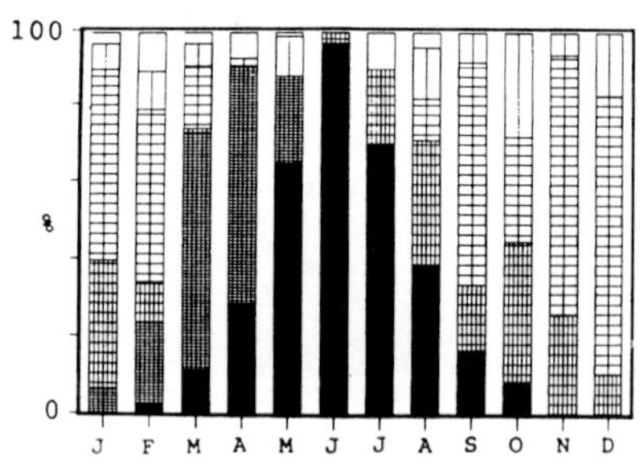

Majority of flowers fully open
Majority of flowers in bud
Majority of flowers recently finished
Flowers finished, fruiting cones present
Neither flowers nor fruiting cones present
Unspecified

Possible pollinators

Pollinator type 305 records: bird (13%) bees, wasps, ants (4%) mammals (1%) no pollinator observed (82%).

Specific pollinator 75 records. honeyeater (9) New Holland Honeyeater (15), White-cheeked Honeyeater (3), White-eared Honeyeater (2) White-plumed Honeyeater (1) Yellow-faced Honeyeater (1) Crescent Honeyeater (3) Noisy Miner (2) Red Wattlebird (10) Little Wattlebird (5) Eastern Spinebill (10) friarbird (1) Rainbow Lorikeet (1) bee (8) European honey bee (4).

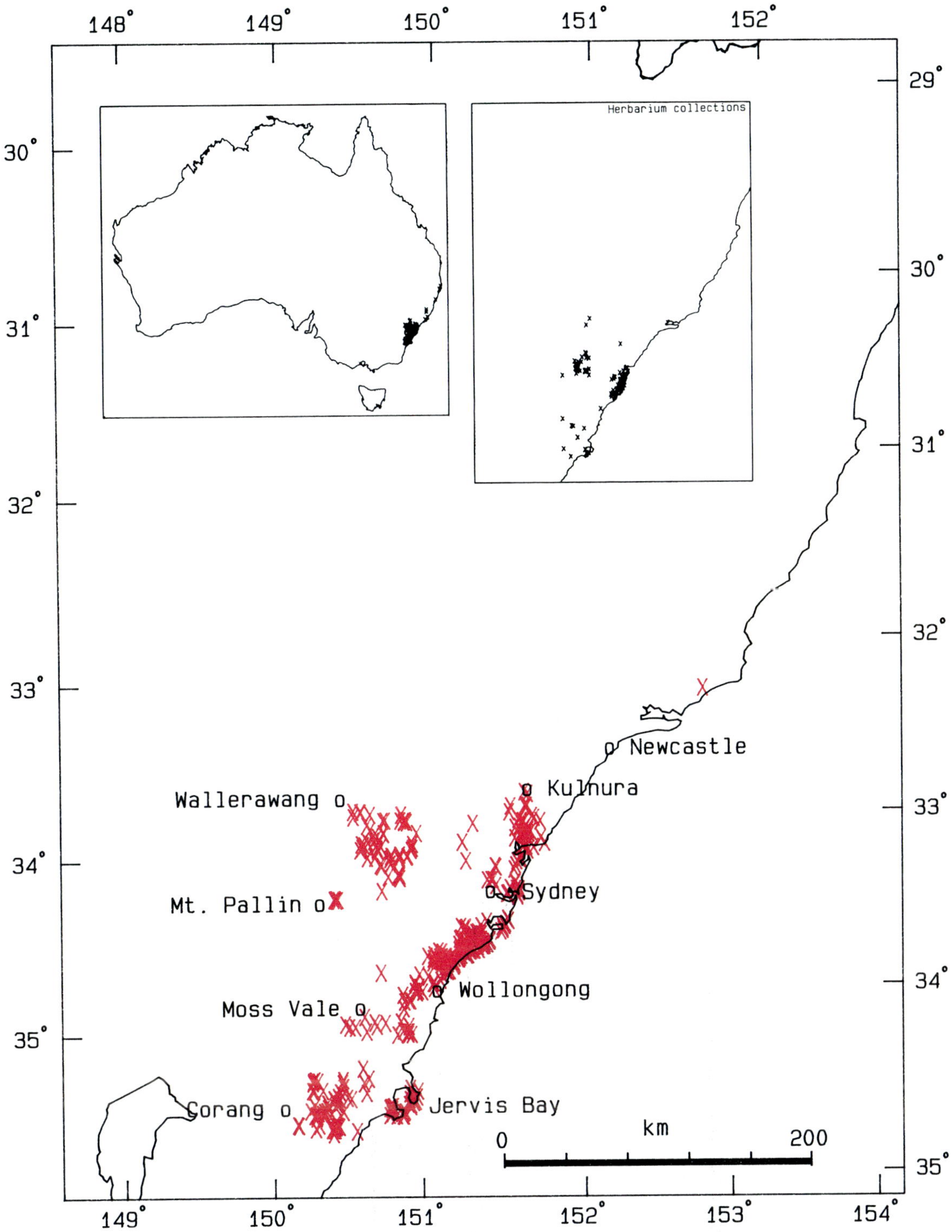

DISCUSSION

Atlas records are similar to herbarium records in indicating a distribution of B. ericifolia var. ericifolia along the New South Wales central coast between Avoca Beach and Lake Conjola (south of Jervis Bay), extending inland to adjacent parts of the Great Dividing Range. A 1966 herbarium collection of D. McGillivray's from Mt Coricudgy was not recorded by Atlas contributors. Neither was a collection from Collaroy made by H. Mair in 1930 and cited by George (1981). However, Atlas contributors did show this variety to be widespread in the Budawang Range and Morton National Park, extending inland almost as far as Corang. There is a single record from Myall Lakes, which, if correctly identified would considerably extend the range of this variety and also bring it to within close proximity of that of var. macrantha.

B. ericifolia var. ericifolia grows as a large shrub or small tree generally up to 4 m in height, but occasionally larger. One-fifth of the populations recorded were within 2 km of the coast, where the variety is a common component of sandy coastal heaths. Inland it is often found in woodland, on mountain slopes or hilltops up to 1 000 m above sea level. Here it is associated with sandstone areas, growing in deep sands or sands over rock. Flowering is from autumn through to spring with the peak flowering period between May and July. Conversely, new shoot growth is low in winter being predominantly in late spring and summer. The flowers are very attractive to honey-eating birds with the most frequently reported visitors being the New Holland Honeyeater, Red Wattlebird and Eastern Spinebill. B. ericifolia var. ericifolia is non-lignotuberous and plants are generally killed by fire and regenerate from seed. Occasionally, fires can be so hot that they kill both mature plants and their stock of seed. Alternatively, large mature plants may survive a relatively cool burn whilst smaller specimens succumb. The few Atlas records of lignotuber resprouts following fire should be investigated. A presumed hybrid of B. ericifolia var. ericifolia x B. spinulosa var. cunninghamii was recorded inland from Wollongong (see Chapter 3 for details).

BANKSIA ERICIFOLIA Linnaeus f.
var. MACRANTHA A.S.George (1981)

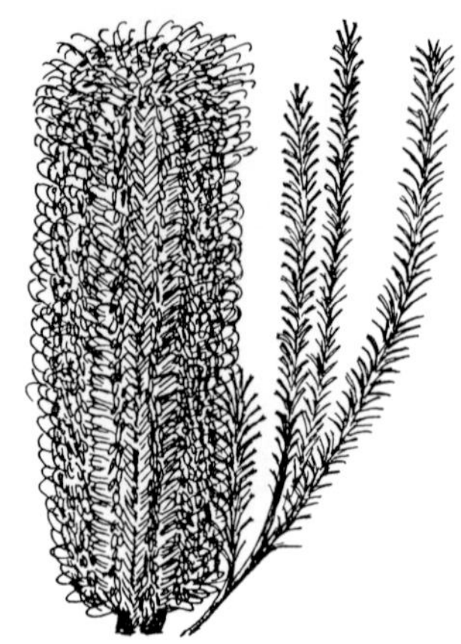

76 RECORDS: Jan (7) Feb (0) Mar (0) Apr (12) May (5) Jun (2) Jul (6) Aug (18) Sep (7) Oct (12) Nov (7) Dec (0).

Population Size: 1-10 (7) 10-100 (14) >100 (52) Unspecified (3).

Conservation Status: Restricted to road verge (5%) Not (95%) Unspecified (0%). In conservation reserve (67%) Not (24%) Unspecified (9%).

Tree/Shrub: Tree form (26%) Shrub form (71%) Unspecified (3%).

Height (metres)

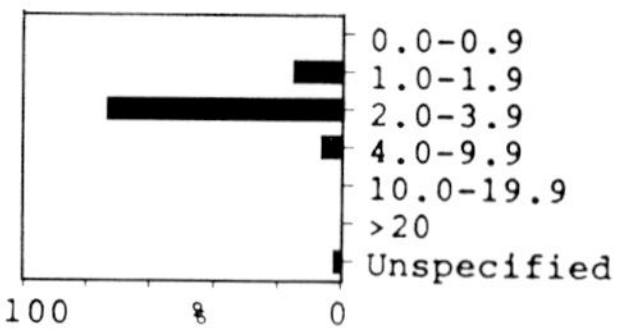

New Shoot Growth

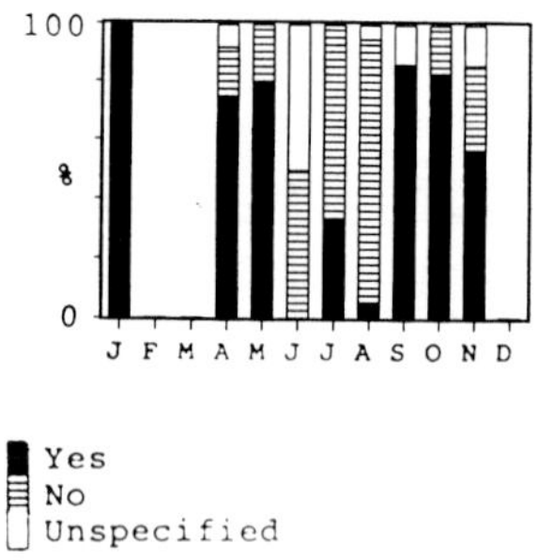

HABITAT

Within 2 km of coast: Yes (76%) No (22%) Unspecified (1%).

Altitude (metres)

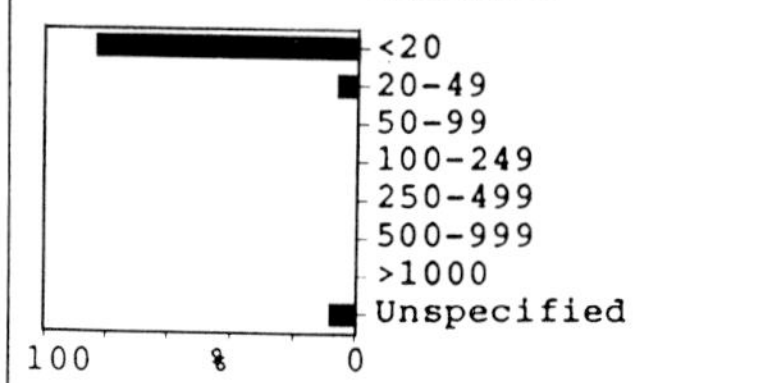

Vegetation

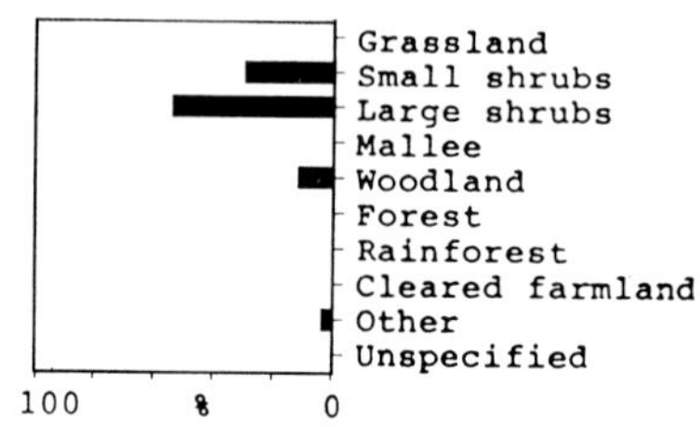

Soil Type

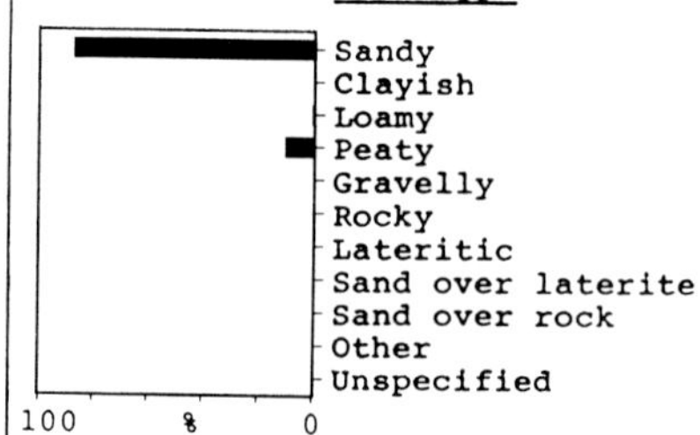

Landform

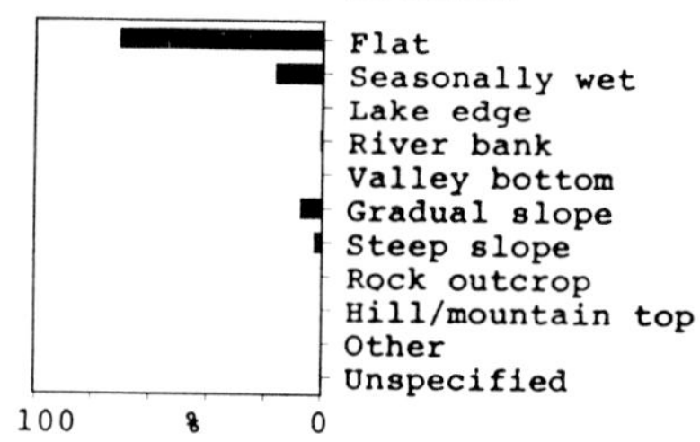

Response to Fire: 10 records

Growth response: Killed, new seedlings (80%) killed, no seedlings (10%) unspecified (10%).

Flowering response: Number of records where flowering had occurred after fire (2). Median reported time to flowering (1-12 months); minimum reported time to flowering (<1 month). Number of records where flowering had not occurred after fire (6). Median reported time to no flowering (1-12 months); maximum reported time to no flowering (25-36 months).

Flowering

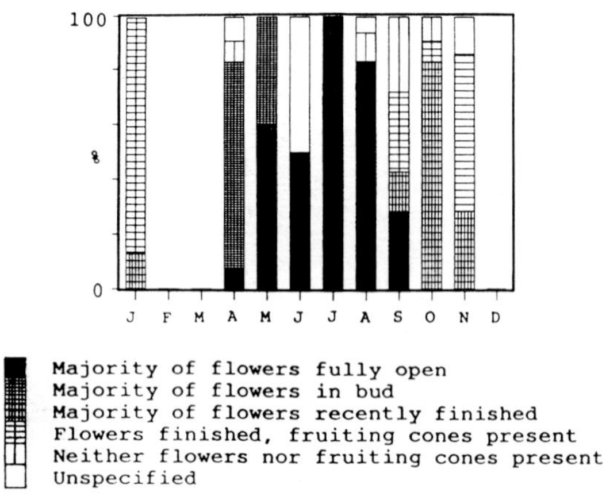

Possible pollinators

Pollinator type 29 records: bird (34%) butterflies, moths (3%) bees, wasps, ants (21%) no pollinator observed (41%).

Specific pollinator 18 records: honeyeater (2) Brown Honeyeater (1) White-cheeked Honeyeater (6) Tawny-crowned Honeyeater (1) Little Wattlebird (3) Eastern Spinebill (1) Black-faced Cuckoo-shrike (1) bee (2) European honey bee (1).

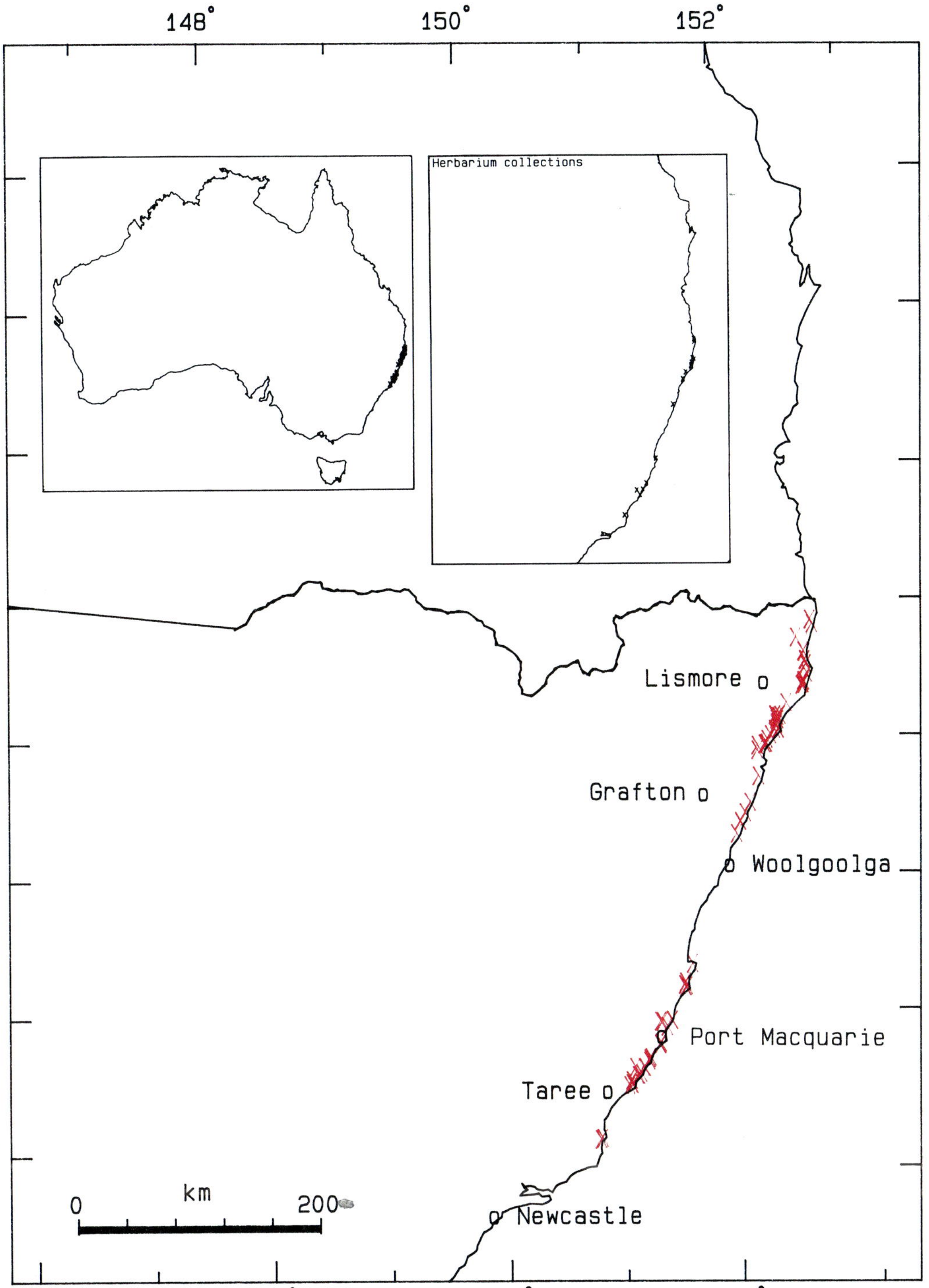

DISCUSSION

Atlas data confirmed herbarium records for B. ericifolia var. macrantha in showing that it occurs in two disjunct populations along the northern New South Wales coast. One population ranges from Kingscliff to Yuraygir National Park, north of Woolgoolga. The other is from Hat Head National Park, north of Port Macquarie as far south as Cape Hawke. The variety is locally common with more than 50% of its recorded populations being of more than 100 plants. It is well represented in conservation reserves, largely those of Limeburners Creek Nature Reserve and Crowdy Bay and Broadwater National Parks, as well as those reserves already mentioned.

The variety occurs as a large shrub or small tree up to 4 m in height, occasionally larger. It has a strongly coastal distribution with more than 75% of its populations within 2 km of the coast where it occurs on low lying sandy flats. Surrounding vegetation is usually low or tall shrubland. The main flowering period is between May and September with a peak in July. The flower spikes are attractive to honey-eating birds, with Atlas contributors recording the White-cheeked Honeyeater as the most frequent visitor. Flower spikes are sometimes very large with lengths of up to 26 cm being recorded from several spikes in Crowdy Bay National Park. Available data suggest a period of new shoot growth from early spring through to late autumn. The variety is non-lignotuberous and regenerates from seed after fire. A very hot fire, however, may sometimes kill seed as well as mature plants. Seedlings may need at least three years to reach flowering stage although at Crowdy Bay some two year old plants were observed with their first flower spikes present.

BANKSIA GARDNERI A.S.George (1981)

Prostrate Banksia

329 RECORDS: Jan (11) Feb (15) Mar (15) Apr (17) May (12) Jun (47) Jul (36) Aug (42) Sep (43) Oct (31) Nov (36) Dec (19).

Population Size: 1-10 (81) 10-100 (150) >100 (83) Unspecified (15).

Conservation Status: Restricted to road verge (14%) Not (81%) Unspecified (5%). In conservation reserve (55%) Not (40%) Unspecified (5%).

Tree/Shrub: Tree form (0%) Shrub form (95%) Unspecified (5%).

Height (metres)

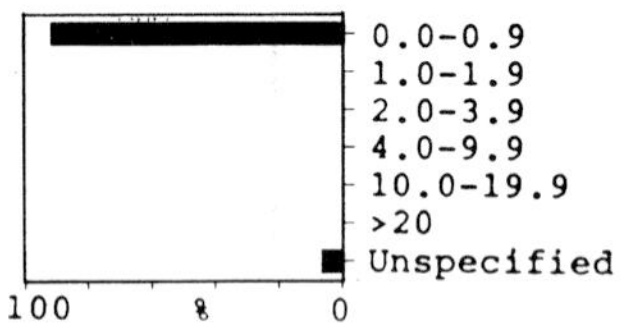

New Shoot Growth

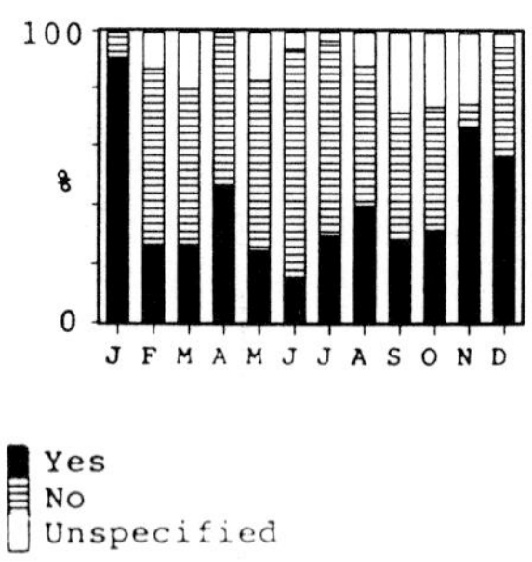

HABITAT

Within 2 km of coast: Yes (6%) No (94%) Unspecified (0%).

Altitude (metres)

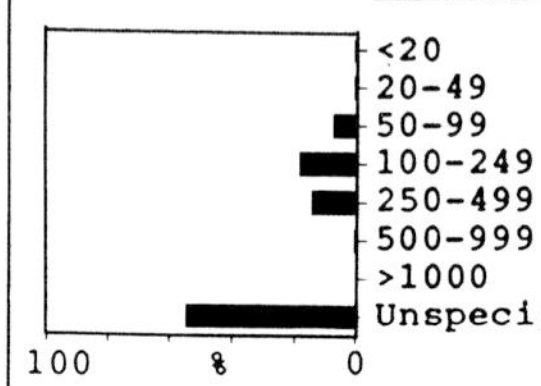

Vegetation

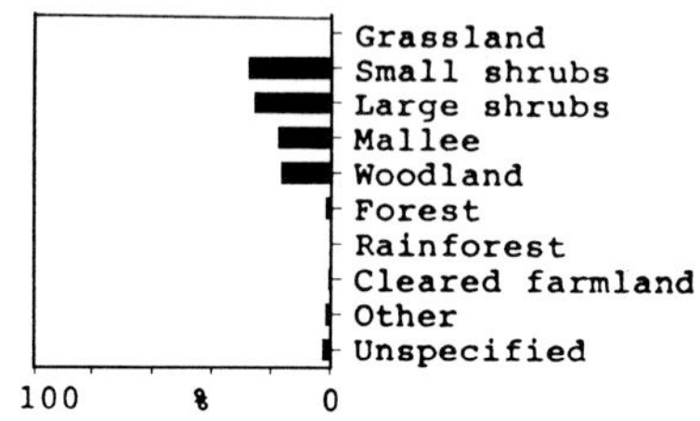

Soil Type

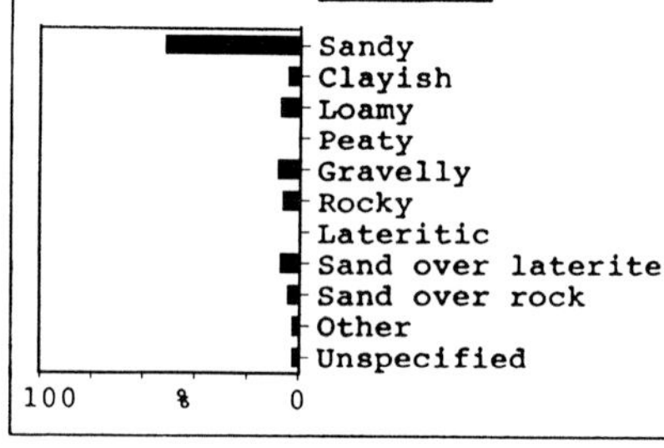

Landform

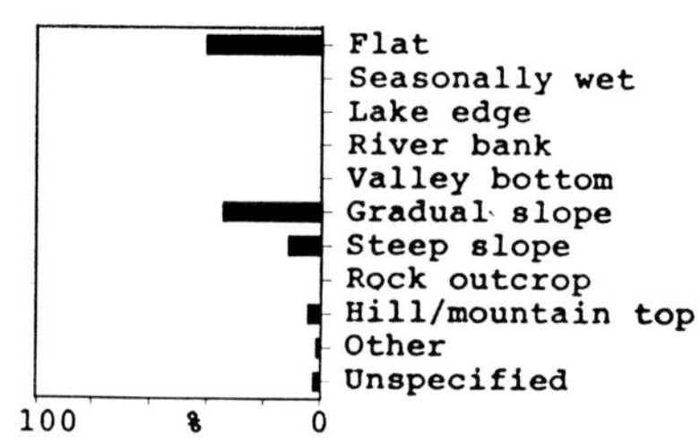

Response to Fire: 46 records

Growth response: Ground resprout (43%) ground resprout and killed, new seedlings (4%) unspecified (52%).

Flowering response: Number of records where flowering had occurred after fire (36). Median reported time to flowering (13-24 months); minimum reported time to flowering (<1 month).
Number of records where flowering had not occurred after fire (6). Median reported time to no flowering (25-36 months); maximum reported time to no flowering (73-84 months).

Flowering

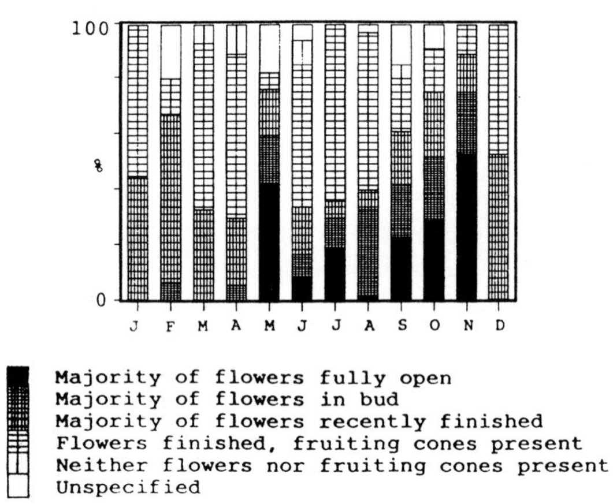

Possible pollinators

Pollinator type 88 records: bees, wasps, ants (5%) beetles (1%) mammals (1%) no pollinator observed (93%).

Specific pollinator 5 records: ant (3) Honey-possum (1) cockroach (1).

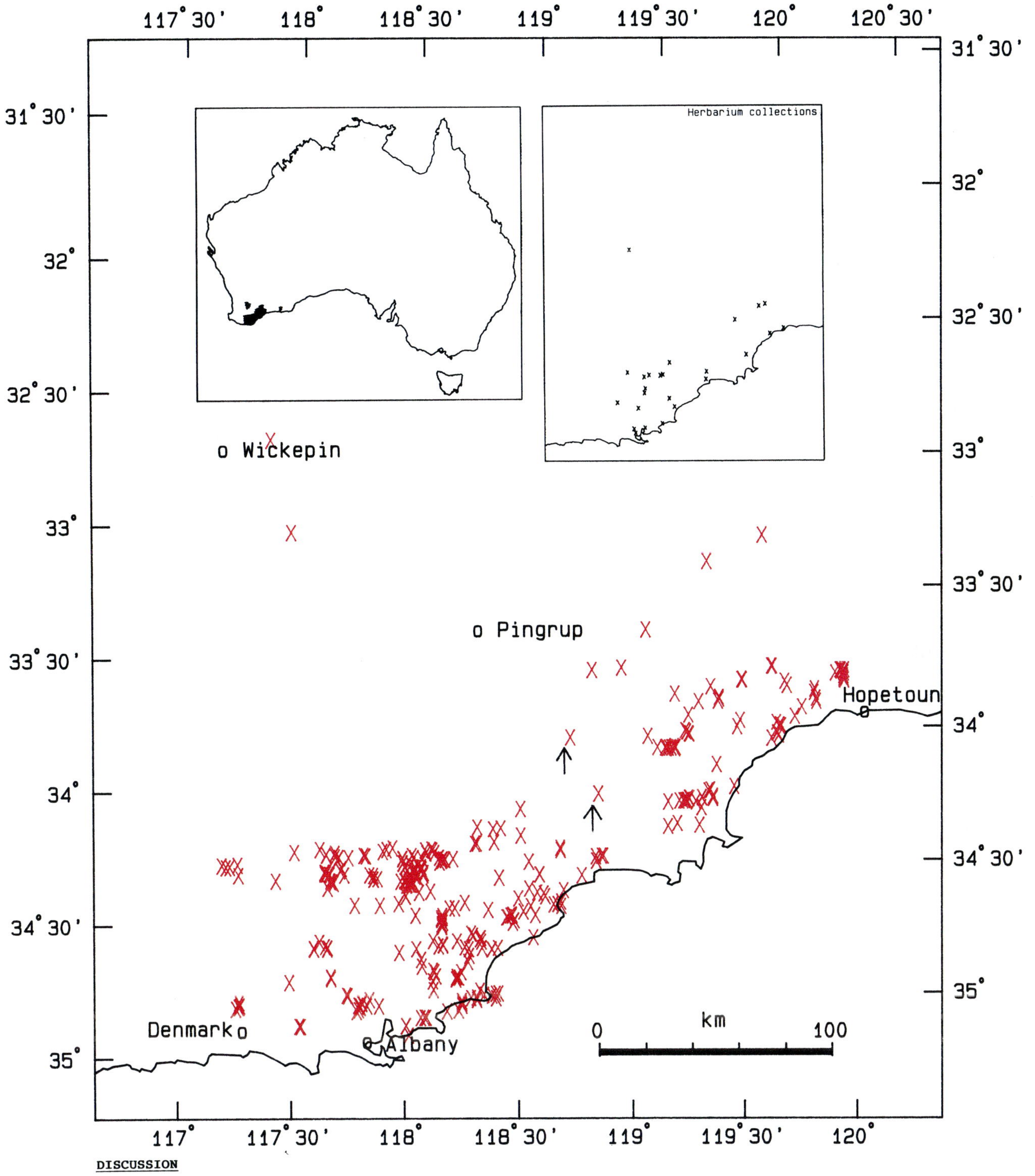

DISCUSSION

B. gardneri is a prostrate shrub with flower spikes and fruiting cones at ground level. Horizontal stems give rise to erect leaves up to 40 cm long. The species has three varieties distinguished by leaf shape, flower colour and flowering period. Var. hiemalis and var. gardneri have similar leaves but are identified by flower colour and flowering time. If flowers were absent, contributors found it difficult to distinguish between these two varieties. There may also have been some confusion with some low-growing dryandras (e.g. D. pteridifolia, D. calophylla) whose leaves are very similar to those of var. gardneri and var. hiemalis. B. gardneri var. brevidentata has a distinctive leaf shape and did not cause any identification problems.

In their distribution, there is an overlap between the varieties brevidentata and gardneri in the Stirling Range. Also, var. gardneri may overlap with var. hiemalis near the western boundary of Fitzgerald River National Park (see discussion under varieties).

The 329 records of B. gardneri include 290 records where a varietal name was specified as well as 39 records where only the species name, B. gardneri, was noted. Of the latter, two in particular (arrowed on the map) are of considerable interest occurring mid-way between the main ranges of var. gardneri and var. hiemalis. It is interesting to wonder which of the two varieties these records belong to. If the former, they represent a range extension eastwards, if the latter a range extension westwards. Either way, they bring the ranges of the two varieties closer together. The two records were both made in the month of September when the plants were noted to be in full flower. According to George (1981) this flowering time would suggest var. gardneri. However, the flowering time of var. hiemalis may not be as uniformly early as George has indicated since Atlas contributors noted several occurrences of later flowering plants (see discussion under var. hiemalis).

BANKSIA GARDNERI A.S.George var. BREVIDENTATA A.S.George (1981)

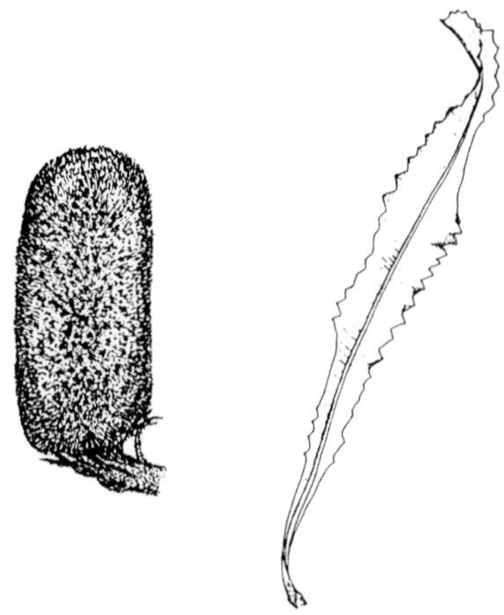

29 RECORDS: Jan (1) Feb (2) Mar (0) Apr (1) May (4) Jun (13) Jul (1) Aug (2) Sep (2) Oct (3) Nov (0) Dec (0).

Population Size: 1-10 (5) 10-100 (15) >100 (9) Unspecified (0).

Conservation Status: Restricted to road verge (0%) Not (100%) Unspecified (0%). In conservation reserve (100%) Not (0%) Unspecified (0%).

Tree/Shrub: Tree form (0%) Shrub form (100%) Unspecified (0%).

Height (metres)

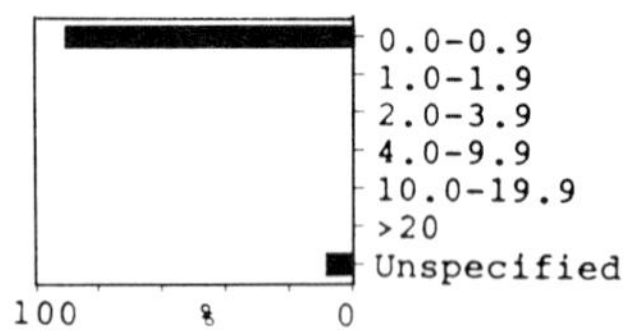

New Shoot Growth

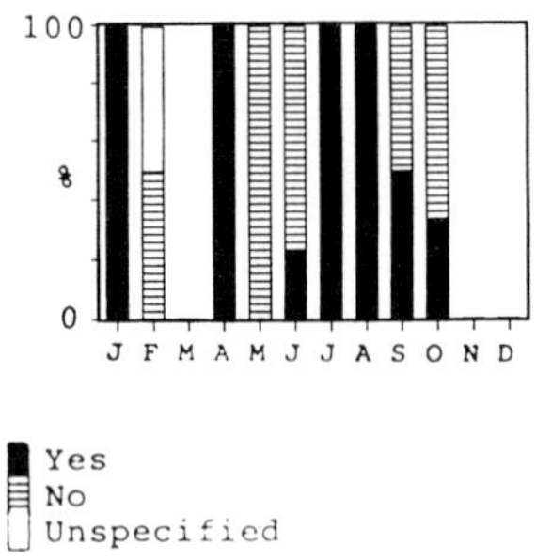

HABITAT

Within 2 km of coast: Yes (0%) No (100%) Unspecified (0%).

Altitude (metres)

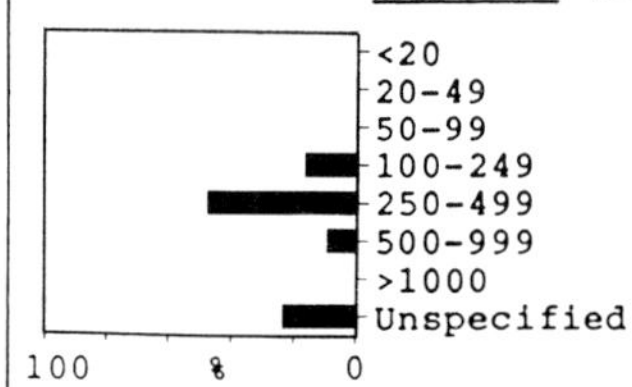

Vegetation

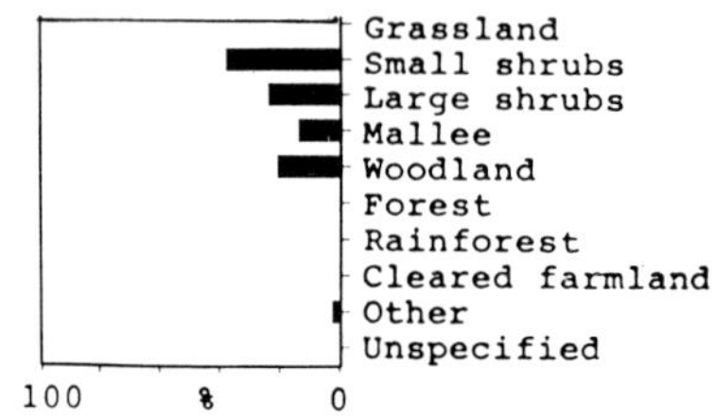

Soil Type

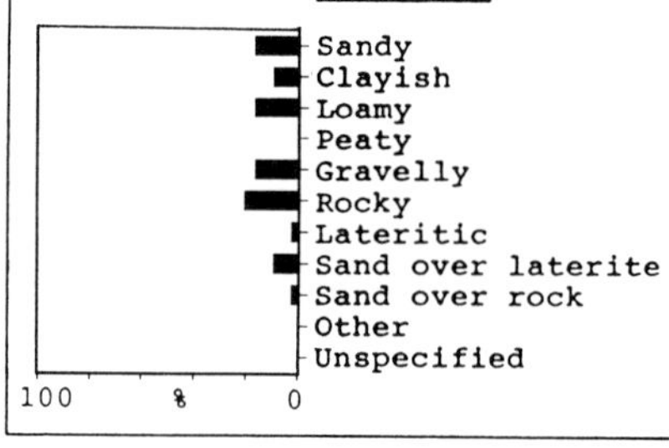

Landform

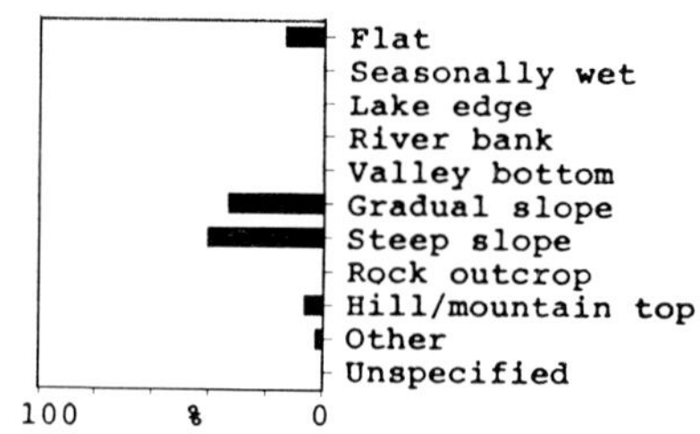

Response to Fire: 8 records

Growth response: Ground resprout (50%) ground resprout and killed, new seedlings (12%) unspecified (37%).

Flowering response: Number of records where flowering had occurred after fire (7). Median reported time to flowering (13-24 months); minimum reported time to flowering (<1 month). Number of records where flowering had not occurred after fire (0).

Flowering

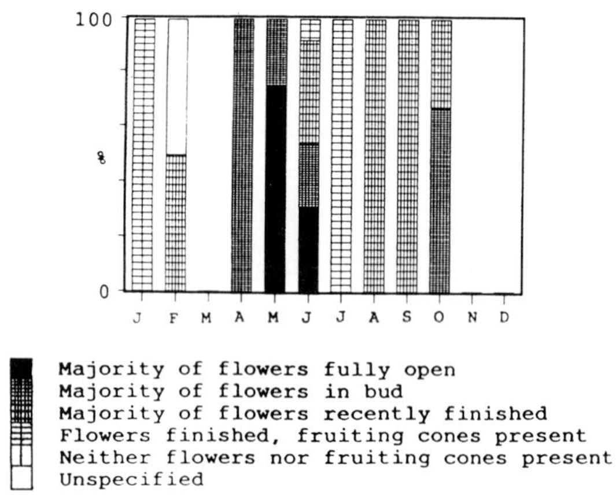

Possible pollinators

Pollinator type 11 records: bees, wasps, ants (9%) no pollinator observed (91%).

Specific pollinator 0 records.

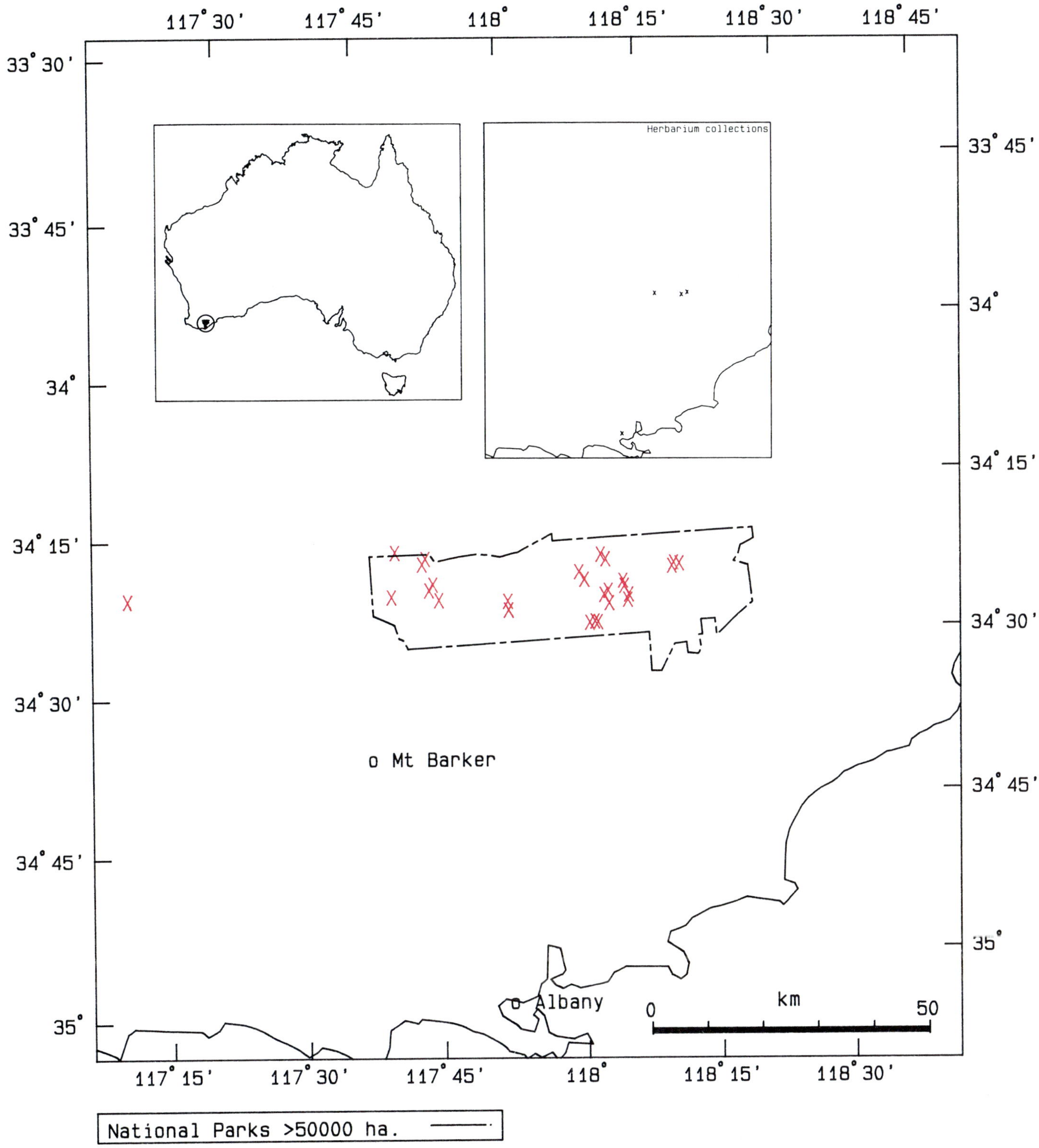

DISCUSSION

Atlas records show B. gardneri var. brevidentata to be well distributed throughout the Stirling Range National Park from Bluff Knoll in the east to Donnelly Track in the west. A significant new recording was made ca. 30 km west of Cranbrook on Warrinup Nature Reserve. A specimen of this population is retained at the herbarium of the Katanning office, Dept of Conservation and Land Management. The 1963 herbarium collection of F. Humphrey's from Millbrook Road, north of Albany, was not recorded by Atlas contributors. Although George (1981) stated that B. gardneri var. brevidentata and B. gardneri var. gardneri did not occur together, on a few occasions Atlas contributors recorded these two varieties from the same locality (e.g. along Yungermere Track, 1 km and 6 km east of Chester Pass Road, also at Warrinup Nature Reserve.

B. gardneri var. brevidentata is a prostrate shrub with stems running horizontally. Its erect leaves are up to 40 cm long. It grows in a diversity of soils generally on the slopes and tops of foothills of the Stirling Range in shrubland, low woodland or amongst mallee eucalypts. The population west of Cranbrook occurs on laterite gravel in eucalypt woodland. Flowering is from autumn to early winter. There were few Atlas records of either pollinators or new shoot growth and, for both these categories, further investigations are needed before reliable conclusions can be drawn. The variety is lignotuberous and may flower in one or two years after fire.

BANKSIA GARDNERI A.S.George (1981) var. GARDNERI

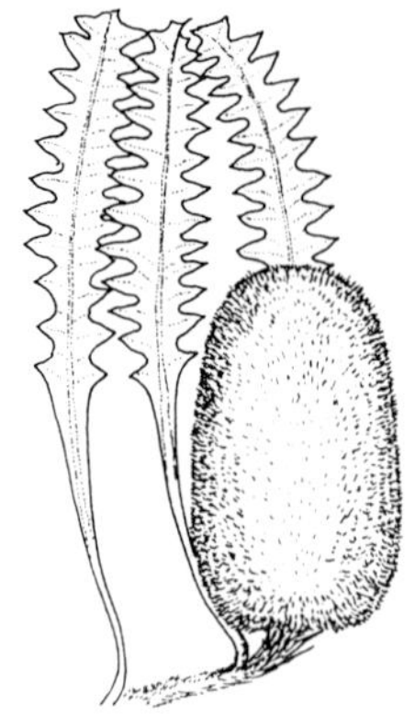

Flowers rusty-brown

180 RECORDS: Jan (7) Feb (8) Mar (7) Apr (15) May (3) Jun (29) Jul (22) Aug (28) Sep (19) Oct (10) Nov (24) Dec (7).

Population Size: 1-10 (49) 10-100 (83) >100 (45) Unspecified (3).

Conservation Status: Restricted to road verge (21%) Not (74%) Unspecified (6%). In conservation reserve (44%) Not (49%) Unspecified (7%).

Tree/Shrub: Tree form (0%) Shrub form (99%) Unspecified (1%).

Height (metres)

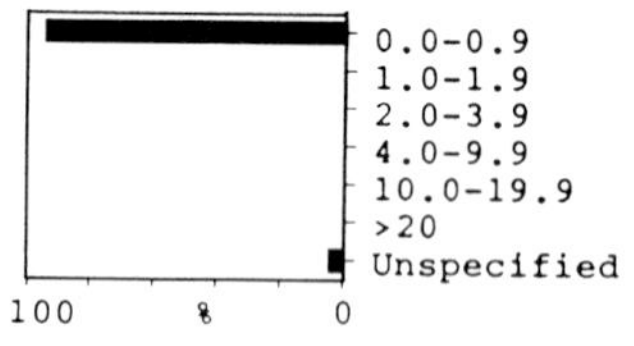

New Shoot Growth

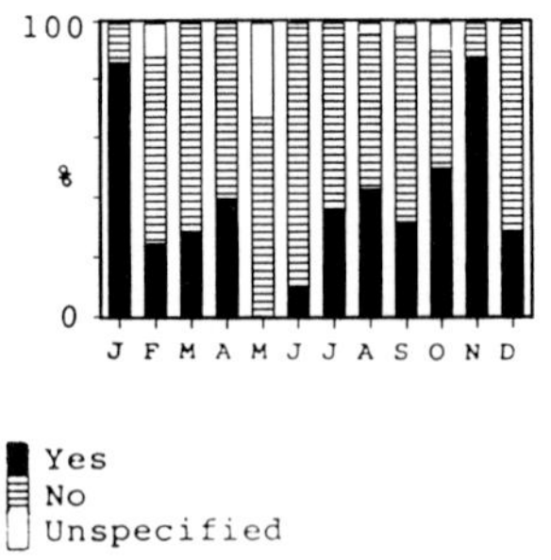

HABITAT

Within 2 km of coast: Yes (8%) No (92%) Unspecified (0%).

Altitude (metres)

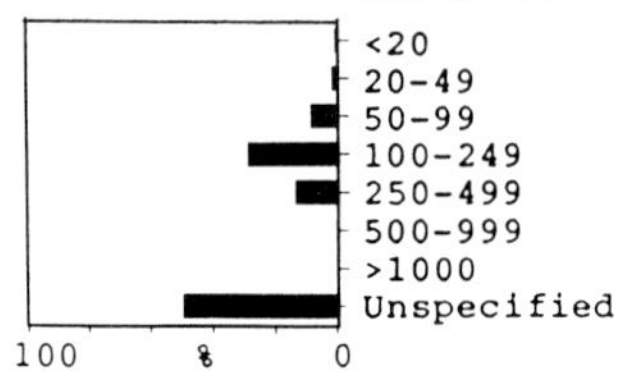

Vegetation

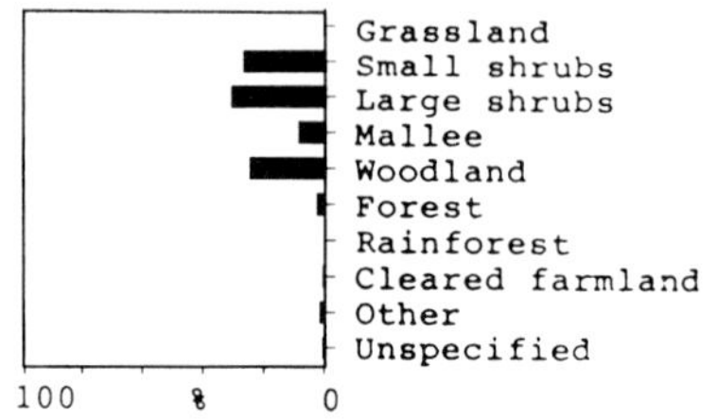

Soil Type

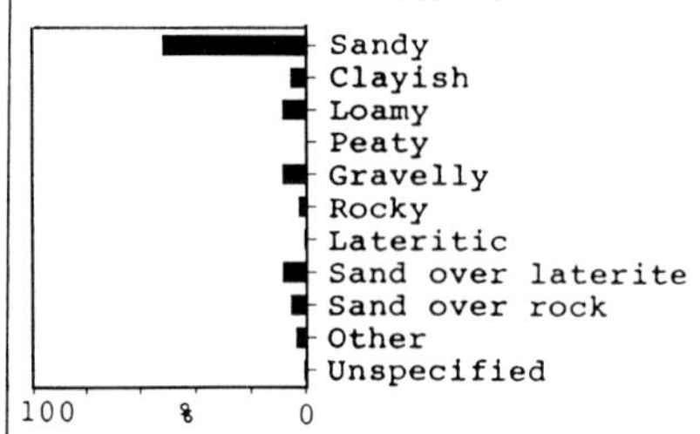

Landform

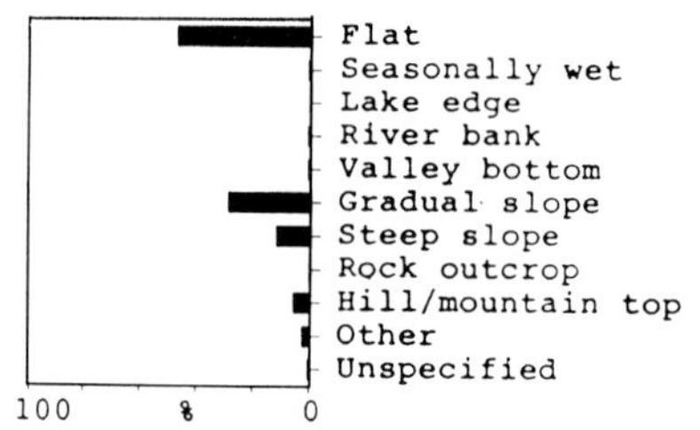

Response to Fire: 27 records

Growth response: Ground resprout (37%) unspecified (63%).

Flowering response: Number of records where flowering had occurred after fire (21). Median reported time to flowering (13-24 months); minimum reported time to flowering (<1 month).
Number of records where flowering had not occurred after fire (4). Median reported time to no flowering (24-25 months); maximum reported time to no flowering (73-84 months).

Flowering

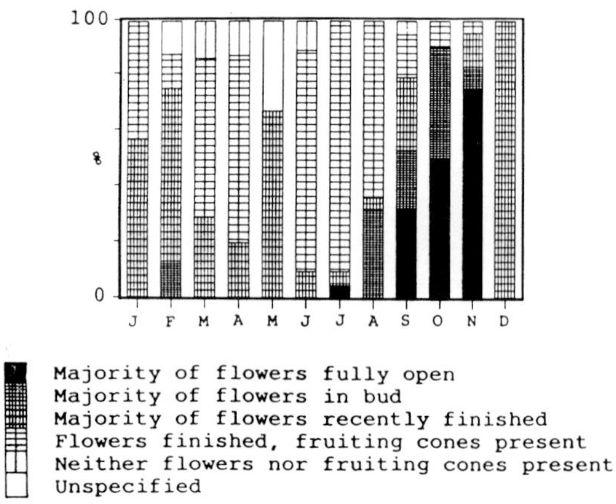

Possible pollinators

Pollinator type 55 records: bees, wasps, ants (5%) no pollinator observed (95%).

Specific pollinator 3 records: ant (3).

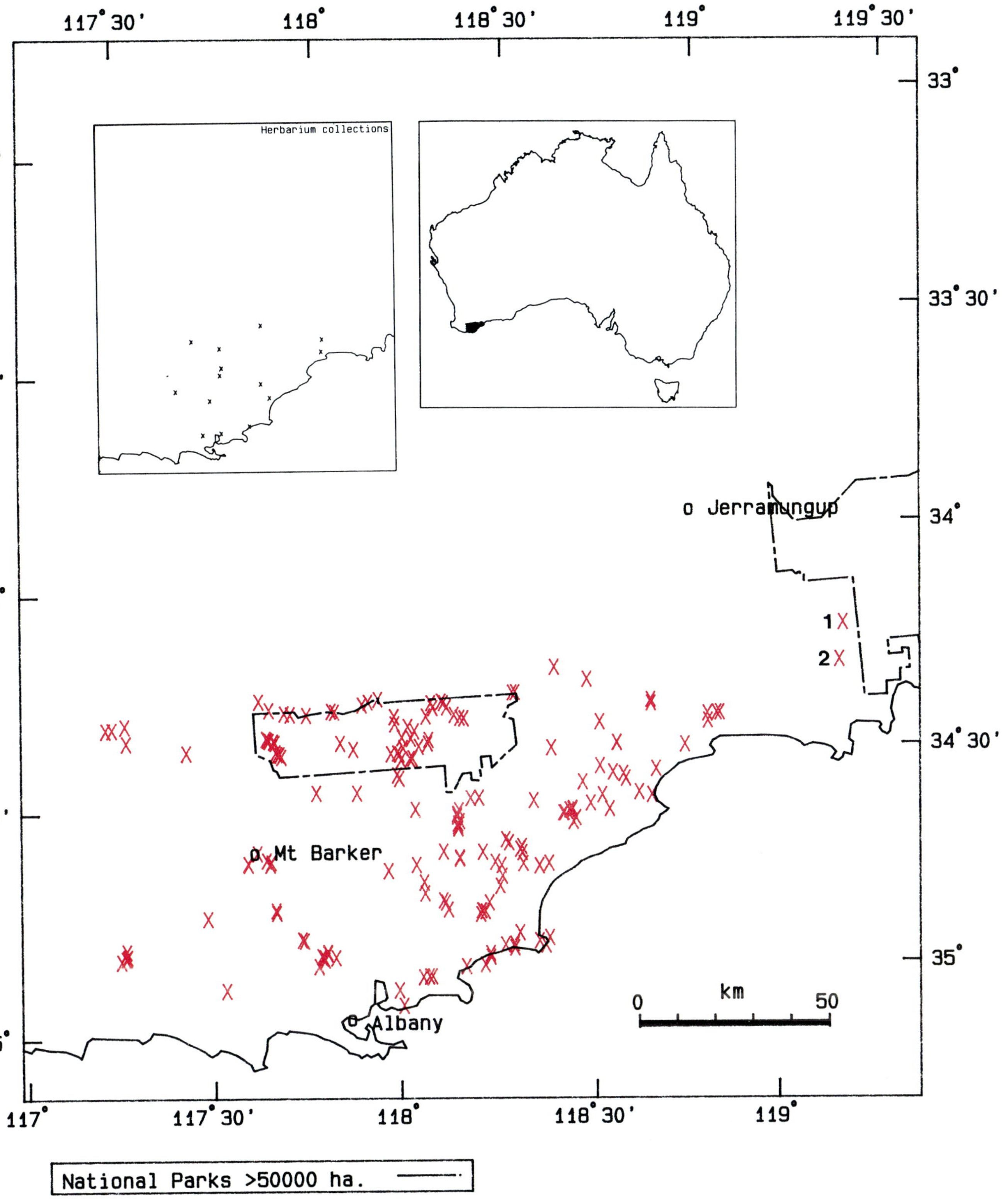

DISCUSSION

Western Australian Herbarium collections indicated a distribution for B. gardneri var. gardneri from the Stirling Range to Albany extending westwards to Mt Barker and almost to Cranbrook and east to Beaufort Inlet. Sight records have confirmed this distribution and extended it further west to Mt Lindesay north of Denmark and to approximately 25 km west of Cranbrook. Eastwards, there are two new recordings from near the Fitzgerald River National Park boundary. If correctly identified, these populations would mean that var. gardneri occurs further east than was previously known and also that its distribution overlaps with that of var. hiemalis. Record 1 (see map) was made in February when the plants were reported to be in "late flower, still with some colour". Old flowers were described as reddish brown, and the leaves dark green. This description conforms with that of B. gardneri var. gardneri. Record 2 was also made in February, but here the plants were "in bud" which suggests a possibility of var. hiemalis. In the Stirling Range, var. gardneri is frequently recorded, occasionally in the same locality as var. brevidentata. George (1981), however, stated that the two varieties never occurred together. At least two such recordings have been checked and both varieties verified.

B. gardneri var. gardneri is a prostrate shrub with leaves and flower spikes at ground level. It prefers sandy soil though sometimes occurring in loamy, clayish or gravelly soils. It generally grows on flat or gently undulating country, but occasionally on steeper mountain slopes. Surrounding vegetation is usually shrubland, though it also occurs amongst low open woodland and sometimes in mallee. The main flowering period is from September to November. Some new shoot growth was reported from almost all months of the year, though apparently peaking in late spring and early summer.

BANKSIA GARDNERI A.S.George
var. HIEMALIS A.S.George (1981)

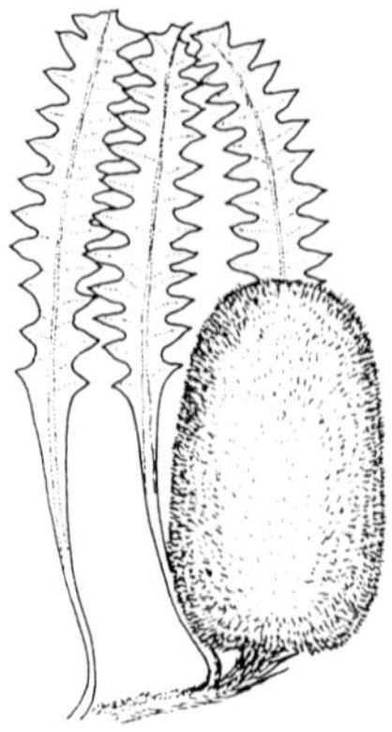

Flowers pale brown-pink

81 RECORDS: Jan (1) Feb (3) Mar (7) Apr (0) May (2) Jun (4) Jul (11) Aug (9) Sep (14) Oct (14) Nov (11) Dec (1).

Population Size: 1-10 (13) 10-100 (36) >100 (20) Unspecified (12).

Conservation Status: Restricted to road verge (4%) Not (86%) Unspecified (10%). In conservation reserve (72%) Not (28%) Unspecified (0%).

Tree/Shrub: Tree form (0%) Shrub form (98%) Unspecified (2%).

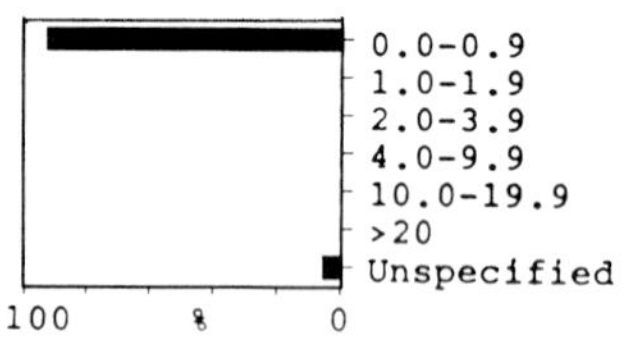

New Shoot Growth

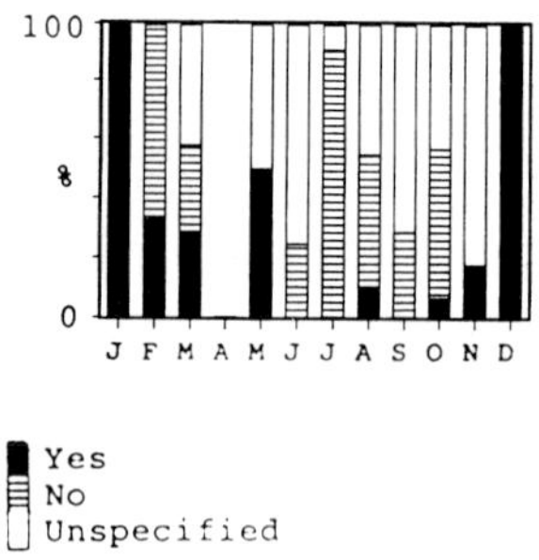

HABITAT

Within 2 km of coast: Yes (1%) No (99%) unspecified (0%)

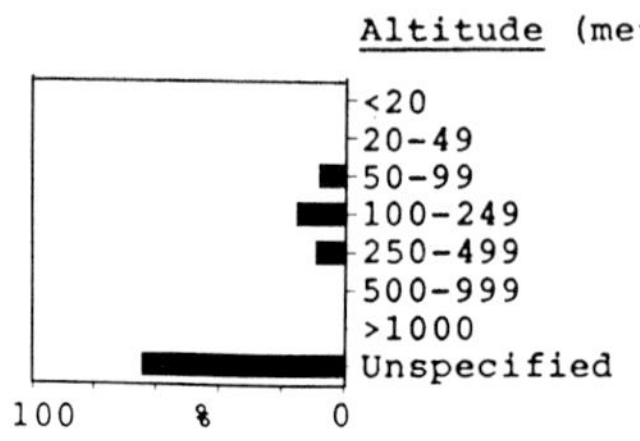

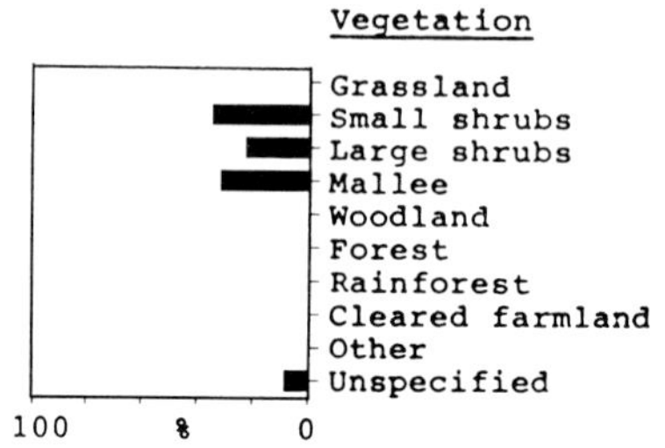

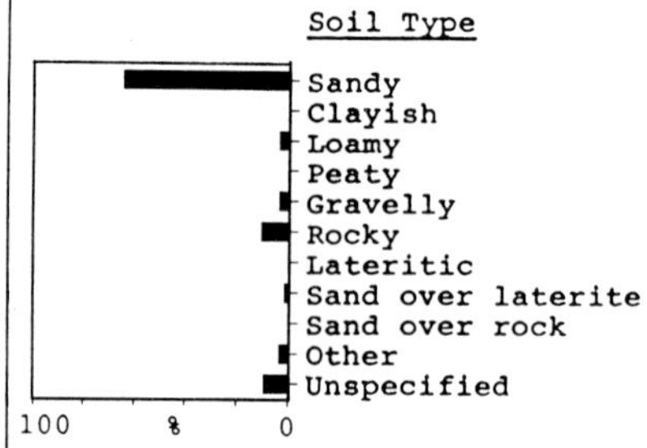

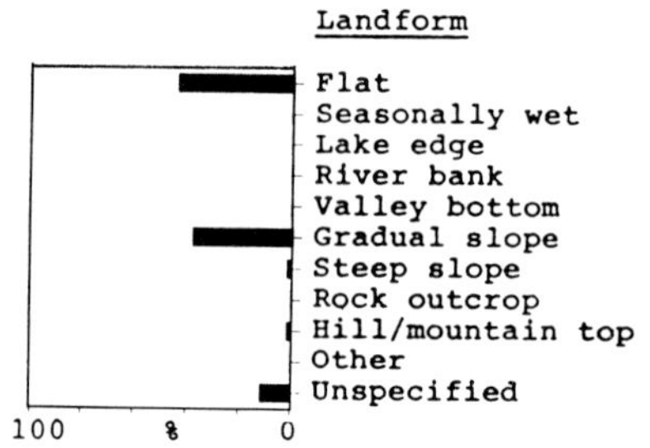

Response to Fire: 2 records

Growth response: Ground resprout (100%)

Flowering response: Number of records where flowering had occurred after fire (2). Median reported time to flowering (13-24 months); minimum reported time to flowering (1-12 months). Number of records where flowering had not occurred after fire (0).

Flowering

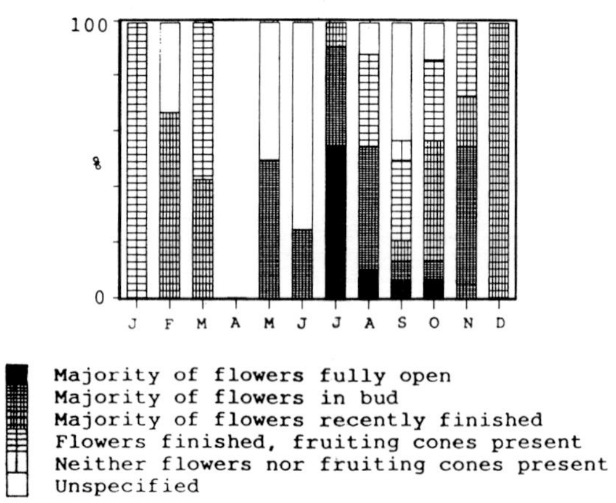

Possible pollinators

Pollinator type 6 records: mammals (17%) no pollinator observed (83%).

Specific pollinator 2 records: Honey-possum (1) cockroach (1).

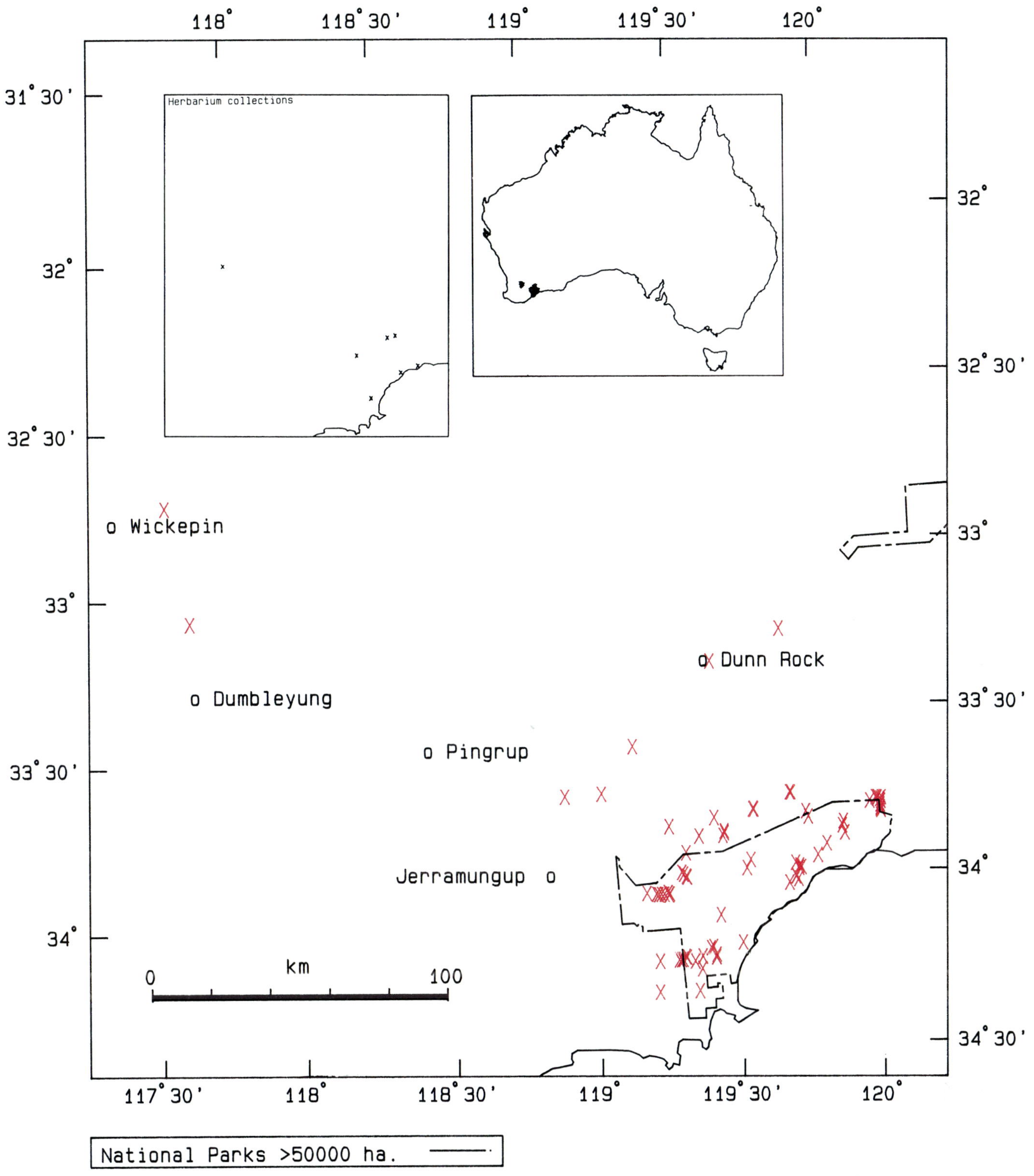

DISCUSSION

B. gardneri var. hiemalis occurs further east than either var. brevidentata or var. gardneri, centred mainly on the Fitzgerald River National Park with outlying populations east of Wickepin, and on Dongolocking, Lake Magenta, Dunn Rock and Pallarup Lake Nature Reserves. The latter two records represent range extensions of up to 60 km. In its southern populations, the variety appears to occur further west than herbarium collections indicated. In fact, Atlas records suggest an overlap between var. hiemalis and the easternmost populations of var. gardneri. However, these two varieties are very similar in appearance, their main difference being in flower colour and time of flowering. George (1981) stated that June-August is the flowering period of var. hiemalis and September-November that of var. gardneri. Flowering data recorded by Atlas contributors for var. hiemalis suggest a peak flowering time in July but with some plants flowering as late as October, whilst in November, 6 out of 11 recordings were stated to be in bud. This suggests that either var. hiemalis flowers later than previously thought, or that some plants recorded as var. hiemalis for the Atlas were wrongly identified. An investigation is needed of those populations near the south-western boundary of the Fitzgerald River National Park to determine whether the two varieties do overlap in their distribution. Also, further studies on the flowering time of var. hiemalis are required.

B. gardneri var. hiemalis occurs mainly in sandy soils on flat to undulating country. Surrounding vegetation is usually small shrubs often with emergent scattered mallee eucalypts.

BANKSIA GOODII R.Brown (1830)

Good's Banksia

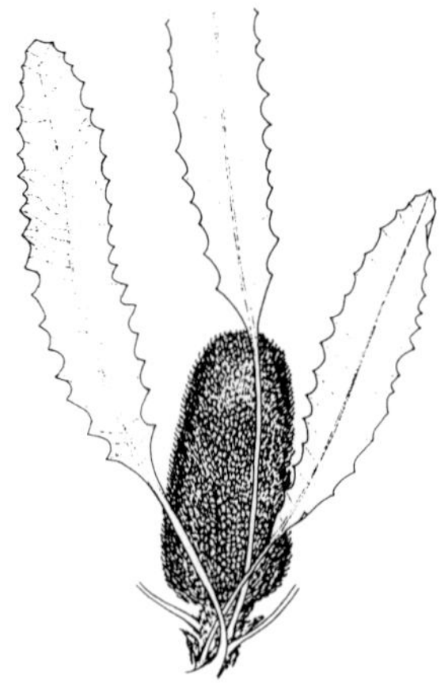

17 RECORDS: Jan (1) Feb (1) Mar (0) Apr (2) May (0) Jun (0) Jul (0) Aug (2) Sep (1) Oct (2) Nov (6) Dec (1).

Population Size: 1-10 (6) 10-100 (6) >100 (5) Unspecified (0).

Conservation Status: Restricted to road verge (41%) Not (53%) Unspecified (6%). In conservation reserve (18%) Not (82%) Unspecified (0%).

Tree/Shrub: Tree form (0%) Shrub form (94%) Unspecified (6%).

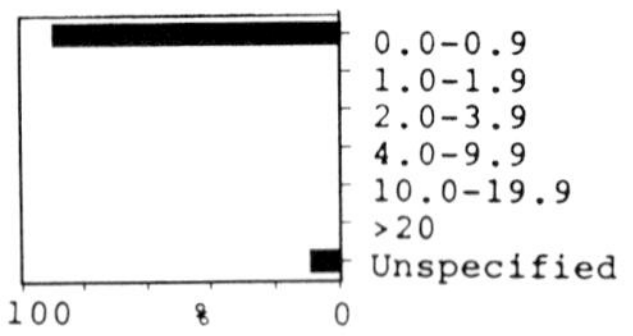

New Shoot Growth

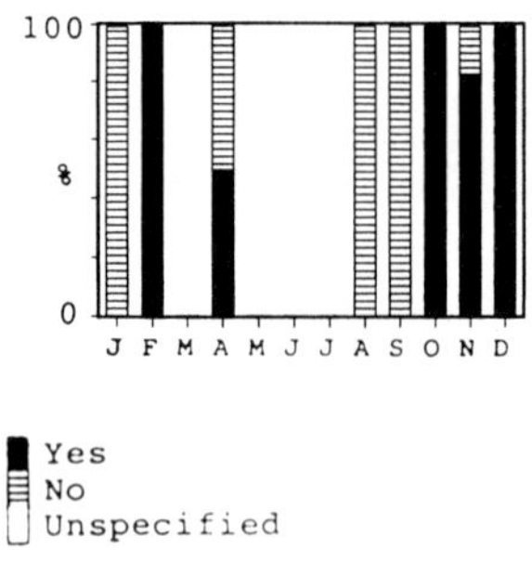

HABITAT

Within 2 km of coast: Yes (0%) No (94%) Unspecified (6%).

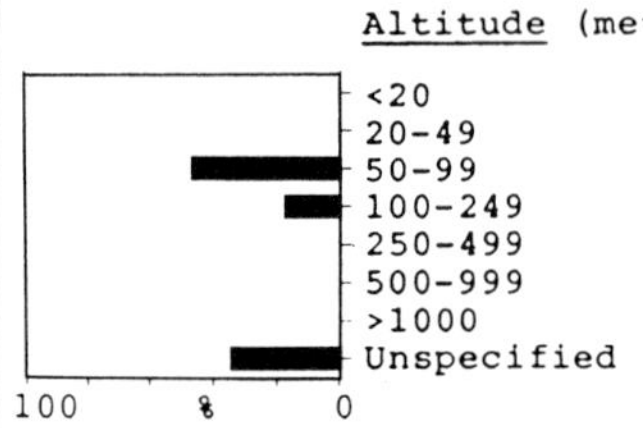

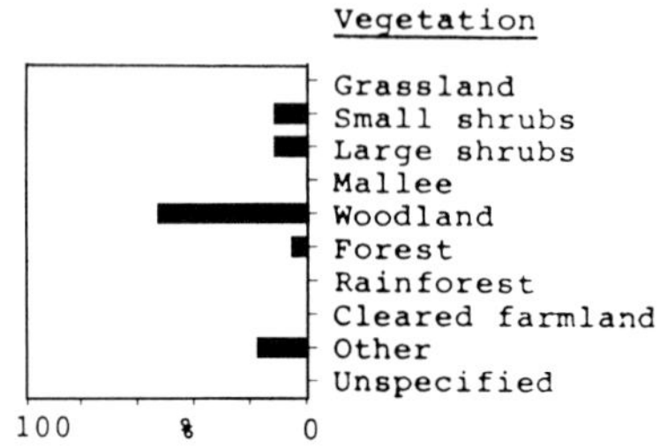

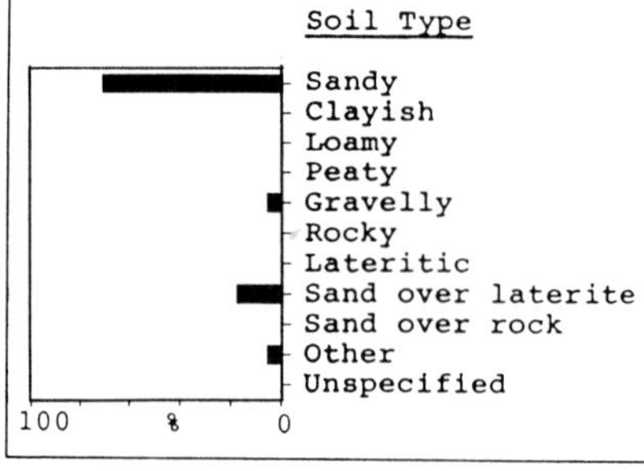

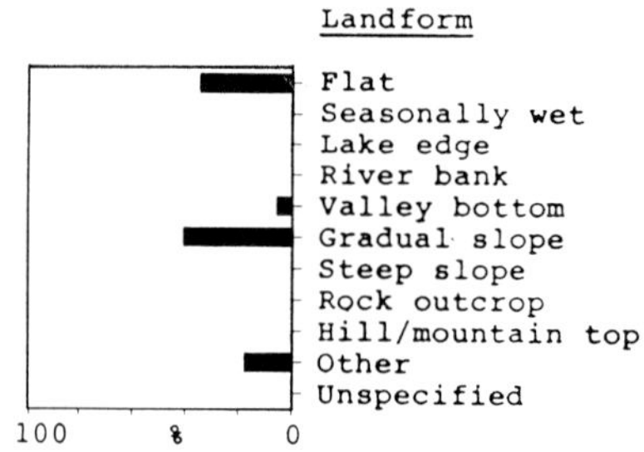

Response to Fire: 5 records

Growth response: Ground resprout (40%) killed, new seedlings (20%) unspecified (40%).

Flowering response: Number of records where flowering had occurred after fire (3). Median reported time to flowering (13-24 months); minimum reported time to flowering (1-12 months). Number of records where flowering had not occurred after fire (1). Maximum reported time to no flowering (13-24 months).

Flowering

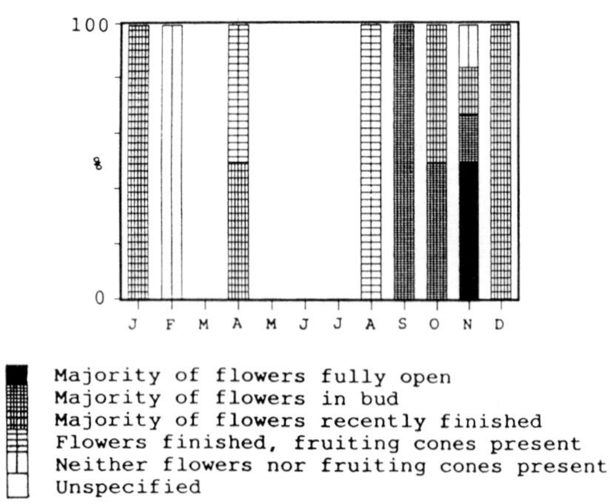

Possible pollinators

Pollinator type 3 records: no pollinator observed (100%).

Specific pollinator 0 records.

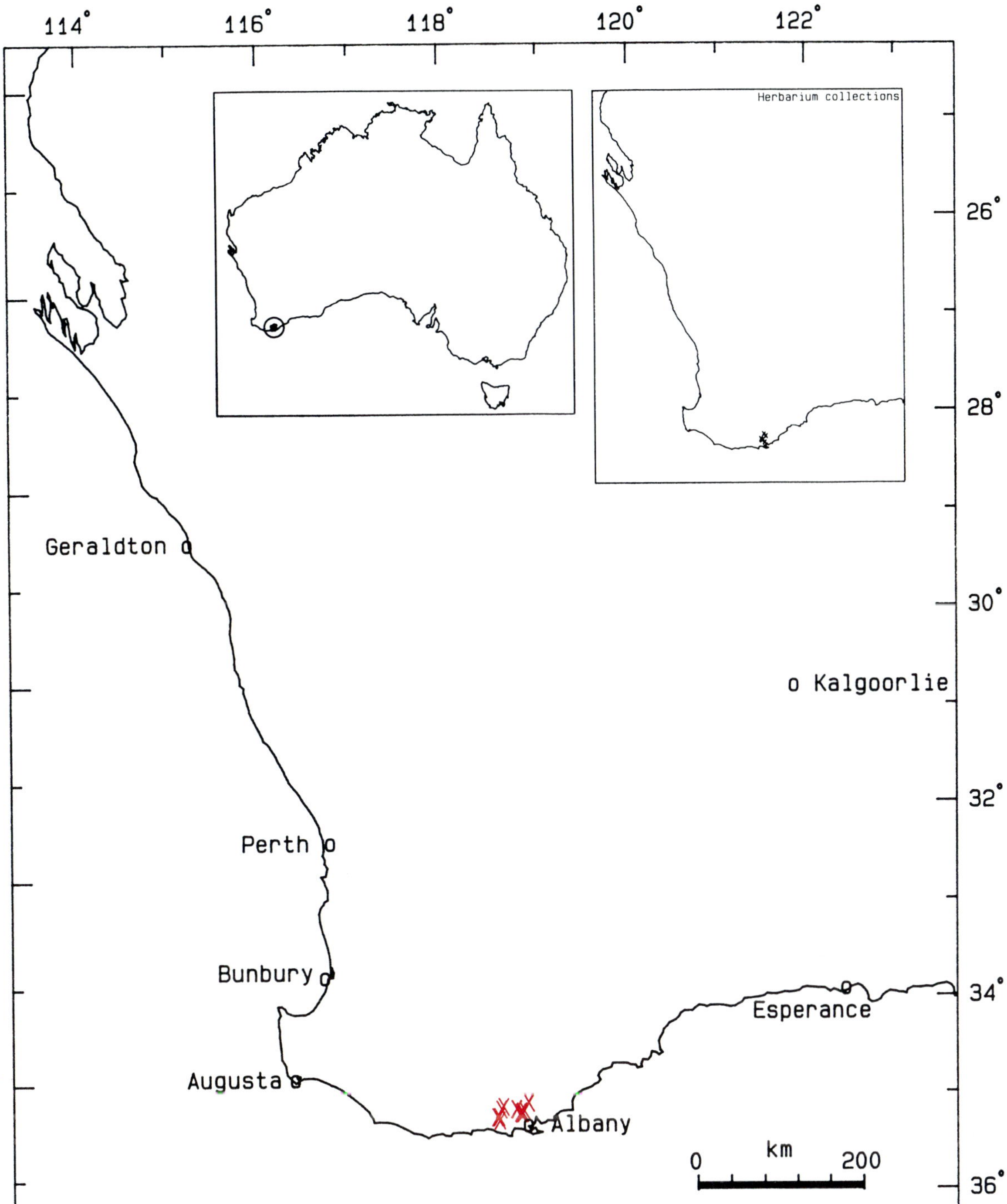

DISCUSSION

B. goodii is a rare banksia previously recorded only between Albany and the Porongurup Range. Atlas records have confirmed and extended this distribution with five new sites from west of the Albany Highway. Of these, two have populations of between 10 and 100 plants, and three of less than 10 plants. East of the Albany Highway there appear to be at least five populations, with three having in excess of 100 plants. The remaining 7 records are duplications of already recorded sites. Of the ten recorded populations of B. goodii, four are on road verges and only one is in a conservation reserve. The latter has approximately 300 plants. It is likely that B. goodii used to be more widespread but the area in which it grows is now largely cleared for agriculture.

B. goodii is a prostrate shrub with leaves and flower spikes at ground level. It grows in sand or sand over laterite, generally in woodland or forest of Eucalyptus marginata and Casuarina fraseriana. The two records of shrubland as surrounding vegetation both include stunted E. marginata. One site is regrowing after clearing for agriculture, the other is possibly post-fire regrowth. Landform is usually flat to gently undulating. The flowering season is short being mainly in November though plants were reported "in bud" as early as September. New shoot growth appears to be in late spring and summer. The species is thought to be lignotuberous and regenerates well after fire. A single Atlas record of plants being killed by fire and regenerating from seed should be investigated.

BANKSIA GRANDIS Willdenow (1798)

Mangite, Bull Banksia, Giant Banksia

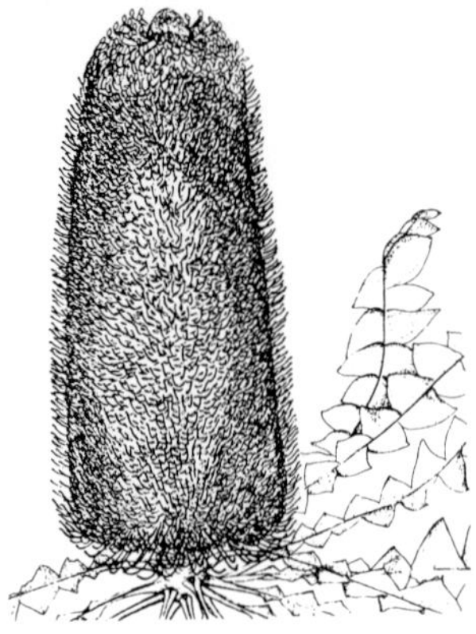

1952 RECORDS: Jan (293) Feb (87) Mar (302) Apr (222) May (125) Jun (136) Jul (72) Aug (129) Sep (182) Oct (126) Nov (143) Dec (133).

Population Size: 1-10 (445) 10-100 (928) >100 (569) Unspecified (10).

Conservation Status: Restricted to road verge (10%) Not (87%) Unspecified (3%). In conservation reserve (19%) Not (76%) Unspecified (5%).

Tree/Shrub: Tree form (90%) Shrub form (9%) Unspecified (1%).

Height (metres)

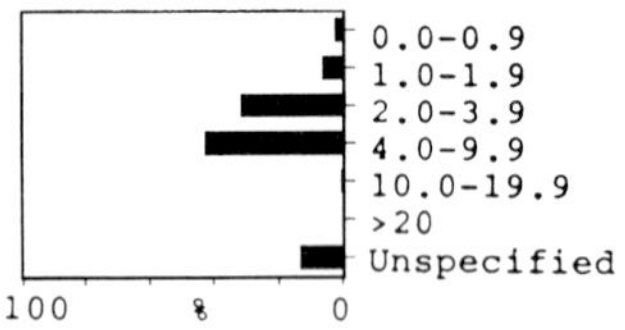

New Shoot Growth

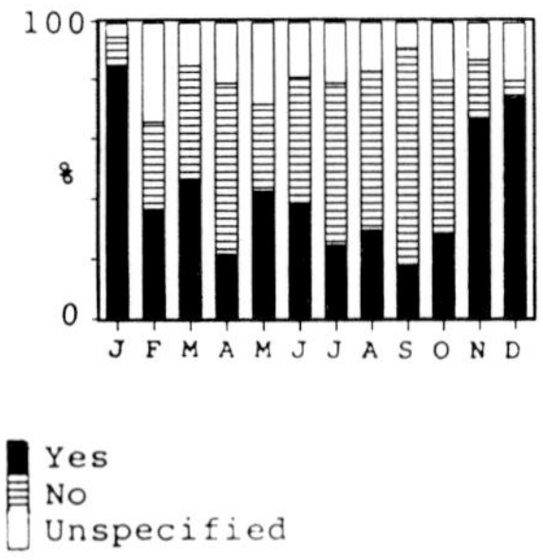

HABITAT

Within 2 km of coast: Yes (7%) No (92%) Unspecified (1%).

Altitude (metres)

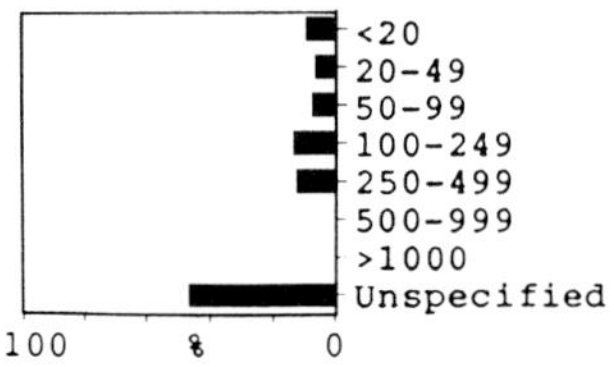

Vegetation

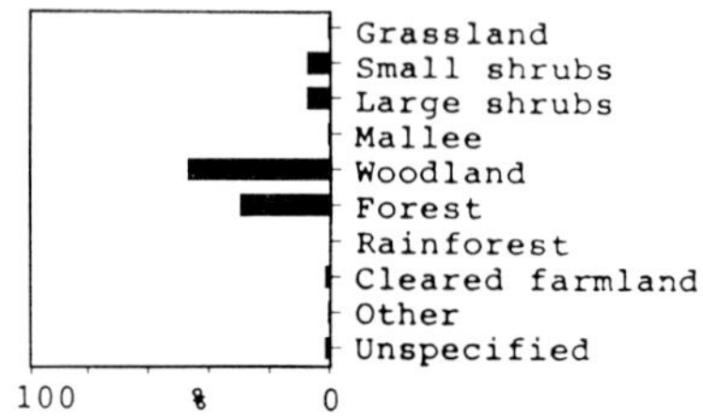

Soil Type

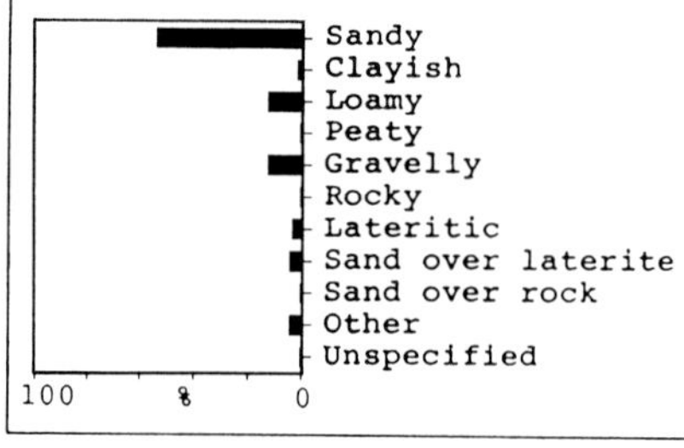

Landform

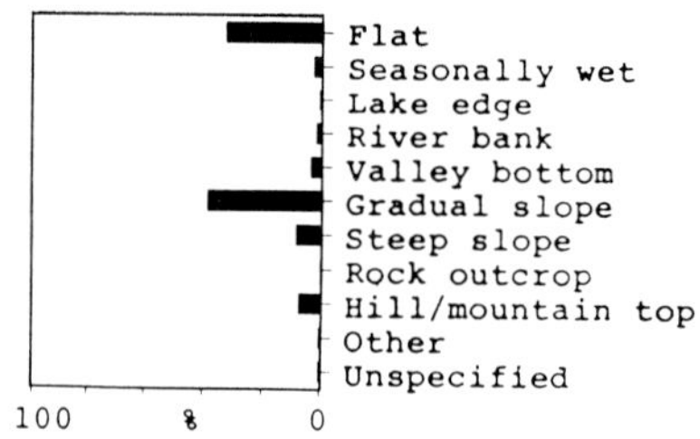

Response to Fire: 385 records

Growth response: Ground resprout (9%) ground resprout and killed, new seedlings (2%) ground resprout and trunk resprout (16%) killed, new seedlings (5%) killed, new seedlings and trunk resprout (2%) trunk resprout (25%) killed, no seedlings and trunk resprout (1%) killed, no seedlings and killed, new seedlings (1%) unspecified (40%).

Flowering response: Number of records where flowering had occurred after fire (197). Median reported time to flowering (1-12 months); minimum reported time to flowering (<1 month). Number of records where flowering had not occurred after fire (76). Median reported time to no flowering (13-24 months); maximum reported time to no flowering (61-72 months).

Flowering

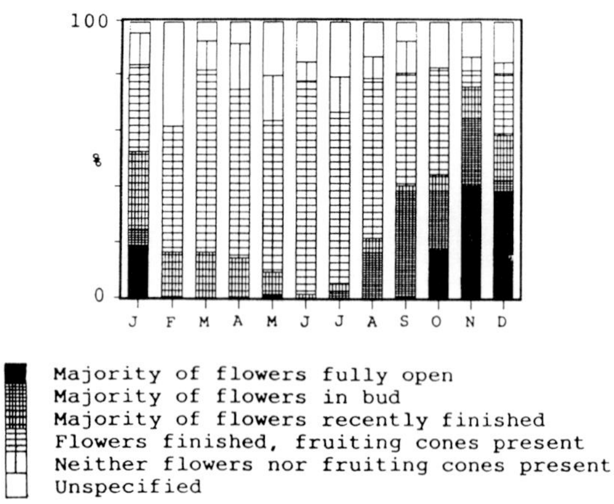

Possible pollinators

Pollinator type 914 records: bird (6%) butteflies/moths (0.4%) bees, wasps, ants (7%) flies (1%) beetles (0.3%) no pollinator observed (85%).

Specific pollinator 123 records: honeyeater (7) New Holland Honeyeater (13) Spiny-cheeked Honeyeater (1) wattlebird (1) Red Wattlebird (13) Western Spinebill (6) Red-capped Parrot (2) Port Lincoln Parrot (4) Purple-crowned Lorikeet (3) Rufous Treecreeper (1) ant (19) bee (43) native bee (3) European honey bee (6) blowfly (1).

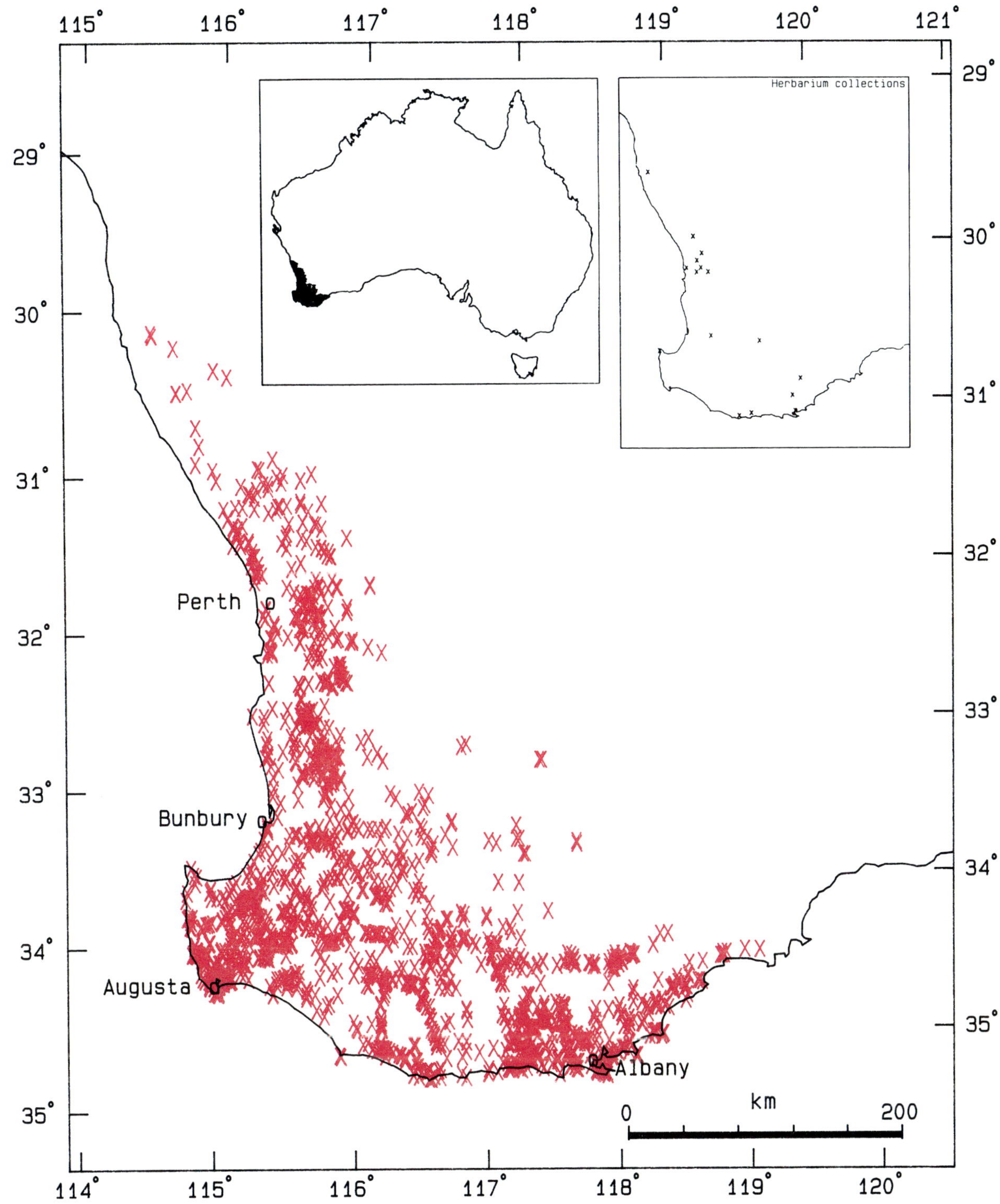

DISCUSSION

The 1952 records of *B. grandis* show it to be a common species on both the coastal plain and Darling Plateau extending from Mt Lesueur southwards to Augusta and eastwards almost to Bremer Bay. Clearly, the relatively few herbarium collections for this species provide an inadequate picture of both its range and abundance. The most inland sight records come from Badgebup and from Dongolocking Nature Reserve. The species is also recorded from Greaves Hill north-east of the Stirling Range, from Narrogin, from near Boyagarring Hill on the Brookton Highway, and from north of Bold Hill between Badgingarra and Moora.

B. grandis usually grows as a small tree up to 10 m in height. However, along the south coast it occurs as a low spreading shrub. Interestingly, seed from the south coast produces plants which retain their low spreading habit when grown elsewhere. On the western coastal plain, *B. grandis* grows in deep brown-grey sands in woodlands of *Eucalyptus marginata*, *B. attenuata*, *B. menziesii* and *B. ilicifolia*. On the Darling Plateau, it is a common understorey tree in forest and woodland of *E. marginata* and *E. calophylla* where the soil types are lateritic gravels and loams. In the extreme south-west it is an understorey tree in the tall forests of *E. diversicolor*. Along the south coast, the species is an emergent shrub above low coastal heath on generally sandy soils.

B. grandis flowers and produces new shoot growth from August to January. The conspicuous flowers are frequently visited by honey-eating birds, and also by bees and ants. Following fire, *B. grandis* resprouts from both epicormic buds and lignotuber and was observed to then flower 1-2 years later on average. Occasionally, mature plants are killed by fire and regenerate from seed.

BANKSIA GROSSA A.S.George (1981)

53 RECORDS: Jan (2) Feb (0) Mar (2) Apr (11) May (12) Jun (2) Jul (3) Aug (6) Sep (6) Oct (4) Nov (0) Dec (5).

Population Size: 1-10 (5) 10-100 (22) >100 (24) Unspecified (2).

Conservation Status: Restricted to road verge (42%) Not (58%) Unspecified (0%). In conservation reserve (19%) Not (81%) Unspecified (0%).

Tree/Shrub: Tree form (0%) Shrub form (98%) Unspecified (2%).

Height (metres)

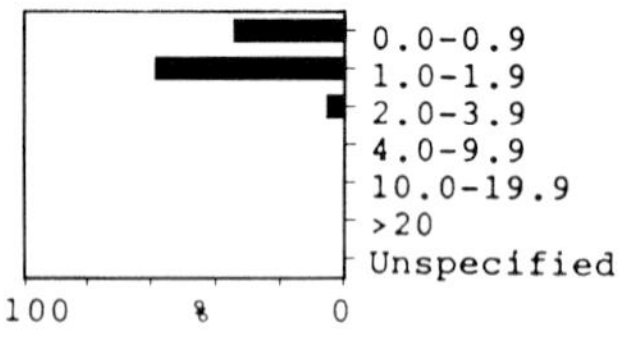

New Shoot Growth

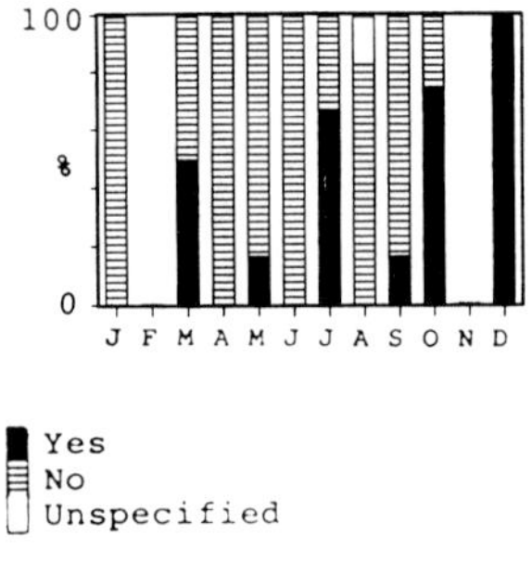

HABITAT

Within 2 km of coast: Yes (0%) No (98%) Unspecified (2%).

Altitude (metres)

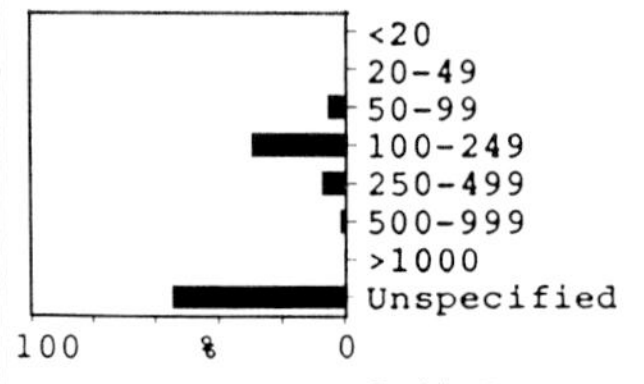

Vegetation

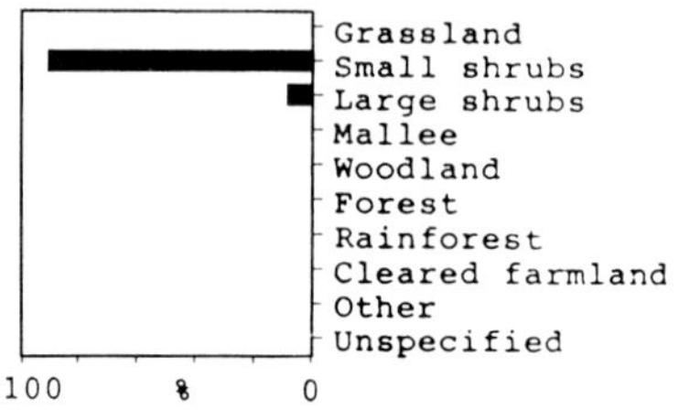

Soil Type

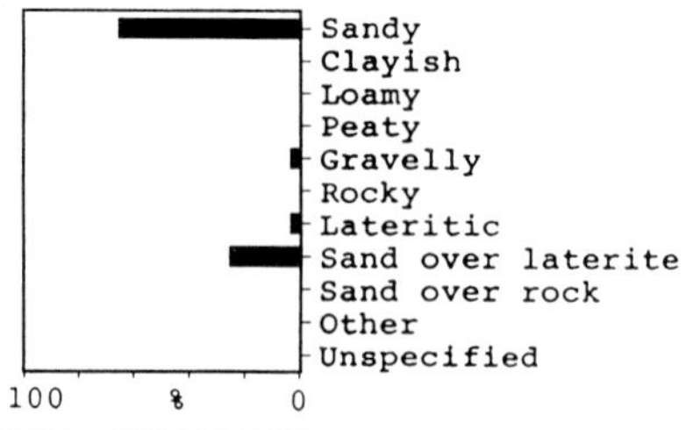

Landform

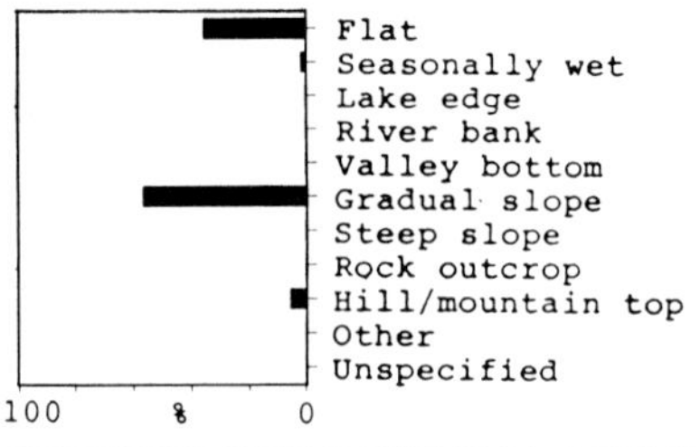

Response to Fire: 0 records

Flowering

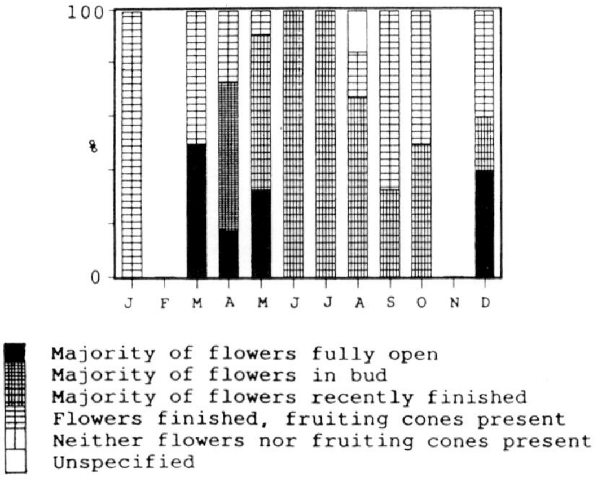

Possible pollinators

Pollinator type 14 records: no pollinator observed (100%).

Specific pollinator 0 records.

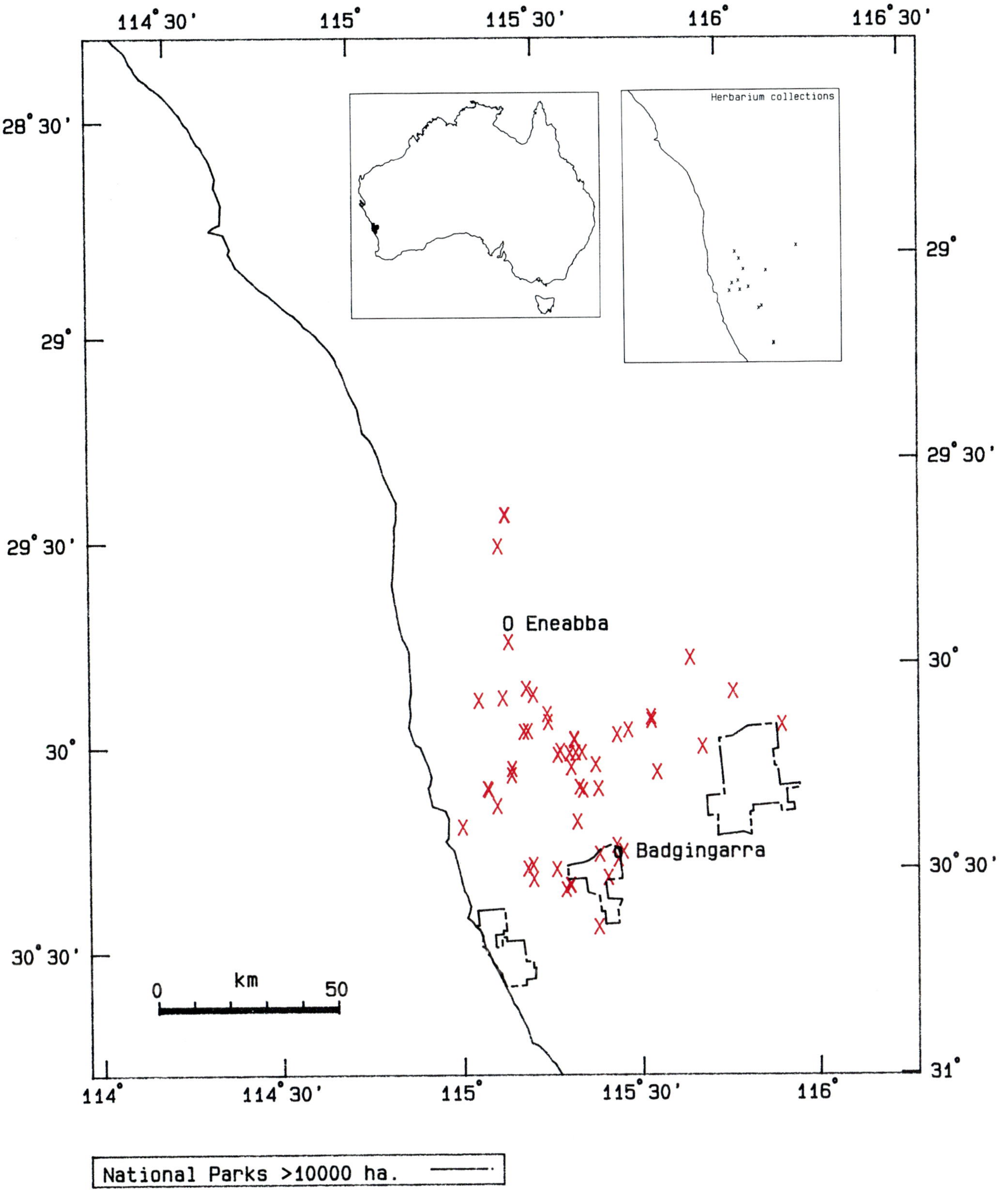

DISCUSSION

Western Australian Herbarium records showed that B. grossa occurred between Eneabba and Badgingarra with outlying populations at Coorow Reserve (C. Chapman's 1967 collection) and Cataby Brook, north of Regans Ford (A.S. George's 1977 collection). Atlas records have confirmed this general distribution though Chapman and George's collections were not refound and recorded. (George, pers. comm., confirmed that the Cataby Brook population was still extant in October 1986, after Atlas contributions had ceased.) However, contributors did extend the range northwards to an area west of Yandanooka Hill. Inland, there were additional sightings from the north-east corner of Watheroo National Park, from Alexander Morrison National Park and from areas south-west of Coorow. Although less than 20% of populations are on conservation reserves and more than 40% are restricted to road verges, the species is well distributed throughout its range and does not appear to be under any immediate threat.

B. grossa is typically a shrub up to 1.5 m in height. It grows in sands and sand over laterite on flat to gently undulating land, occasionally on the tops of lateritic rises. Surrounding vegetation is predominantly low heath. Flowering is mainly in autumn, though some flowers were recorded as early as December. New shoot growth appears to be in spring and early summer. There were no observations of either pollinators or post-fire responses, but George (1981) recorded the species as lignotuberous.

BANKSIA HOOKERIANA Meissner (1855)

Hooker's Banksia

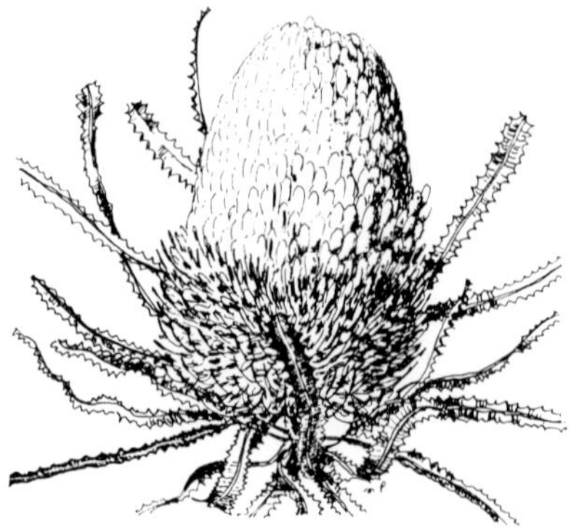

36 RECORDS: Jan (1) Feb (1) Mar (2) Apr (0) May (3) Jun (2) Jul (5) Aug (3) Sep (5) Oct (3) Nov (1) Dec (7).

Population Size: 1-10 (3) 10-100 (5) >100 (26) Unspecified (2).

Conservation Status: Restricted to road verge (3%) Not (97%) Unspecified (0%). In conservation reserve (11%) Not (83%) Unspecified (6%).

Tree/Shrub: Tree form (3%) Shrub form (94%) Unspecified (3%).

Height (metres)

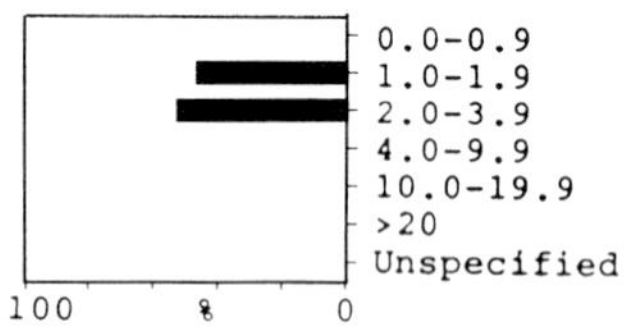

New Shoot Growth

HABITAT

Within 2 km of coast: Yes (0%) No (100%) Unspecified (0%).

Altitude (metres)

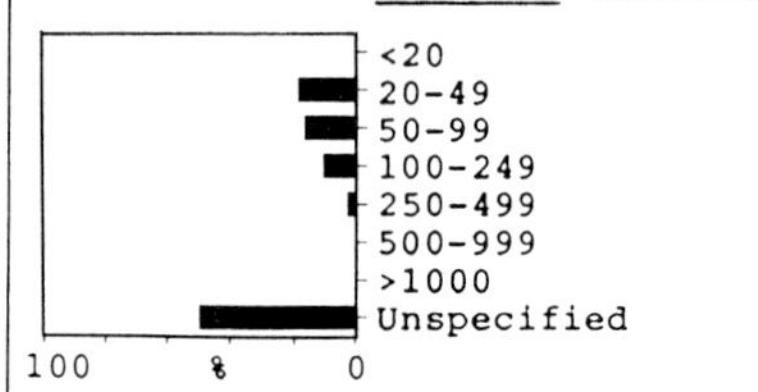

Vegetation

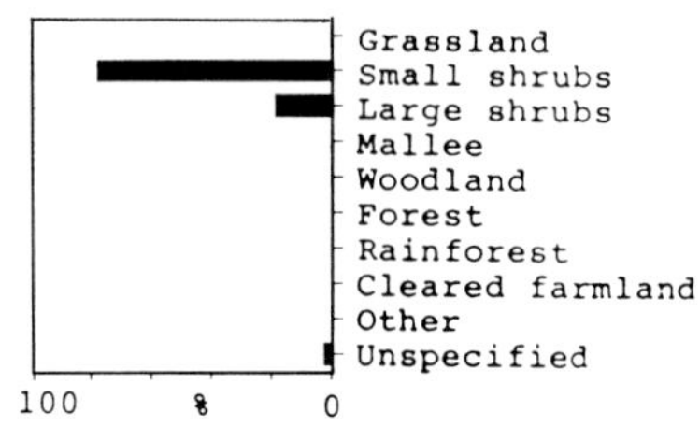

Soil Type

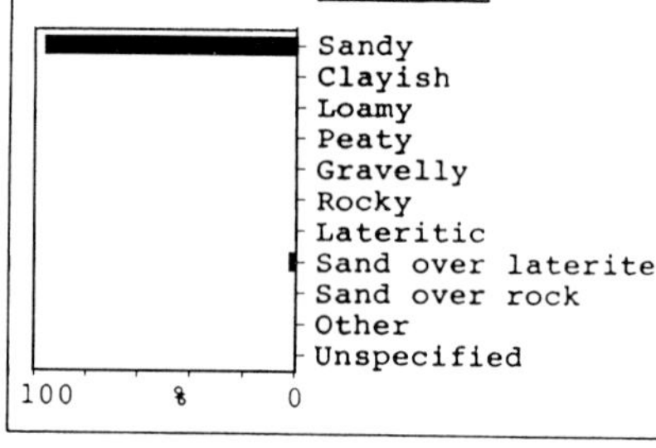

Landform

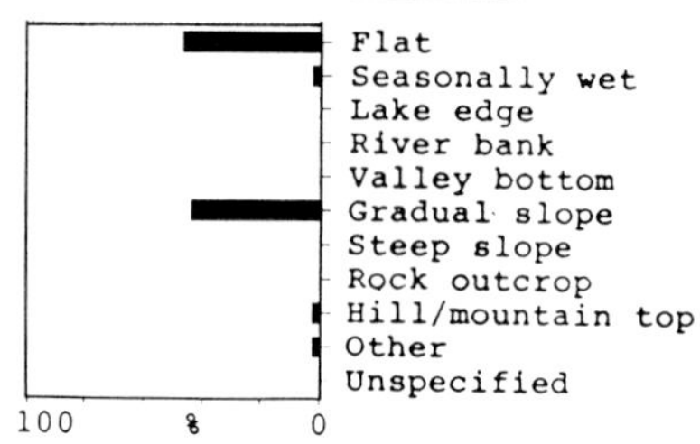

Response to Fire: 15 records

Growth response: Ground resprout (13%) ground resprout and killed, new seedlings (13%) killed, new seedlings (33%) killed, no seedlings (7%) unspecified (33%).

Flowering response: Number of records where flowering had occurred after fire (9). Median reported time to flowering (13-24 months); minimum reported time to flowering (1-12 months). Number of records where flowering had not occurred after fire (4). Median reported time to no flowering (13-24 months); maximum reported time to no flowering (13-24 months).

Flowering

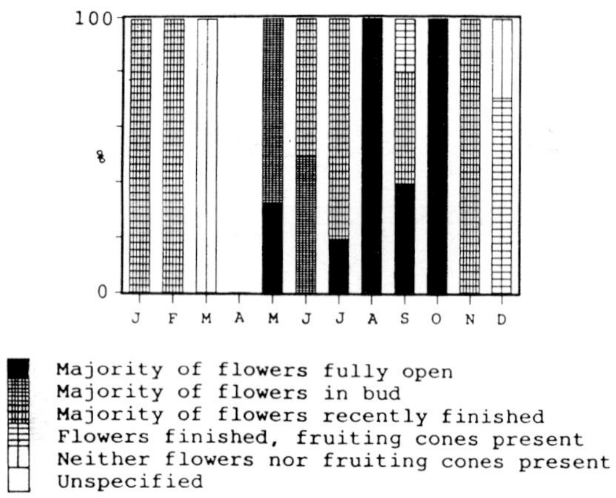

Possible pollinators

Pollinator type 12 records: bird (33%) bees, wasps, ants (25%) no pollinator observed (42%).

Specific pollinator 7 records: New Holland Honeyeater (3) Brown Honeyeater (1) ant (2) bee (1).

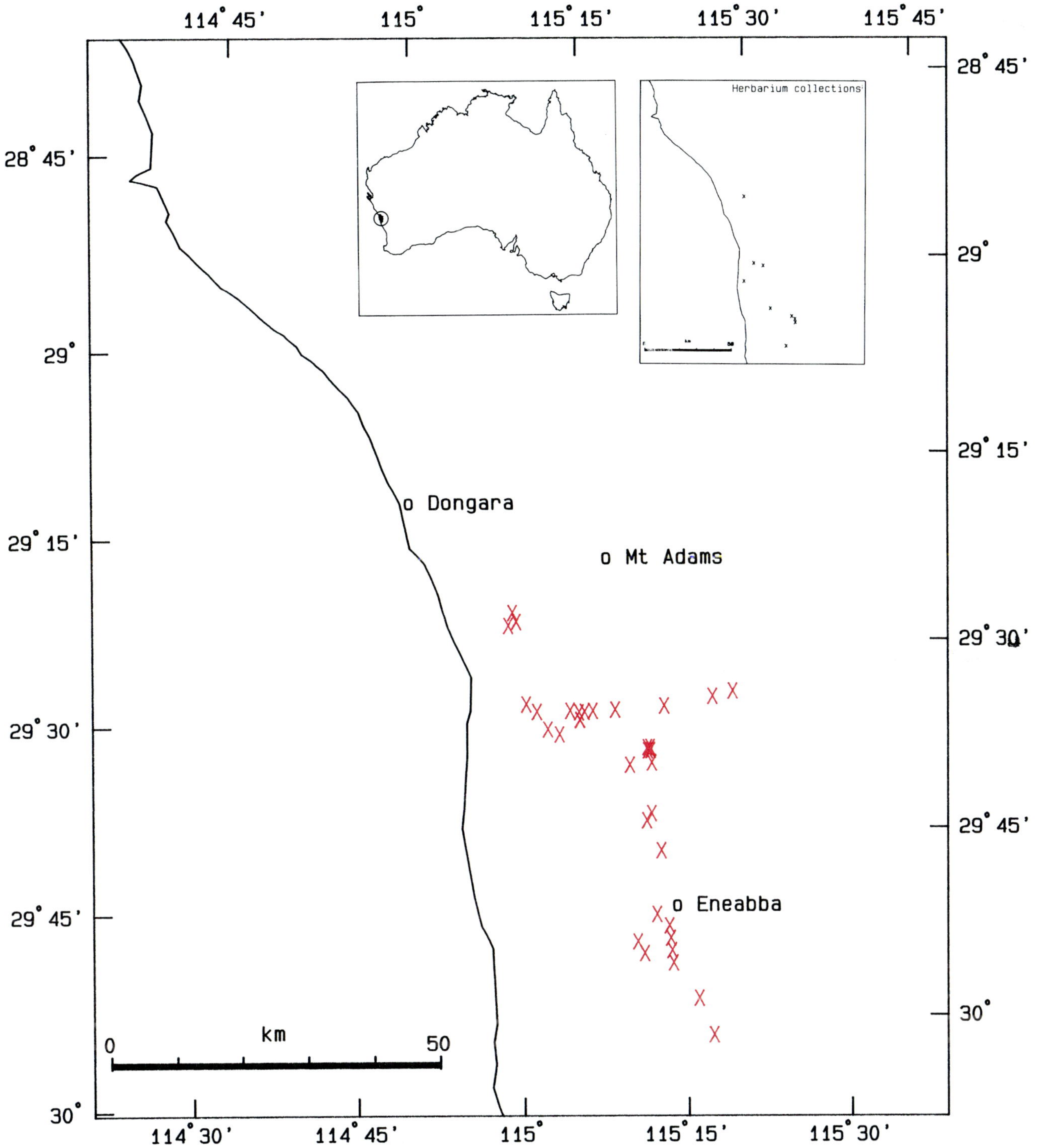

DISCUSSION

B. hookeriana has a restricted range chiefly between Arrowsmith and up to 20 km south of Eneabba. K. Newbey's 1965 collection from east of Dongara was not recorded but new sightings were noted from an area about 15 km east of White Point and from near Yandanooka Hill. B. hookeriana is locally common with most populations having in excess of 100 plants. However, the area in which it occurs is being increasingly cleared for agriculture and mining development. It is also heavily used in the cut wildflower trade (Burgman and Hopper 1982), is highly susceptible to dieback disease, and is not well represented in conservation reserves. Its conservation status therefore needs monitoring.

B. hookeriana is typically a shrub between 1 and 4 m in height growing on flat or gently undulating sandplain amongst small and large shrubs. Flowering is from late autumn to spring when the conspicuous flowers are attractive to honey-eating birds, bees and ants. New shoot growth appears to be in spring and summer. The species was previously thought to be killed by fire and to regenerate from seed. This is confirmed by some Atlas records, but several records suggest that some plants may be lignotuberous. Possible hybrids between B. hookeriana and B. prionotes have been reported from the Lake Indoon and Mt Adams areas. The latter is also the site of a presumed hybrid between B. hookeriana and B. attenuata. Details of both these hybrids appear in Chapter 3.

BANKSIA ILICIFOLIA R.Brown (1810)

Holly-leaved Banksia

557 RECORDS: Jan (56) Feb (17) Mar (135) Apr (64) May (36) Jun (16) Jul (20) Aug (49) Sep (39) Oct (58) Nov (43) Dec (24).

Population Size: 1-10 (122) 10-100 (293) >100 (134) Unspecified (8).

Conservation Status: Restricted to road verge (3%) Not (92%) Unspecified (5%). In conservation reserve (32%) Not (64%) Unspecified (4%).

Tree/Shrub: Tree form (96%) Shrub form (3%) Unspecified (1%).

Height (metres)

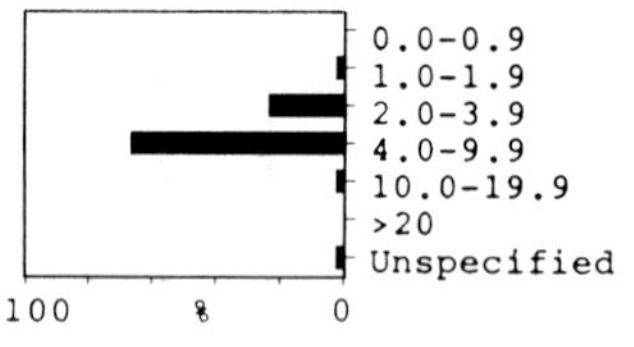

New Shoot Growth

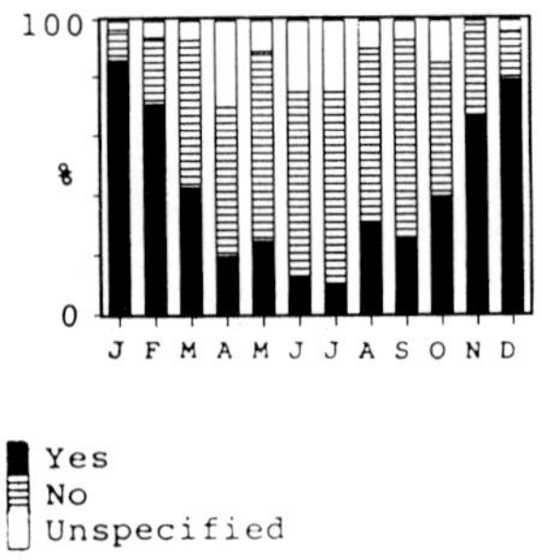

HABITAT

Within 2 km of coast: Yes (15%) No (83%) Unspecified (1%).

Altitude (metres)

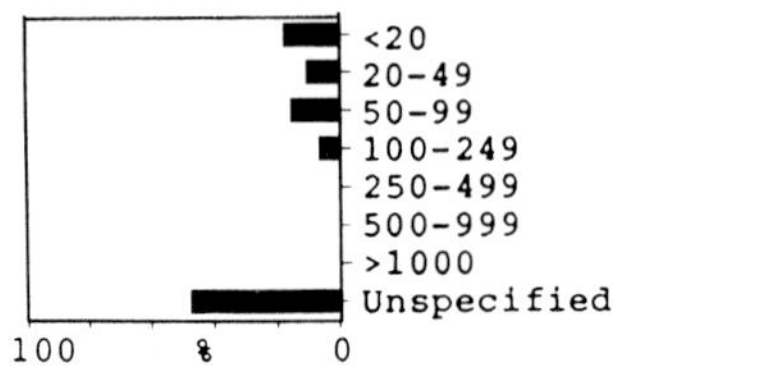

Vegetation

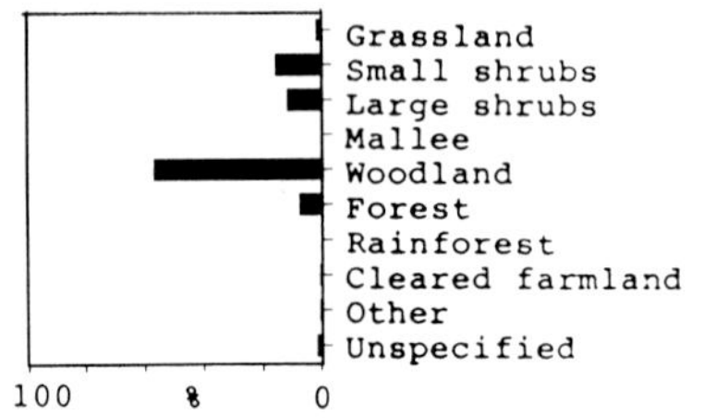

Soil Type

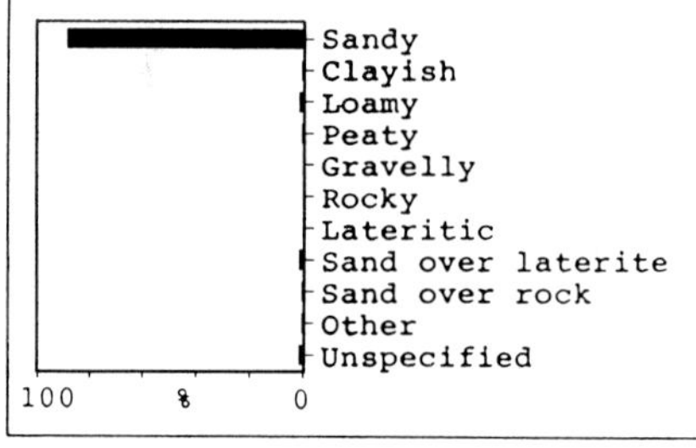

Landform

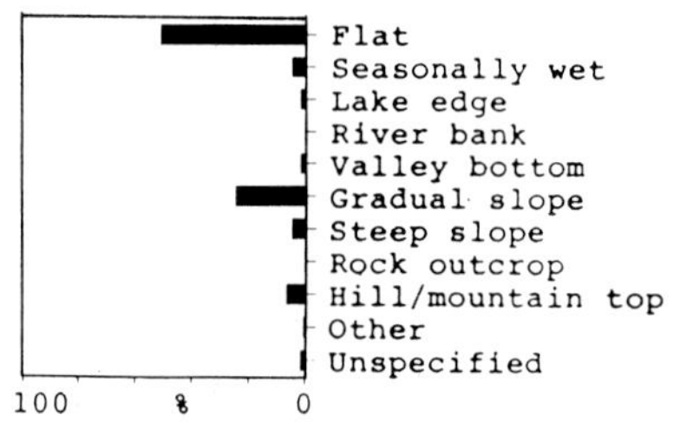

Response to Fire: 99 records

Growth response: Ground resprout (4%) ground resprout and killed, new seedlings (2%) ground resprout and trunk resprout (15%) killed, new seedlings (4%) trunk resprout (31%) killed, no seedlings and trunk resprout (1%) unspecified (42%).

Flowering response: Number of records where flowering had occurred after fire (65). Median reported time to flowering (13-24 months); minimum reported time to flowering (<1 month). Number of records where flowering had not occurred after fire (18). Median reported time to no flowering (13-24 months); maximum reported time to no flowering (37-48 months).

Flowering

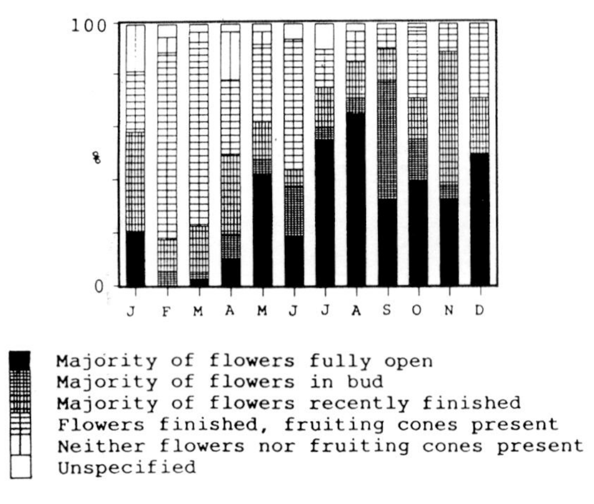

Possible pollinators

Pollinator type 255 records: bird (16%) bees, wasps, ants (10%) flies (1%) beetles (2%) no pollinator observed (71%) .

Specific pollinator 71 records: honeyeater (6) New Holland Honeyeater (20) Brown Honeyeater (5) White-cheeked Honeyeater (1) Singing Honeyeater (2) Silvereye (1) Red Wattlebird (1) Western Spinebill (8) Port Lincoln Parrot (4) ant (4) bee (16) native bee (1) European honey bee (2).

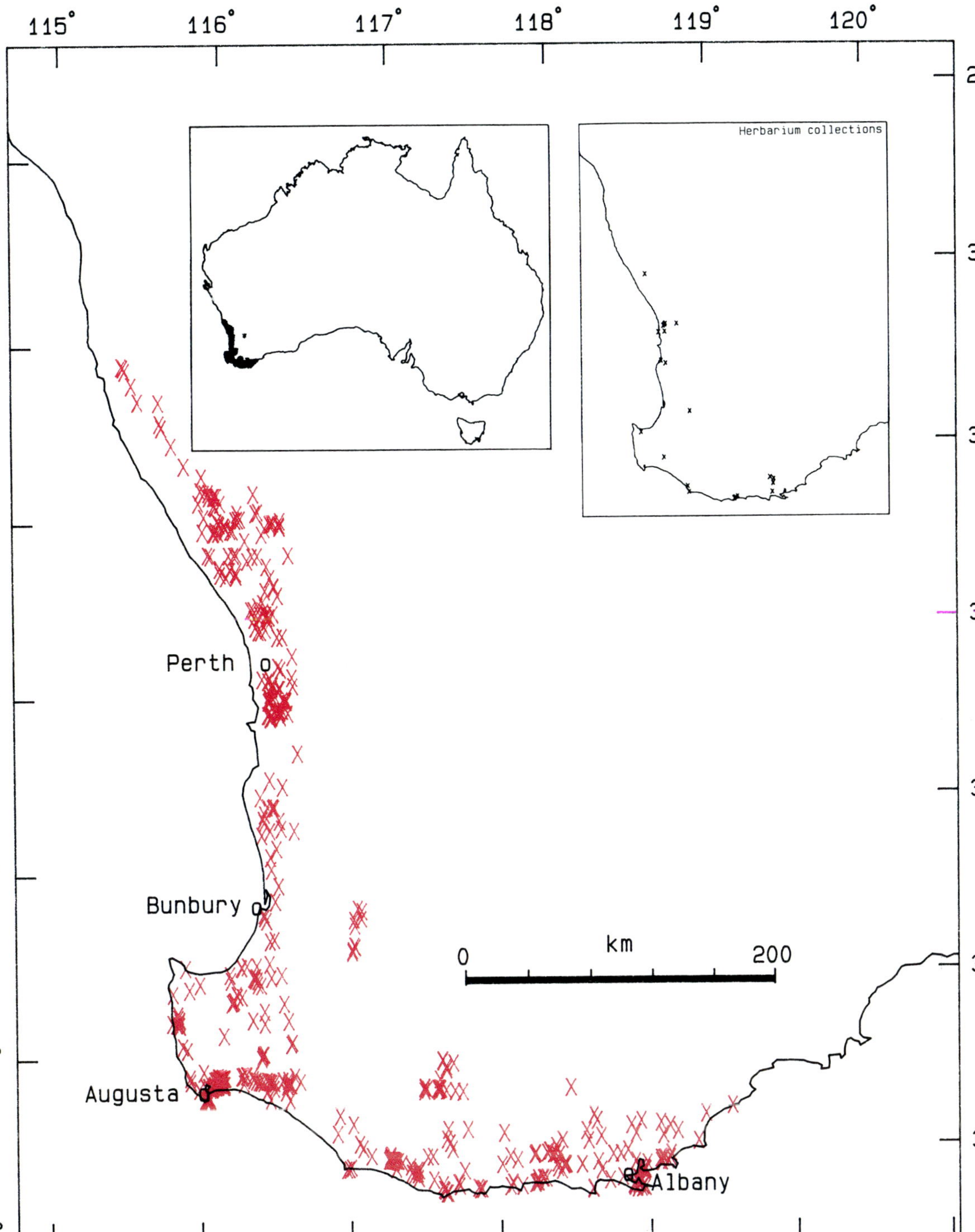

DISCUSSION

The 557 records of B. ilicifolia show it to be relatively common on the sandy coastal plain from Mt Lesueur south to Augusta and east as far as the Cordinup River between Albany and Bremer Bay. Herbarium records had indicated an eastern limit at Albany. Therefore, Atlas records have extended the range by about 90 km eastwards. Heavy soils constitute a significant barrier to the distribution of B. ilicifolia. Thus it does not appear east of the Darling Scarp except in the Collie area east of Bunbury and in the Tonebridge-Lake Muir area east of Manjimup, where there are re-occurrences of sandplain. Along the south coast it is also found mainly on sandy soils. North-west of Albany there is an outlying record from Sheepwash Creek Nature Reserve in the Narrikup area.

B. ilicifolia is typically an erect tree up to 10 m in height growing in open woodland with such species as Banksia attenuata, B. menziesii, and Eucalyptus marginata. Towards Mt Lesueur it forms small patches of woodland amongst shrubland. 15% of its populations are within 2 km of the coast. These are mainly along the south coast where the species occurs as a small spreading tree in coastal heath, sometimes on cliff tops. B. ilicifolia flowers intermittently in most months of the year with an apparent peak in July, August and September. The flowers are attractive to honey-eating birds with the New Holland Honeyeater being the most commonly observed. Bees were also frequent visitors. New shoot growth is predominantly in summer. Following fire, the species resprouts from both its epicormic buds and lignotuber. There is some evidence to suggest that very hot fires may kill mature plants.

BANKSIA INCANA A.S.George (1981)

38 RECORDS: Jan (4) Feb (1) Mar (7) Apr (4) May (8) Jun (0) Jul (2) Aug (3) Sep (6) Oct (0) Nov (1) Dec (2).

Population Size: 1-10 (7) 10-100 (16) >100 (14) Unspecified (1).

Conservation Status: Restricted to road verge (24%) Not (76%) Unspecified (0%). In conservation reserve (32%) Not (5%) Unspecified (0%).

Tree/Shrub: Tree form (0%) Shrub form (100%) Unspecified (0%).

Height (metres)

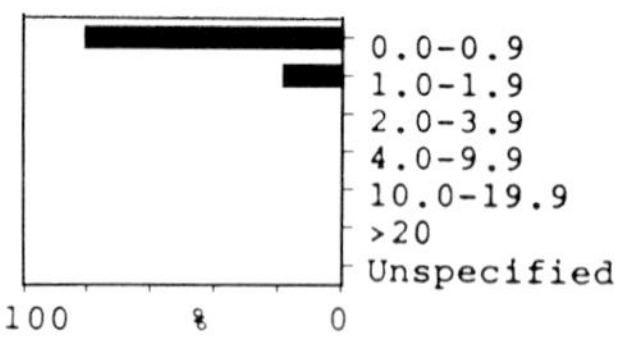

New Shoot Growth

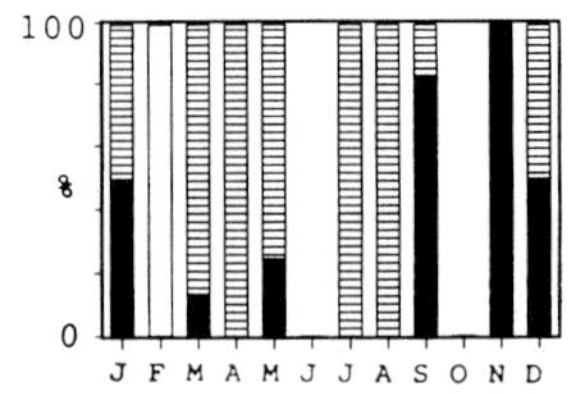

Yes
No
Unspecified

HABITAT

Within 2 km of coast: Yes (0%) No (100%) Unspecified (0%).

Altitude (metres)

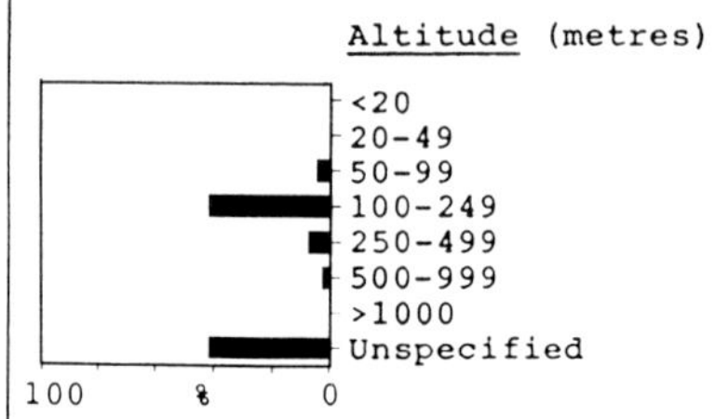

Vegetation

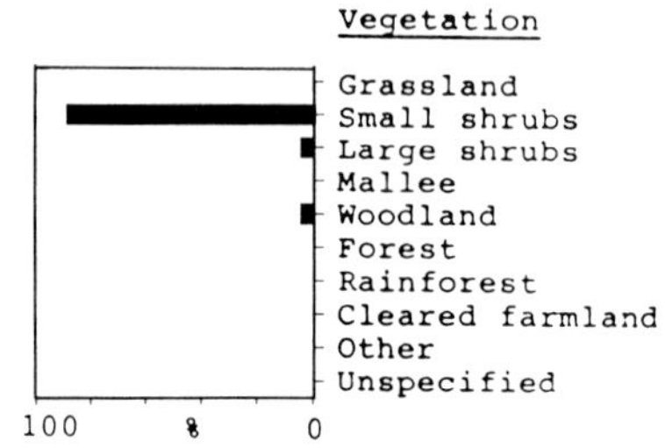

Soil Type

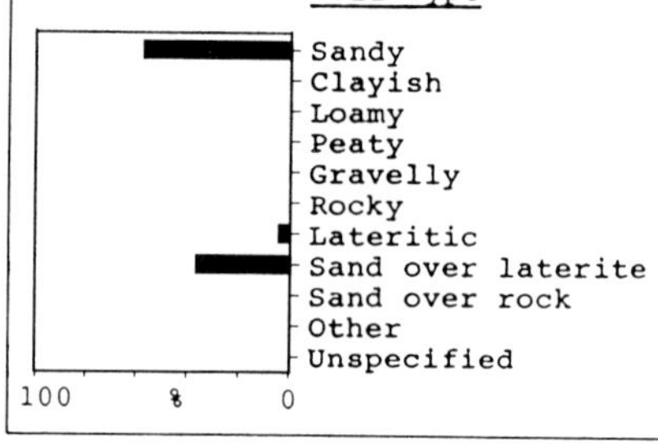

Landform

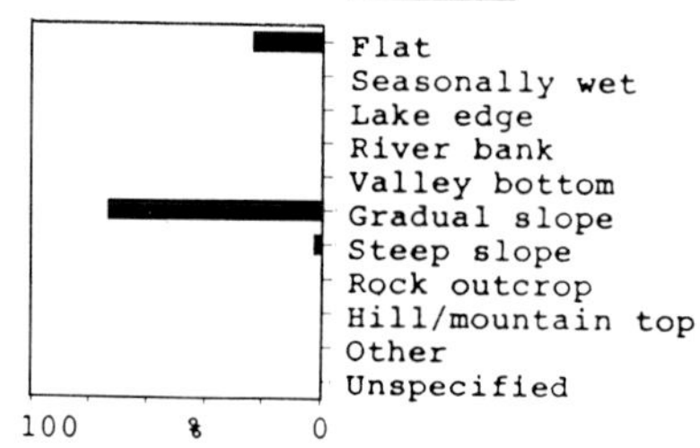

Response to Fire: 0 records

Flowering

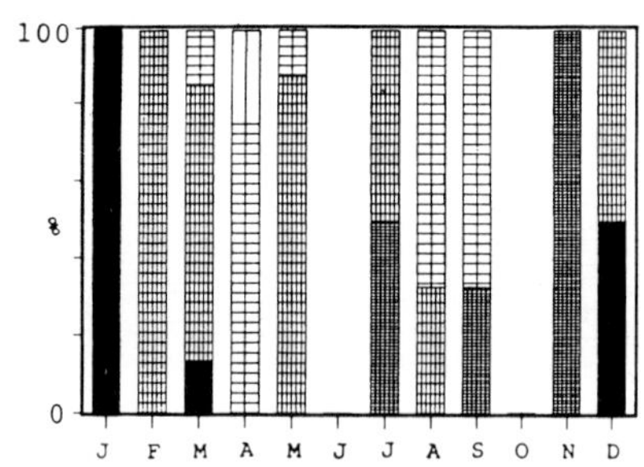

Majority of flowers fully open
Majority of flowers in bud
Majority of flowers recently finished
Flowers finished, fruiting cones present
Neither flowers nor fruiting cones present
Unspecified

Possible pollinators

Pollinator type 12 records: no pollinator observed (100%).

Specific pollinator 0 records.

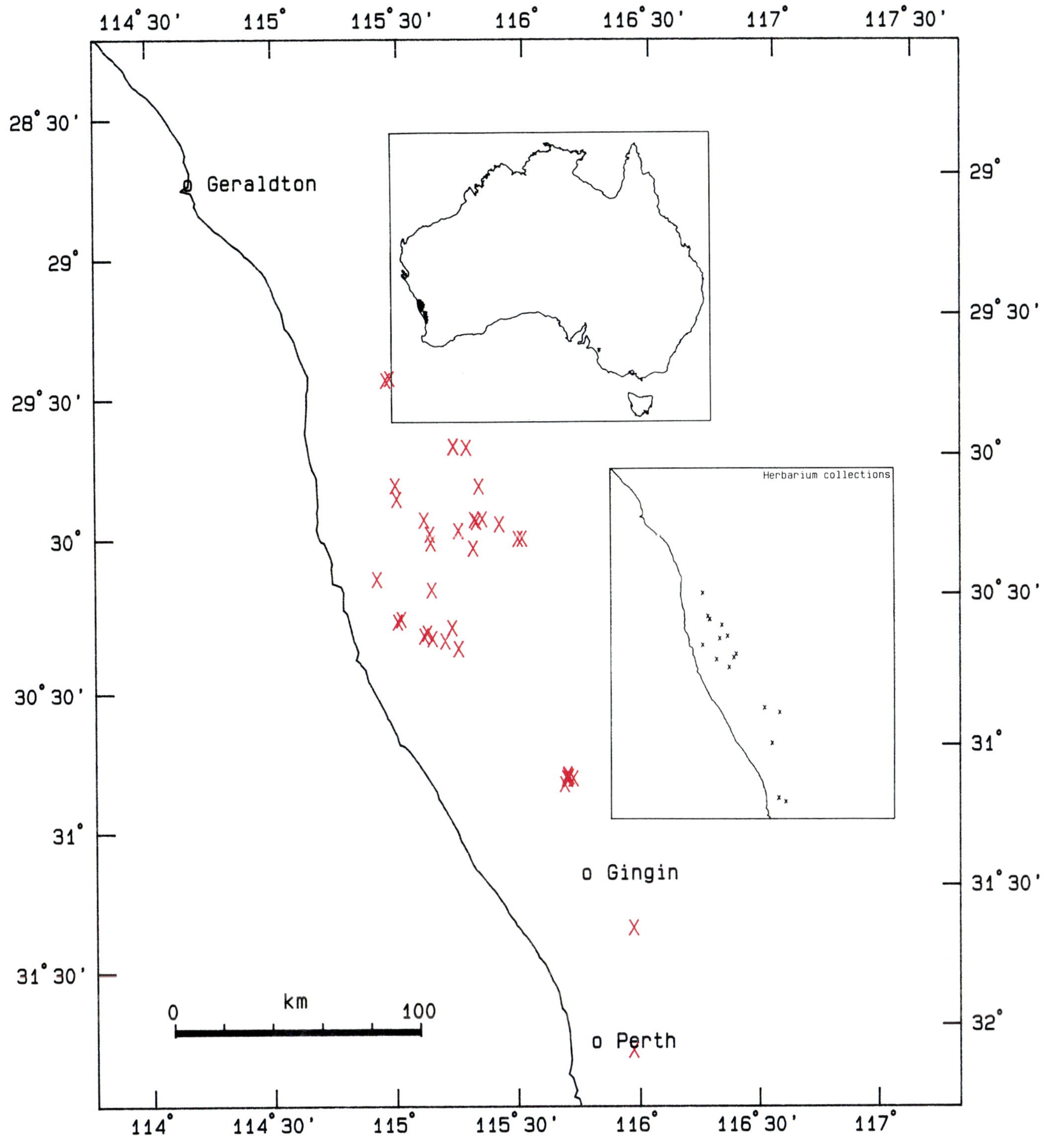

DISCUSSION

Atlas data have confirmed that the main range of B. incana is between Badgingarra National Park and an area north-west of Yandanooka Hill (ca. 35 km north of Eneabba). The latter record extended the northernmost known location of B. incana by about 20 km. Within this main group, the known range has also been extended further eastwards e.g. to the north-west corner of Watheroo National Park. South of Badgingarra National Park there appears to be a break in its distribution until a population is encountered on an area of private bushland to the west of Mogumber. This area is currently threatened with clearance for agriculture. Further south, there are outlying records from Muchea and from Forrestfield in Perth.

B. incana is a recently named taxon which, until 1981, was included within B. sphaerocarpa. It is one of the easiest of these new taxa to recognise due to its distinctive fruiting cone with very large follicles and deciduous old flowers. It is a small shrub generally less than 1 m in height. B. incana grows in deep sand or sand over laterite on flat to gently sloping land. Surrounding vegetation is usually low heath. The single Perth record was in woodland of E. marginata and banksias. Atlas records suggest a flowering season in summer. According to A.S. George (1981) the season may extend into autumn, with flowers at late as April. New shoot growth appears to be in spring and early summer. There were no records of the species' respone to fire. However, A.S. George (1981) stated that it resprouts from its lignotuber when burnt.

BANKSIA INTEGRIFOLIA Linnaeus f. (1782)

Coast Banksia

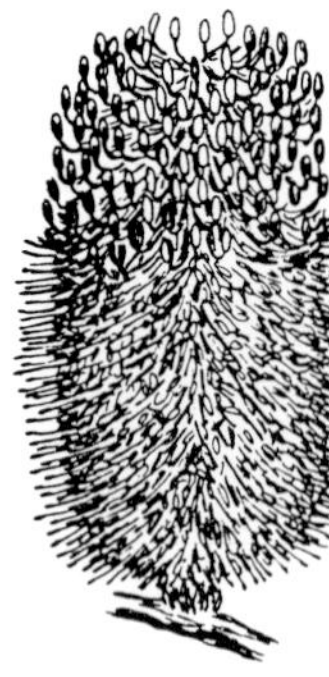

1736 RECORDS: Jan (163) Feb (141) Mar (149) Apr (230) May (267) Jun (222) Jul (175) Aug (126) Sep (56) Oct (53) Nov (46) Dec (101).

Population Size: 1-10 (342) 10-100 (614) >100 (748) Unspecified (32).

Conservation Status: Restricted to road verge (9%) Not (87%) Unspecified (4%). In conservation reserve (44%) Not (47%) Unspecified (9%).

Tree/Shrub: Tree form (75%) Shrub form (22%) Unspecified (3%).

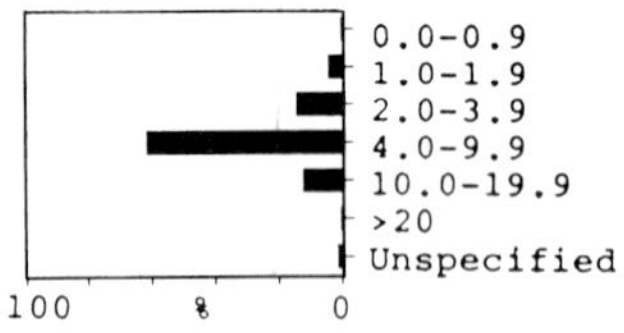

New Shoot Growth

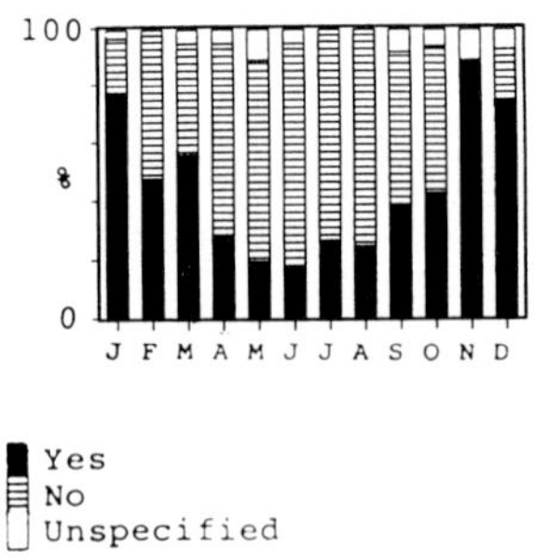

HABITAT

Within 2 km of coast: Yes (65%) No (34%) Unspecified (1%).

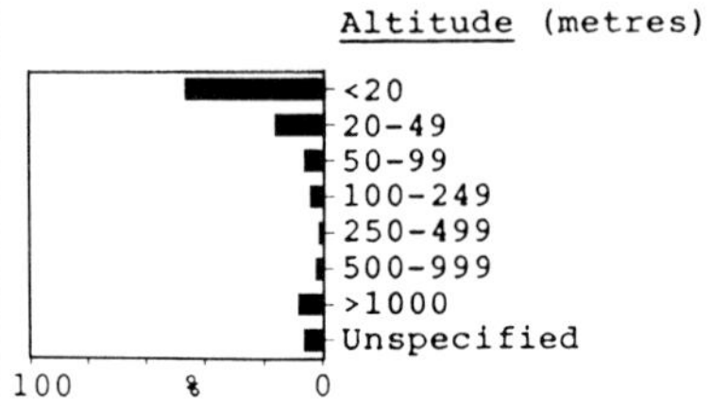

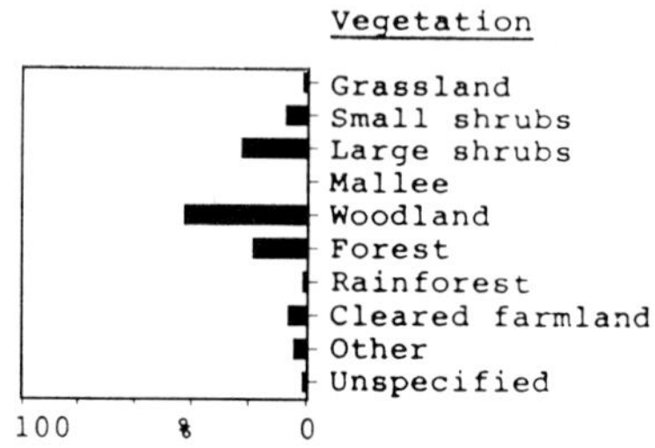

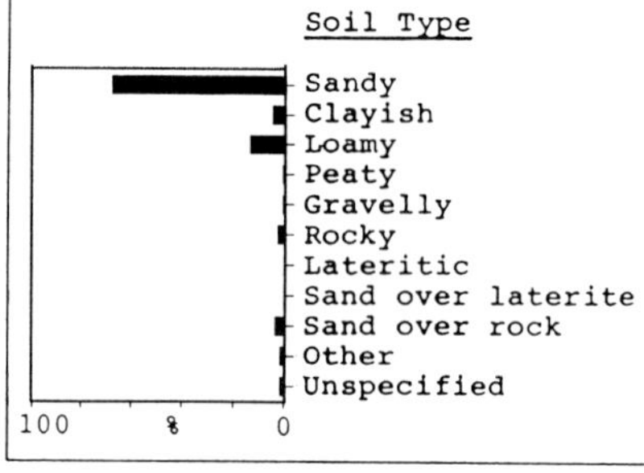

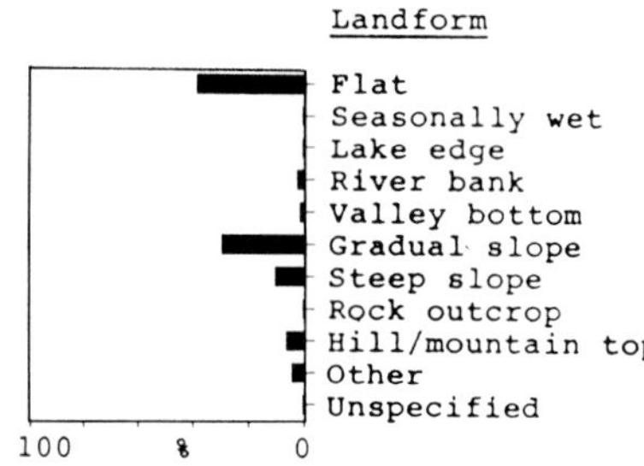

Response to Fire: 122 records

Growth response: Ground resprout (13%) ground resprout and killed, new seedlings (2%) ground resprout and trunk resprout (7%) killed, new seedlings (7%) killed, new seedlings and trunk resprout (1%) trunk resprout (17%) trunk resprout and sucker resprout (1%) killed, no seedlings (2%) killed, no seedlings and ground resprout (1%) unspecified (50%).

Flowering response: Number of records where flowering had occurred after fire (70). Median reported time to flowering (13-24 months); minimum reported time to flowering (<1 month).
Number of records where flowering had not occurred after fire (20). Median reported time to no flowering (1-12 months); maximum reported time to no flowering (85-96 months).

Flowering

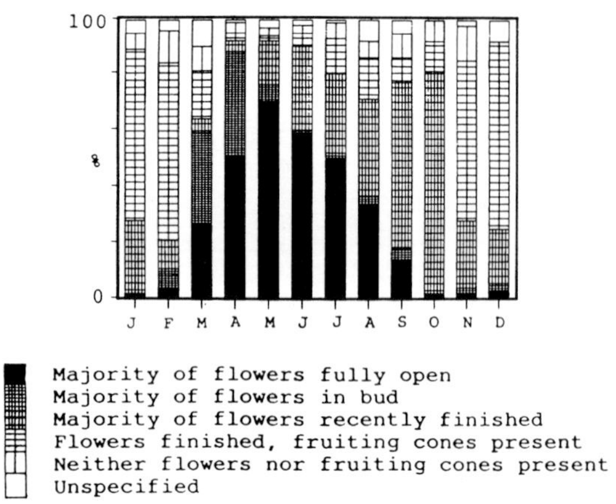

Possible pollinators

Pollinator type 920 records: bird (19%) moths/ butterflies (1%) bees, wasps, ants (5%) beetles (0.3%) mammals (0.3%) no pollinator observed (75%).

Specific pollinator 275 records: Honeyeaters - honeyeater (26) New Holland (25) White-eared (5) White-naped (1) Brown (13) White-cheeked (6) Singing (1) Tawny-crowned (1) Lewin's (4) White-naped (3) White-plumed (2) Yellow-faced (4) Spiny-cheeked (2) Blue-faced (2) Dusky (1) White-throated (2) Scarlet (1) Fuscous (1) Bridled (1) Noisy Miner (9) Silvereye (7) Wattlebird (14) Red Wattlebird (21) Little Wattlebird (26) Eastern Spinebill (16) friarbird (7) Noisy Friarbird (6) Little friarbird (3) rosella (2) lorikeet (2) Rainbow Lorikeet (14) Black-faced Cuckoo-shrike (1) weavers and allies (1) Striated Pardalote (1) thornbill (1) Brown Thornbill (1) ant (4) bee (18) native bee (2) European honey bee (14) european wasp (1) spider (1) possum (2).

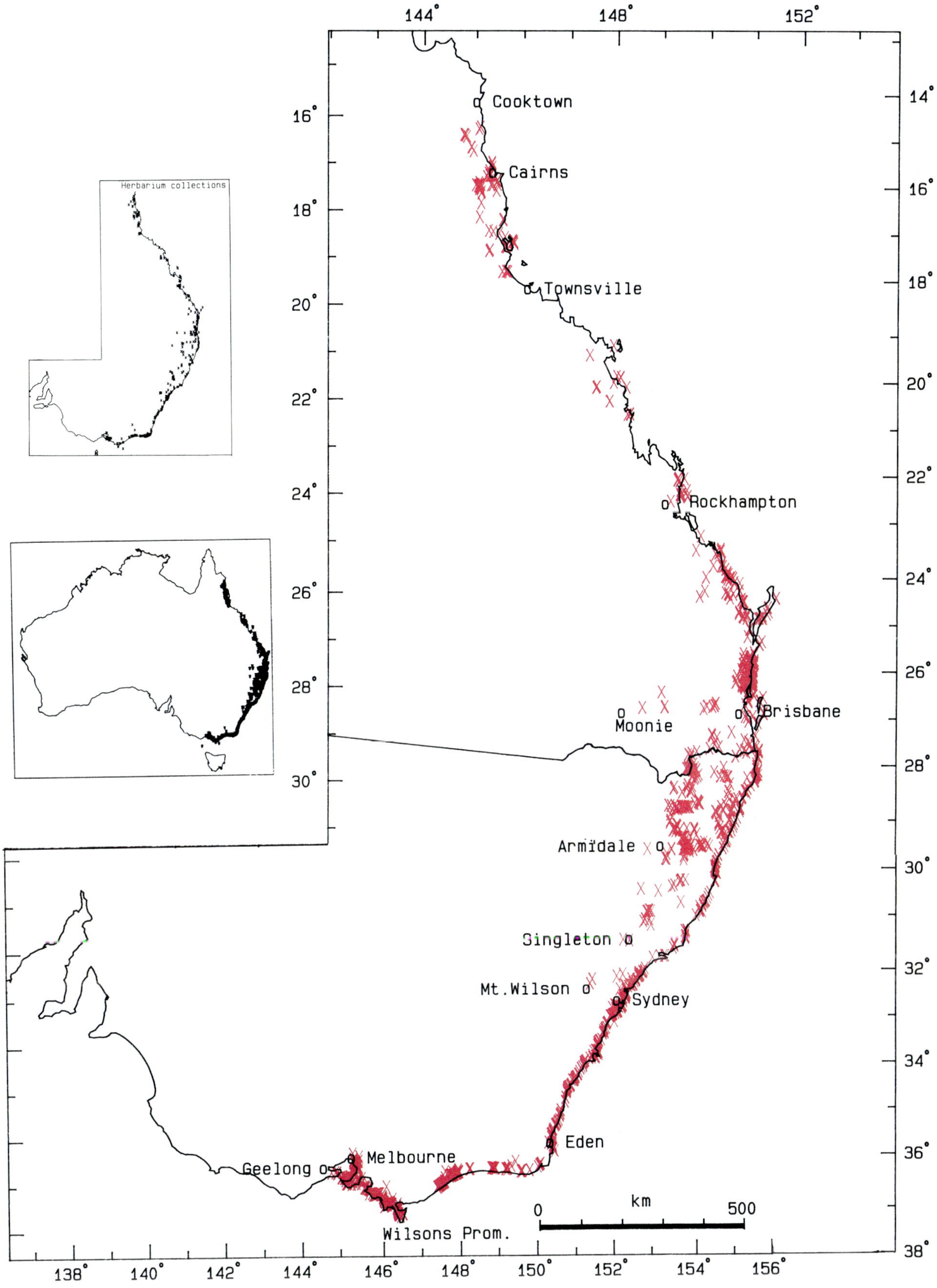

DISCUSSION

Of the 1 736 sight records for B. integrifolia, 1 676 had a varietal name specified and 60 (3%) were recorded by the species name only. The species is divided into three varieties on the basis of leaf size, shape, and arrangement on the stem. However, Atlas contributors found great variability in these measurements with specimens often falling outside the varietal limits specified by George (1981) or being intermediate between two varieties. This was particularly so with var. integrifolia and var. compar especially where their ranges overlap in coastal parts of southern Queensland and northern New South Wales. In these areas, the co-existence of these two varieties also appears more common than was previously thought.

B. integrifolia is one of the most common eastern banksias. Indeed, there are few sections of coast between Melbourne and Rockhampton where it does not occur. It is a predominantly autumn- and winter- flowering species though some flowers have been recorded in all months of the year. Summer is the main season of new shoot growth. The species usually occurs as a tree between 4 m and 10 m in height though larger specimens of all three varieties are not uncommon. Specimens of B. integrifolia var. compar from Washpool National Park are more than 30 m tall with the trunk diameter at breast height measuring almost one metre. These may well be the largest banksia specimens that exist. B. integrifolia generally prefers sandy soils, though particularly with its northern form (var. aquilonia), heavier soils are also tolerated. The species as a whole is highly resistant to dieback disease and is being investigated as a rootstock on which to graft some of the sensitive western species for possible commercial cut-flower production.

BANKSIA INTEGRIFOLIA Linnaeus f.
var. AQUILONIA A.S.George (1981)

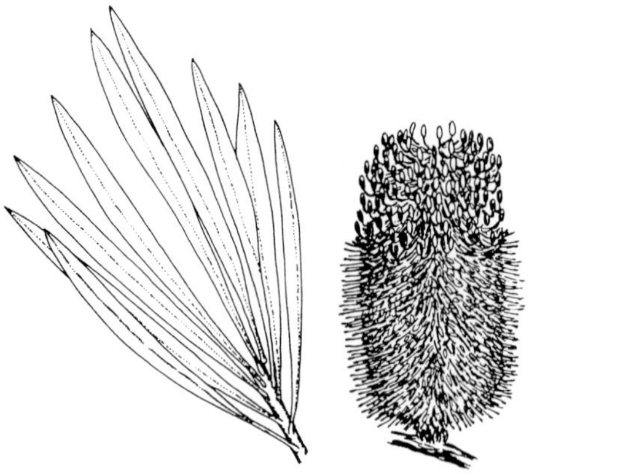

92 RECORDS: Jan (1) Feb (0) Mar (4) Apr (3) May (12) Jun (17) Jul (33) Aug (5) Sep (13) Oct (2) Nov (1) Dec (0).

Population Size: 1-10 (16) 10-100 (65) >100 (11) Unspecified (0).

Conservation Status: Restricted to road verge (5%) Not (91%) Unspecified (3%). In conservation reserve (36%) Not (53%) Unspecified (11%).

Tree/Shrub: Tree form (93%) Shrub form (7%) Unspecified (0%).

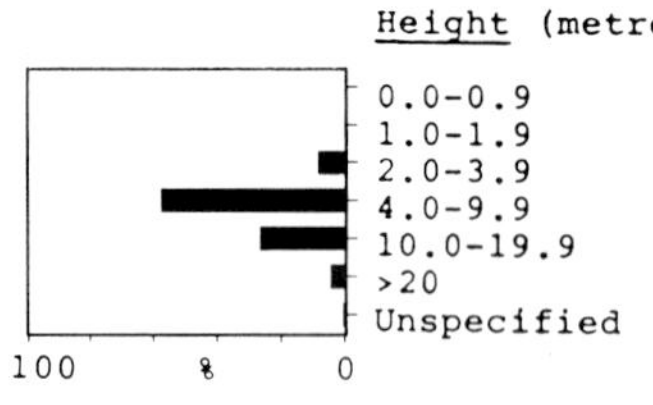

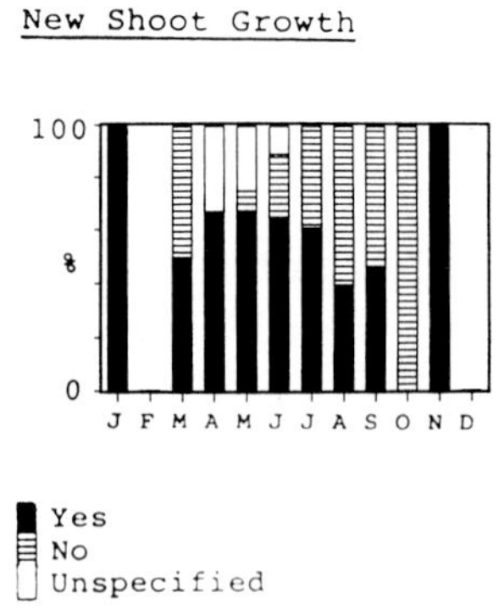

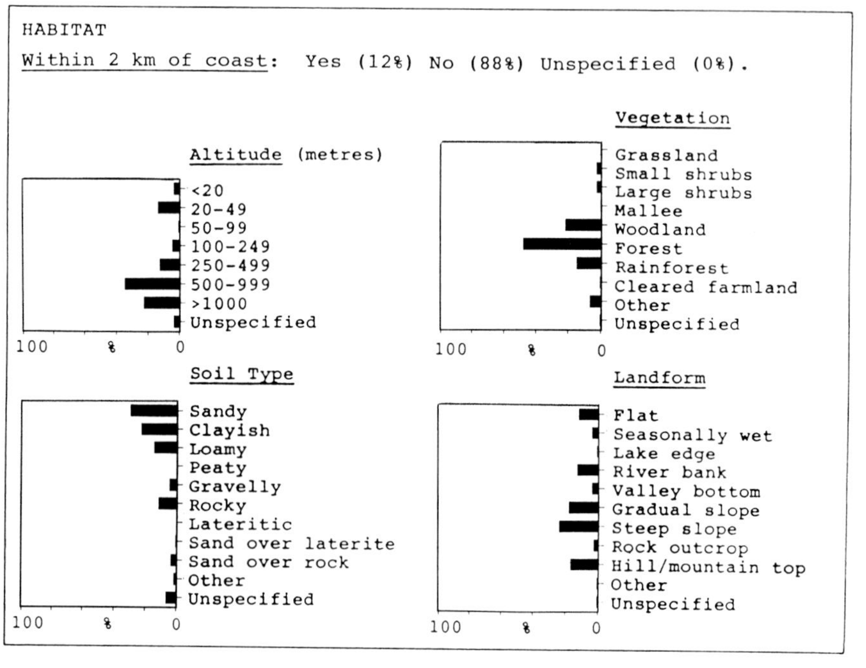

Response to Fire: 14 records

Growth response: Ground resprout (14%) trunk resprout (64%) unspecified (21%).

Flowering response: Number of records where flowering had occurred after fire (10). Median reported time to flowering (13-24 months); minimum reported time to flowering (1-12 months). Number of records where flowering had not occurred after fire (3). Median reported time to no flowering (13-24 months); maximum reported time to no flowering (13-24 months).

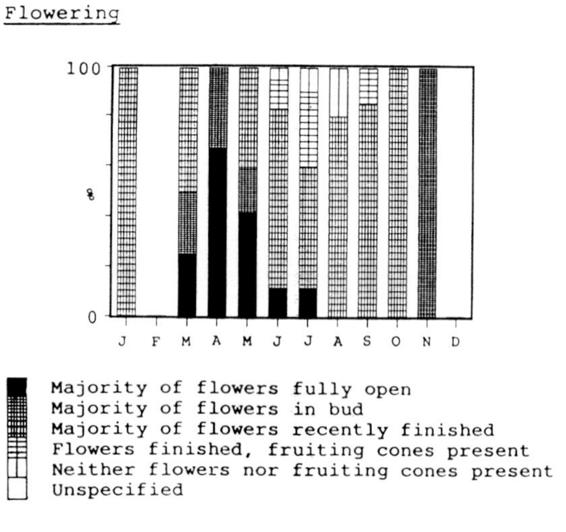

Possible pollinators

Pollinator type 23 records: bird (9%) no pollinator observed (91%).

Specific pollinator 4 records. White-cheeked Honeyeater (1) Bridled Honeyeater (1) Eastern Spinebill (1) Rainbow Lorikeet (1).

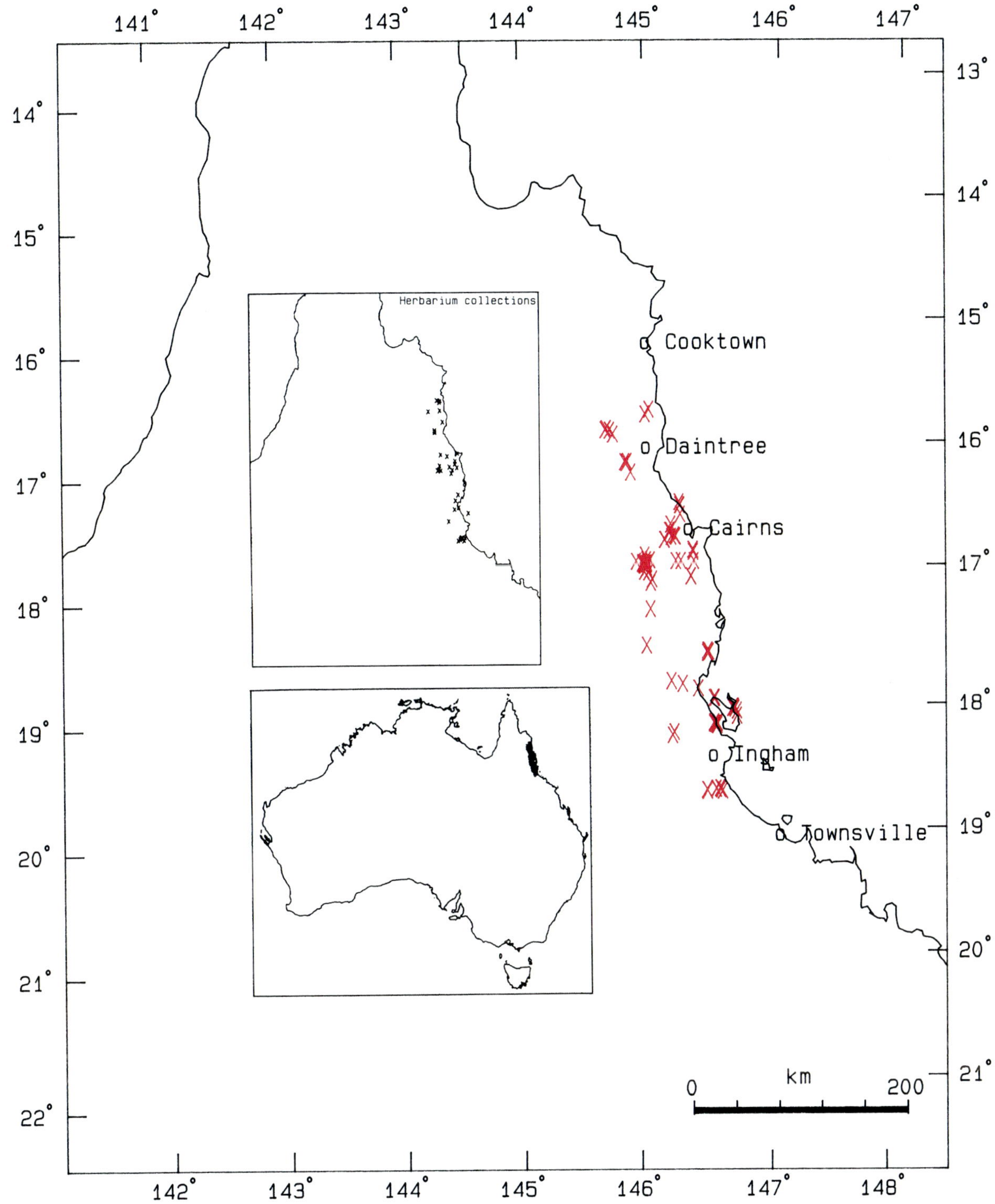

DISCUSSION

Atlas records confirmed a distribution of B. integrifolia var. aquilonia in northern Queensland between the Paluma Range south of Ingham almost to Finnigan National Park south of Cooktown. The variety also was on Hinchinbrook Island.

The variety occurs in a wide range of habitats from coastal sometimes swampy flats and sand dunes to mountain slopes, summits and plateaus exceeding 1 000 m. Soil types vary from sands to loams and clays, and are occasionally rocky. Surrounding vegetation is usually eucalypt woodland or forest and sometimes rainforest.

B. integrifolia var. aquilonia is usually a tree up to 15 m in height. Flowering is from March to July. The main season of new shoot growth may be in summer though further data are needed to confirm this. Two interesting variations on the typical leaf size and shape were noted by Atlas contributors. Along the Tully to Mission Beach Road are large-leaved plants with both adult and juvenile leaves up to 38 cm long and juvenile leaves up to 25 mm wide. Near the summit of Coronation Lookout a small-leaved form appears. Leaves up to 13 cm long and only 4 mm wide grow on 3 m high trees. Taller trees with 'normal' sized leaves occur lower down the mountain.

BANKSIA INTEGRIFOLIA Linnaeus f.
var. COMPAR (R.Brown) Bailey (1913)

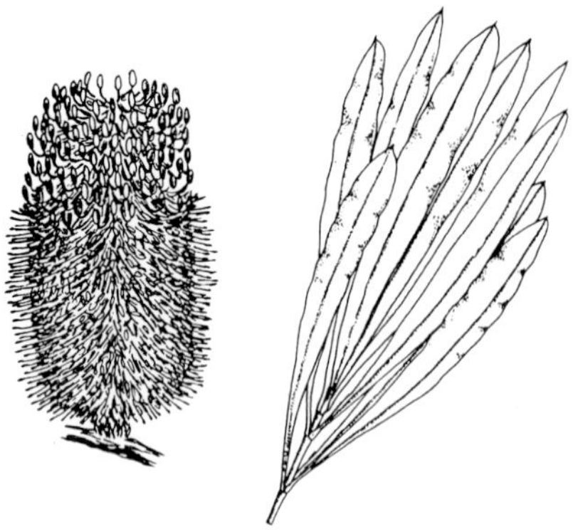

437 RECORDS: Jan (69) Feb (20) Mar (21) Apr (40) May (69) Jun (27) Jul (66) Aug (39) Sep (19) Oct (12) Nov (13) Dec (40).

Population Size: 1-10 (112) 10-100 (221) >100 (94) Unspecified (10).

Conservation Status: Restricted to road verge (13%) Not (85%) Unspecified (2%). In conservation reserve (27%) Not (59%) Unspecified (14%).

Tree/Shrub: Tree form (91%) Shrub form (9%) Unspecified (1%).

Height (metres)

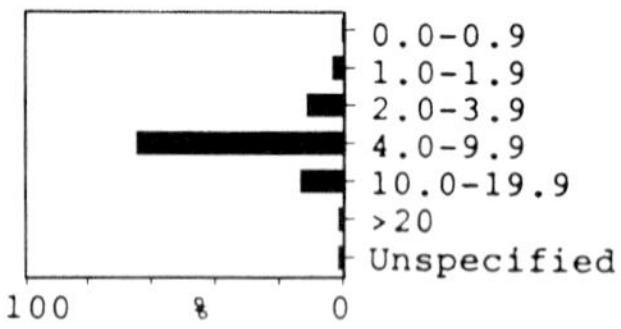

New Shoot Growth

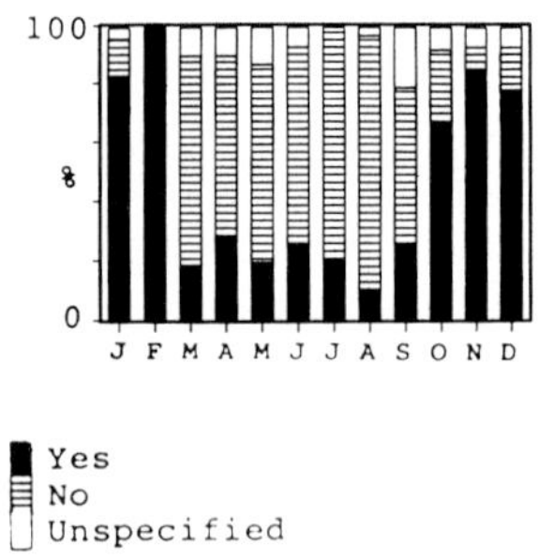

HABITAT

Within 2 km of coast: Yes (15%) No (83%) Unspecified (2%).

Altitude (metres)

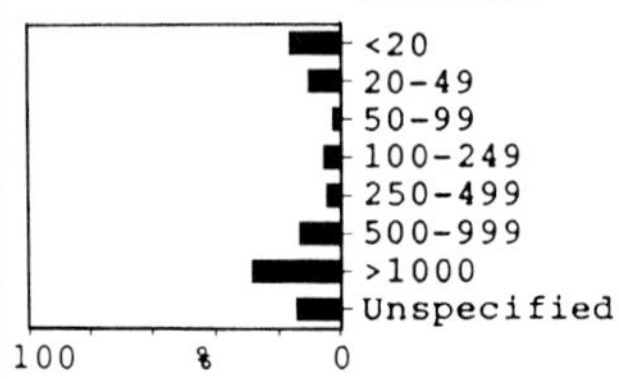

Vegetation

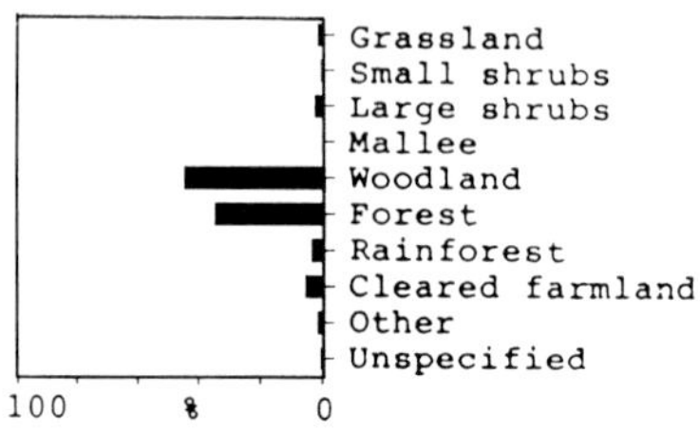

Soil Type

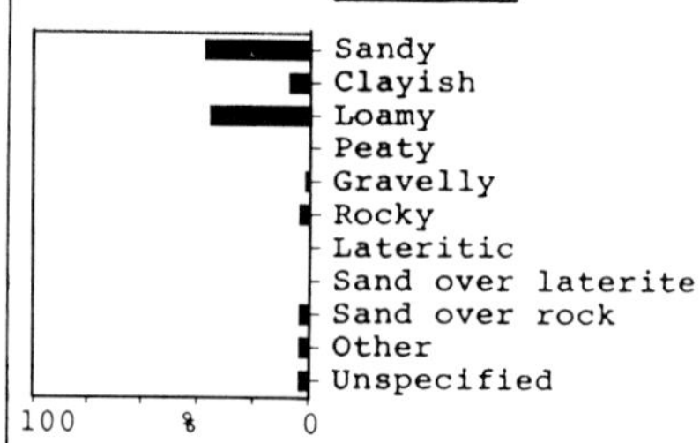

Landform

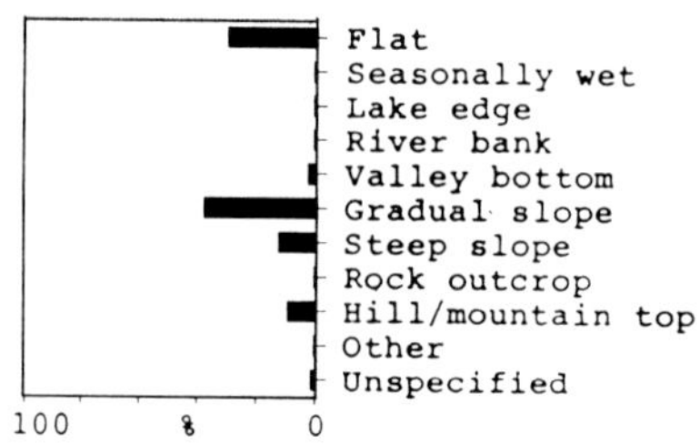

Response to Fire: 34 records

Growth response: Ground resprout (15%) ground resprout and killed, new seedlings (3%) killed, new seedlings (6%) trunk resprout (21%) unspecified (56%).

Flowering response: Number of records where flowering had occurred after fire (22). Median reported time to flowering (12-13 months); minimum reported time to flowering (<1 month). Number of records where flowering had not occurred after fire (4). Median reported time to no flowering (1-12 months); maximum reported time to no flowering (13-24 months).

Flowering

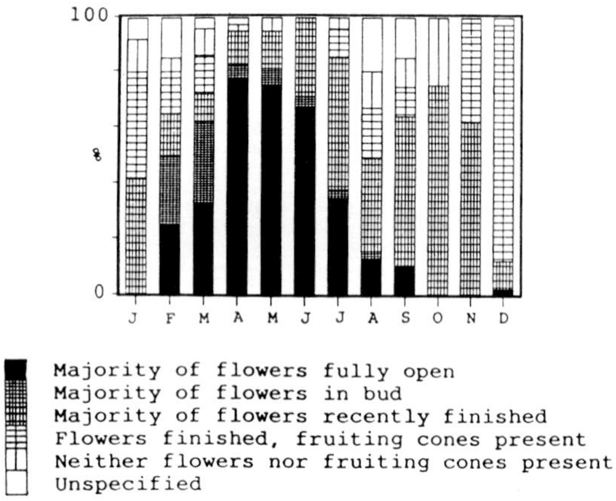

Possible pollinators

Pollinator type 187 records: bird (28%) butterflies, moths (3%) bees, wasps, ants (9%) mammal (2%) no pollinator observed (59%).

Specific pollinator 76 records: honeyeater (11) New Holland Honeyeater (1) Brown Honeyeater (4) Lewin's Honeyeater (2) Yellow-faced Honeyeater (1) Noisy Miner (6) Dusky Honeyeater (1) Spiny-cheeked Honeyeater (1) White-throated Honeyeater (2) Scarlet Honeyeater (1) Fuscous Honeyeater (1) Red Wattlebird (2) Eastern Spinebill (9) rosella (2) friarbird (1) Noisy Friarbird (3) Little Friarbird (3) lorikeets and allies (1) Rainbow Lorikeet (5) weavers and allies (1) Striated Pardalote (1) thornbill (1) Brown Thornbill (1) ant (2) bee (5) native bee (2) European honey bee (4) possum (2).

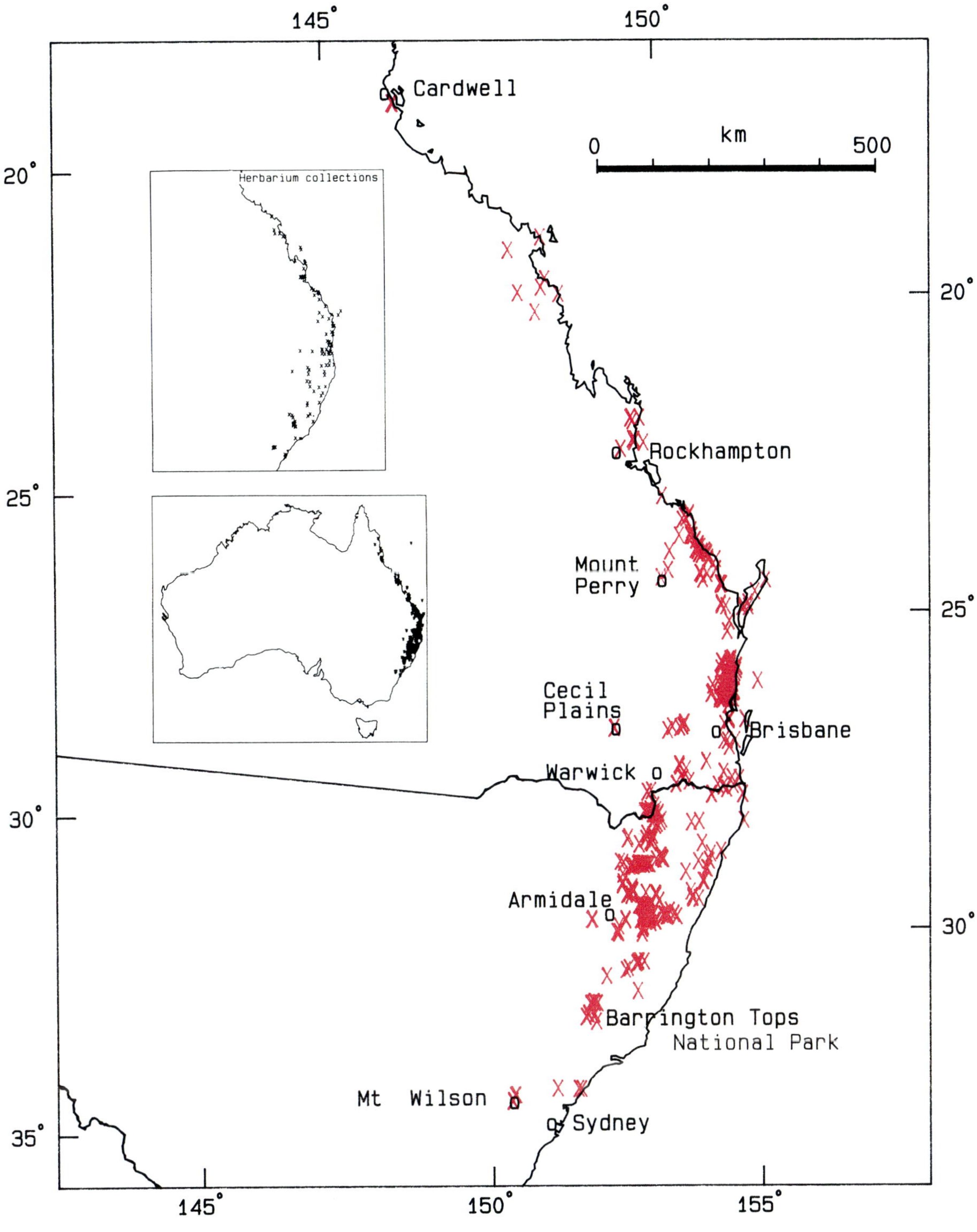

DISCUSSION

B. integrifolia var. compar was recorded between Mt Wilson (N.S.W.) and Proserpine (Queensland) with a possible outlier near Bishops Peak, Cardwell (north Queensland). If the latter record is correct it will extend the range of this variety by about 400 km. Mature leaves from Bishops Peak plants have a shape and size like that of var. compar (up to 28 cm long and 2.8 cm wide) but they have dark hairs along the underside of the midrib - a characteristic of var. aquilonia. Further investigations are needed to determine their true status. In New South Wales, var. compar usually occurs on the Great Dividing Range, extending to the coastal plain between Maclean and Coffs Harbour. The most inland record is from ca. 40 km west of Armidale. In Queensland the variety is more coastal and in southern parts often co-exists with var. integrifolia (e.g. around Caloundra). It is in coastal parts of southern Queensland and northern New South Wales that the ranges of the two varieties overlap and they are often difficult to distinguish.

In mountainous parts of northern New South Wales where var. compar grows at altitudes in excess of 1 000 m, it appears different from the typical form from lower altitudes. Adult leaves are longer and narrower, fruiting cones are black rather than grey and follicles remain closed either until the following summer or until they are burnt. Whether these are adaptations to a cold climate or are significant morphological differences is not known. In some mountain areas, var. compar grows in rainforest usually on the edges of firebreaks or roads or within areas that have been disturbed e.g. by fire or logging. Some specimens are very large e.g. up to 30 m tall from Washpool National Park. More commonly, the variety grows in eucalypt forests and woodlands. In Queensland, the variety ranges from coastal sand dunes to more inland mountain sites e.g. Mt Perry and Mt Tibrogargon. Flowering is from February to September and summer is the main season of new shoot growth. Following fire, the variety may resprout from epicormic buds or its lignotuber. Some plants may also be killed by fire.

BANKSIA INTEGRIFOLIA Linnaeus f. (1782) var. INTEGRIFOLIA

Coast Banksia

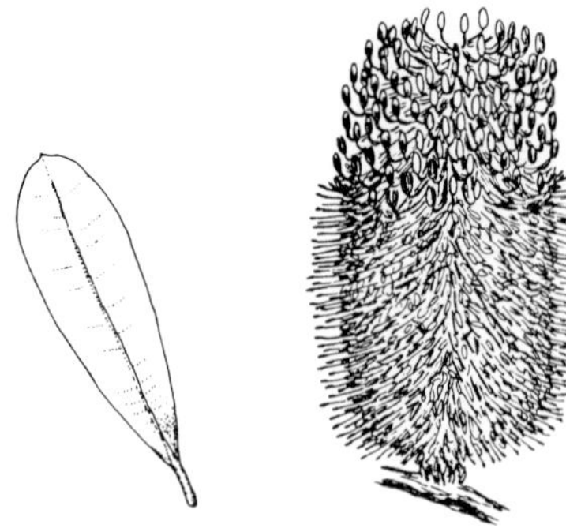

1147 RECORDS: Jan (87) Feb (117) Mar (120) Apr (182) May (175) Jun (172) Jul (68) Aug (80) Sep (24) Oct (35) Nov (29) Dec (57).

Population Size: 1-10 (200) 10-100 (300) >100 (629) Unspecified (18).

Conservation Status: Restricted to road verge (8%) Not (86%) Unspecified (6%). In conservation reserve (50%) Not (42%) Unspecified (7%).

Tree/Shrub: Tree form (54%) Shrub form (45%) Unspecified (1%).

Height (metres)

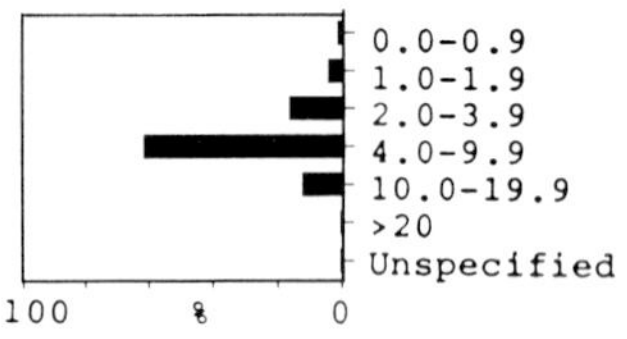

New Shoot Growth

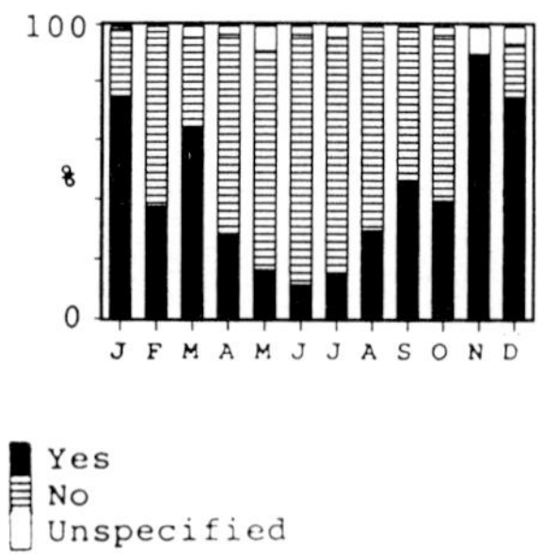

HABITAT

Within 2 km of coast: Yes (89%) No (10%) Unspecified (1%).

Altitude (metres)

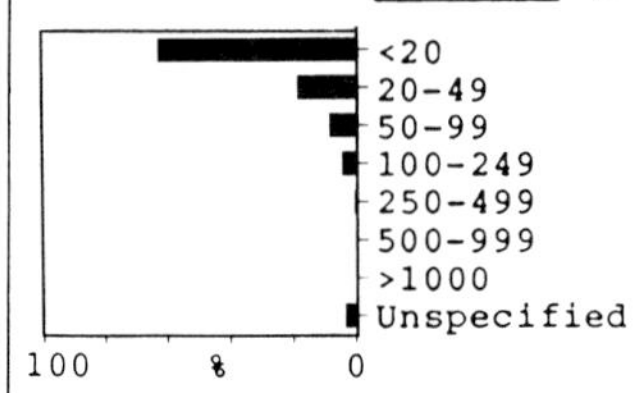

Vegetation

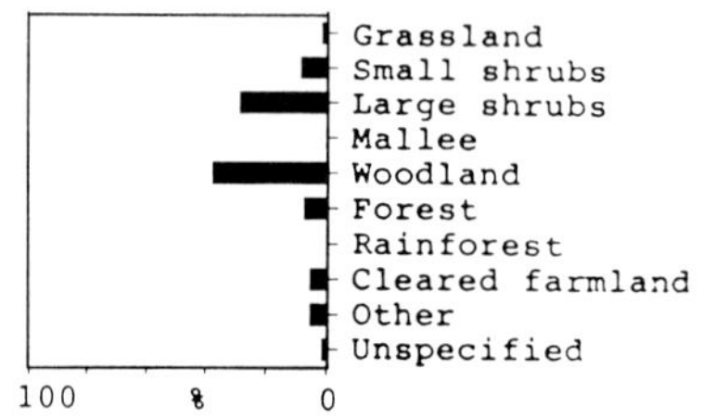

Soil Type

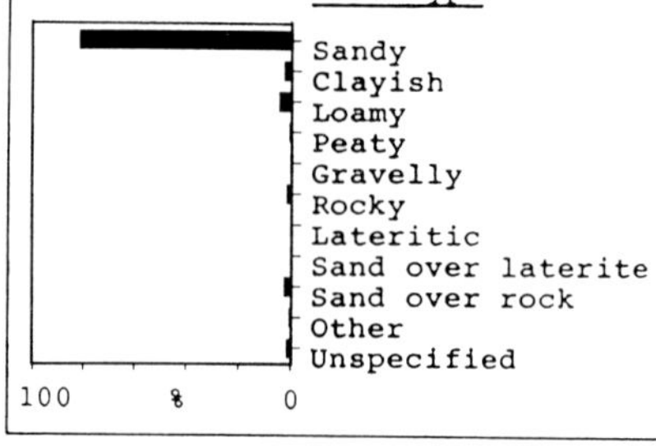

Landform

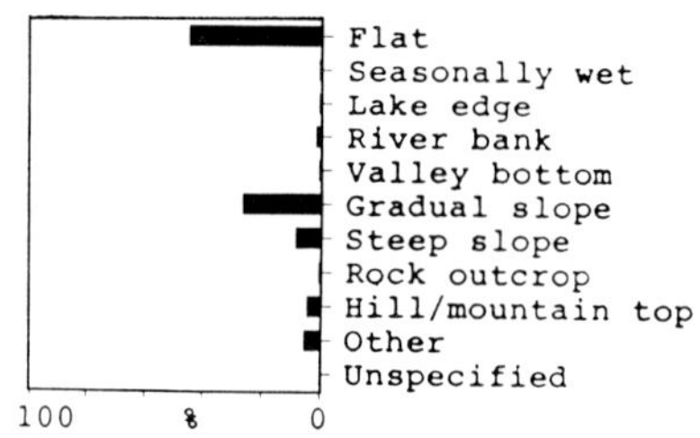

Response to Fire: 69 records

Growth response: Ground resprout (10%) ground resprout and killed, new seedlings (1%) ground resprout and trunk resprout (12%) killed, new seedlings (10%) killed, new seedlings and trunk resprout (1%) trunk resprout (7%) trunk resprout and sucker resprout (1%) killed, no seedlings (4%) killed, no seedlings and ground resprout (1%) unspecified (51%).

Flowering response: Number of records where flowering had occurred after fire (38). Median reported time to flowering (13-24 months); minimum reported time to flowering (<1 month).
Number of records where flowering had not occurred after fire (11). Median reported time to no flowering (1-12 months); maximum reported time to no flowering (85-96 months).

Flowering

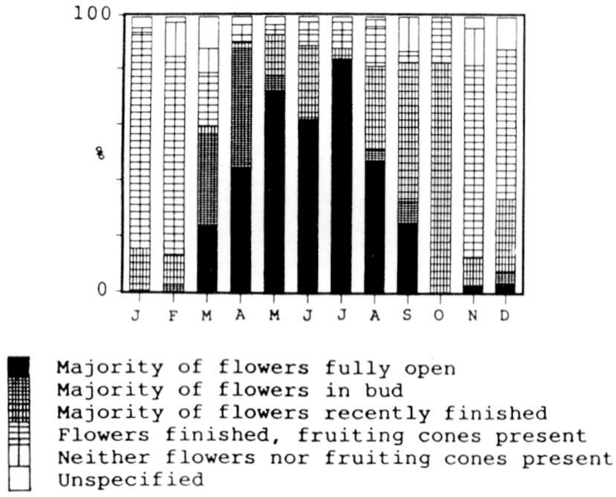

Possible pollinators

Pollinator type 671 records: bird (17%) butterflies/moths (0.4%) bees, wasps, ants (3%) beetles (0.4%) no pollinator observed (79%).

Specific pollinator 177 records: honeyeater (13) New Holland Honeyeater (21) Brown Honeyeater (9) White-cheeked Honeyeater (5) Singing Honeyeater (1) White-eared Honeyeater (4) Tawny-crowned Honeyeater (1) Lewin's Honeyeater (2) White-naped Honeyeater (3) White-plumed Honeyeater (2) Yellow-faced Honeyeater (3) Noisy Miner (3) Spiny-cheeked Honeyeater (1) Blue-faced Honeyeater (2) Silvereye (7) wattlebird (14) Red Wattlebird (18) Little Wattlebird (26) Eastern Spinebill (5) friarbird (5) Noisy Friarbird (3) lorikeets and allies (1) Rainbow Lorikeet (8) Black-faced Cuckoo-shrike (1) ant (1) bee (7) European honey bee (9) European wasp (1) spider (1).

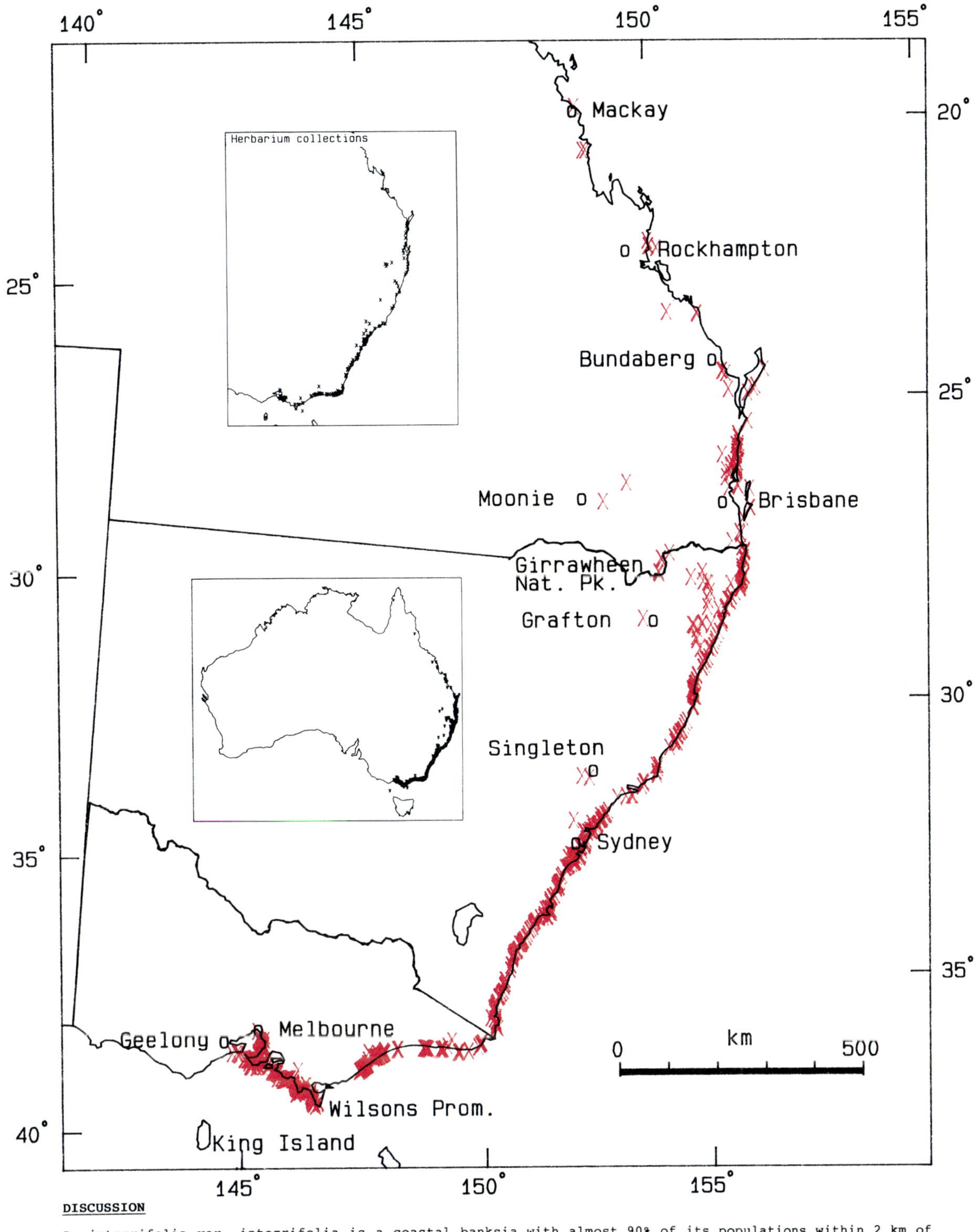

DISCUSSION

B. integrifolia var. integrifolia is a coastal banksia with almost 90% of its populations within 2 km of the coast. Herbarium records indicated a distribution between the eastern side of Port Phillip Bay (Victoria) to Wide Bay and Fraser Island (Queensland) including King and Long Islands in Bass Strait. Apart from these two island collections which were not recorded by Atlas contributors, the 1 147 sight records have confirmed this general distribution, and extended it to the western side of Port Phillip Bay and to several inland localities. The most inland of these are from the Weir River east of Moonie and from 7 km west of Lake Broadwater (Queensland). Specimens of both these records are in the Queensland Herbarium (B. Ballingall). Other inland records are from Girraween National Park (Queensland), west of Grafton (N.S.W.) and from Warkworth near Singleton (N.S.W.). North of Fraser Island, populations which had been identified as var. compar by A. George (1981) were submitted as var. integrifolia by A. Salkin, who stated that the only difference between these plants and those of var. integrifolia further south was in the leaves being slightly longer.

B. integrifolia var. integrifolia occurs as both a tree and shrub generally up to 10 m in height, but occasionally larger. It has a strong preference for sandy soils and commonly occurs on consolidated coastal sand dunes amongst tall shrubland or woodland. Inland it may occur on low-lying, riverine, sandy flats. It is a good coloniser often occurring on the edge of disturbed habitats. Thus in northern New South Wales it is found on the edges of rainforest wherever a man-made clearing (e.g. a road) allows the entry of more light. Possibly the highest altitude at which it occurs is 460 m on North Brother Mountain, south of Laurieton (N.S.W).

B. integrifolia var. integrifolia flowers from March to September with a few flowers recorded in November and December. There were over 100 records of birds visiting the flower with the most common being the Little Wattlebird, New Holland Honeyeater and Red Wattlebird. New shoot growth is predominantly in summer. Following fire, the variety resprouts both from its lignotuber and from its epicormic buds. Some plants are also killed by fire and regenerate from seed. There are also records of plants resprouting from suckers which have spread underground up to 4 m from the parent trunk.

BANKSIA LAEVIGATA Meissner (1856)

59 RECORDS: Jan (1) Feb (4) Mar (1) Apr (17) May (1) Jun (0) Jul (0) Aug (2) Sep (12) Oct (9) Nov (8) Dec (3).

Population Size: 1-10 (16) 10-100 (31) >100 (11) Unspecified (1).

Conservation Status: Restricted to road verge (15%) Not (82%) Unspecified (3%). In conservation reserve (25%) Not (68%) Unspecified (7%).

Tree/Shrub: Tree form (2%) Shrub form (96%) Unspecified (2%).

Height (metres)

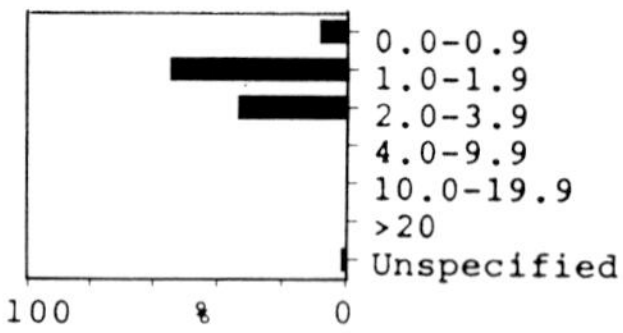

New Shoot Growth

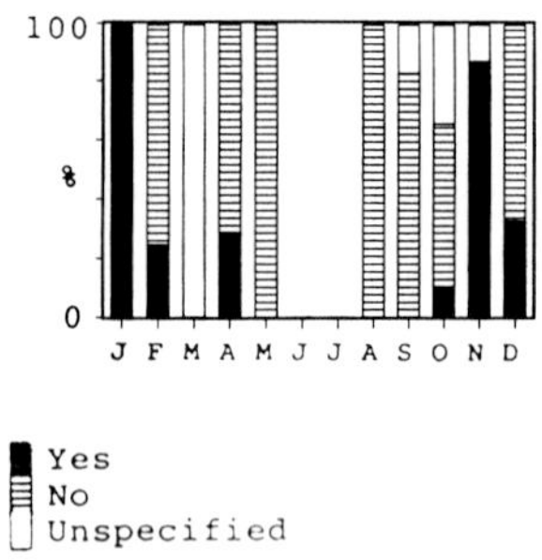

HABITAT

Within 2 km of coast: Yes (0%) No (100%) Unspecified (0%).

Altitude (metres)

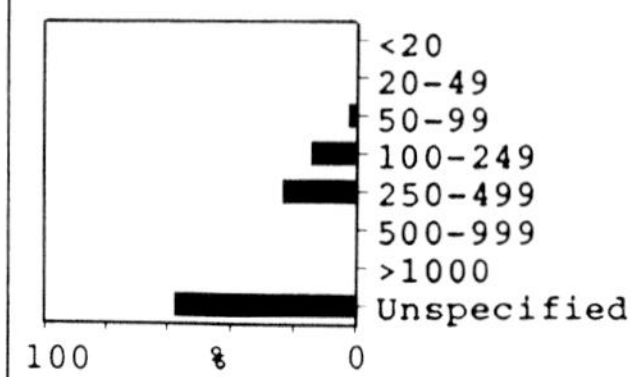

Vegetation

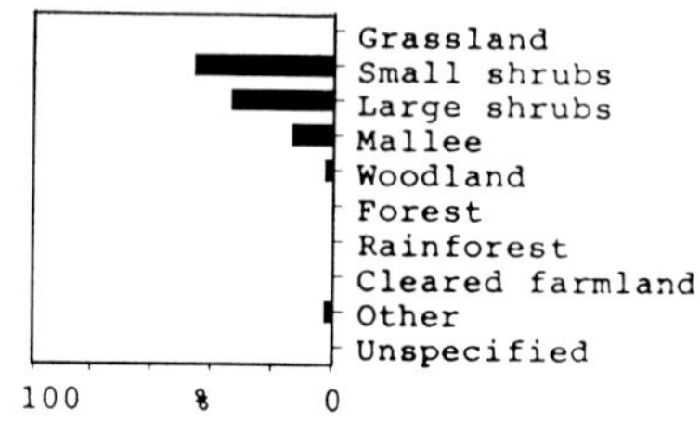

Soil Type

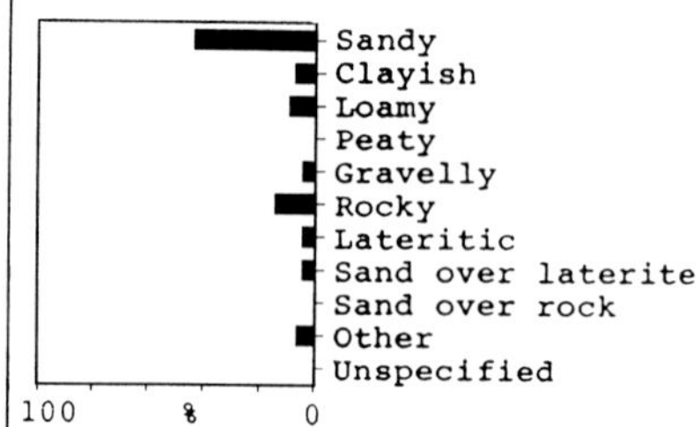

Landform

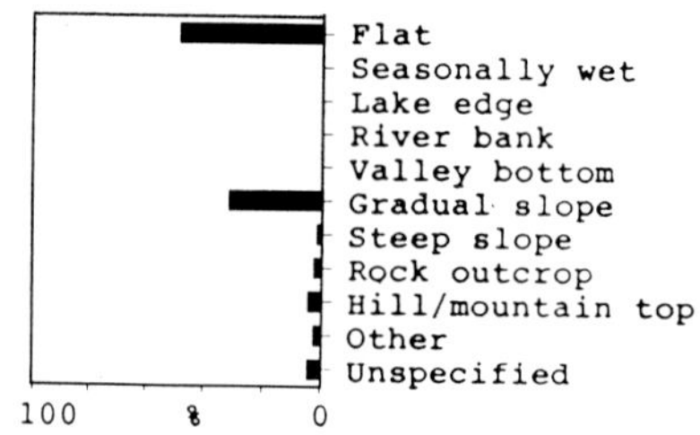

Response to Fire: 8 records

Growth response: Killed, new seedlings (50%) killed, no seedlings and killed, new seedlings (12%) unspecified (37%).

Flowering response: Number of records where flowering had occurred after fire (2). Median reported time to flowering (36-37 months); minimum reported time to flowering (13-24 months). Number of records where flowering had not occurred after fire (4). Median reported time to no flowering (12-13 months); maximum reported time to no flowering (25-36 months).

Flowering

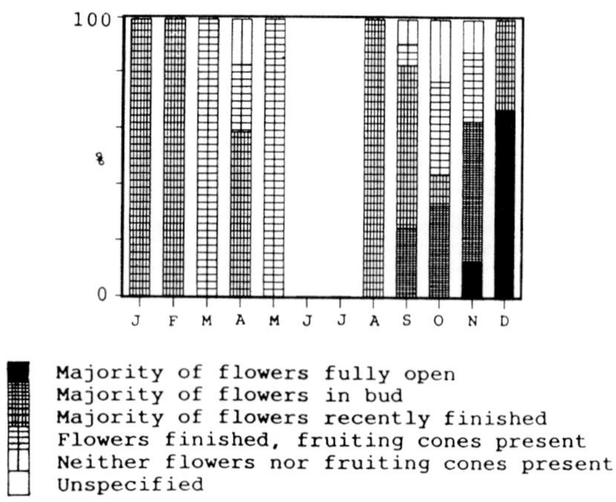

Possible pollinators

Pollinator type 9 records: bees, wasps, ants (11%) no pollinator observed (89%).

Specific pollinator 0 records.

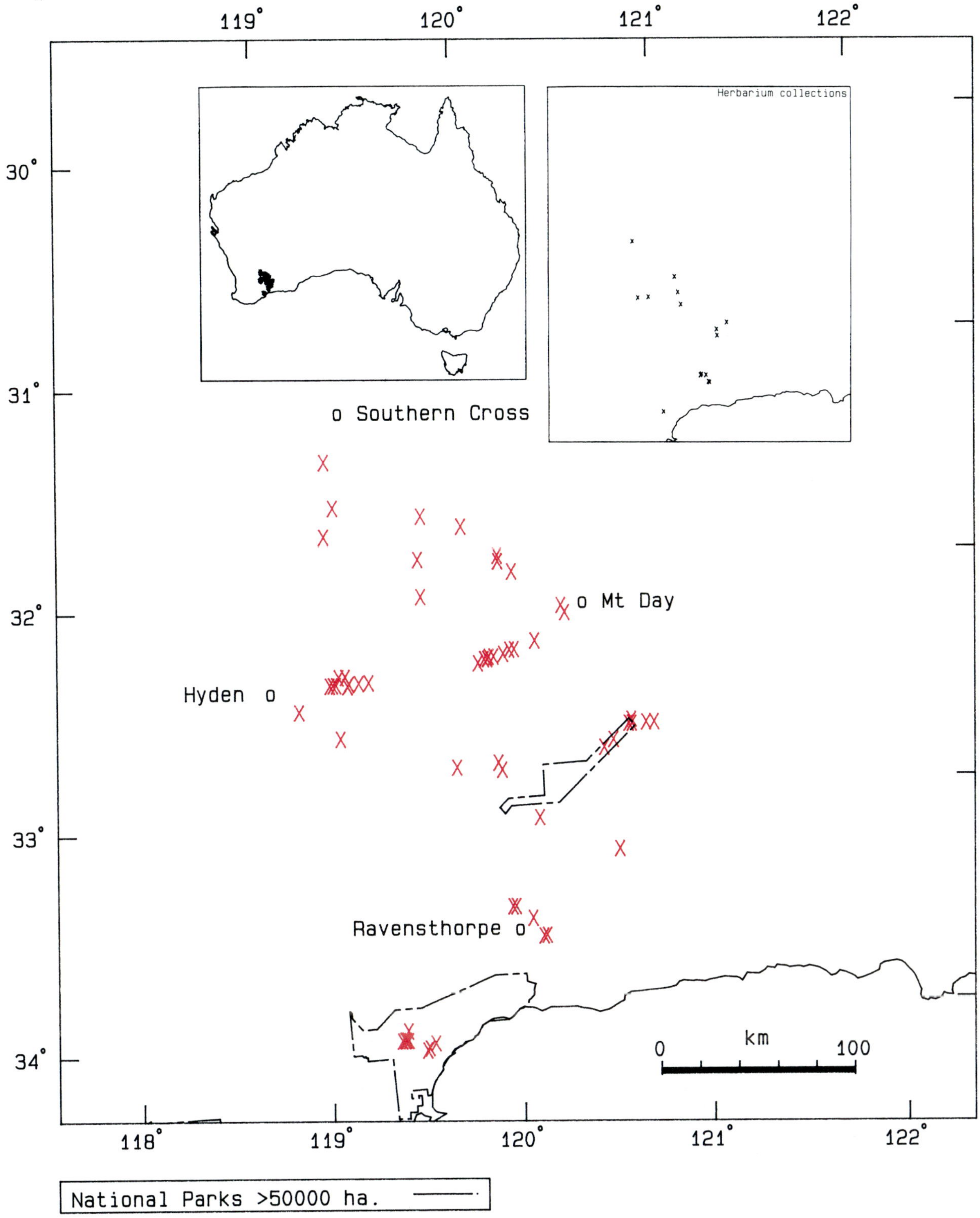

DISCUSSION

B. laevigata occurs from near Southern Cross south to the Fitzgerald River National Park in Western Australia. It grows in heath and mallee communities, usually in gently undulating terrain. Northern populations (subsp. fuscolutea) favour sandy soils, while populations near Ravensthorpe and south-westwards (subsp. laevigata) are found on rocky hills and breakaways. The species occurs mainly beyond agricultural land, and seems fairly secure. However, subsp. laevigata is rare and requires monitoring in the future (see below).

The two subspecies are distinguished on floral features. Subsp. fuscolutea has bright yellow flowers with rusty brown hairs, while subsp. laevigata has pale yellow to cream flowers with grey hairs. Only four of the 59 sight records of the species did not have the subspecies specified, indicating that contributors had little difficulty with identification in this area.

BANKSIA LAEVIGATA Meissner subsp. FUSCOLUTEA A.S.George (1966)

Flowers bright yellow with rusty brown hairs.

41 RECORDS: Jan (0) Feb (2) Mar (1) Apr (16) May (0) Jun (0) Jul (0) Aug (2) Sep (6) Oct (4) Nov (6) Dec (3).

Population Size: 1-10 (13) 10-100 (23) >100 (4) Unspecified (1).

Conservation Status: Restricted to road verge (20%) Not (80%) Unspecified (0%). In conservation reserve (15%) Not (78%) Unspecified (7%).

Tree/Shrub: Tree form (0%) Shrub form (95%) Unspecified (5%).

Height (metres)

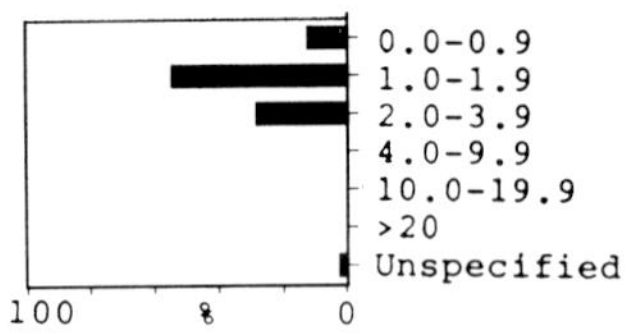

New Shoot Growth

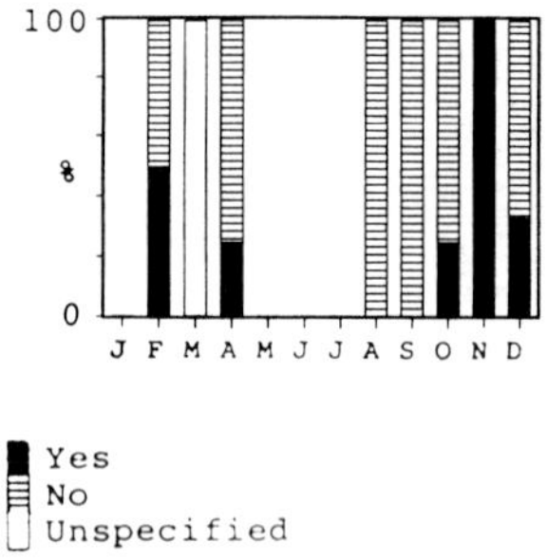

HABITAT

Within 2 km of coast: Yes (0%) No (100%) Unspecified (0%).

Altitude (metres)

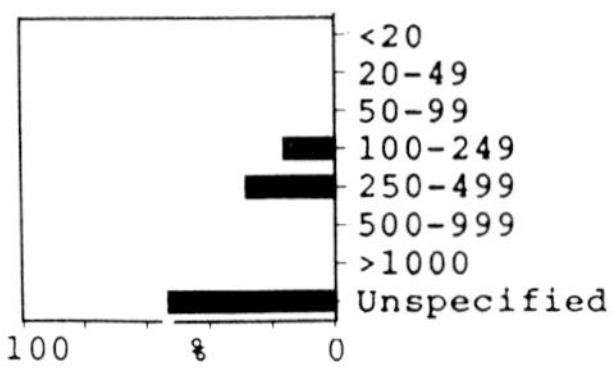

Vegetation

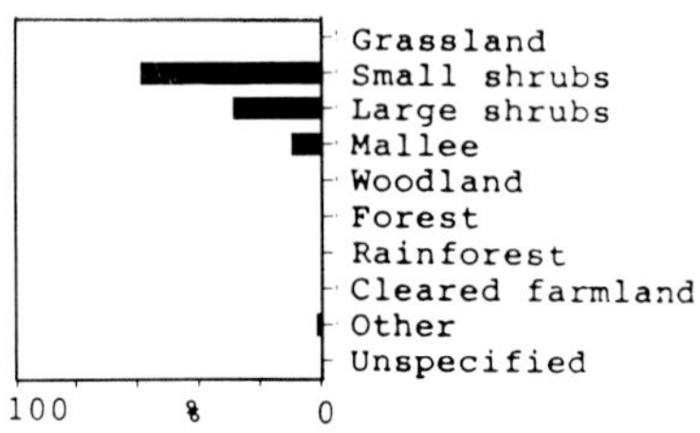

Soil Type

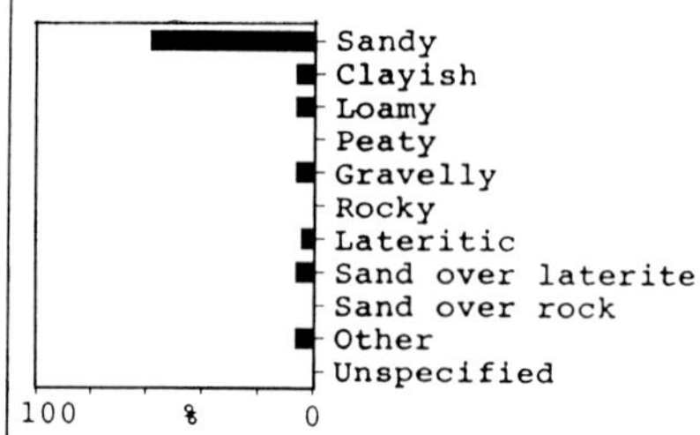

Landform

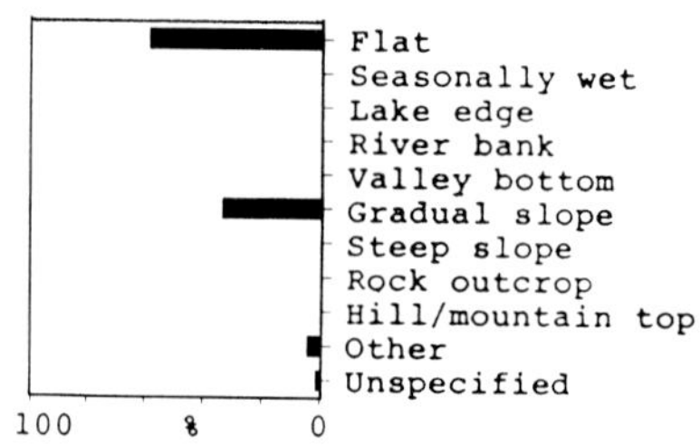

Response to Fire: 8 records

Growth response: Killed, new seedlings (50%) killed, no seedlings and killed, new seedlings (12%) unspecified (37%).

Flowering response: Number of records where flowering had occurred after fire (2). Median reported time to flowering (25-36 months); minimum reported time to flowering (13-24 months). Number of records where flowering had not occurred after fire (4). Median reported time to no flowering (12-13 months); maximum reported time to no flowering (25-36 months).

Flowering

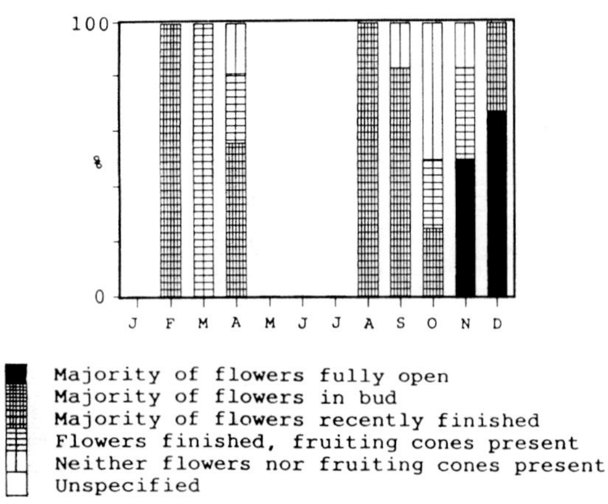

Possible pollinators

Pollinator type 5 records: no pollinator observed (100%).

Specific pollinator 0 records.

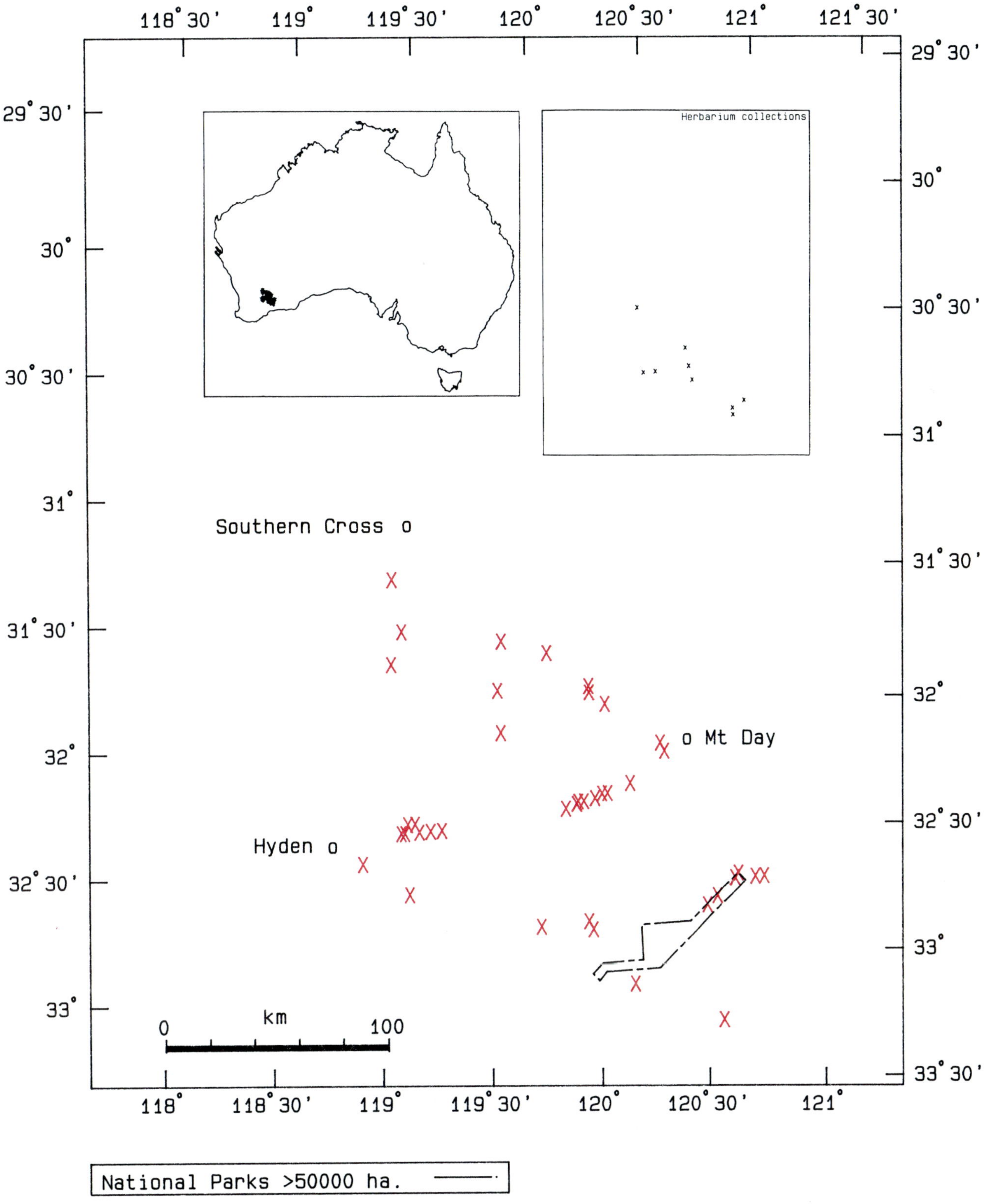

DISCUSSION

Atlas records confirm that B. laevigata subsp. fuscolutea occurs in scattered populations between Southern Cross, Mt Day, Hyden and Frank Hann National Park. It is only along the Hyden - Norseman track and in the eastern section of Frank Hann National Park that it becomes relatively common. Its range has been extended about 50 km further south by an Atlas record near the headwaters of the Young River. B. laevigata subsp. fuscolutea may be under-recorded since access is difficult in much of its habitat. Thus, the apparent linear distribution in parts of its range is due to records conforming with existing roads and tracks. Most populations were of less than 100 plants and only 15% were recorded from conservation reserves (notably Frank Hann National Park). The range of the subspecies is mainly uncleared land to the east and north of the Wheatbelt. Therefore, its conservation status seems relatively secure.

The subspecies grows as a shrub up to 4 m in height generally on flat to gently undulating sandy soil, which may sometimes overlie laterite. Surrounding vegetation is usually shrubland, possibly with emergent mallees. George (1981) described the flowering period as December to January. Atlas records show it to be flowering in November and December. However, there were no Atlas records from January and only two from February so the flowering period may last longer than was recorded. New shoot growth was strongly evident in November and to a lesser extent in December and February. The subspecies is killed by fire and regenerates from seed.

BANKSIA LAEVIGATA Meissner (1856) subsp. LAEVIGATA

Flowers pale yellow with grey hairs. Styles cream.

14 RECORDS: Jan (1) Feb (0) Mar (0) Apr (0) May (1) Jun (0) Jul (0) Aug (0) Sep (5) Oct (5) Nov (2) Dec (0).

Population Size: 1-10 (3) 10-100 (6) >100 (5) Unspecified (0).

Conservation Status: Restricted to road verge (0%) Not (86%) Unspecified (14%). In conservation reserve (64%) Not (29%) Unspecified (7%).

Tree/Shrub: Tree form (7%) Shrub form (93%) Unspecified (0%).

Height (metres)

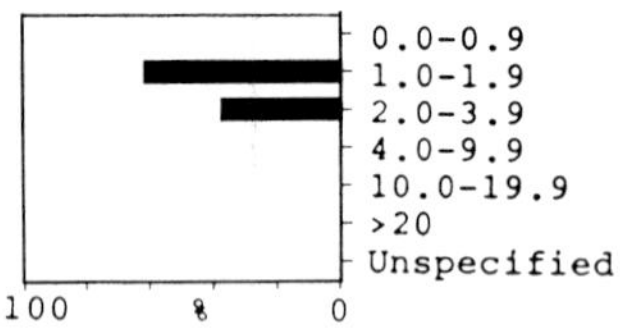

New Shoot Growth

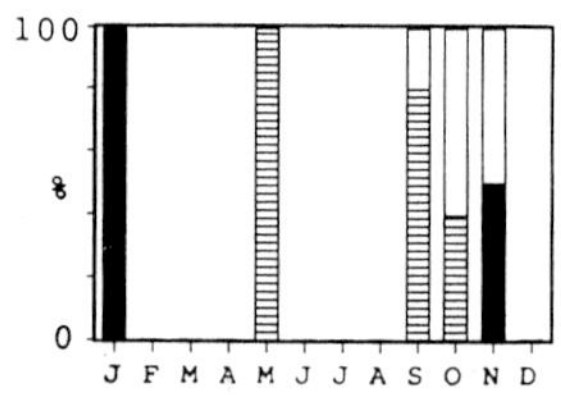

Yes
No
Unspecified

HABITAT

Within 2 km of coast: Yes (0%) No (100%) Unspecified (0%).

Altitude (metres)

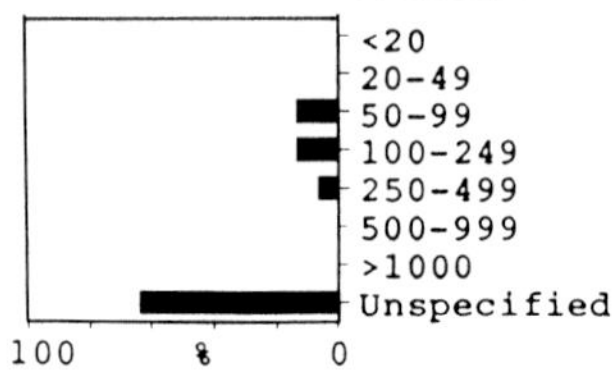

Vegetation

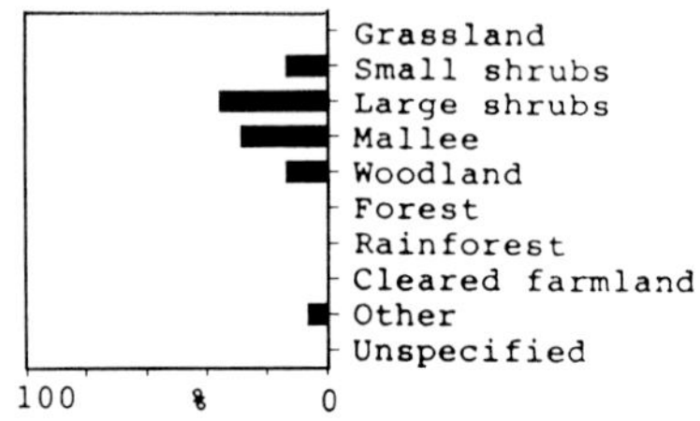

Soil Type

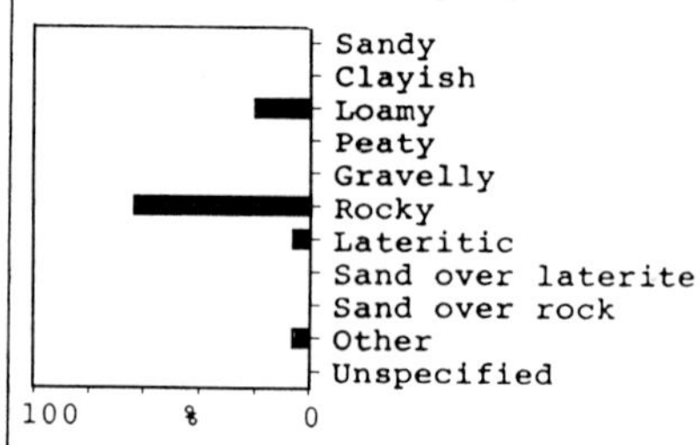

Landform

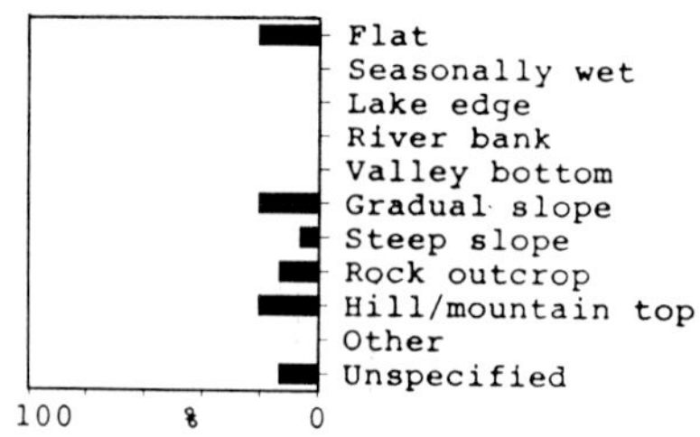

Response to Fire: 0 records

Flowering

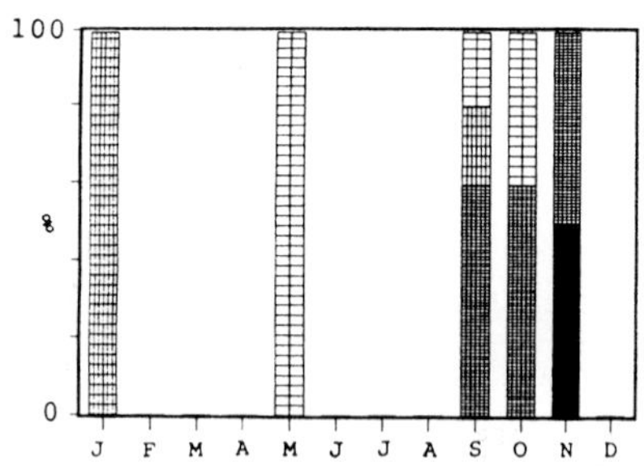

Majority of flowers fully open
Majority of flowers in bud
Majority of flowers recently finished
Flowers finished, fruiting cones present
Neither flowers nor fruiting cones present
Unspecified

Possible pollinators

Pollinator type 3 records: bees, wasps, ants (33%) no pollinator observed (67%).

Specific pollinator 0 records.

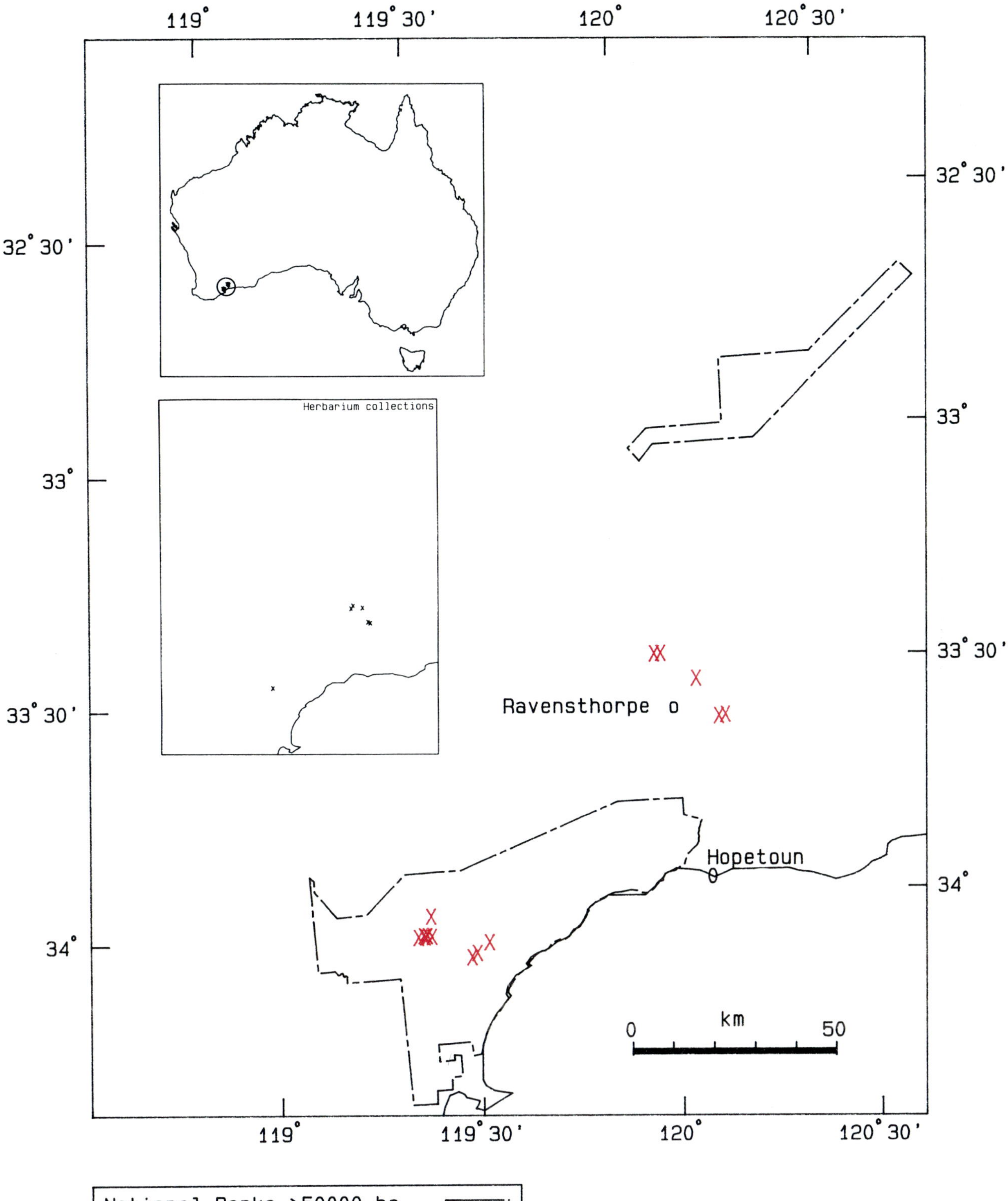

DISCUSSION

B. laevigata subsp. laevigata is a relatively rare banksia confined to the Ravensthorpe Range and a few sites in the Fitzgerald River National Park (generally near to the Fitzgerald River). The southern-most records are all new sites, not represented in herbarium collections. The subspecies may be even more rare than it appears on the map since 5 of the 14 records represent one site close to the Twertup Field Studies Centre. If this is the case there would be only 10 known sites with 50% in conservation reserves. Most populations are of less than 100 plants. Moreover, the species is fairly susceptible to dieback disease. For all these reasons, populations of this subspecies should be closely monitored.

B. laevigata subsp. laevigata occurs as a shrub or small tree up to 4 m in height. Typically, it grows on the rocky slopes and tops of hills and breakaways. Surrounding vegetation is generally shrubland and mallee eucalypts e.g. E. tetragona, E. falcata. Frequently, it is associated with B. lemanniana. Occasionally, B. laevigata subsp. laevigata occurs in woodland e.g. of E. gardneri or E. lehmannii. George (1981) described the flowering period of B. laevigata subsp. laevigata as November to January. Atlas records show it to be largely in bud in September/October and flowering in November. There are no records in December and a single record in January is of "flowers mainly finished still with some colour". New shoot growth appears to be in summer. There are no Atlas records of the subspecies' response to fire but George (1981) stated that it is non-lignotuberous and dependent on seed germination for regeneration.

BANKSIA LANATA A.S.George (1981)

17 RECORDS: Jan (0) Feb (0) Mar (2) Apr (5) May (1) Jun (0) Jul (1) Aug (2) Sep (2) Oct (3) Nov (0) Dec (1).

Population Size: 1-10 (2) 10-100 (4) >100 (9) Unspecified (2).

Conservation Status: Restricted to road verge (41%) Not (59%) Unspecified (0%). In conservation reserve (18%) Not (82%) Unspecified (0%).

Tree/Shrub: Tree form (0%) Shrub form (94%) Unspecified (6%).

Height (metres)

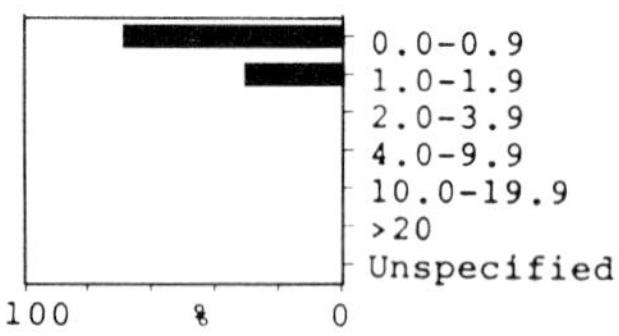

New Shoot Growth

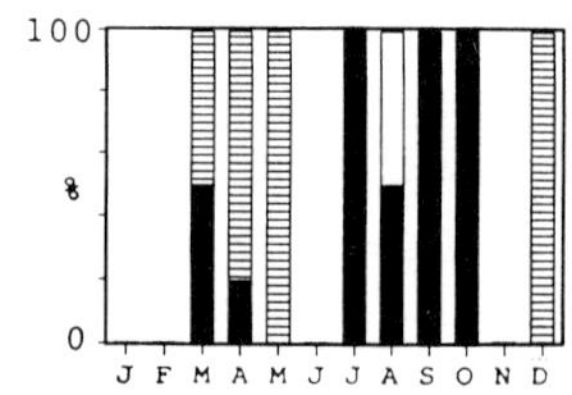

Yes
No
Unspecified

HABITAT

Within 2 km of coast: Yes (0%) No (94%) Unspecified (6%).

Altitude (metres)

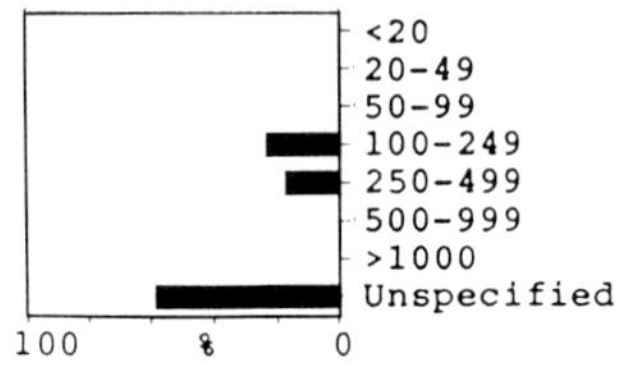

Vegetation

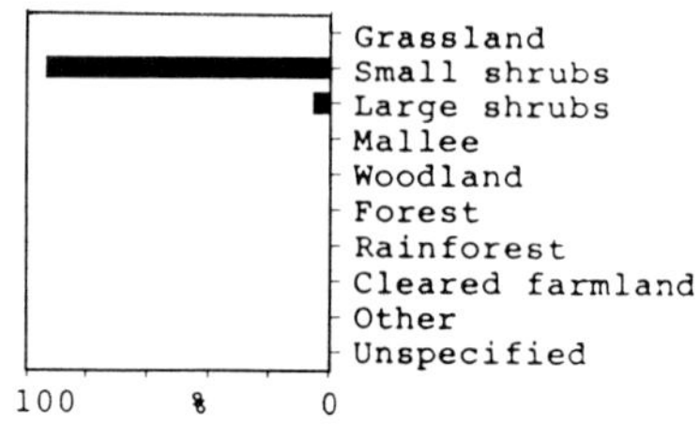

Soil Type

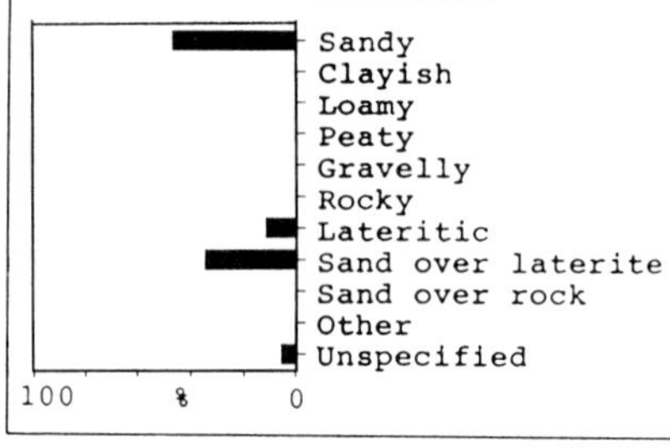

Landform

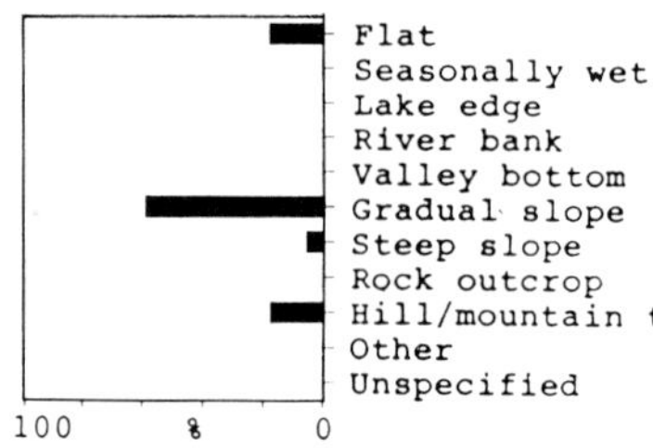

Response to Fire: 0 records

Flowering

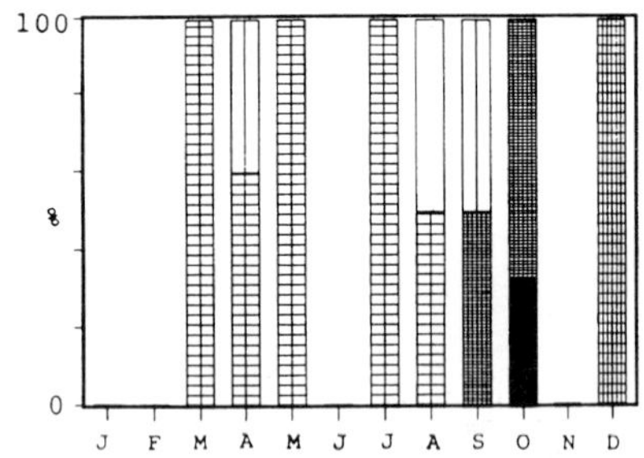

Majority of flowers fully open
Majority of flowers in bud
Majority of flowers recently finished
Flowers finished, fruiting cones present
Neither flowers nor fruiting cones present
Unspecified

Possible pollinators

Pollinator type 4 records: no pollinator observed (100%).

Specific pollinator 0 records.

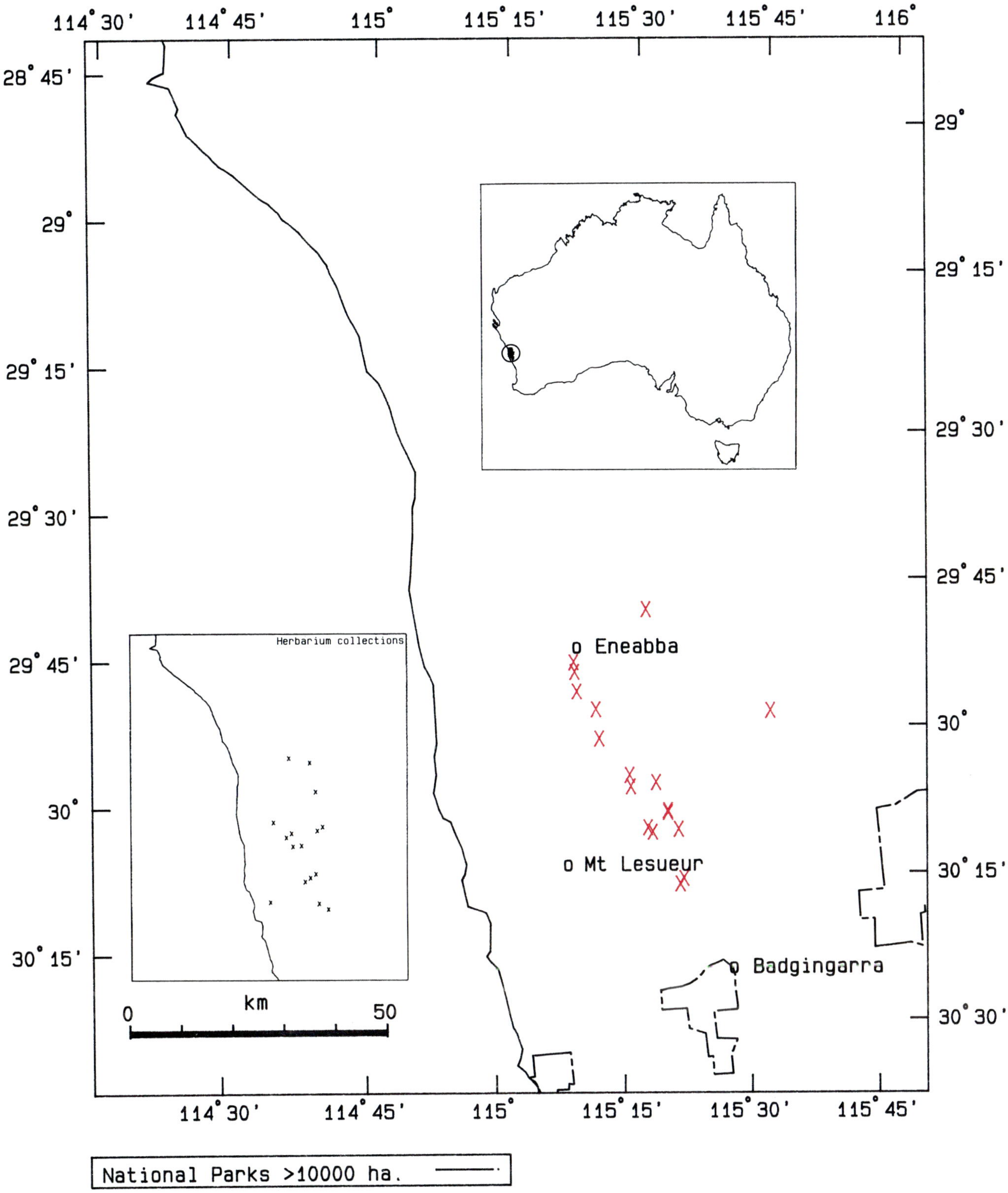

DISCUSSION

Western Australian Herbarium records indicated that B. lanata ranged between Coomallo Creek and a location 32 km west of Arrino with an eastern boundary at Tathra National Park and the most western record near Cockleshell Gully. Atlas records convey a broadly similar picture although the majority form a linear distribution between Coomallo Creek and Eneabba, always within 10 km of the Brand Highway. An apparent new Atlas record is from Willmott Road north of Alexander Morrison National Park. However, this could be doubtful as the recorder was not certain of his identification. The species has a fairly small geographic range of less than 100 km. It is not well represented on conservation reserves and has a fairly high percentage of its populations restricted to road verges. It is, however, locally common and does not appear to be under any immediate threat.

B. lanata is one of several species that until 1981 was included within B. sphaerocarpa. It is identified by its bright pink new growth and its floral bracts which are covered with white, woolly hairs. Typically it is a small shrub of 1 m or less growing in deep white sand or sand over laterite. It occurs on gently undulating to flat land, rarely on lateritic hill tops. Surrounding vegetation is predominantly low heath. Atlas data on flowering are poor, as with many Western Australian summer-flowering species. George (1981) described the flowering period as late October to January. New shoot growth appears to be in winter and spring. However, there are no records from summer months when new shoot growth might be expected to occur. There are no Atlas records of the species' response to fire, but George (1981) stated that it is non-lignotuberous and relies on seed germination for regeneration.

BANKSIA LARICINA C.Gardner (1964)

Rose Banksia, Rose-fruited Banksia

23 RECORDS: Jan (0) Feb (0) Mar (2) Apr (2) May (7) Jun (0) Jul (3) Aug (3) Sep (2) Oct (0) Nov (1) Dec (3).

Population Size: 1-10 (1) 10-100 (6) >100 (16) Unspecified (0).

Conservation Status: Restricted to road verge (9%) Not (91%) Unspecified (0%). In conservation reserve (65%) Not (35%) Unspecified (0%).

Tree/Shrub: Tree form (0%) Shrub form (100%) Unspecified (0%).

Height (metres)

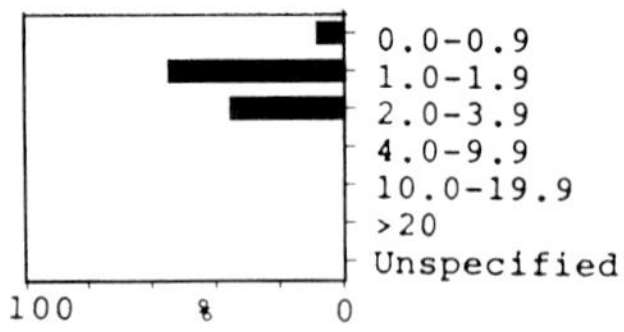

HABITAT

Within 2 km of coast: Yes (0%) No (100%) Unspecified (0%).

Altitude (metres)

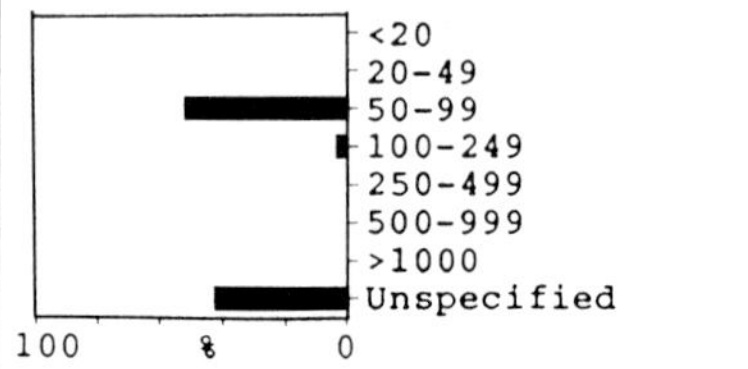

Vegetation

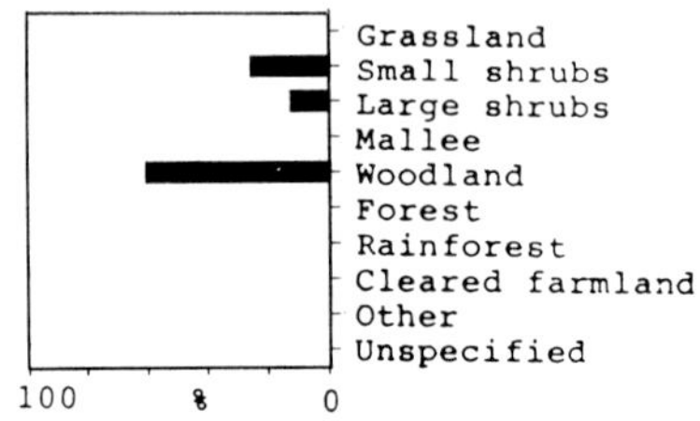

Soil Type

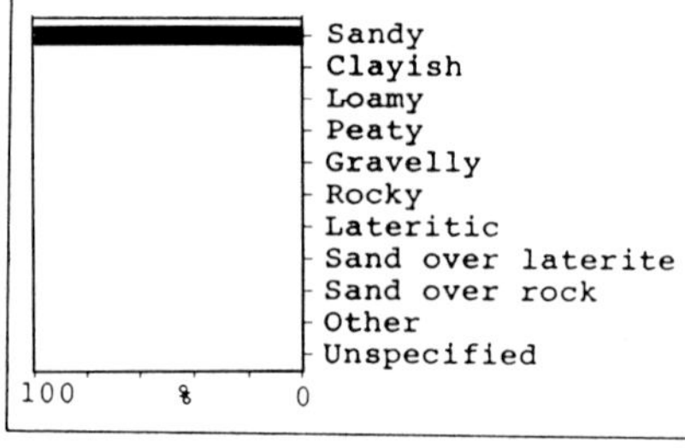

Landform

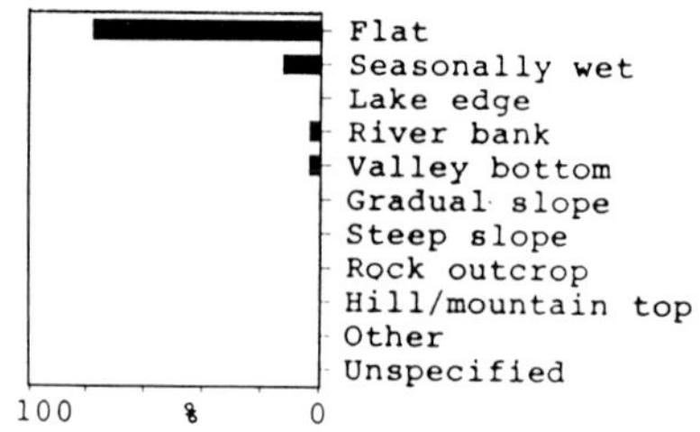

New Shoot Growth

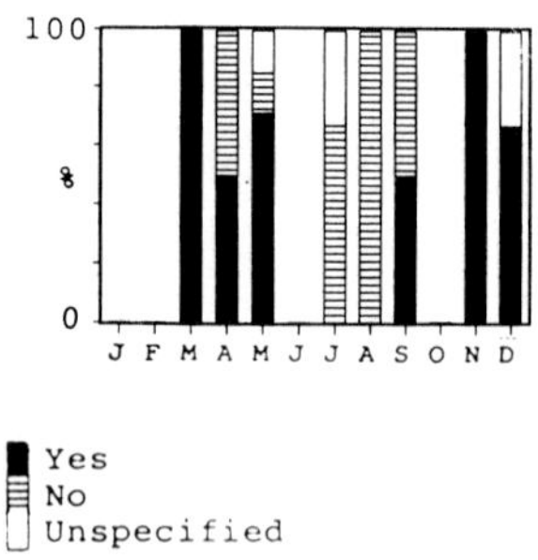

Response to Fire: 0 records

Flowering

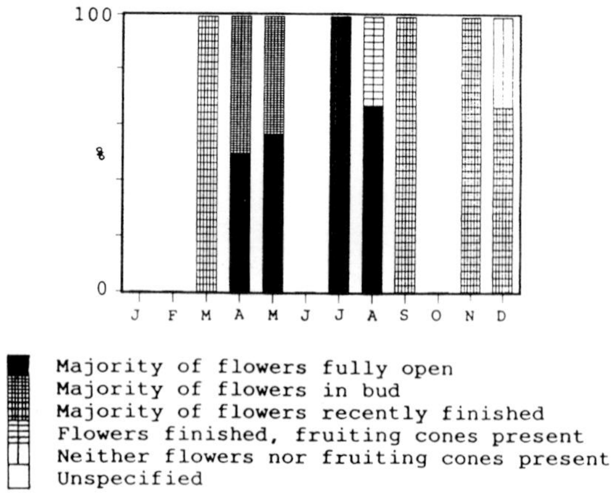

Possible pollinators

Pollinator type 4 records: bees, wasps, ants (25%) no pollinator observed (75%).

Specific pollinator 2 records: ant (1) bee (1)

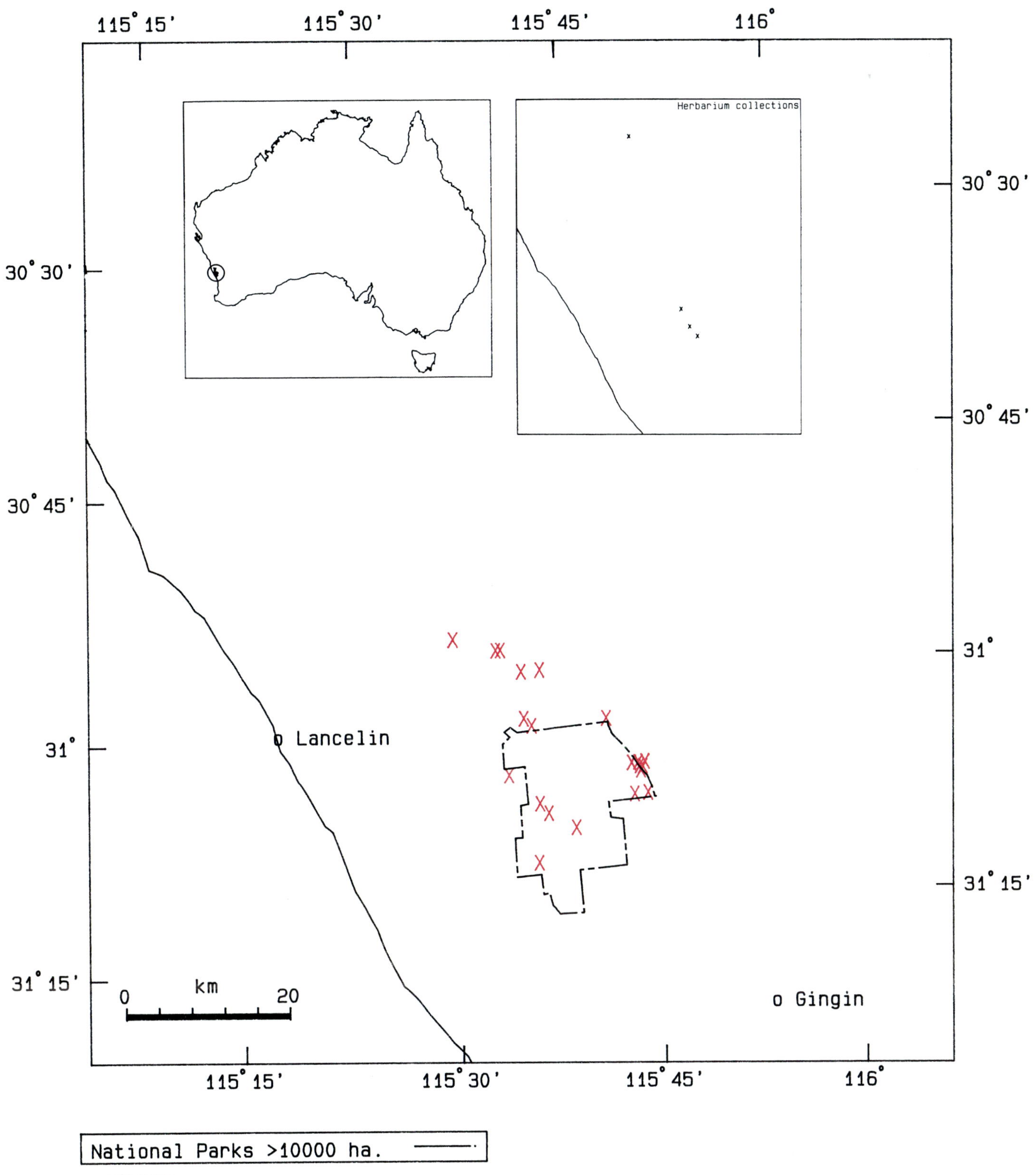

DISCUSSION

B. laricina has a restricted distribution with a geographic range of only 35 km. It occurs both north and south of the Moore River, from Namming Nature Reserve in the north to the south-west corner of Moore River National Park. A 1965 collection of R. Roberts from Badgingarra was not recorded. Atlas records suggest that the conservation status of the species is better than was previously thought. It is well represented in conservation reserves where it is locally common with most if its populations being over 100 plants. In parts of Moore River National Park it was described as the dominant species and one area was reported to have "thousands" of plants. However, its attractive fruits continue to be harvested for the commercial wildflower trade, and it is also one of the most susceptible banksias to dieback disease. Therefore, its status should continue to be monitored.

B. laricina is a shrub between 1 and 2 m high. It occurs only on deep white or grey sands where the landform is flat and sometimes seasonally wet. Surrounding vegetation is usually open woodland with B. menziesii, B. attenuata and B. ilicifolia. In wetter areas the species may be associated with B. littoralis and B. telmatiaea. B. laricina is also found in shrubland, often dominated by Adenanthos cygnorum. Flowering is from April to August and new shoot growth appears to be in summer though further data are needed to confirm this. There were no observations of the species' response to fire, although George (1981) stated that it is non-lignotuberous and it is therefore dependent on seed germination for regeneration. An interesting form of fruiting cone with wavy follicle margins was found on a single plant on Moore River National Park (see figure elsewhere in main text). Such an unusual fruit would probably have great horticultural potential. It may not be as rare as it would seem since similar cones have been noted from other areas (G. Collison, pers. comm.).

BANKSIA LEMANNIANA Meissner (1856)

Lemann's Banksia

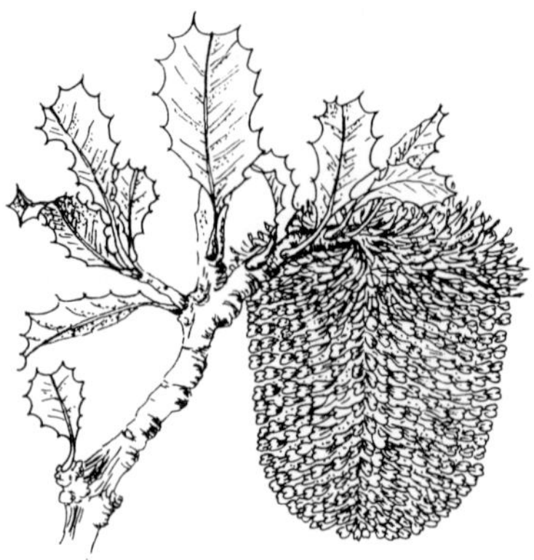

69 RECORDS Jan (9) Feb (3) Mar (5) Apr (2) May (0) Jun (2) Jul (3) Aug (4) Sep (14) Oct (13) Nov (7) Dec (5).

Population Size: 1-10 (13) 10-100 (29) >100 (24) Unspecified (3).

Conservation Status: Restricted to road verge (1%) Not (94%) Unspecified (4%). In conservation reserve (74%) Not (25%) Unspecified (1%).

Tree/Shrub: Tree form (7%) Shrub form (88%) Unspecified (4%).

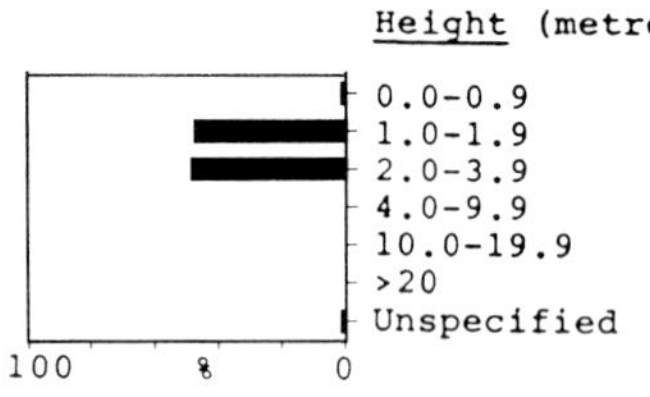

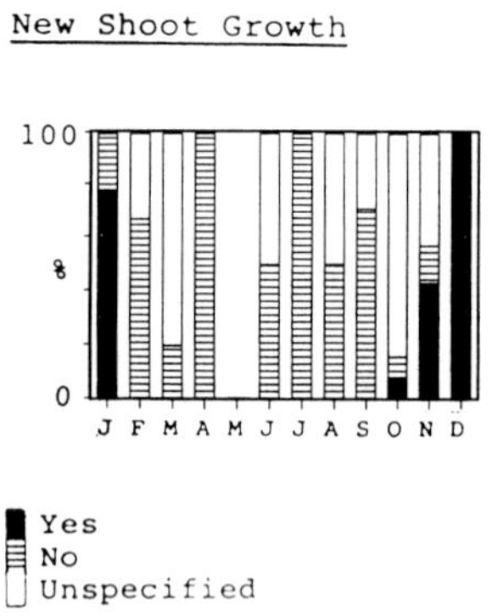

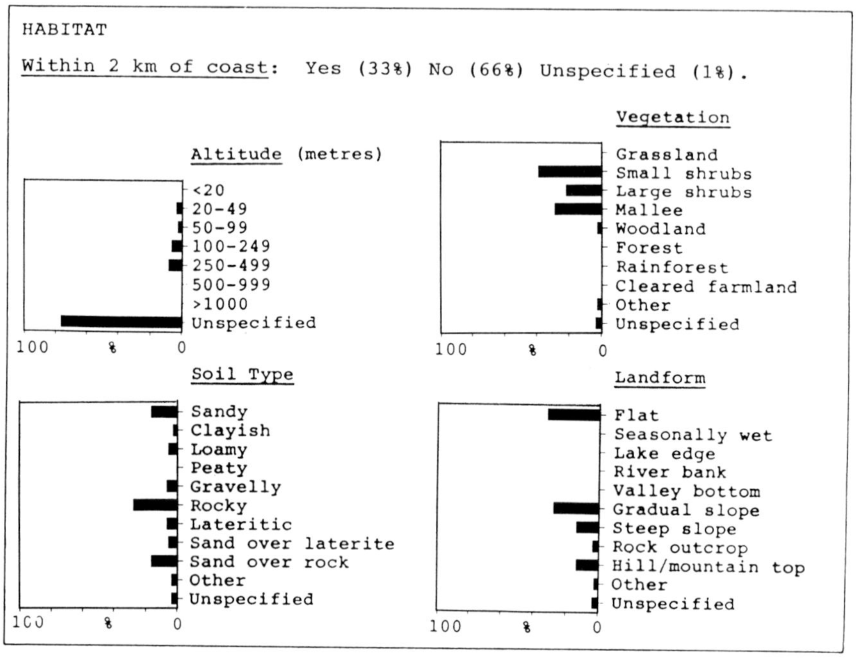

Response to Fire: 4 records

Growth response: Killed, new seedlings (50%) unspecified (50%).

Flowering response: Number of records where flowering had occurred after fire (3). Median reported time to flowering (13-24 months); minimum reported time to flowering (13-24 month). Number of records where flowering had not occurred after fire (0).

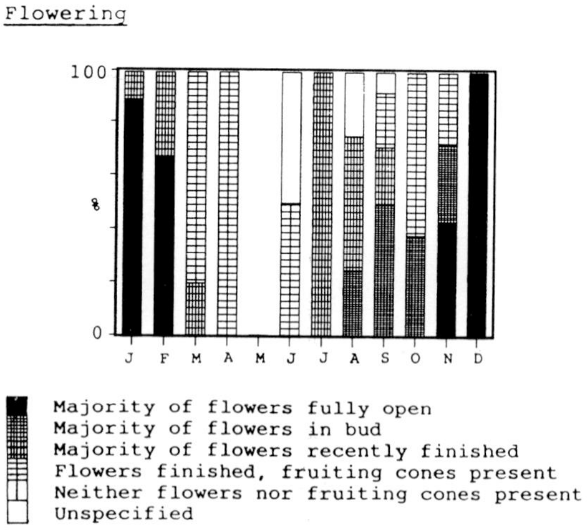

Possible pollinators

Pollinator type 19 records: bird (16%) bees, wasps, ants (16%) no pollinator observed (68%).

Specific pollinator 6 records: New Holland Honeyeater (1) Red Wattlebird (1) ant (2) native bee (1) European honey bee (1)

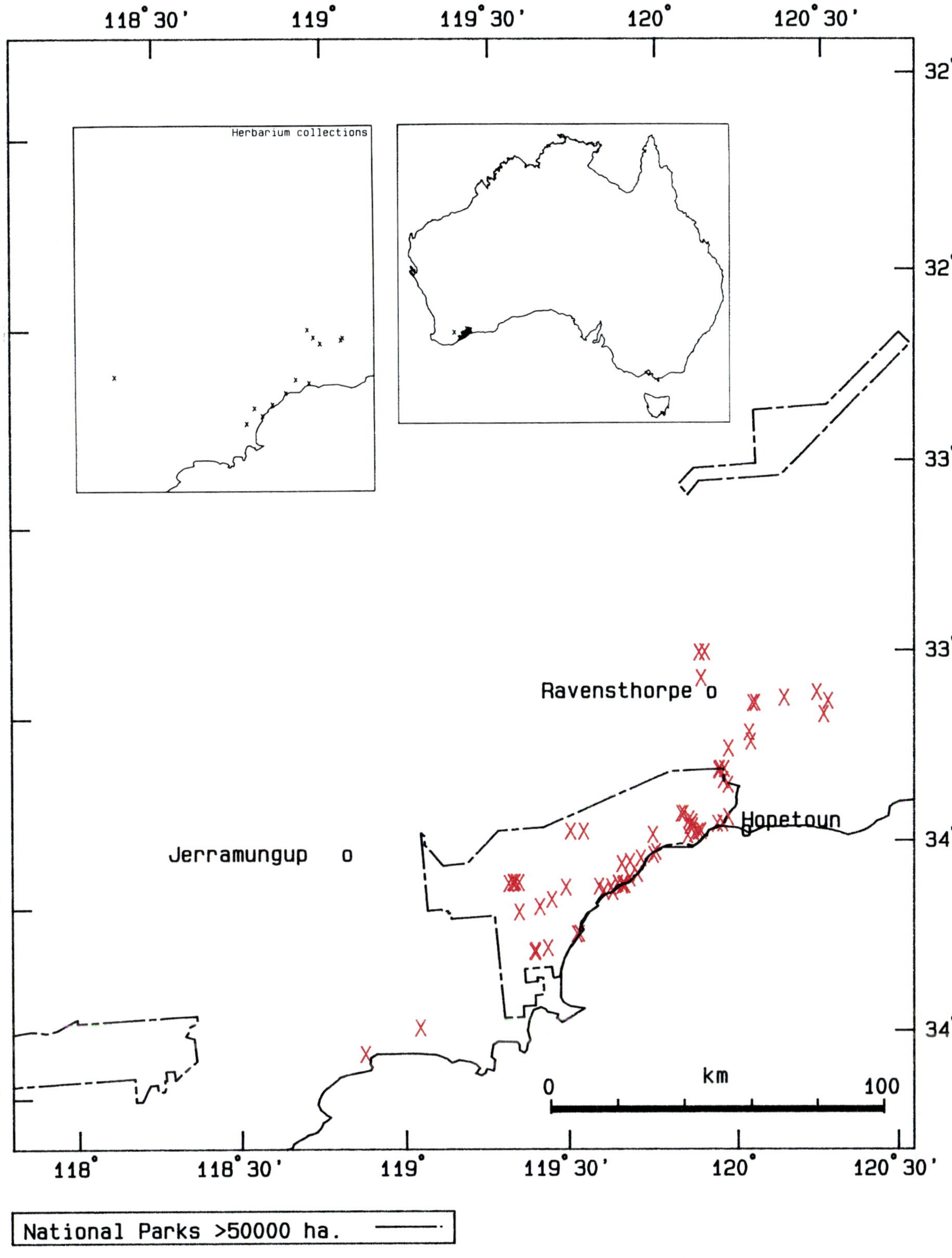

DISCUSSION

Atlas data confirm that the main range of B. lemanniana is within the Fitzgerald River National Park, extending to the Ravensthorpe Range and as far east as the Rabbit Proof Fence. Westwards there are two new records from outside the national park. The recording from Pallinup estuary extends the previously known range by some 60 km. B. lemanniana is relatively secure with almost three-quarters of its recorded populations being within the Fitzgerald River National Park. It is only slightly susceptible to dieback disease and is not utilised by the commercial wildflower trade.

The species is generally a tall shrub up to 4 m in height. Tree-like forms are also recorded. It grows predominantly in rocky soil which is sometimes lateritic. It also occurs in sand or sand overlying laterite. B. lemanniana may be found on hilltops, mountain slopes, and flat plains amongst heath, tall shrubland or mallee eucalypts. It flowers in summer, when birds such as the New Holland Honeyeater and Red Wattlebird were observed on the inflorescences. Summer is also the season of new shoot growth. The species is killed by fire and regenerates from seed. There are insufficient data to indicate when seedlings reach maturity and flower, but George (1984) stated that 5-6 years are required.

BANKSIA LEPTOPHYLLA A.S.George (1981)

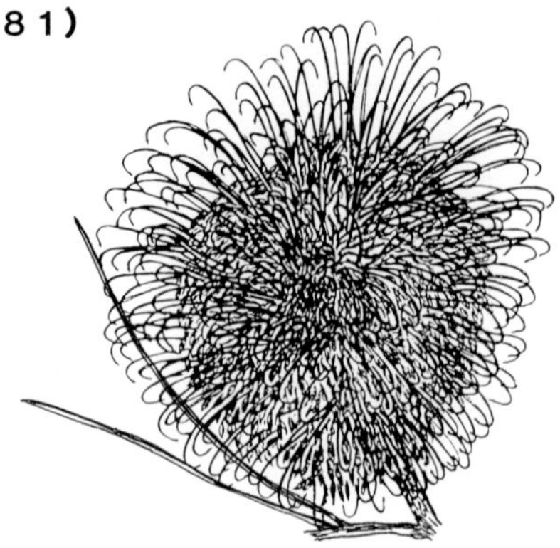

104 RECORDS: Jan (18) Feb (8) Mar (12) Apr (3) May (9) Jun (2) Jul (3) Aug (11) Sep (19) Oct (7) Nov (4) Dec (6).

Population Size: 1-10 (15) 10-100 (34) >100 (54) Unspecified (1).

Conservation Status: Restricted to road verge (25%) Not (74%) Unspecified (1%). In conservation reserve (30%) Not (67%) Unspecified (3%).

Tree/Shrub: Tree form (0%) Shrub form (99%) Unspecified (1%).

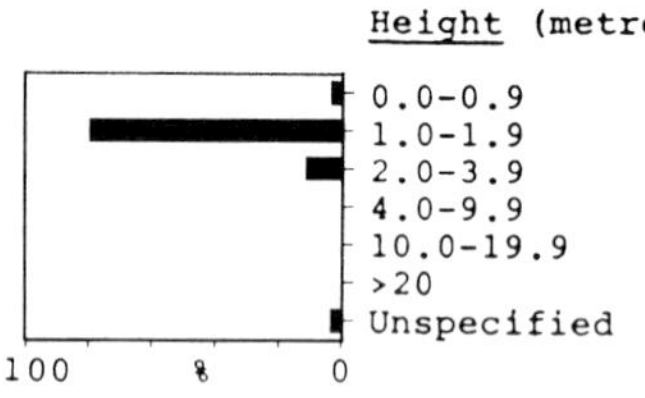

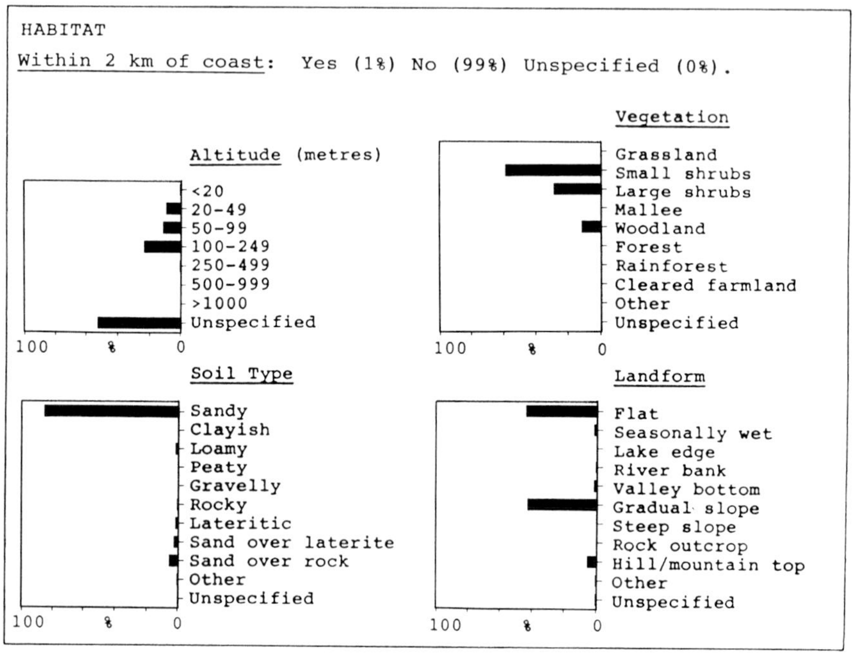

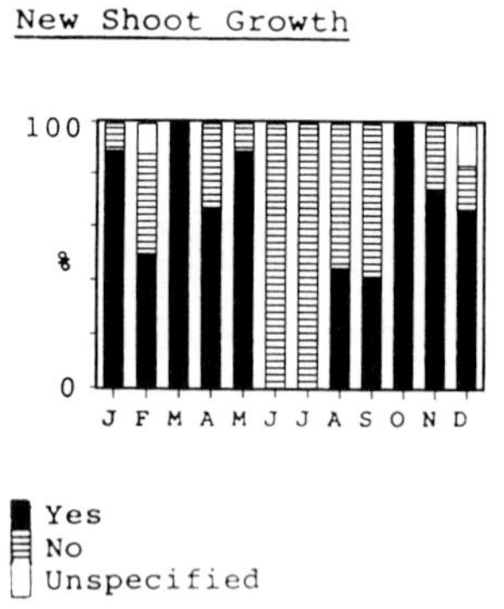

Response to Fire: 13 records

Growth response: Killed, new seedlings (38%) killed, no seedlings (15%) unspecified (46%).

Flowering response: Number of records where flowering had occurred after fire (6). Median reported time to flowering (13-24 months); minimum reported time to flowering (<1 month). Number of records where flowering had not occurred after fire (5). Median reported time to no flowering (13-24 months); maximum reported time to no flowering (37-48 months).

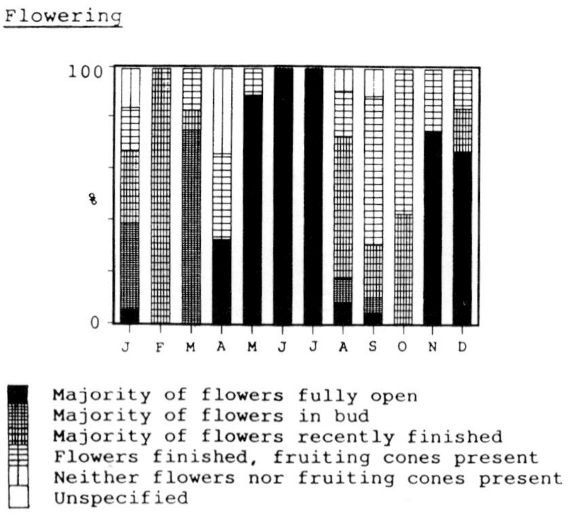

Possible pollinators

Pollinator type 19 records: bees, wasps, ants (11%) no pollinator observed (89%).

Specific pollinator 1 record: ant (1).

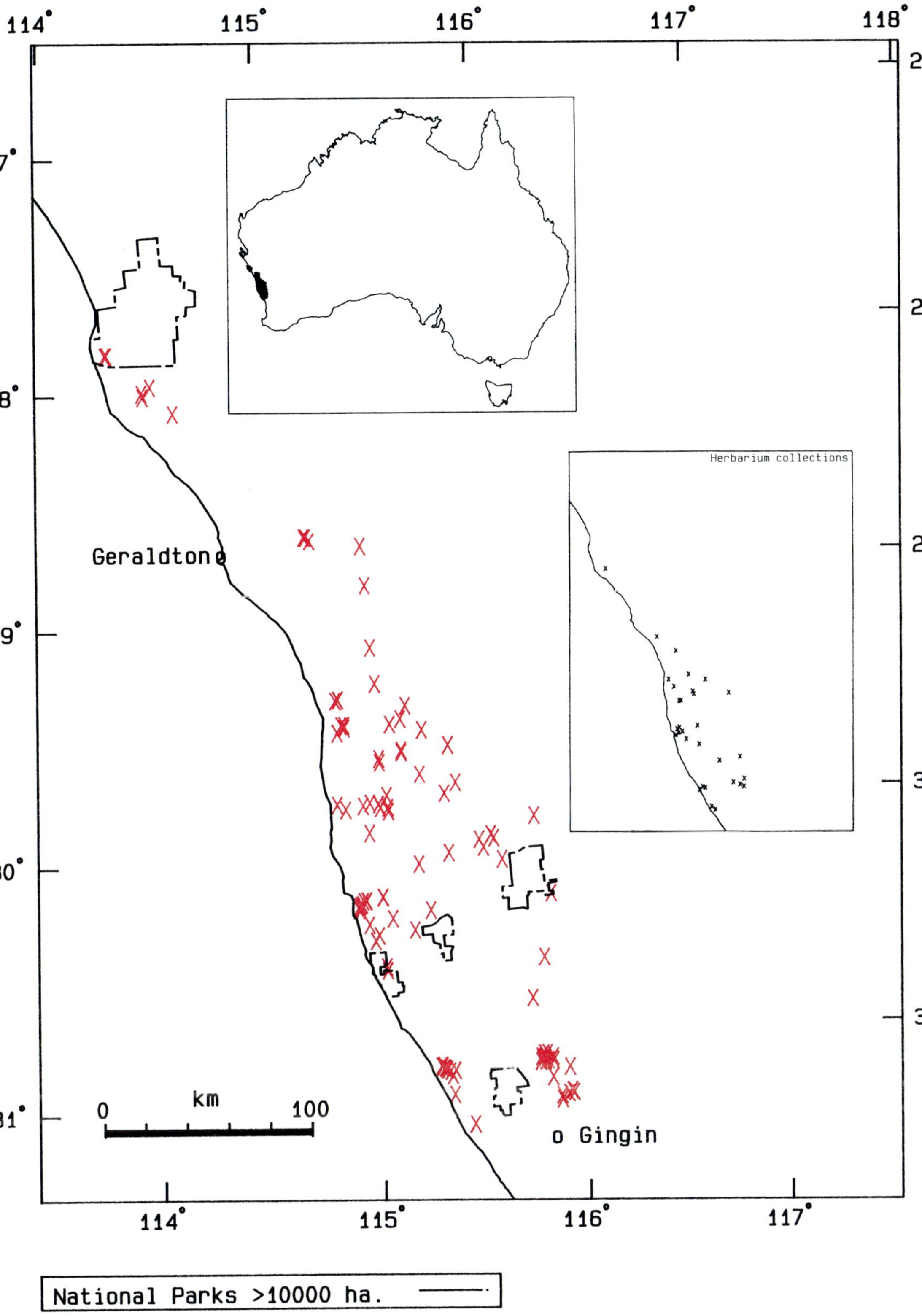

DISCUSSION

B. leptophylla occurs from Kalbarri southwards to Guilderton and almost to Gingin. Both Atlas and herbarium records suggest that the distribution may not be continuous. A northern population from Kalbarri south to Hutt River appears to be separated by a gap of approximately 90 km from the main southern group. The furthest inland records are from 15 km south-south-west of Coorow and the south-east corner of Watheroo National Park. The species is locally common with most if its recorded populations being of more than 100 plants. It is also well represented in conservation reserves.

B. leptophylla is typically a shrub less than 2 m in height, though sometimes larger. It occurs predominantly in sandy soil which, in near-coastal areas, may sometimes overlie limestone. Landforms are usually flat or gently undulating. Surrounding vegetation is generally shrubland but occasionally woodland. Although the species is often found near the coast it is rarely within 2 km of the coastline. Flowering occurs in two distinct peaks, one in November - December, the other from April to July. The fact that summer-flowering populations have large flowers and inflorescences and winter-flowering populations have small flowers and inflorescences has led George (1987) to now recognise two varieties. B. leptophylla var. leptophylla now refers to the form with large flowers appearing in summer. It generally occurs in the eastern parts of the species' range. B. leptophylla var. melletica now refers to plants with small flowers produced in winter and with a generally western distribution. The species shows new shoot growth throughout much of the year (except winter) with apparent peaks in October, January, March and May. B. leptophylla is non-lignotuberous, being killed by fire and regenerating from seed.

BANKSIA LINDLEYANA Meissner (1855)

Porcupine Banksia

28 RECORDS: Jan (0) Feb (2) Mar (0) Apr (3) May (0) Jun (0) Jul (0) Aug (3) Sep (20) Oct (0) Nov (0) Dec (0).

Population Size: 1-10 (7) 10-100 (16) >100 (5) Unspecified (0).

Conservation Status: Restricted to road verge (0%) Not (89%) Unspecified (11%). In conservation reserve (54%) Not (43%) Unspecified (4%).

Tree/Shrub: Tree form (0%) Shrub form (100%) Unspecified (0%).

Height (metres)

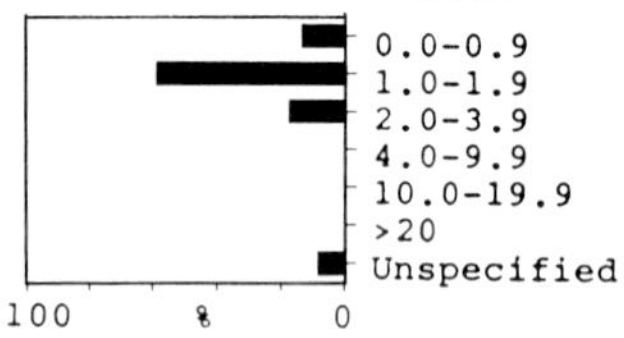

New Shoot Growth

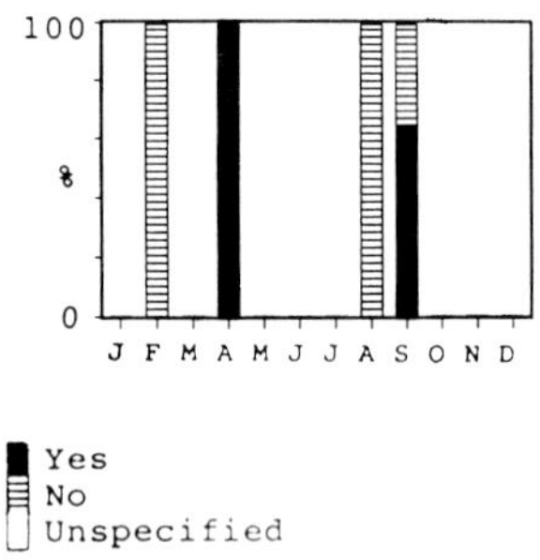

HABITAT

Within 2 km of coast: Yes (14%) No (86%) Unspecified (0%).

Altitude (metres)

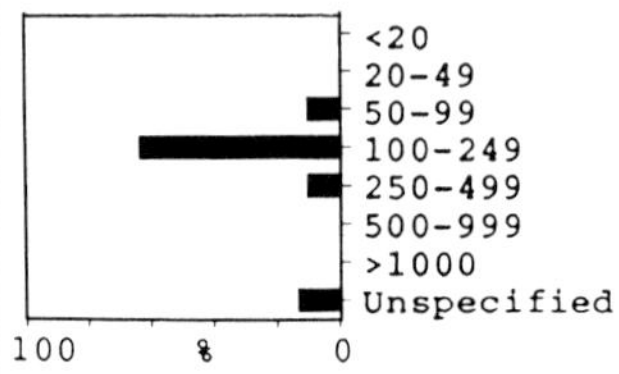

Vegetation

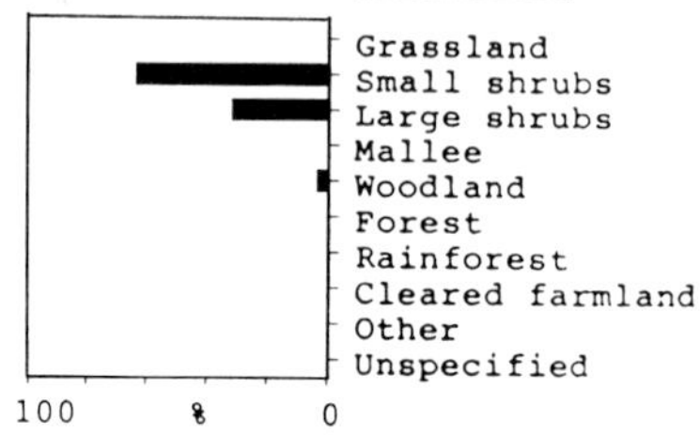

Soil Type

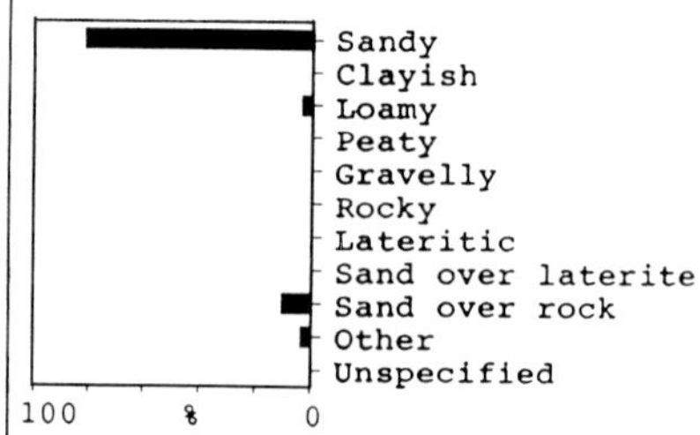

Landform

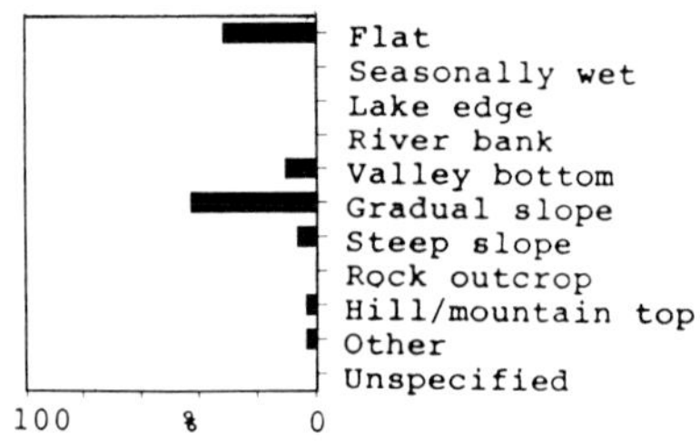

Response to Fire: 4 records

Growth response: Ground resprout (75%) unspecified (25%).

Flowering response: Number of records where flowering had occurred after fire (2). Median reported time to flowering (37-48 months); minimum reported time to flowering (37-48 month). Number of records where flowering had not occurred after fire (0).

Flowering

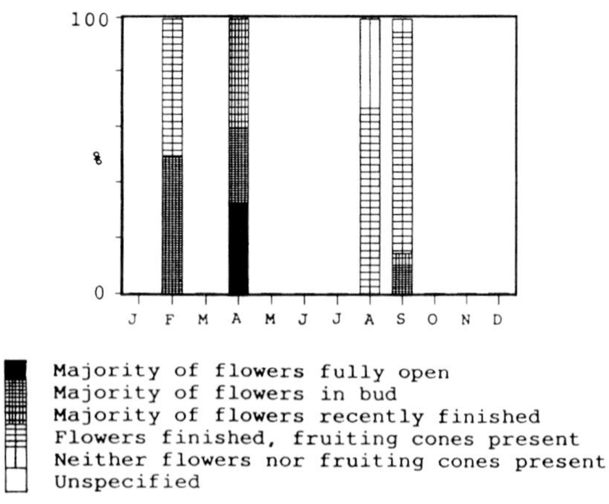

Possible pollinators

Pollinator type 2 records: bees, wasps, ants (50%) no pollinator observed (50%).

Specific pollinator 1 record: ant (1).

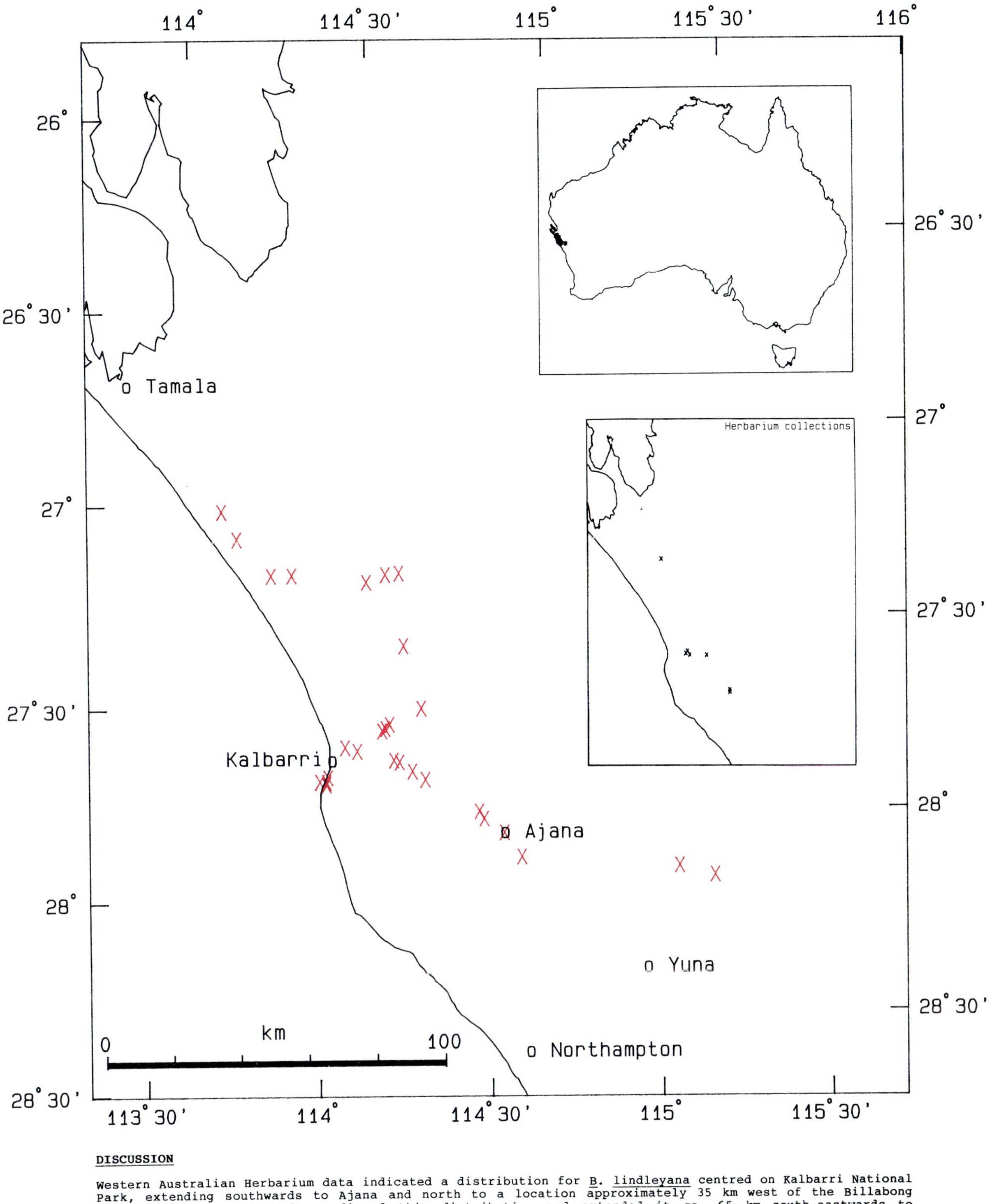

DISCUSSION

Western Australian Herbarium data indicated a distribution for B. lindleyana centred on Kalbarri National Park, extending southwards to Ajana and north to a location approximately 35 km west of the Billabong Roadhouse. Atlas records confirmed this distribution and extended it ca. 65 km south-eastwards to Wandana Nature Reserve. Only the western boundary of this large reserve was explored, so it is possible that the species extends even further inland. There are also new Atlas records from along the State barrier fence north of Kalbarri National Park, and from within and 15 km north of Zuytdorp National Park. Although most of the area in which B. lindleyana occurs was sparsely covered by contributors, Atlas records suggest that it is probably more common than was previously thought. Also, more than 50% of recorded populations were in conservation reserves. However, most populations were of less than 100 plants.

The species is typically a shrub less than 2 m in height though sometimes up to 4 m. It favours deep yellow sandy soils and generally occurs in shrubland sometimes associated with B. sceptrum, B. prionotes or B. ashbyi. Near the coast it often grows in swales between sand dunes, whilst inland it occurs on flat sandy plains. Flowering was sparsely recorded by Atlas contributors, though George (1981) stated that it is from January to March. At this time of year field work is unpleasant due to the very high temperatures experienced. Data on new shoot growth are inconclusive. Following fire, the species regenerates from its lignotuber. An interesting B. lindleyana x B. prionotes hybrid plant was recorded from north of Kalbarri National Park (see Fig. 14 in Chapter 3).

BANKSIA LITTORALIS R.Brown (1810)

Western Swamp Banksia, Swamp Banksia

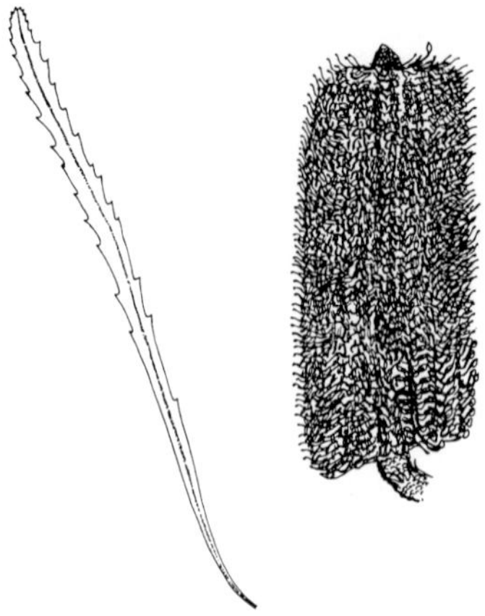

726 RECORDS: Jan (99) Feb (29) Mar (146) Apr (86) May (53) Jun (44) Jul (28) Aug (52) Sep (71) Oct (53) Nov (38) Dec (25).

Population Size: 1-10 (267) 10-100 (353) >100 (98) Unspecified (8).

Conservation Status: Restricted to road verge (11%) Not (84%) Unspecified (5%). In conservation reserve (26%) Not (69%) Unspecified (5%).

Tree/Shrub: Tree form (92%) Shrub form (7%) Unspecified (0%).

Height (metres)

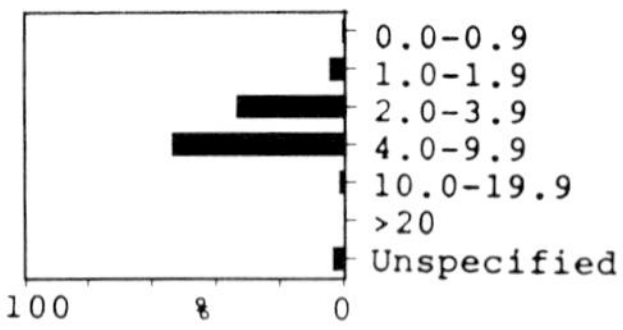

New Shoot Growth

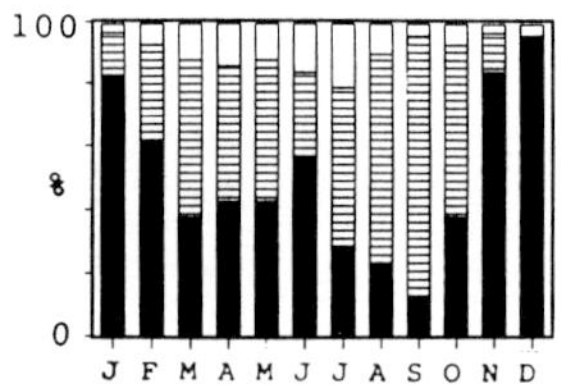

Yes
No
Unspecified

HABITAT

Within 2 km of coast: Yes (11%) No (88%) Unspecified (1%).

Altitude (metres)

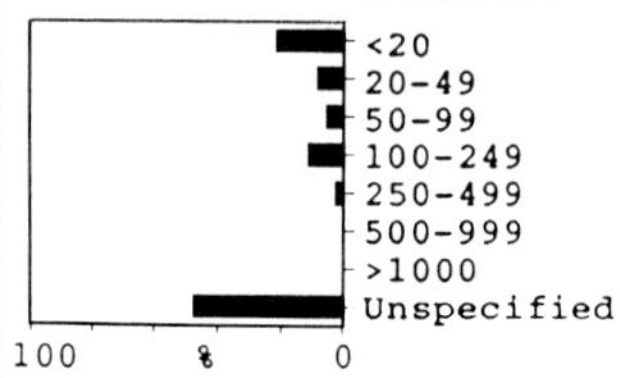

Vegetation

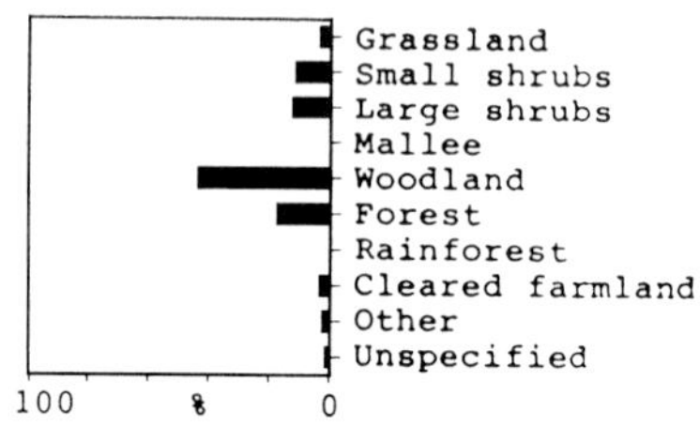

Soil Type

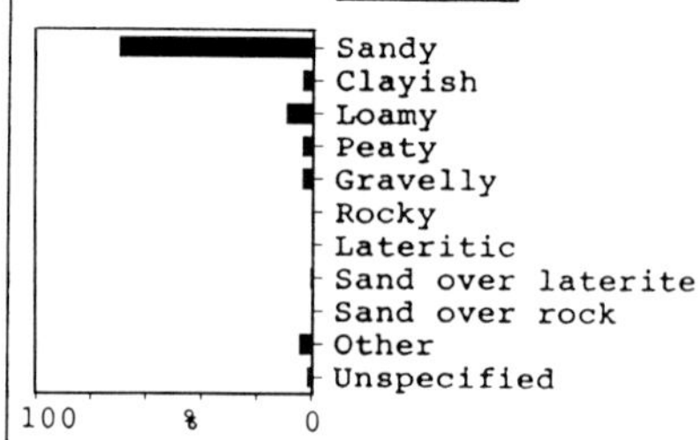

Landform

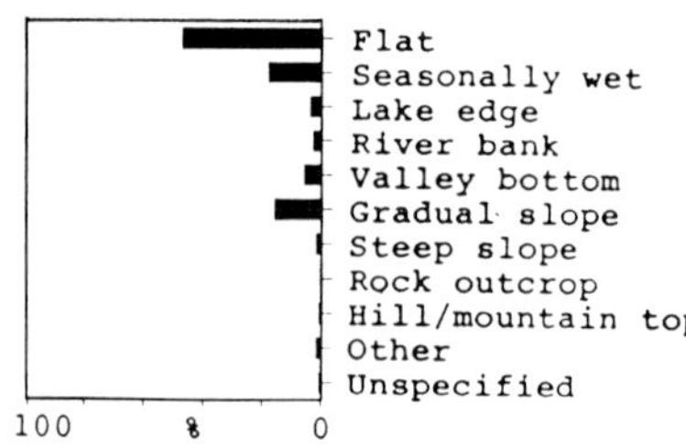

Response to Fire: 125 records

Growth response: Ground resprout (5%) ground resprout and trunk resprout (10%) killed, new seedlings (2%) killed, new seedlings and trunk resprout (1%) trunk resprout (30%) unspecified (52%).

Flowering response: Number of records where flowering had occurred after fire (70). Median reported time to flowering (13-24 months); minimum reported time to flowering (<1 month). Number of records where flowering had not occurred after fire (23). Median reported time to no flowering (1-12 months); maximum reported time to no flowering (61-72 months).

Flowering

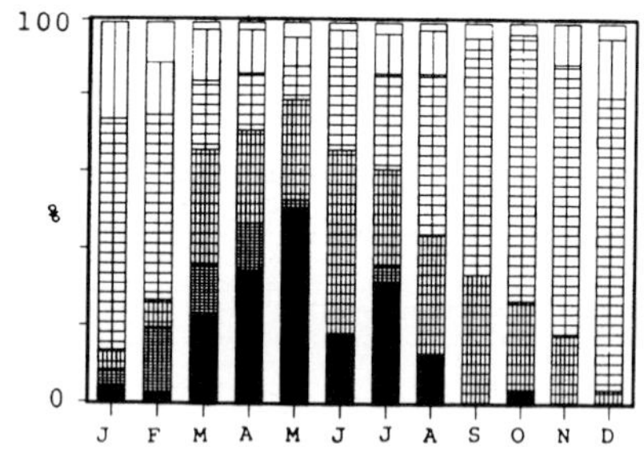

Majority of flowers fully open
Majority of flowers in bud
Majority of flowers recently finished
Flowers finished, fruiting cones present
Neither flowers nor fruiting cones present
Unspecified

Possible pollinators

Pollinator type 382 records: bird (10%) bees, wasps, ants (7%) beetle (1%) mammal (1%) no pollinator observed (82%).

Specific pollinator 70 records: honeyeater (6) New Holland Honeyeater (16) Brown Honeyeater (1) Silvereye (2) wattlebird (1) Red Wattlebird (5) Little Wattlebird (4) Red-tailed Black-Cockatoo (1) ant (13) bee (14) native bee (1) European honey bee (4) Honey-possum (1) jewel beetle (1)

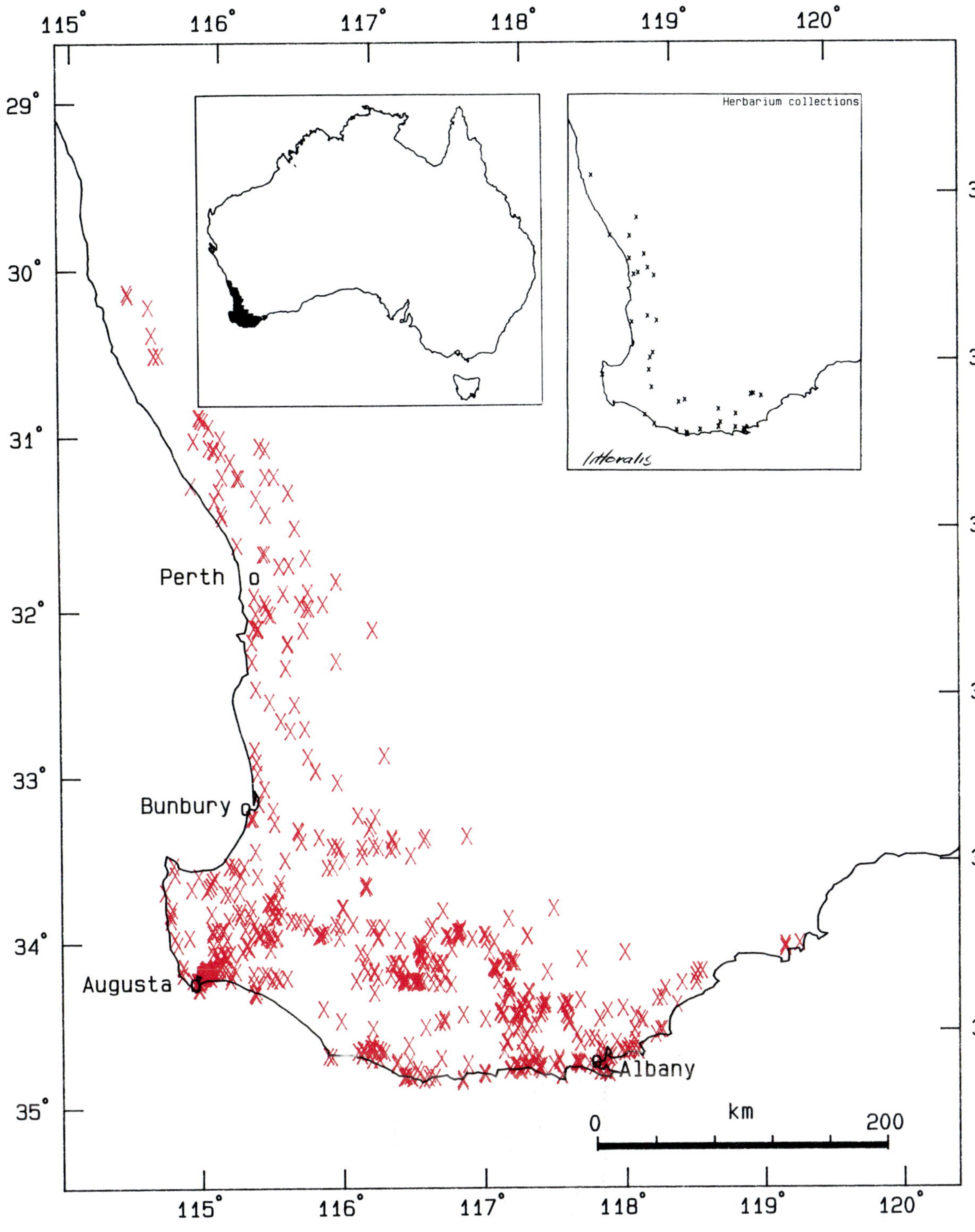

DISCUSSION

Western Australian Herbarium collections indicated a distribution of B. littoralis from Mt Lesueur southwards to the Leeuwin-Naturaliste National Park and eastwards to the Stirling Range and Two Peoples Bay, always within 80 km of the coast. Atlas contributors have extended the range approximately 95 km eastwards with new recordings at both Bremer Bay and Dillon Bay. B. littoralis has also been shown to extend considerably further inland with records from Tambellup, west of Woodanilling, and near Moodiarup. The distribution of B. littoralis now appears fairly continuous with several records from the Leeuwin - Naturaliste area and south of Mt Lesueur, areas not well respresented by herbarium collections.

B. littoralis is usually a tree less than 10 m in height growing on flats, in seasonally wet areas and around lake margins. Soil types are usually sandy, sometimes loamy or peaty. In the Darling Range it may grow on clayish or gravelly soils. Surrounding vegetation is usually woodland or forest. More rarely it occurs as a shrub on coastal dunes with surrounding vegetation of both tall and low shrubs. Populations of B. littoralis are generally small, often being restricted to seasonally damp patches in an otherwise drier environment. 37% of recorded populations are of less than 10 plants, and 85% are of less than 100 plants. 26% of populations are in conservation areas. The main flowering period is from March to August though flowers have also been recorded as early as January and as late as October. The most commonly recorded animals visiting flower spikes were birds and ants, the former particularly well represented by the New Holland Honeyeater. The main period of new shoot growth is in late spring and summer. Following fire, the species resprouts from both its trunk and lignotuber. There are also three records of its being killed by fire and regenerating from seed.

BANKSIA LULLFITZII C.Gardner (1966)

7 RECORDS: Jan (2) Feb (0) Mar (0) Apr (1) May (1) Jun (0) Jul (0) Aug (2) Sep (1) Oct (0) Nov (0) Dec (0).

Population Size: 1-10 (2) 10-100 (3) >100 (2) Unspecified (0).

Conservation Status: Restricted to road verge (0%) Not (100%) Unspecified (0%). In conservation reserve (71%) Not (29%) Unspecified (0%).

Tree/Shrub: Tree form (0%) Shrub form (100%) Unspecified (0%).

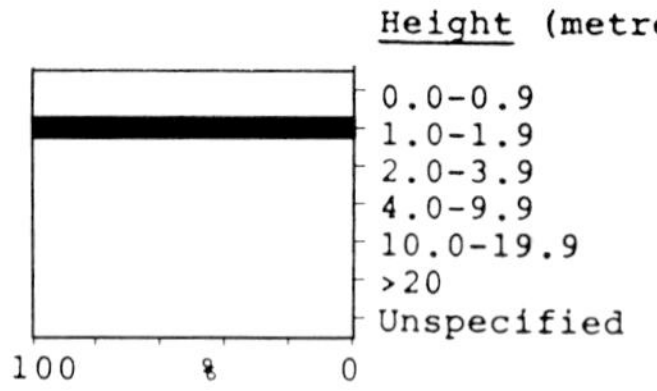

New Shoot Growth

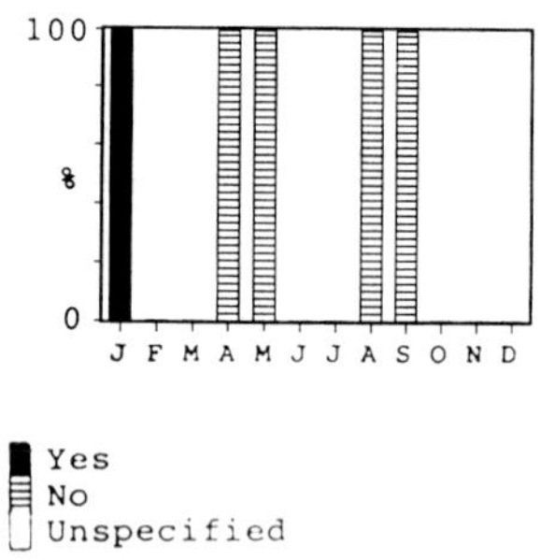

HABITAT

Within 2 km of coast: Yes (0%) No (100%) Unspecified (0%).

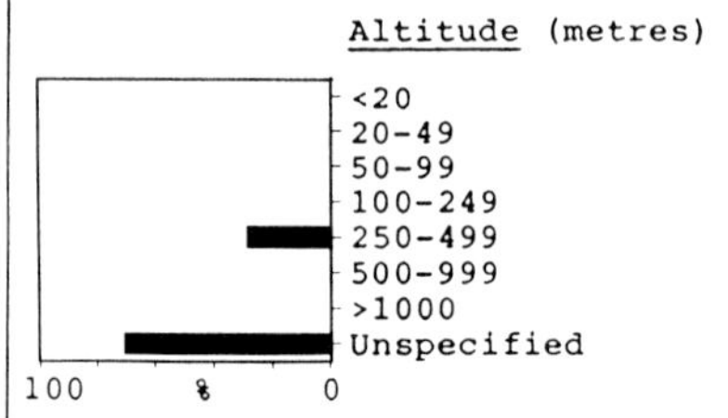

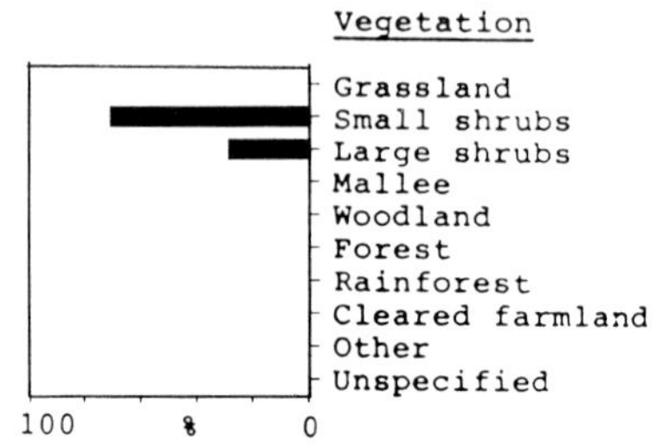

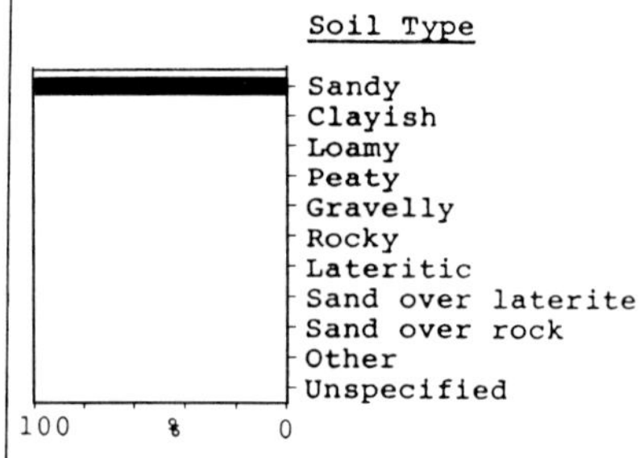

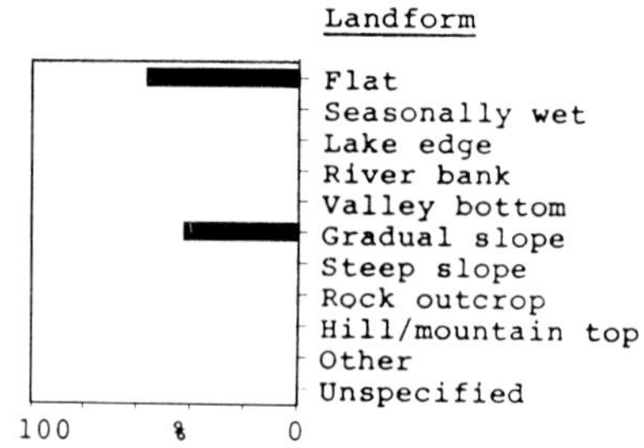

Response to Fire: 1 record

Growth response: Ground resprout (100%)

Flowering response: Number of records where flowering had occurred after fire (1). Minimum reported time to flowering (25-36 month).
Number of records where flowering had not occurred after fire (0).

Flowering

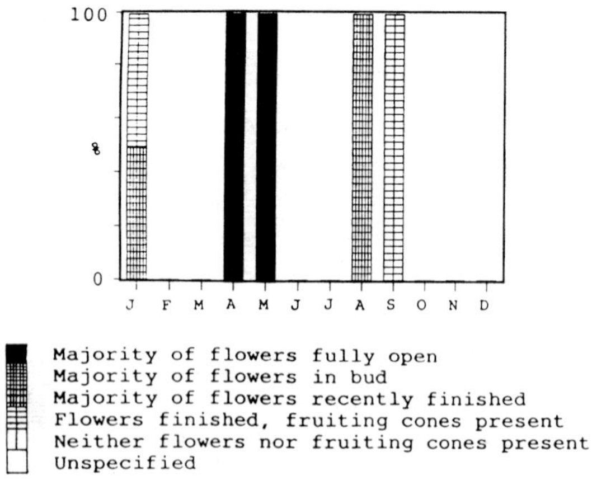

Possible pollinators

Pollinator type 1 record: bird (100%)

Specific pollinator 4 records: Brown Honeyeater (1) White-eared Honeyeater (1) Tawny-crowned Honeyeater (1) Red Wattlebird (1)

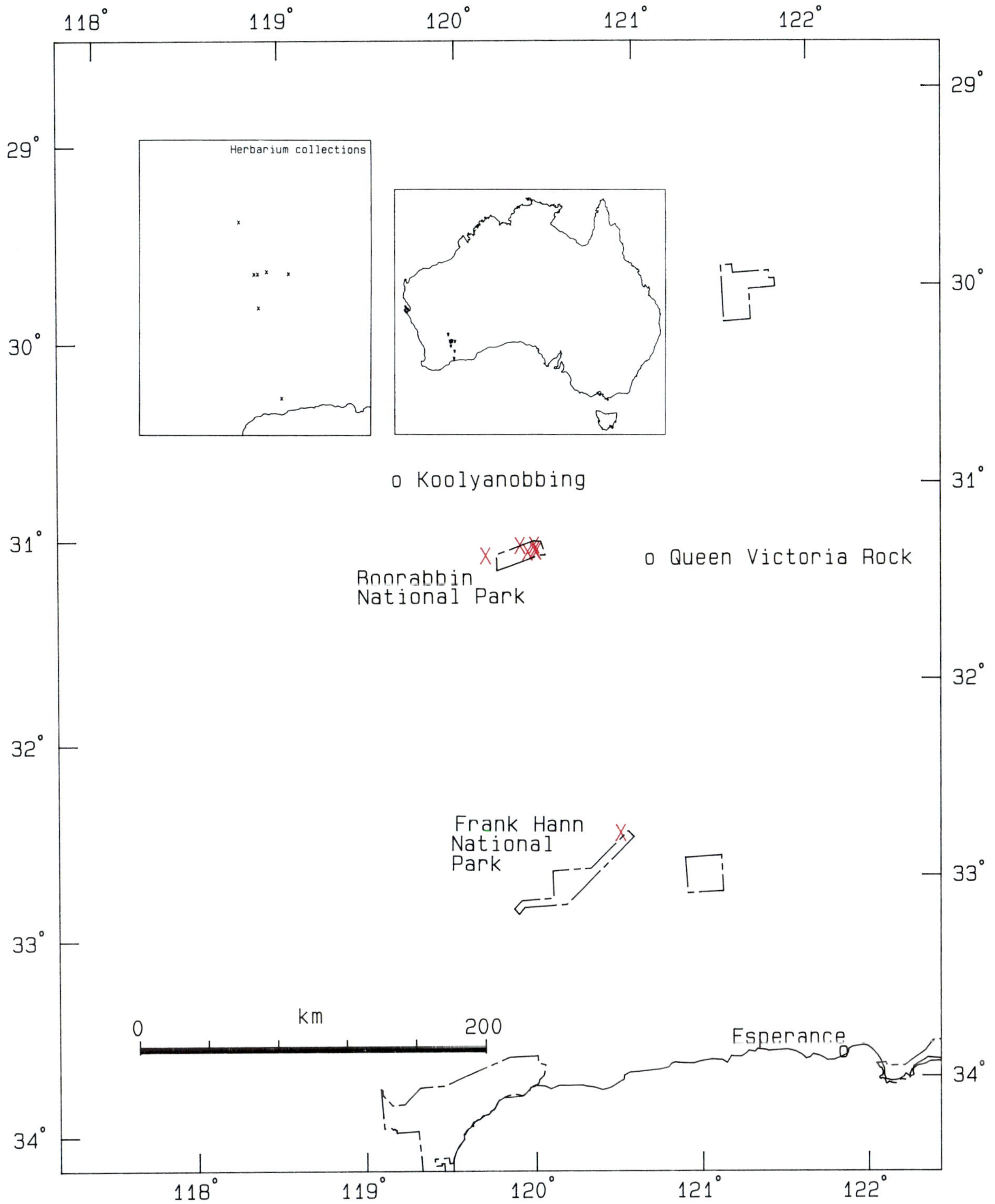

National Parks >10000 ha.

DISCUSSION

Western Australian Herbarium records indicated that B. lullfitzii occurred in scattered populations between the Ravensthrope to Esperance Road and a location approximately 70 km NNE of Koolyanobbing. A survey of the Eastern Goldfields (Newbey and Hnatiuk 1985) had also located it near Queen Victoria Rocks south-west of Coolgardie. Atlas contributors recorded five separate sites within Boorabbin National Park and another site just west of the park near Koorarawalyee. A new recording was also made from near the eastern boundary of Frank Hann National Park. The species is most probably under-recorded since access is difficult in much of its habitat and there are large areas still unexplored. Also it has a similar appearance to B. elderiana, differing mainly in its narrower leaves and erect orange flower spikes. When not in flower the two can be mistaken easily. Although there are relatively few populations known and most are of less than 100 plants, the species is probably not under threat. It is well represented in conservation reserves (chiefly Boorabbin National Park) and its range is in land which is unlikely to be cleared for agriculture due to insufficient rainfall.

B. lullfitzii is a small shrub of less than 2 m in height growing in deep yellow sands on flat to gently sloping plains. It is frequently associated with B. audax and/or B. elderiana in heath or mallee-heath with patches of spinifex grassland sometimes interspersed. Emergent trees of Callitris sp. and Casuarina sp. may also be present. Flowering is in autumn when birds such as the White-eared, Tawny-crowned and Brown Honeyeaters, and Red Wattlebird were observed on the flower spikes. New shoot growth was recorded in January, but further data are needed before the full season can be determined. Following fire, B. lullfitzii resprouts from its lignotuber. A single record suggests that it may be flowering within 2-3 years of being burnt.

BANKSIA MARGINATA Cavanilles (1800)

Silver Banksia

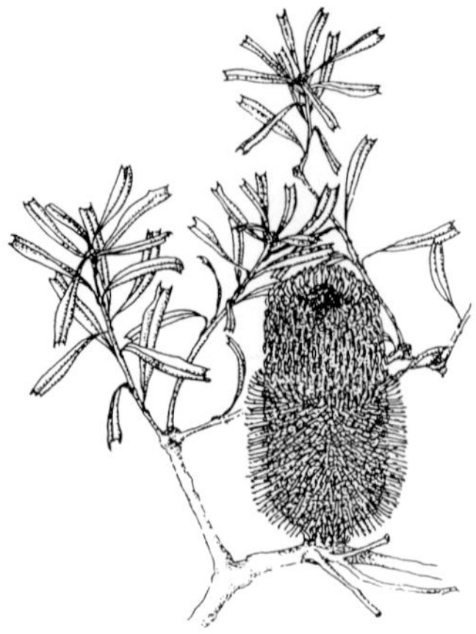

3775 RECORDS: Jan (244) Feb (362) Mar (270) Apr (352) May (527) Jun (333) Jul (364) Aug (175) Sep (301) Oct (339) Nov (250) Dec (218).

Population Size: 1-10 (836) 10-100 (1496) >100 (1302) Unspecified (141).

Conservation Status: Restricted to road verge (29%) Not (66%) Unspecified (5%). In conservation reserve (31%) Not (64%) Unspecified (5%).

Tree/Shrub: Tree form (39%) Shrub form (57%) Unspecified (5%).

Height (metres)

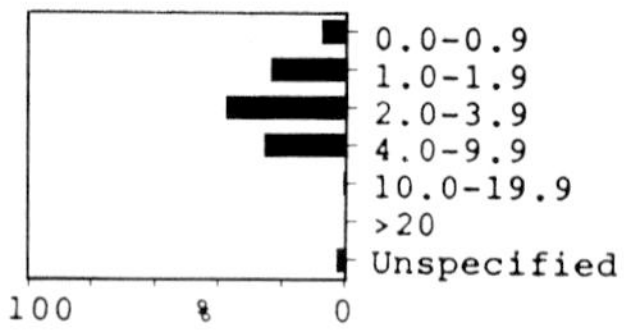

New Shoot Growth

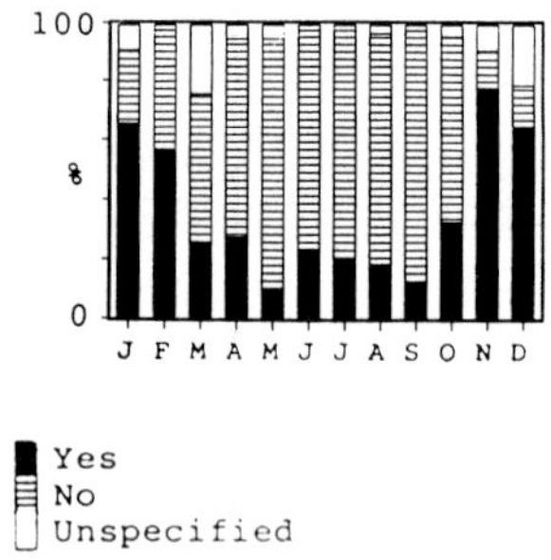

HABITAT

Within 2 km of coast: Yes (16%) No (82%) Unspecified (2%).

Altitude (metres)

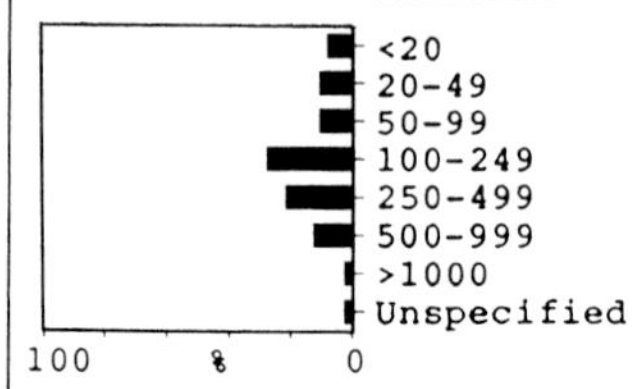

Vegetation

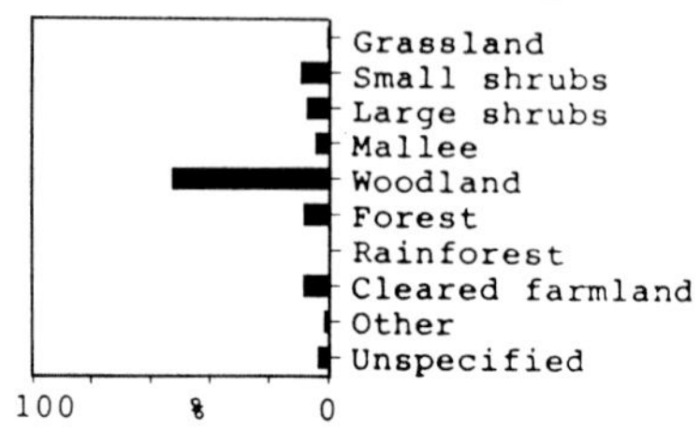

Soil Type

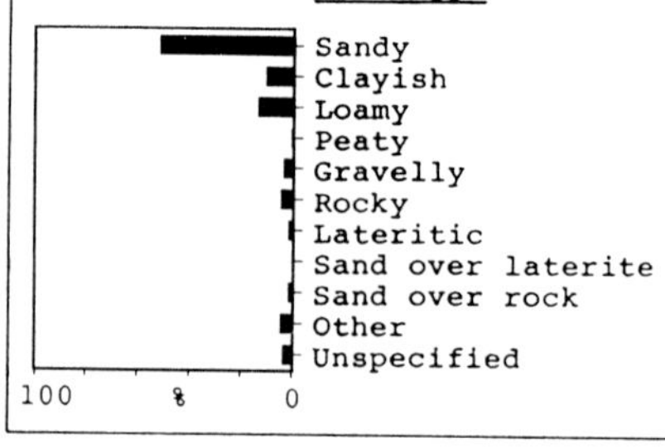

Landform

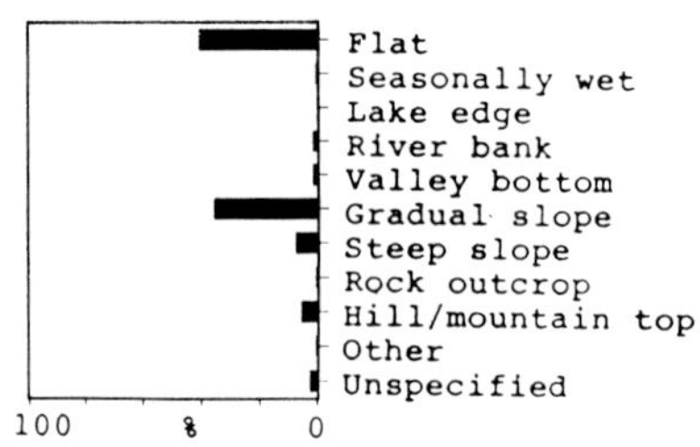

Response to Fire: 408 records

Growth response: Ground resprout (37%) ground resprout and killed, new seedlings (4%) ground resprout and trunk resprout (4%) killed, new seedlings (19%) trunk resprout (8%) killed, no seedlings (1%) unspecified (26%).

Flowering response: Number of records where flowering had occurred after fire (167). Median reported time to flowering (13-24 months); minimum reported time to flowering (<1 month). Number of records where flowering had not occurred after fire (159). Median reported time to no flowering (13-24 months); maximum reported time to no flowering (85-96 months).

Flowering

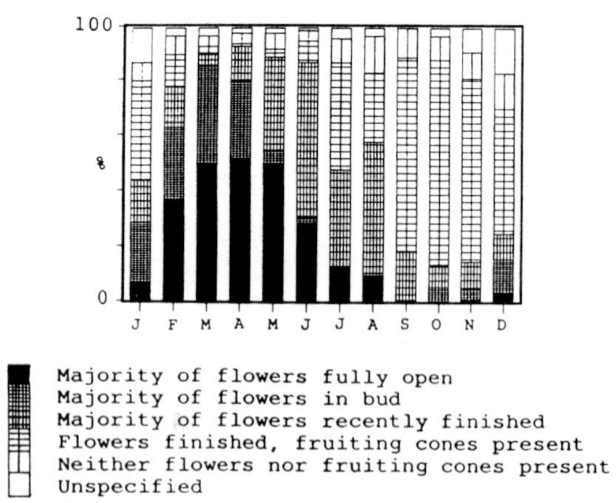

Possible pollinators

Pollinator type 2469 records: bird (8%) butterflies/moths (0.2%) bees, wasps, ants (5%) flies (0.2%) beetles (0.1%) mammals (0.2%) no pollinator observed (86%).

Specific pollinator 321 records: honeyeater (42) New Holland Honeyeater (87) Singing Honeyeater (2) White-eared Honeyeater (2) Tawny-crowned Honeyeater (3) White-naped Honeyeater (3) White-plumed Honeyeater (3) Yellow-faced Honeyeater (3) Crescent Honeyeater (14) Noisy Miner (1) wattlebird (5) Red Wattlebird (27) Little Wattlebird (7) Yellow Wattlebird (1) Eastern Spinebill (17) cockatoos and parrots (1) Musk Lorikeet (1) thornbill (2) ant (31) bee (42) native bee (2) European honey bee (19) blowfly (3) brown butterfly (2) Sugar Glider (1)

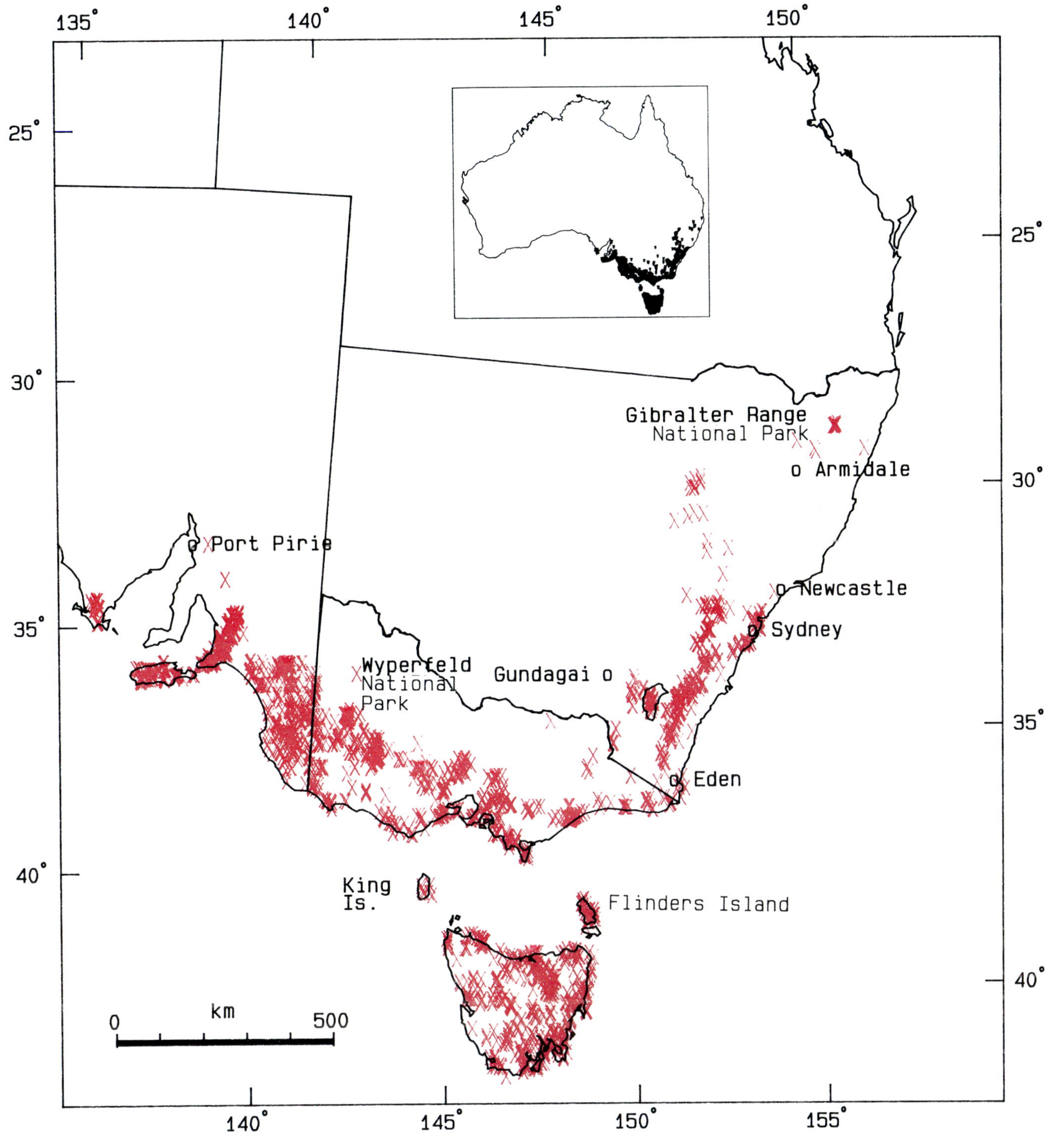

DISCUSSION

B. marginata was the most commonly recorded species in the Atlas project, with 3 775 records reported. It is a widespread and often common species in parts of South Australia, Victoria, New South Wales and Tasmania. The above map shows its entire distribution. This is then subdivided into three sections and larger scale maps of each of these sections appear on the following pages.

In South Australia, B. marginata occurs in the southern tip of the Eyre Peninsula and throughout much of Kangaroo Island. A large number of records show it to be widespread in the Mt Lofty Ranges, east of Adelaide, from the tip of Fleurieu Peninsula as far north as Nuriootpa. North of this population are two outliers from near Port Pirie and Clare. Throughout south-eastern South Australia, Atlas records show B. marginata to be widespread, occurring from Ashville and Parrakie south to Mt Gambier. Previously, there had been very few herbarium collections from this area.

In Victoria, the bulk of the populations lie generally south of a line from the Little Desert through Stawell, Maryborough, Metcalfe, the Dandenongs and Walhalla. From Traralgon eastwards the bulk of the populations are generally within 40 km of the coast. Inland, there are outlying populations near the Bowang River, to the north and west of Benambra and west of Albury. In western Victoria, a population in Wyperfeld National Park had not previously been recorded, and extends the species' known distribution by some 80 km in this part of its range. In Tasmania, Atlas contributors increased the known range of B. marginata with several records from the South West Wilderness, the north-west and north-east tips of the island as well as the eastern coastline. The species was abundantly recorded on Bruny Island and the Tasman Peninsula, areas not mapped in herbarium collections. There is also a new recording of B. marginata on Cape Barren island, to the south of Flinders Island.

cont'd...

BANKSIA MARGINATA cont'd

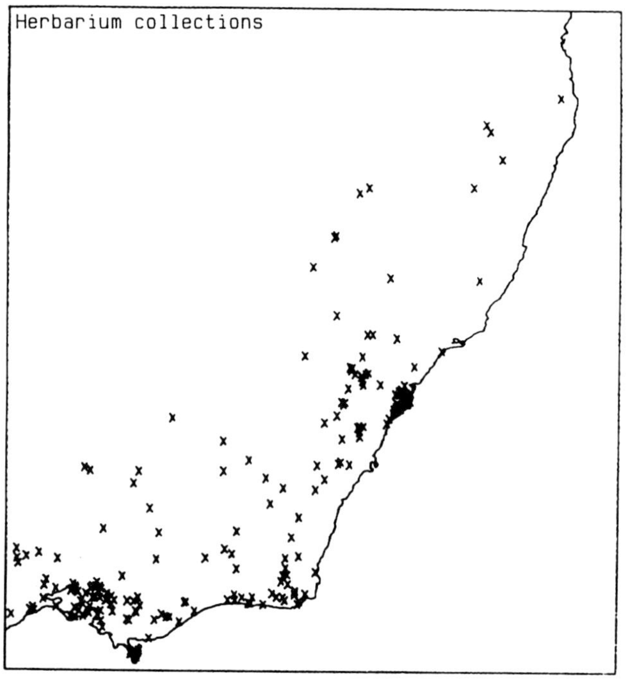

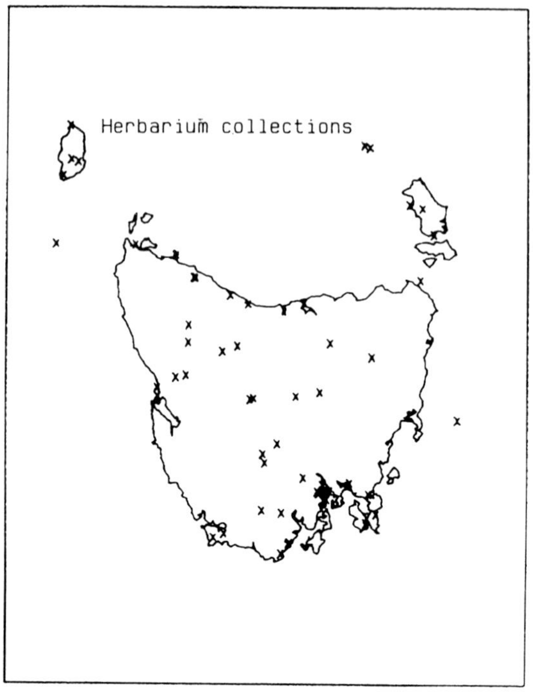

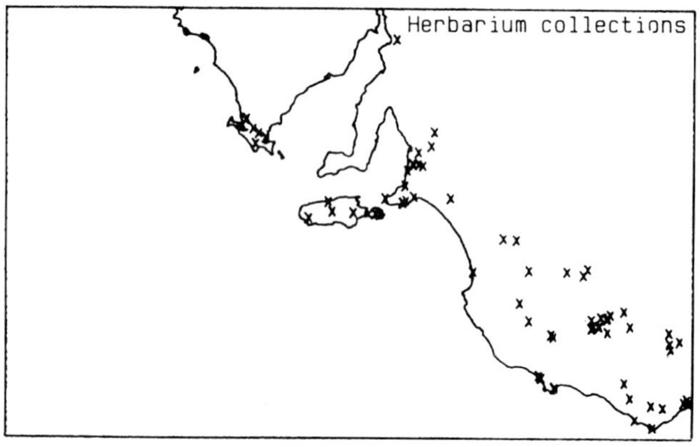

Discussion of Banksia marginata cont'd...

In New South Wales *B. marginata* generally occurs along the Great Dividing Range and further inland. Apart from an extensive area around Sydney, it is largely absent from the coastal plain. There are several herbarium collections of inland populations not recorded by Atlas contributors e.g. the 1879 collection of G. Betche from the Narrandera district and the 1905 collection of O. Wilshire from the Deniliquin district. However, the inland population east of Baradine and Gilgandra was recorded by Atlas contributors. The only Queensland collection of *B. marginata*, that of V. Jaegermann's in 1973, from the Springbrook Mountains, south-south-west of Southport, was not recorded by Atlas contributors.

B. marginata is a highly variable plant ranging from a small shrub less than 0.5 metres in height to a large tree taller than 10 m. Individual specimens with weeping foliage have also been observed. Leaves vary enormously both in size and shape and even in the same general area there can be plants with different leaf types and growth habits. It is not known to what extent these variations are morphological (and therefore hereditary) or are a physiological response to differing environments. The species grows in a range of soil types though sandy soils are predominant. Surrounding vegetation is often woodland, though it may also be shrubland, heath, mallee or forest. Records of a habitat of cleared farmland are a reflection of the general distribution of *B. marginata*. It occurs throughout many cleared agricultural areas and often remains as remnant roadside vegetation and isolated trees in paddocks. The species grows at altitudes varying from sea level to over 1 000 m (e.g. 1 200 m at Mt Field National Park, Tasmania). Response to fire is also variable with some populations being lignotuberous, or resprouting from epicormic buds. Others are killed by fire and are dependent on seed germination for regrowth. Suckering of plants has also been recorded. The extent of variation seen in *B. marginata* suggests that subspecies or varieties may be present. The following presumed hybrids were recorded for the Atlas - *B. marginata* x *B. paludosa*, *B. marginata* x *B. saxicola*, *B. marginata* x *B. integrifolia* var. *integrifolia*. Further details of these (where available) are provided in Chapter 3.

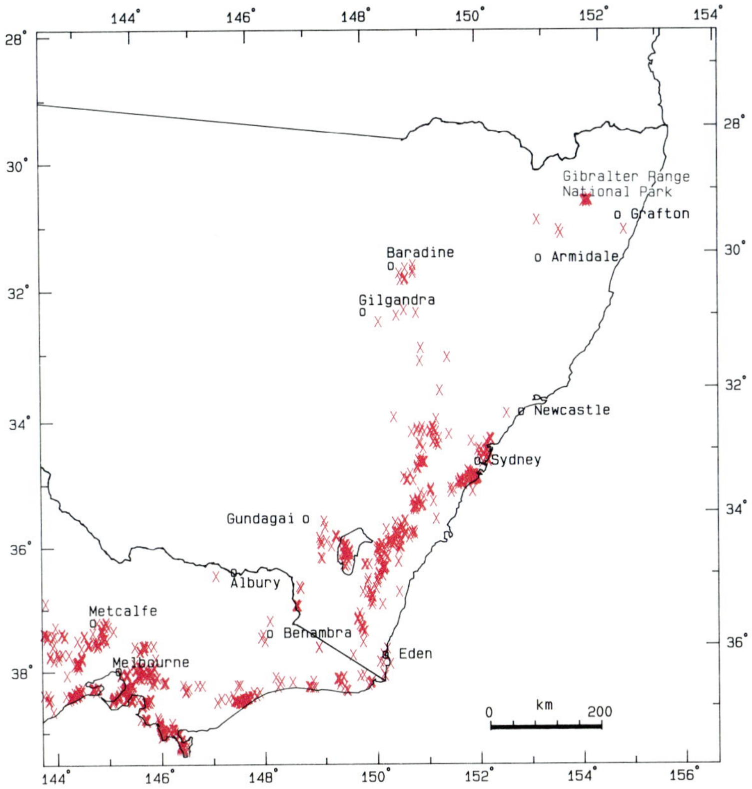
Gibralter Range
National Park
Grafton
Baradine
Armidale
Gilgandra
Newcastle
Sydney
Gundagai
Albury
Metcalfe
Benambra
Melbourne
Eden
0 km 200

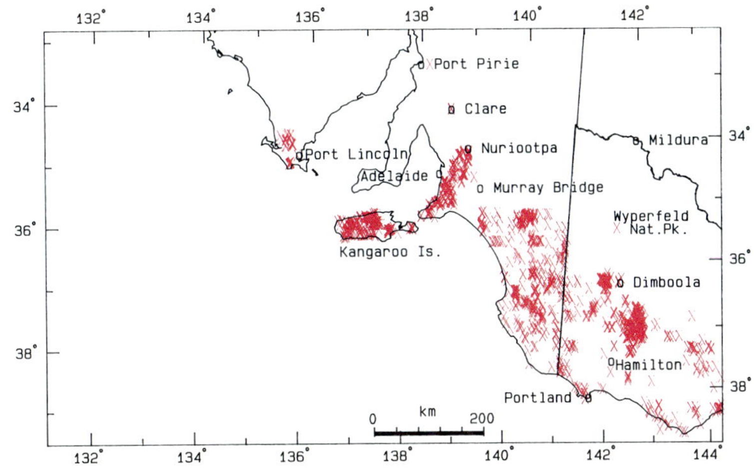
Port Pirie
Clare
Mildura
Nuriootpa
Port Lincoln
Adelaide
Murray Bridge
Wyperfeld
Nat.Pk.
Kangaroo Is.
Dimboola
Hamilton
Portland
0 km 200

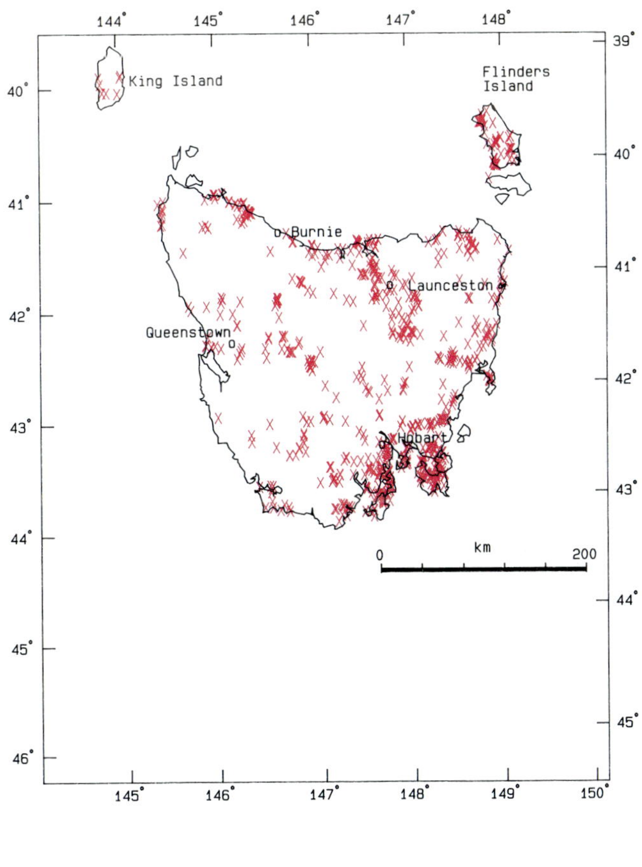
King Island
Flinders
Island
Burnie
Launceston
Queenstown
Hobart
0 km 200

BANKSIA MEDIA R.Brown (1830)

Southern Plains Banksia, Golden Stalk Banksia

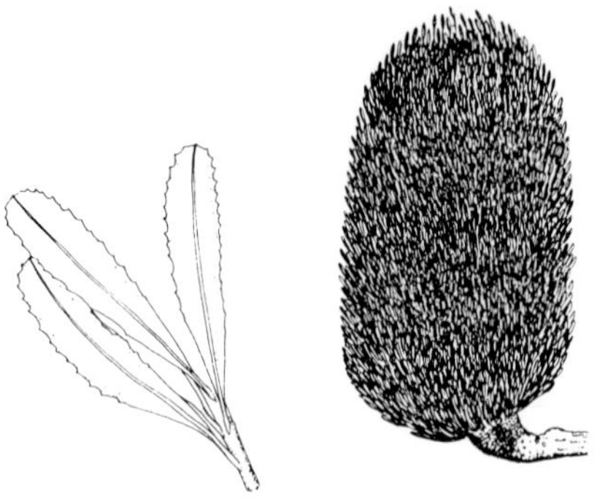

336 RECORDS: Jan (33) Feb (24) Mar (19) Apr (30) May (16) Jun (9) Jul (15) Aug (25) Sep (45) Oct (62) Nov (40) Dec (14).

Population Size: 1-10 (59) 10-100 (136) >100 (130) Unspecified (11).

Conservation Status: Restricted to road verge (18%) Not (76%) Unspecified (6%). In conservation reserve (43%) Not (52%) Unspecified (5%).

Tree/Shrub: Tree form (13%) Shrub form (87%) Unspecified (0%).

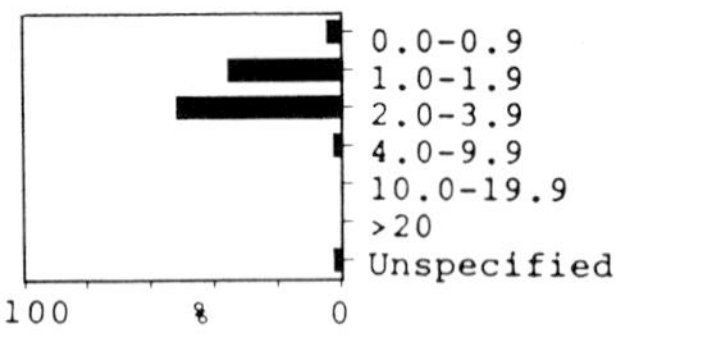

New Shoot Growth

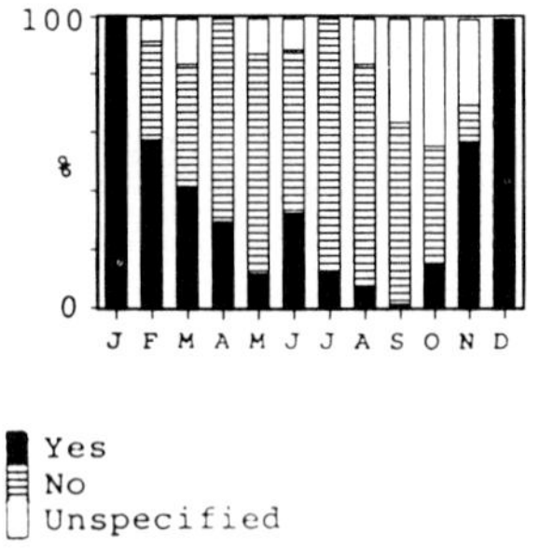

HABITAT

Within 2 km of coast: Yes (14%) No (85%) Unspecified (0%).

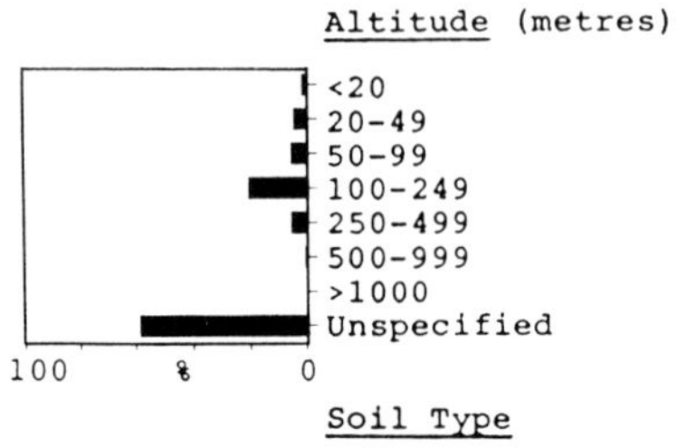

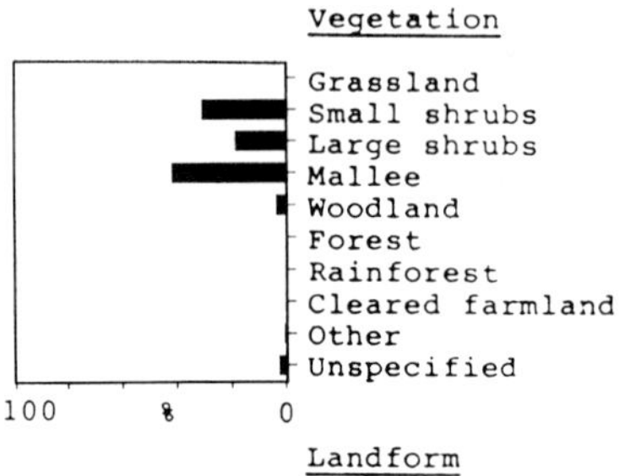

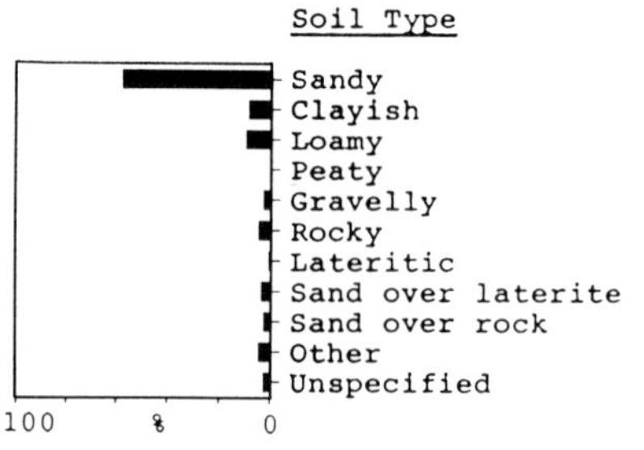

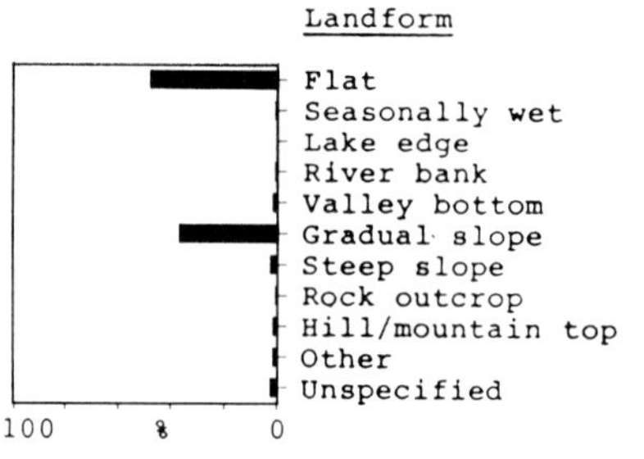

Response to Fire: 26 records

Growth response: Ground resprout (12%) killed, new seedlings (42%) killed, no seedlings (4%) killed, no seedlings and killed, new seedlings (15%) unspecified (27%).

Flowering response: Number of records where flowering had occurred after fire (9). Median reported time to flowering (37-48 months); minimum reported time to flowering (1-12 months). Number of records where flowering had not occurred after fire (10). Median reported time to no flowering (13-24 months); maximum reported time to no flowering (25-36 months).

Flowering

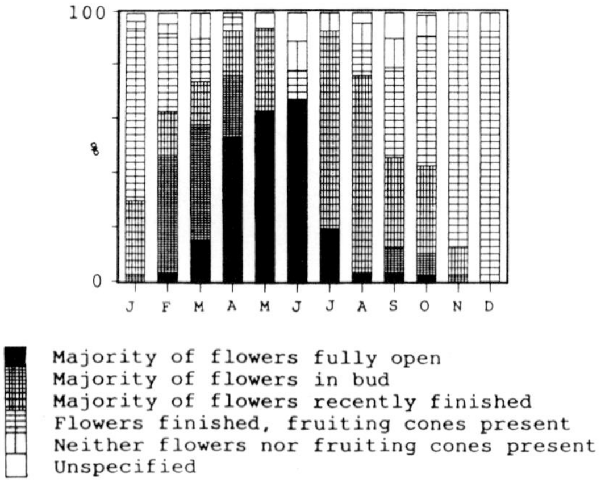

Possible pollinators

Pollinator type 112 records: bird (9%) bees, wasps, ants (14%) flies (4%) beetles (1%) no pollinator observed (71%).

Specific pollinator 28 records: honeyeater (4) New Holland Honeyeater (2) wattlebird (1) Red Wattlebird (1) Western Spinebill (1) thornbill (1) ant (12) meat ant (1) bee (3) European honey bee (2)

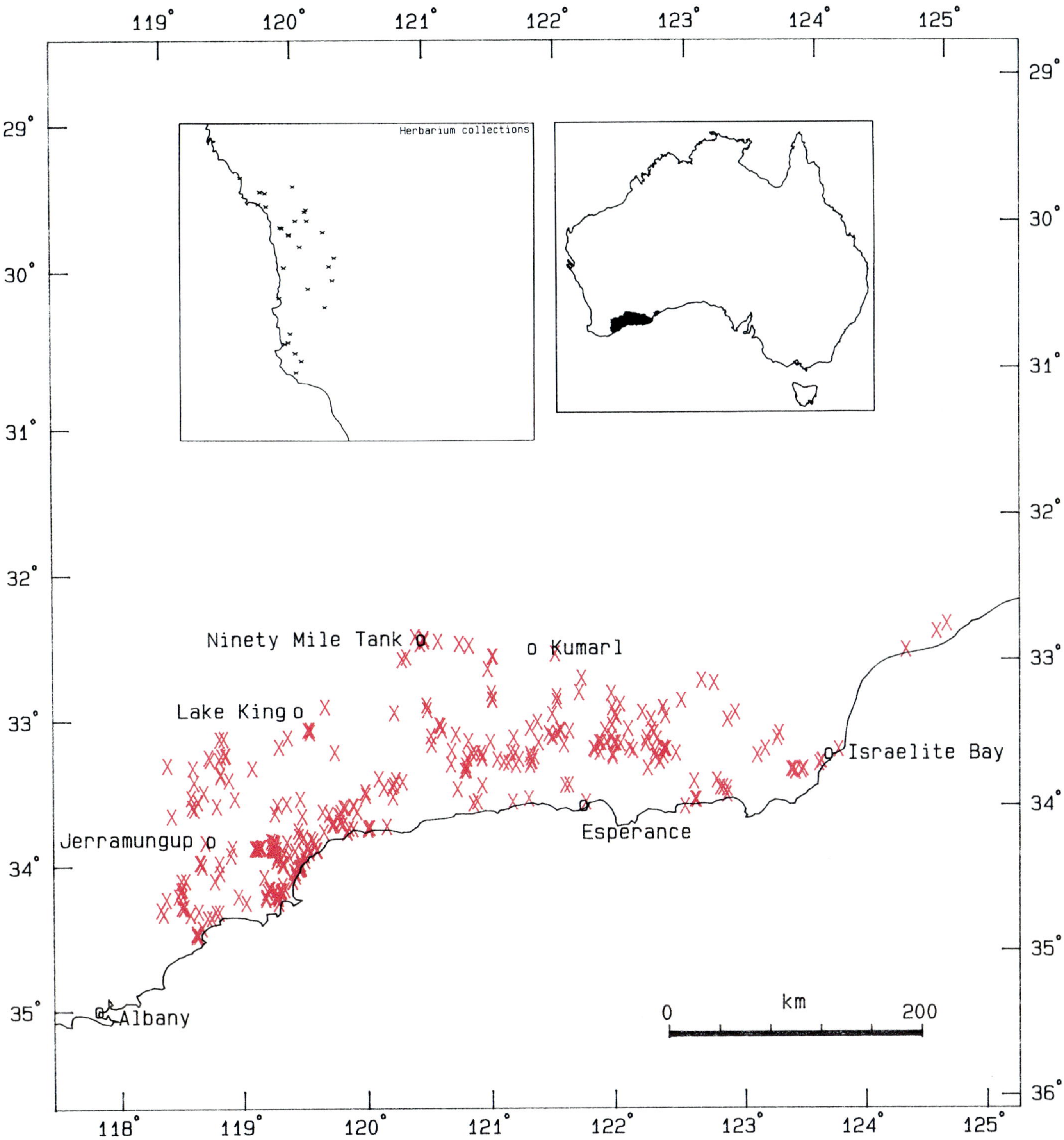

DISCUSSION

Atlas records have confirmed that the main distribution of B. media is between the eastern boundary of Stirling Range National Park and Israelite Bay, extending inland as far as Pingrup, Frank Hann National Park, 15 km east-north-east of Dowak and 35 km north-west of Mt Buraminya. There is an eastern outlier at Point Culver and Toolinna where B. media grows in association with the newly named B. epica. B. media is locally common with many of its populations having more than 100 plants. It is well represented in conservation reserves (particularly the Fitzgerald River National Park), and is only moderately susceptible to dieback disease.

B. media is generally a shrub, sometimes a tree up to 4 m in height. George (1981) noted that it can grow as large as 10 m. It tolerates a mixture of soil types. White sand is preferred though loamy, clayish and rocky soils have been recorded. Surrounding vegetation is shrubland or mallee eucalypts. The main flowering period is from March to July with some flowers recorded in February and some as late as October. Honeyeaters appear to be important pollinators. Ants were also frequently recorded. New shoots grow predominantly in summer. The species is generally non-lignotuberous, regenerating from seed after fire. However, there are three post-fire Atlas records of plants resprouting from lignotubers, so this may be a possibility if fire intensity is only light. Following fire, young B. media plants may require at least 3-4 years before flowering recommences. An interesting population of B. media with very large flower spikes was located north of Mt Ridley. The entire population of less than 100 plants has uniformly large flower spikes up to 26 cm long. This is 60-80% larger than usual.

BANKSIA MEISNERI Lehmann (1845)

Meisner's Banksia

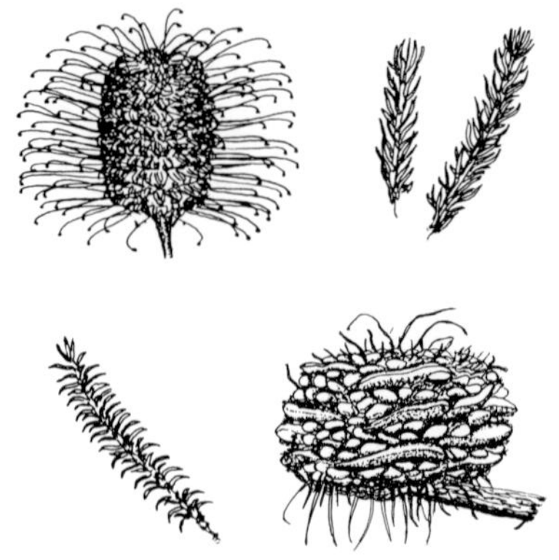

52 RECORDS: Jan (1) Feb (0) Mar (5) Apr (6) May (5) Jun (3) Jul (6) Aug (10) Sep (6) Oct (6) Nov (0) Dec (0).

Population Size: 1-10 (11) 10-100 (24) >100 (16) Unspecified (1).

Conservation Status: Restricted to road verge (8%) Not (86%) Unspecified (6%). In conservation reserve (50%) Not (46%) Unspecified (4%).

Tree/Shrub: Tree form (0%) Shrub form (100%) Unspecified (0%).

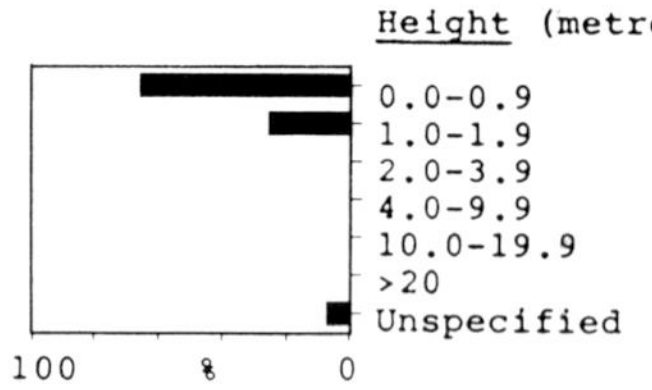

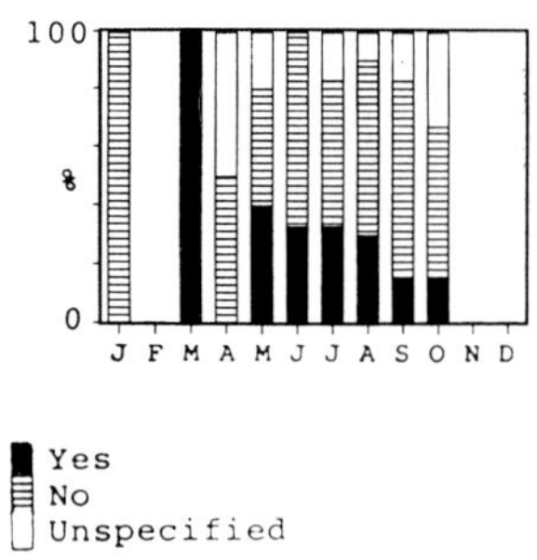

HABITAT

Within 2 km of coast: Yes (8%) No (90%) Unspecified (2%).

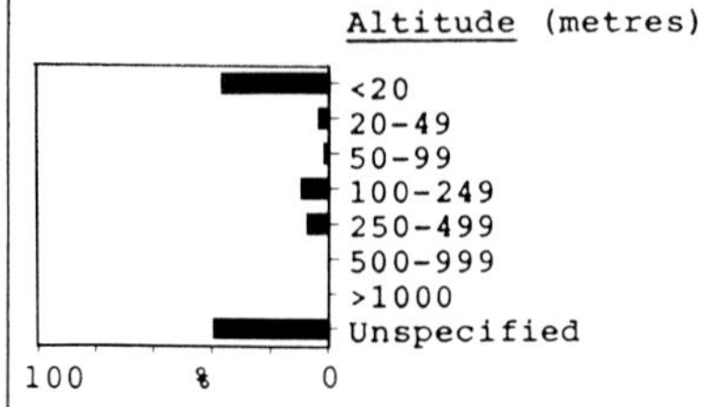

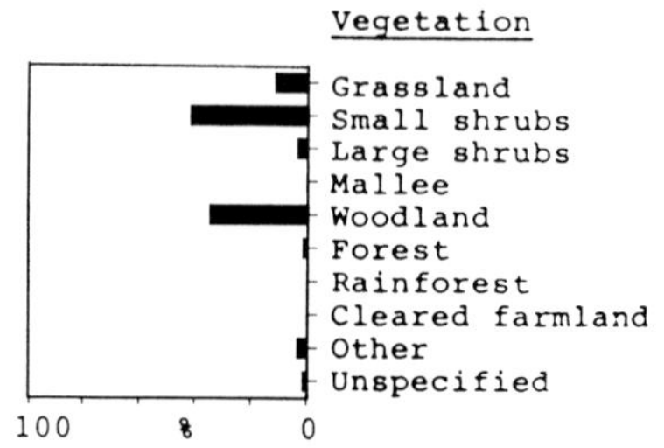

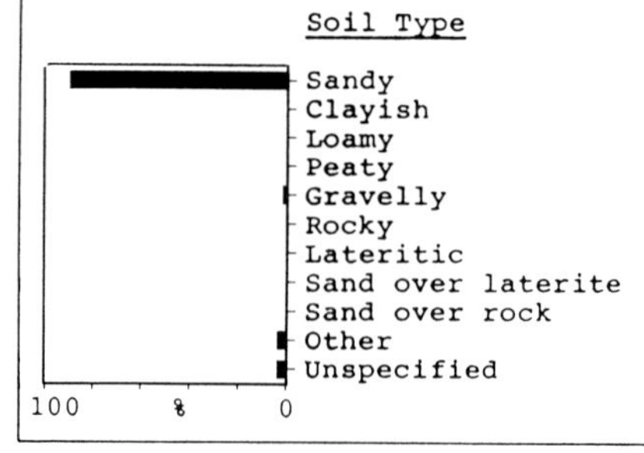

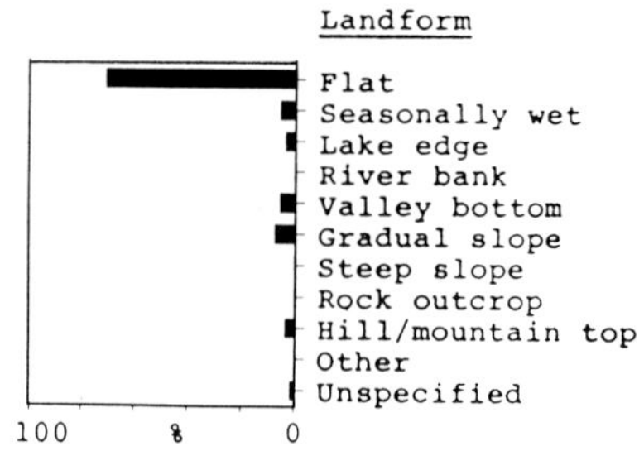

Response to Fire: 10 records

Growth response: Ground resprout and killed, new seedlings (10%) killed, new seedlings (50%) killed, no seedlings (20%) unspecified (20%).

Flowering response: Number of records where flowering had occurred after fire (7). Median reported time to flowering (1-12 months); minimum reported time to flowering (<1 month). Number of records where flowering had not occurred after fire (3). Median reported time to no flowering (13-24 months); maximum reported time to no flowering (25-36 months).

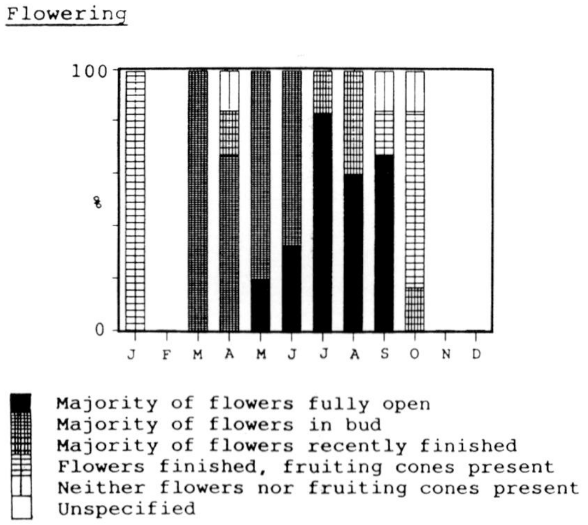

Possible pollinators

Pollinator type 24 records: bees, wasps, ants (25%) no pollinator observed (75%).

Specific pollinator 5 records: bee (1) ant (4)

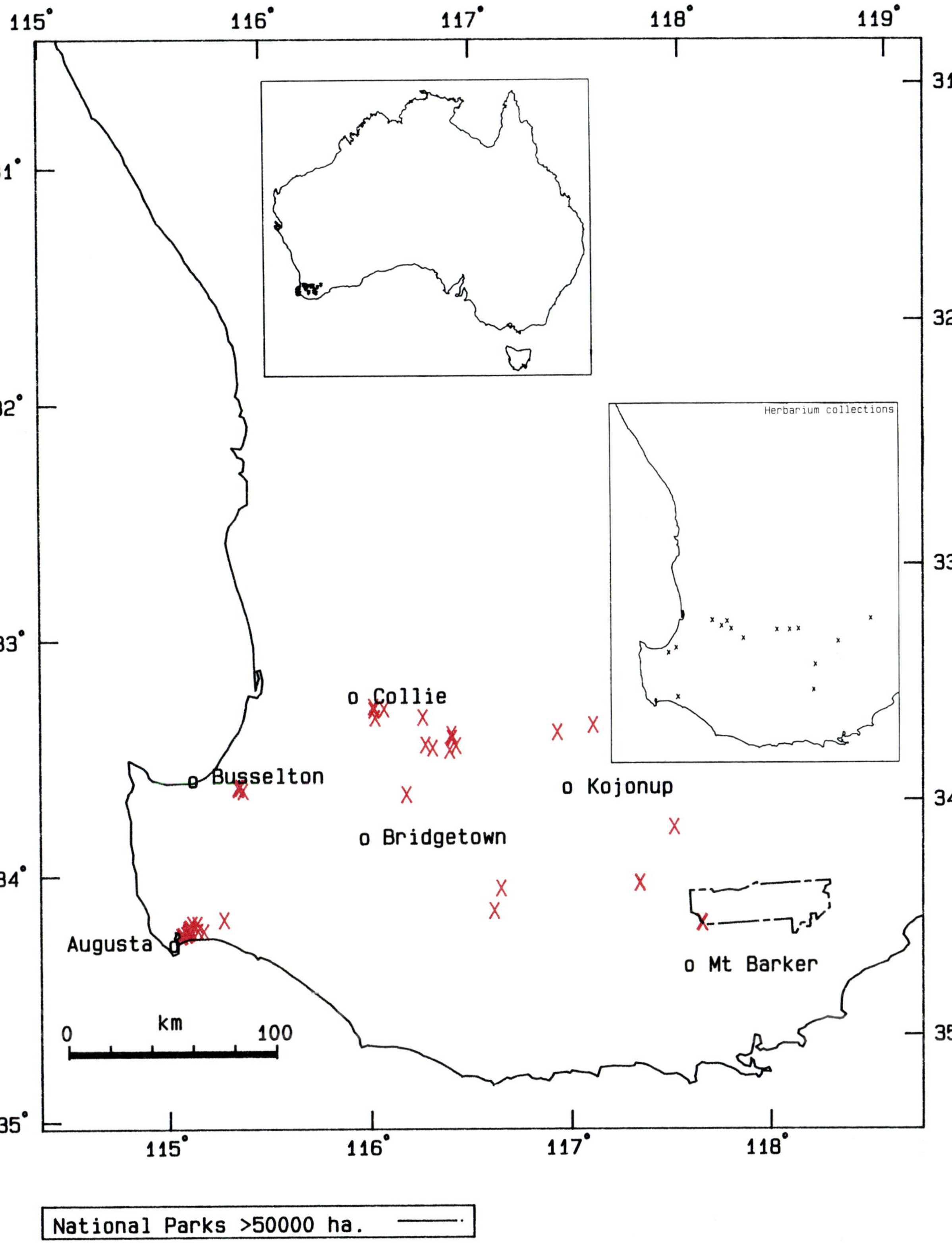

DISCUSSION

B. meisneri is a small uncommon shrub with disjunct populations ranging from near Busselton and Augusta eastwards to the Stirling Range. Typically, it favours sandy soils in heath or woodland, often in flat swampy land. It has small inflorescences and fruits.

George (1981) recognised two varieties. Var. meisneri has down-turned leaves less than 7 mm long, while var. ascendens has erect or spreading leaves that are 8-15 mm long. The two varieties are disjunct, with var. ascendens occupying the westernmost locations. Atlas contributors had no difficulty in identifying the varieties.

BANKSIA MEISNERI Lehmann
var. ASCENDENS A.S.George (1981)
Scott River Banksia

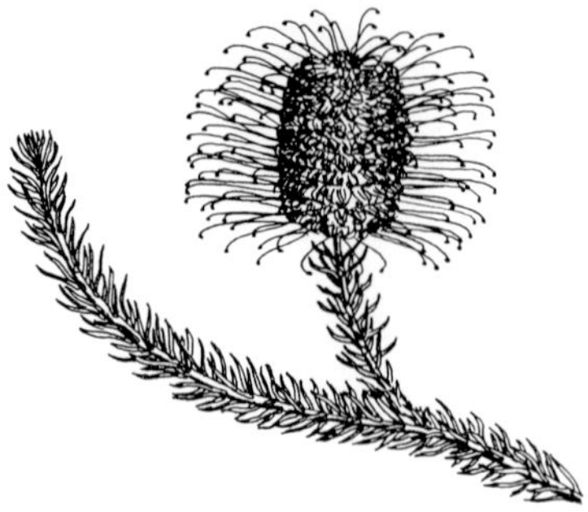

29 RECORDS: Jan (1) Feb (0) Mar (5) Apr (6) May (4) Jun (2) Jul (0) Aug (4) Sep (1) Oct (6) Nov (0) Dec (0).

Population Size: 1-10 (7) 10-100 (11) >100 (11) Unspecified (0).

Conservation Status: Restricted to road verge (10%) Not (86%) Unspecified (3%). In conservation reserve (55%) Not (41%) Unspecified (3%).

Tree/Shrub: Tree form (0%) Shrub form (100%) Unspecified (0%).

Height (metres)

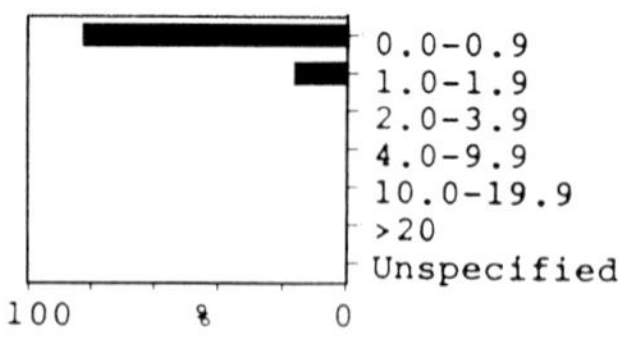

New Shoot Growth

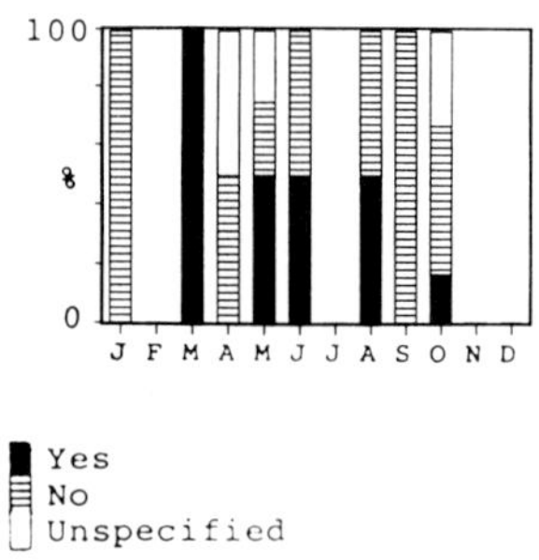

HABITAT

Within 2 km of coast: Yes (14%) No (83%) Unspecified (3%).

Altitude (metres)

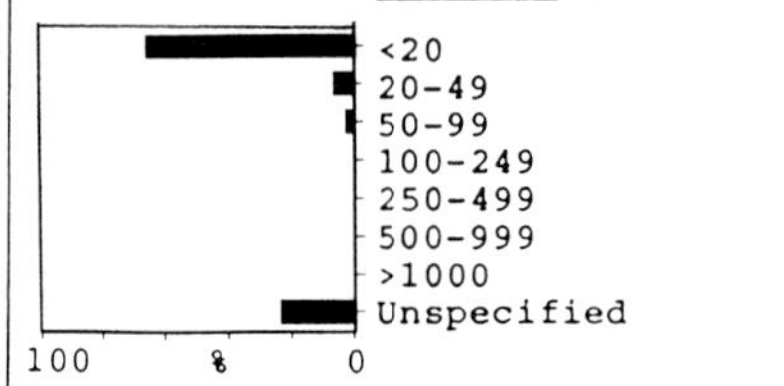

Vegetation

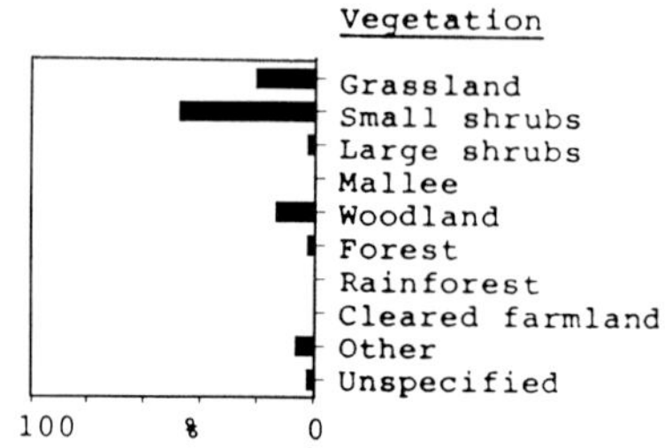

Soil Type

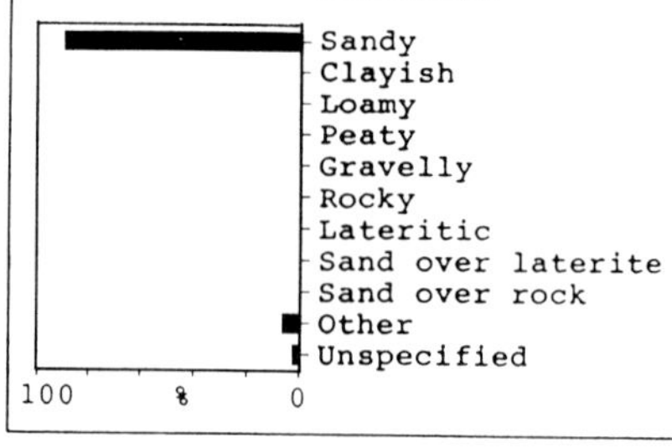

Landform

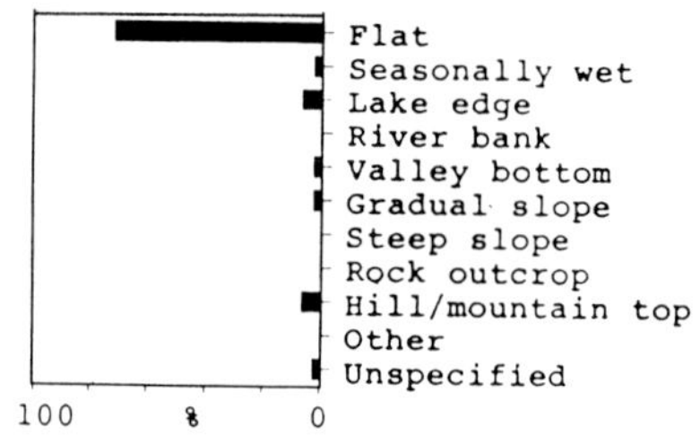

Response to Fire: 4 records

Growth response: Killed, new seedlings (50%) killed, no seedlings (25%) unspecified (25%).

Flowering response: Number of records where flowering had occurred after fire (2). Median reported time to flowering (1 month); minimum reported time to flowering (<1 month). Number of records where flowering had not occurred after fire (2). Median reported time to no flowering (1-12 months); maximum reported time to no flowering (25-36 months).

Flowering

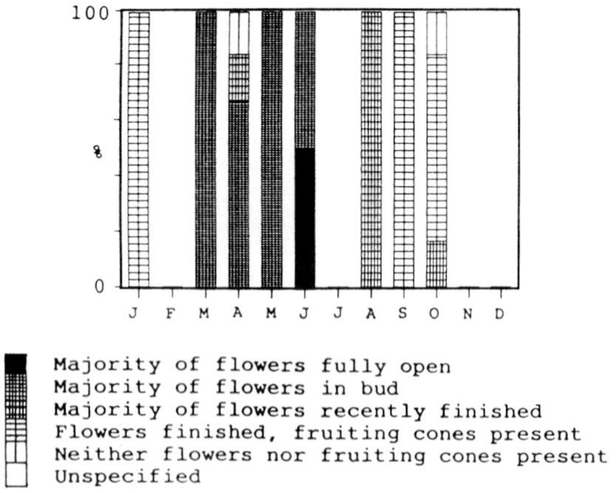

Possible pollinators

Pollinator type 22 records: bees, wasps, ants (23%) no pollinator observed (77%).

Specific pollinator 4 records: ant (4)

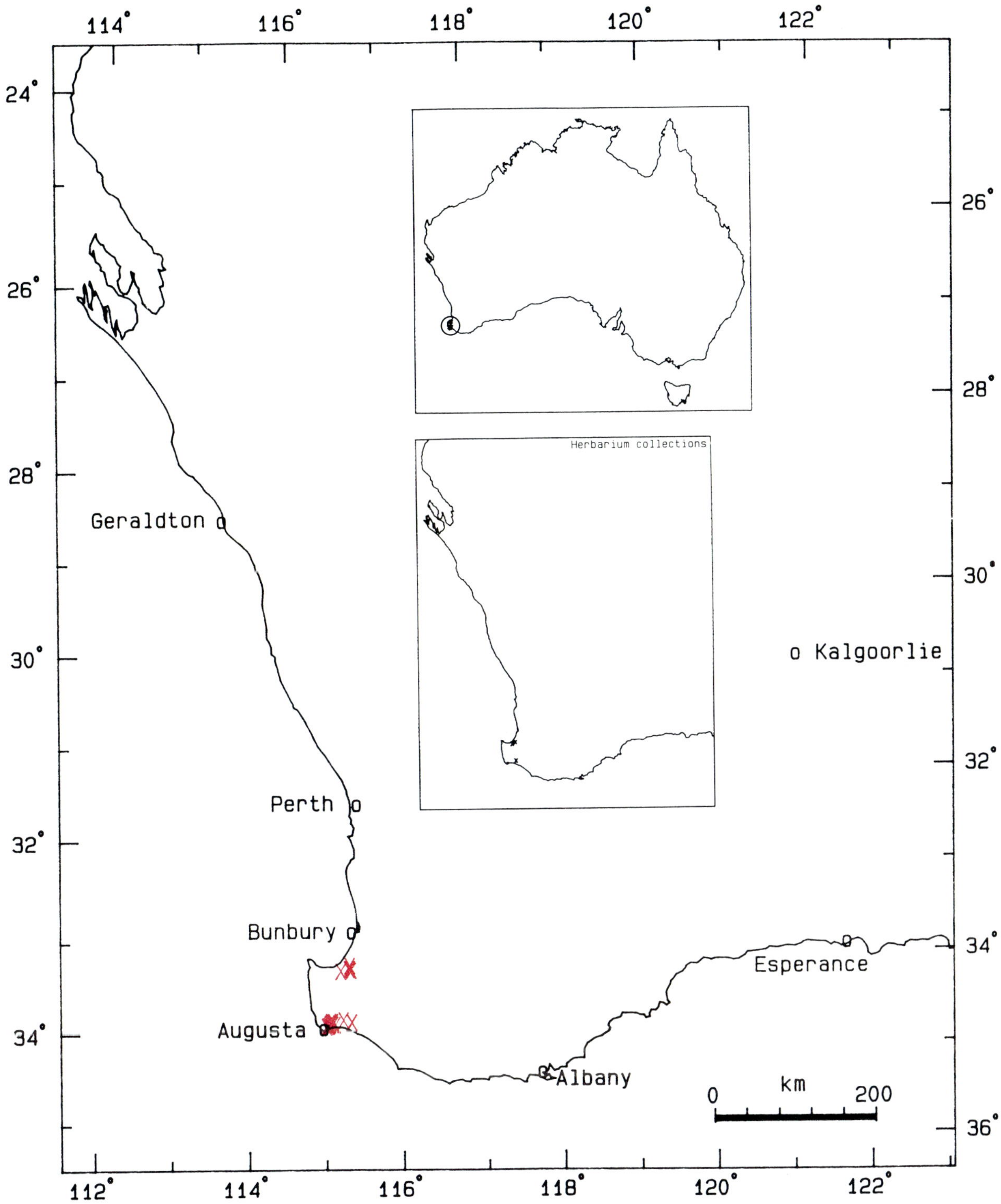

DISCUSSION

B. meisneri var. ascendens is a rare banksia which occurs in two disjunct populations, one near Busselton and the other on the Scott River Plains east of Augusta. The variety is less common than the 29 records suggest owing to duplicated recording at several sites. The northern population was recorded from only two sites. One of these had approximately 100 plants. The other, which had recently been burnt, had only two mature plants, but hundreds of 30 cm high seedlings. In the southern part of its range there are at least 10 500 m x 500 m grids containing B. meisneri var. ascendens. Generally the populations are of more than 100 plants. Most of these southern populations occur within the Scott River National Park. The two northern populations are not in conservation reserves.

The variety is a small shrub, generally less than 1 m in height. It grows in white or grey sands, on flat land often close to swamps. Surrounding vegetation is predominantly small shrubs with native grasses interspersed. Sometimes the latter become dominant. Amongst the shrubs there are often scattered emergent trees of Eucalyptus marginata, Nuytsia floribunda, Banksia ilicifolia, B. attenuata and B. littoralis. Occasionally, such trees become the dominant vegetation forming a low open woodland. Flowering was recorded in June only, whilst buds were recorded as early as March. This is in contrast to the April-August flowering period suggested by George (1981). Further data on flowering time should be sought. New shoot growth occurs in March when it was reported from five out of five records. However, further data are required before a wider season of new shoot growth can be determined. B. meisneri var. ascendens is killed by fire and regenerates from seed. It may take at least four years for seedlings to reach maturity and flower.

BANKSIA MEISNERI Lehmann (1845) var. MEISNERI

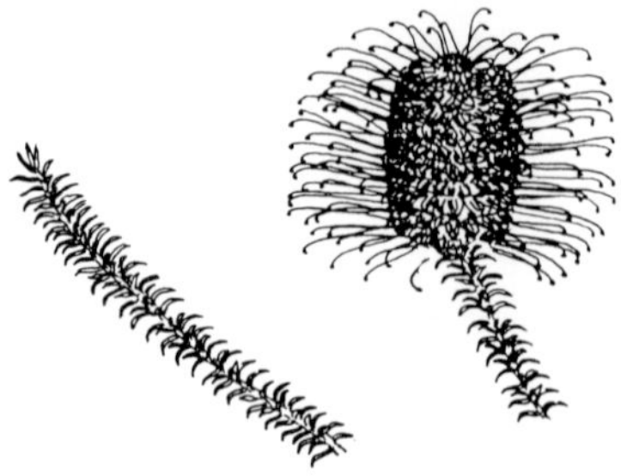

22 RECORDS: Jan (0) Feb (0) Mar (0) Apr (0) May (1) Jun (1) Jul (6) Aug (6) Sep (5) Oct (0) Nov (0) Dec (0).

Population Size: 1-10 (4) 10-100 (12) >100 (5) Unspecified (1).

Conservation Status: Restricted to road verge (5%) Not (86%) Unspecified (9%). In conservation reserve (45%) Not (50%) Unspecified (5%).

Tree/Shrub: Tree form (0%) Shrub form (100%) Unspecified (0%).

Height (metres)

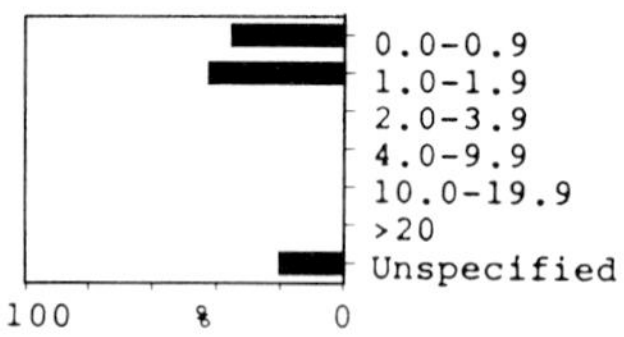

New Shoot Growth

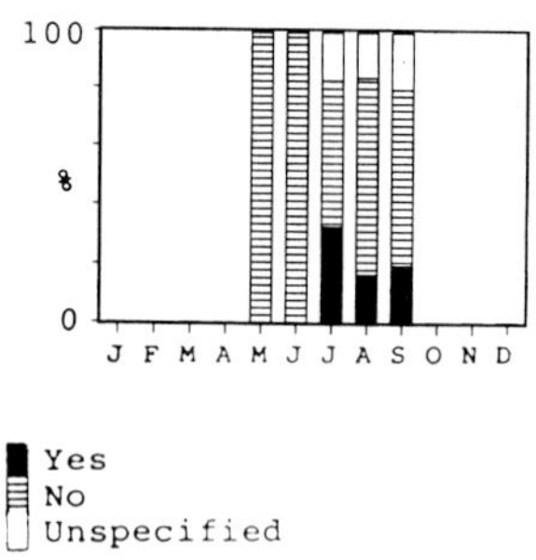

HABITAT

Within 2 km of coast: Yes (0%) No (100%) Unspecified (0%).

Altitude (metres)

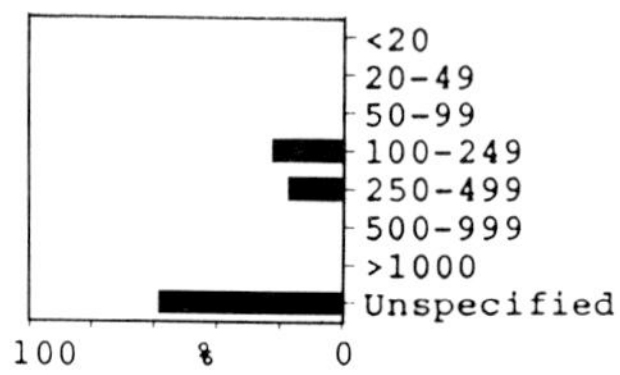

Vegetation

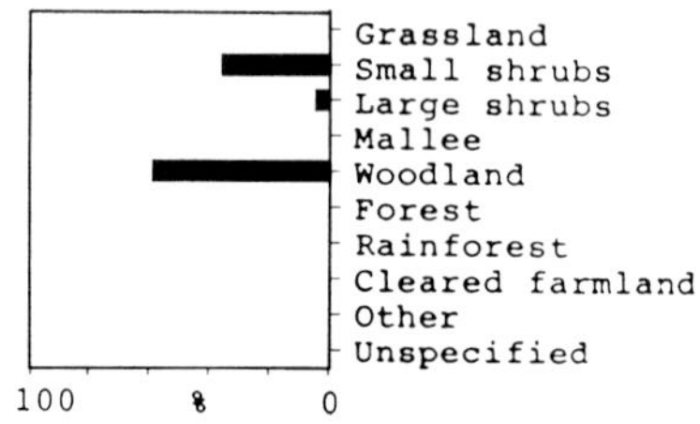

Soil Type

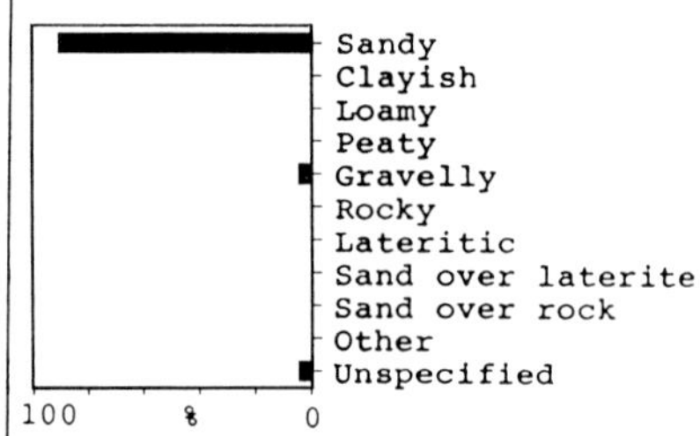

Landform

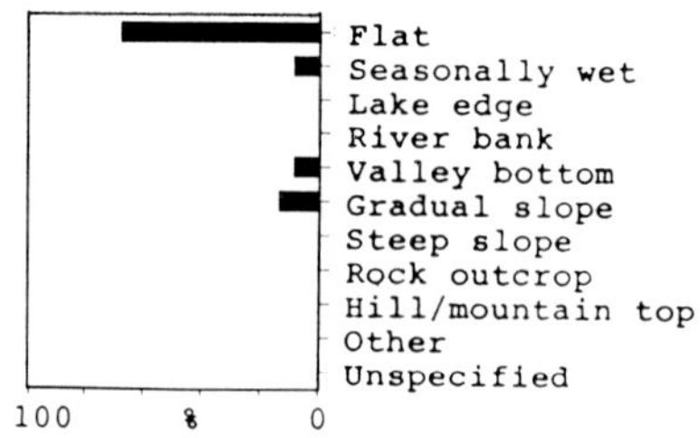

Response to Fire: 6 records

Growth response: Ground resprout and killed, new seedlings (17%) killed, new seedlings (50%) killed, no seedlings (17%) unspecified (17%).

Flowering response: Number of records where flowering had occurred after fire (5). Median reported time to flowering (13-24 months); minimum reported time to flowering (1-12 month). Number of records where flowering had not occurred after fire (1). Maximum reported time to no flowering (13-24 months).

Flowering

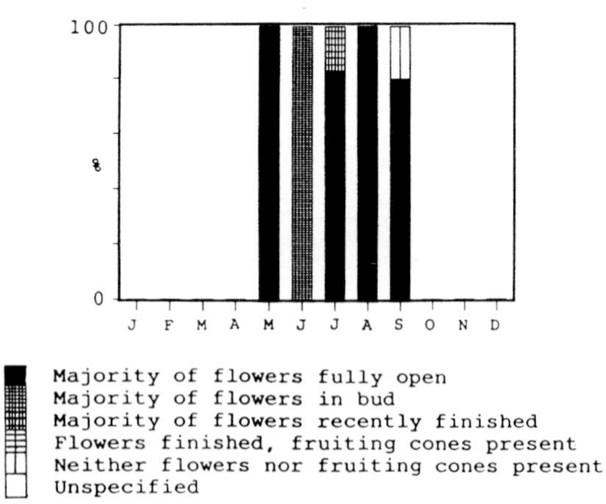

Possible pollinators

Pollinator type 2 records: bees, wasps, ants (50%) no pollinator observed (50%).

Specific pollinator 1 record: bee (1)

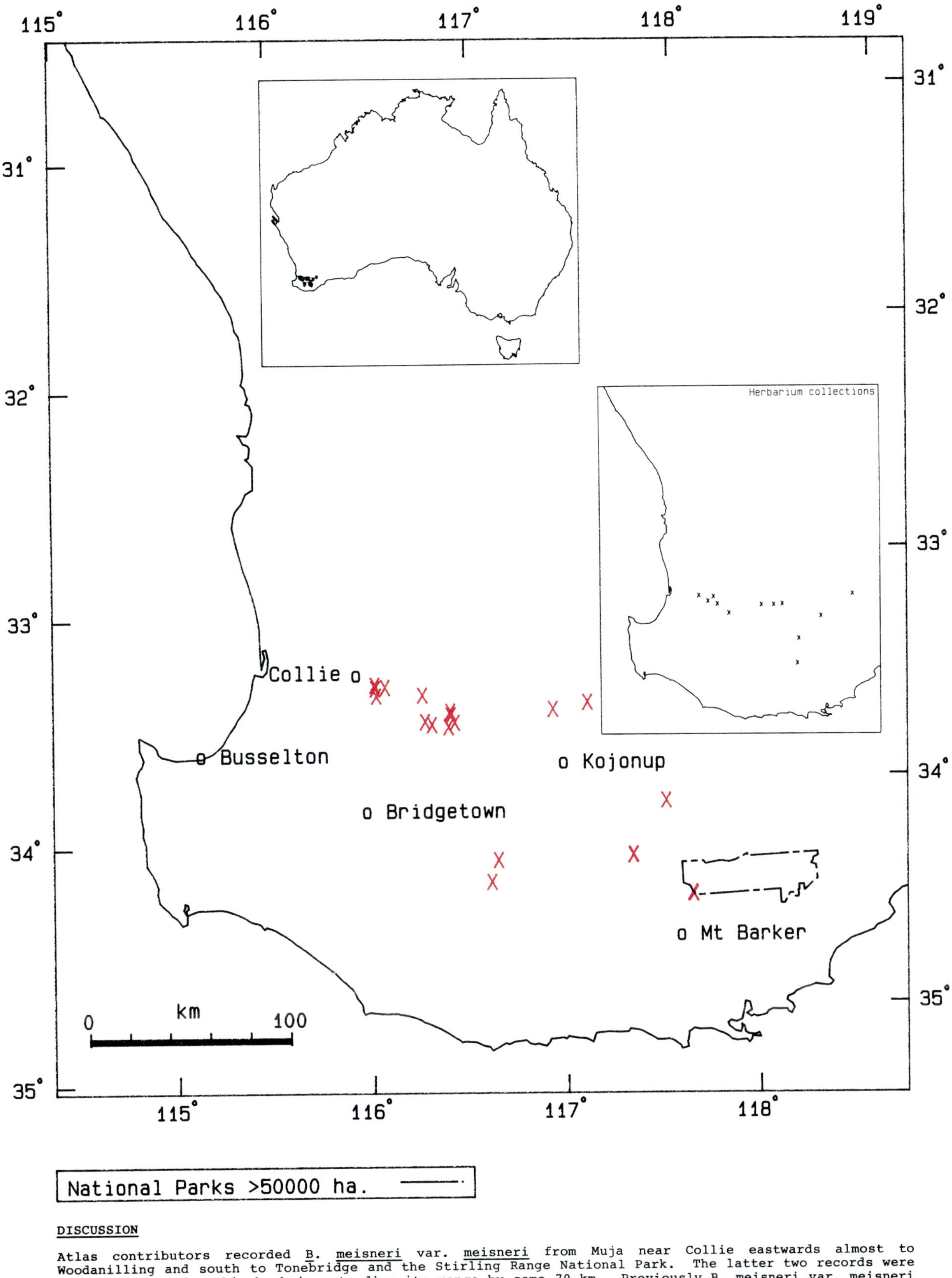

DISCUSSION

Atlas contributors recorded B. meisneri var. meisneri from Muja near Collie eastwards almost to Woodanilling and south to Tonebridge and the Stirling Range National Park. The latter two records were new localities for this banksia extending its range by some 70 km. Previously B. meisneri var. meisneri had not been recorded from within the national park boundary. A 1933 herbarium collection by W.E. Blackall from near Pingrup was not recorded by Atlas contributors. The 1965 collection locality of H. Daniels from Tenterden has now been cleared for agriculture but a new Atlas sighting was made from nearby at Cranbrook. The variety appears less common than was previously thought with the bulk of plants occurring in a relatively small area between Collie, Boyup Brook and Darkan. To the east of this main group there were only a few scattered recordings. However, this banksia is a relatively inconspicuous one and some populations may have been missed.

B. meisneri var. meisneri is typically a small shrub up to 2 m in height. It has a strong preference for white sandy soils and generally on low-lying sometimes swampy flats. Western populations are often in low open woodland of Eucalyptus marginata, Banksia grandis, Melaleuca preissiana, or are occasionally in tall shrubland e.g. with Kunzea ericifolia. Eastern populations are generally in low heath with scattered emergent trees of e.g. Banksia littoralis, B. attenuata. Flowering is from late autumn to early spring. Further data on pollinators and new shoot growth are needed for reliable conclusions to be drawn. B. meisneri var. meisneri is thought to be non-lignotuberous and to regenerate from seed after fire. A single Atlas record from Tonebridge of plants resprouting from lignotubers should be investigated.

BANKSIA MENZIESII R.Brown (1830)

Menzies' Banksia, Firewood Banksia

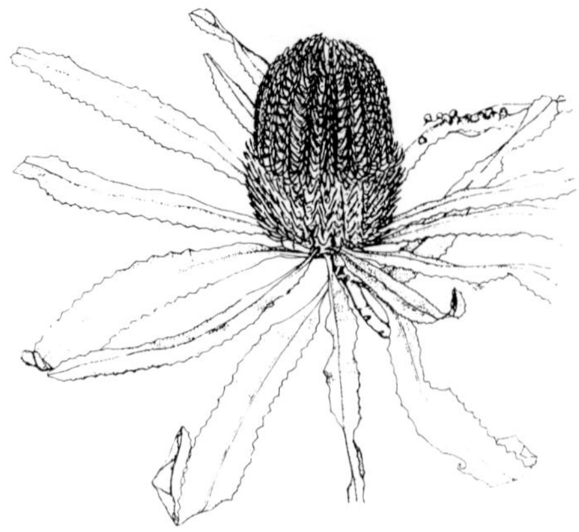

442 RECORDS: Jan (46) Feb (5) Mar (46) Apr (45) May (51) Jun (16) Jul (20) Aug (55) Sep (63) Oct (30) Nov (12) Dec (32).

Population Size: 1-10 (62) 10-100 (182) >100 (189) Unspecified (7).

Conservation Status: Restricted to road verge (9%) Not (88%) Unspecified (3%). In conservation reserve (21%) Not (77%) Unspecified (3%).

Tree/Shrub: Tree form (85%) Shrub form (13%) Unspecified (1%).

Height (metres)

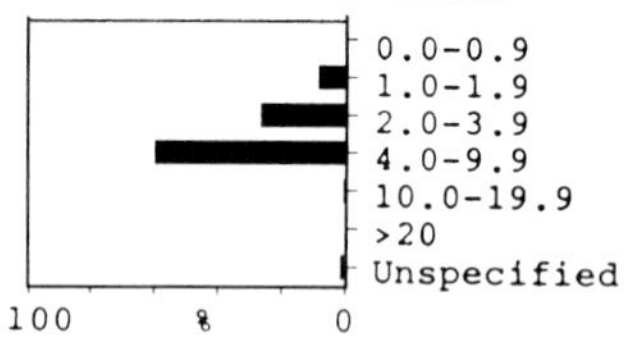

New Shoot Growth

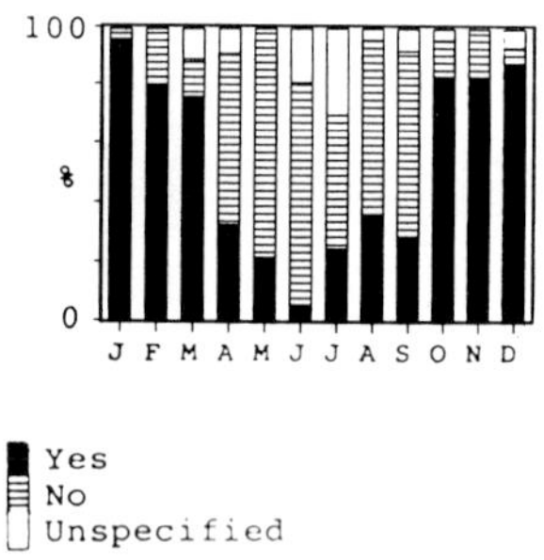

HABITAT

Within 2 km of coast: Yes (2%) No (98%) Unspecified (0%).

Altitude (metres)

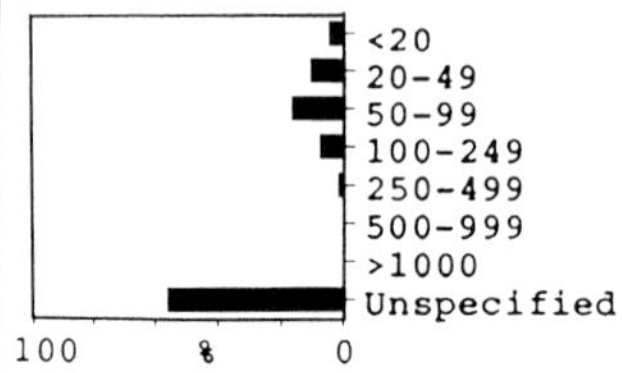

Vegetation

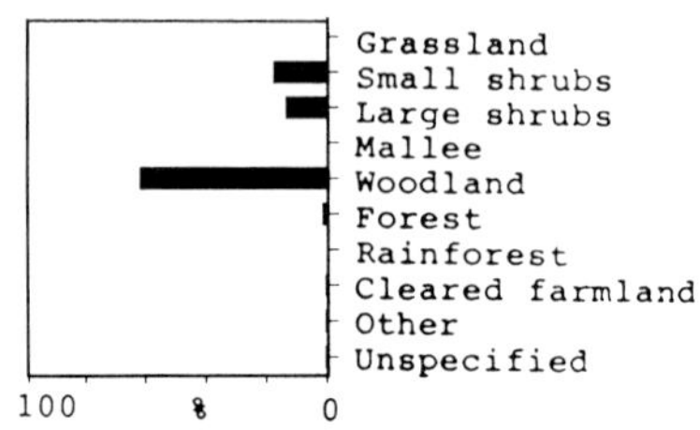

Soil Type

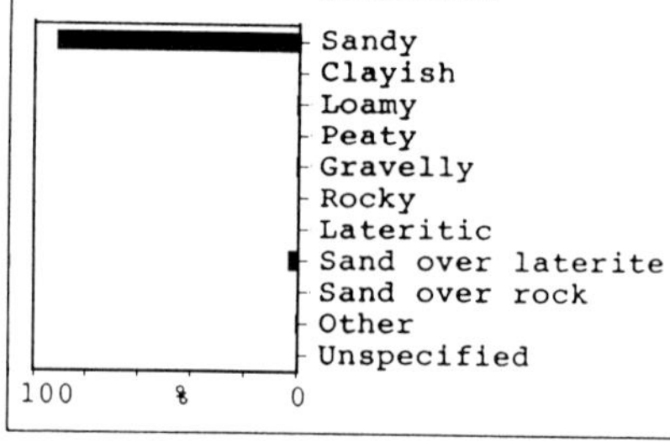

Landform

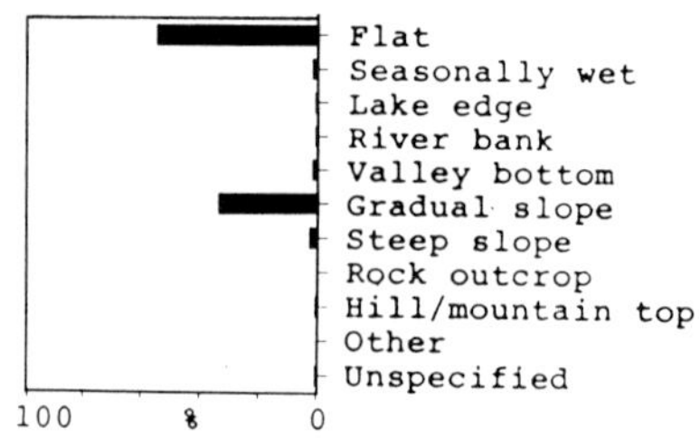

Response to Fire: 80 records

Growth response: Ground resprout (10%) ground resprout and killed, new seedlings (5%) ground resprout and trunk resprout (25%) killed, new seedlings (2%) trunk resprout (31%) killed, no seedlings (2%) unspecified (24%).

Flowering response: Number of records where flowering had occurred after fire (51). Median reported time to flowering (13-24 months); minimum reported time to flowering (<1 month). Number of records where flowering had not occurred after fire (14). Median reported time to no flowering (13-24 months); maximum reported time to no flowering (13-24 months).

Flowering

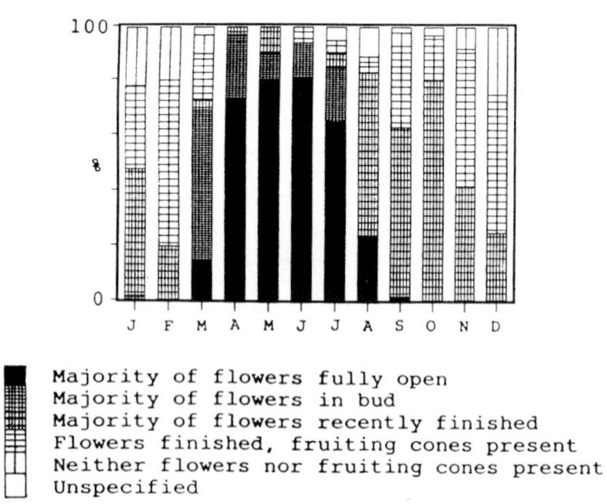

Possible pollinators

Pollinator type 134 records: bird (23%) butterflies, moths (1%) bees, wasps, ants (16%) beetles (1%) no pollinator observed (59%).

Specific pollinator 60 records: honeyeater (3) New Holland Honeyeater (6) Brown Honeyeater (3) White-cheeked Honeyeater (2) Singing Honeyeater (2) Red Wattlebird (3) Little Wattlebird (4) Western Spinebill (4) cockatoos and parrots (1) Port Lincoln Parrot (4) Western Rosella (2) Red-tailed Black-Cockatoo (1) Western Warbler (1) ant (9) bee (12) European honey bee (3)

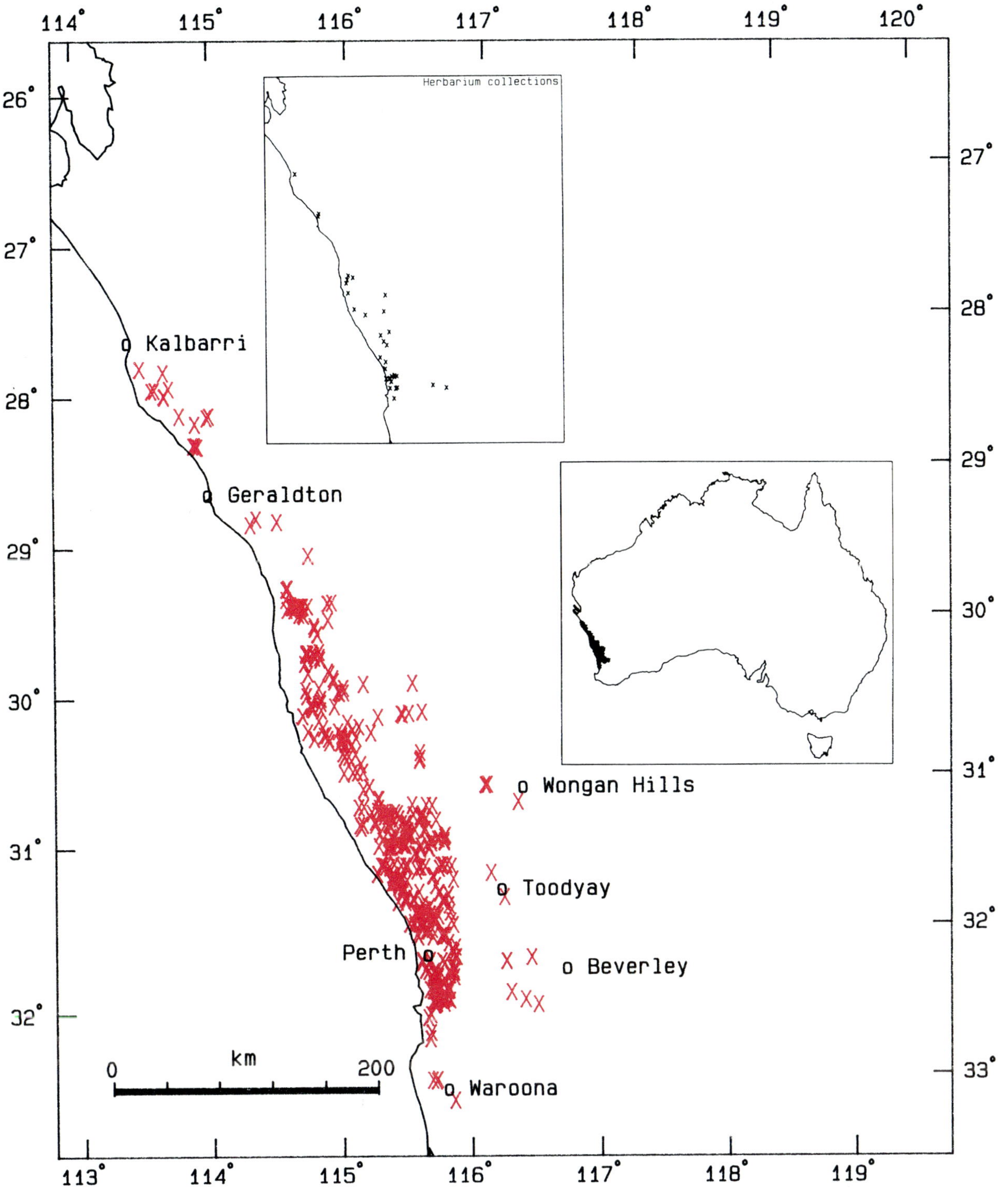

DISCUSSION

B. menziesii is a common component of woodlands and shrublands on deep sandy soils from Kalbarri southwards to Waroona. To the east it is limited by the heavy soils of the Darling Scarp and only reoccurs in isolated pockets of sand e.g. towards Beverley, Toodyay and Wongan Hills. The latter record is the furthest inland being some 120 km from the coast. The most inland herbarium collection of R. Hnatiuk's in 1979 from north-east of Brookton was not recorded by Atlas contributors.

Populations from around Badgingarra northwards are generally lignotuberous shrubs 1-2 m tall. Occasionally they occur as a non-lignotuberous small tree e.g. in Cockleshell Gully west of Mt Lesueur. In both cases the species is usually emergent above an otherwise low heath. South of Badgingarra, B. menziesii is typically a small tree 3-10 m in height, occurring in low open woodland of Banksia attenuata, B. ilicifolia, and Eucalyptus marginata. The species is also variable in leaf size. North of Moore River, leaves are shorter and narrower than those to the south e.g. leaves from Badgingarra, average length is 11-14 cm, from Clackline east of Perth, average length is 21 cm (G. Keighery, pers. comm.). Flower colour is also variable with both bronze and yellow forms occurring occasionally. A greenish-coloured variant has also been observed (G. Keighery, pers. comm.). The normal fruiting cone sheds its flowers at an early stage but some plants have been observed to retain their old flowers. The flowering period is March-August when the conspicuous flowers attract birds such as the New Holland Honeyeater, Brown Honeyeater, Western Spinebill and wattlebirds. New shoot growth is from late spring through summer. Following fire the species resprouts either from its lignotuber or from epicormic buds. There are also a few records of mature plants being killed by fire and regenerating from seed. A rare B. menziesii x B. hookeriana pollen-sterile hybrid has been collected by G. Keighery from north of Badgingarra.

BANKSIA MICRANTHA A.S.George 1981)

31 RECORDS: Jan (0) Feb (1) Mar (3) Apr (9) May (6) Jun (0) Jul (0) Aug (2) Sep (8) Oct (0) Nov (0) Dec (2).

Population Size: 1-10 (5) 10-100 (22) >100 (1) Unspecified (3).

Conservation Status: Restricted to road verge (35%) Not (65%) Unspecified (0%). In conservation reserve (23%) Not (71%) Unspecified (6%).

Tree/Shrub: Tree form (0%) Shrub form (97%) Unspecified (3%).

Height (metres)

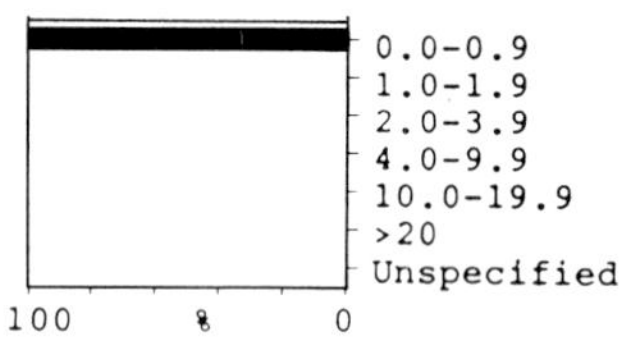

New Shoot Growth

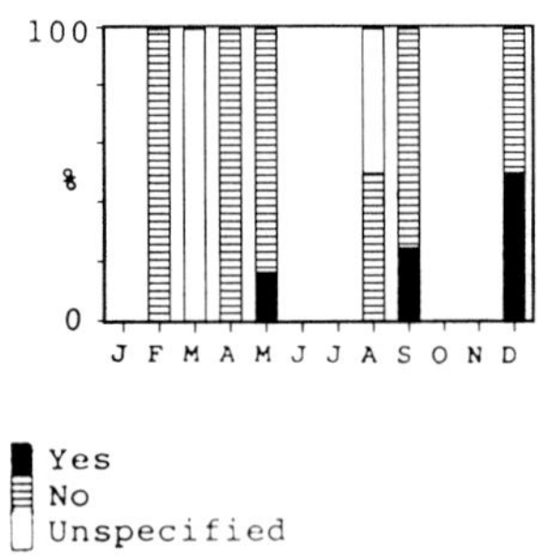

HABITAT

Within 2 km of coast: Yes (0%) No (100%) Unspecified (0%).

Altitude (metres)

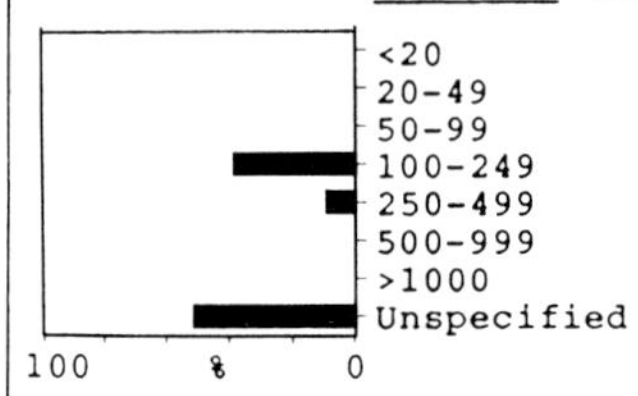

Vegetation

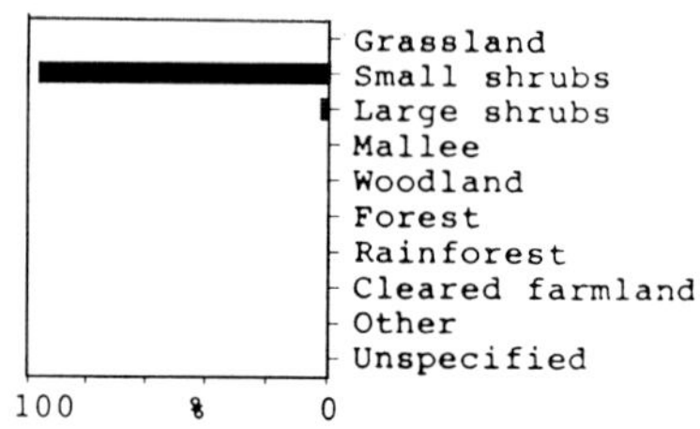

Soil Type

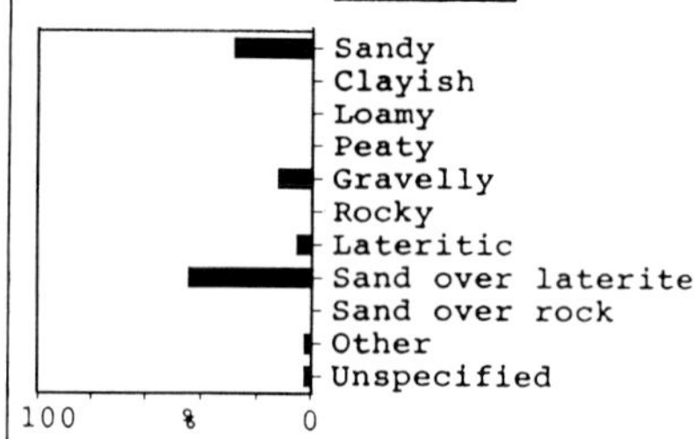

Landform

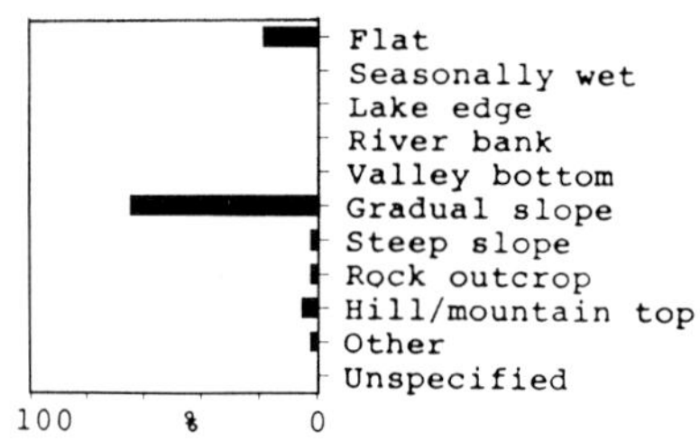

Response to Fire: 3 records

Growth response: Ground resprout (100%)

Flowering response: Number of records where flowering had occurred after fire (2). Median reported time to flowering (12-13 months); minimum reported time to flowering (1-12 months). Number of records where flowering had not occurred after fire (0).

Flowering

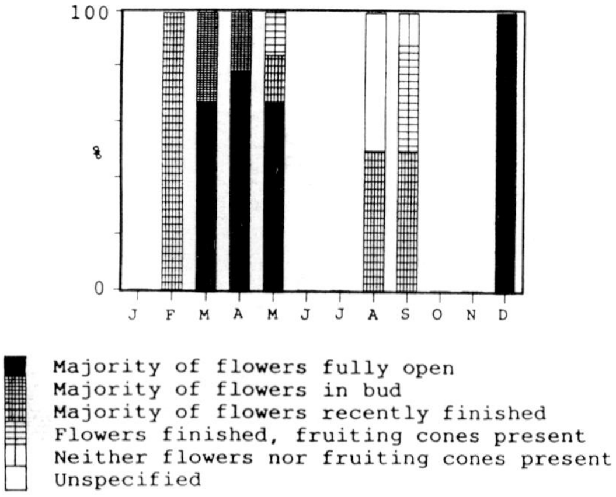

Possible pollinators

Pollinator type 10 records: bees, wasps, ants (10%) no pollinator observed (90%).

Specific pollinator 0 records:

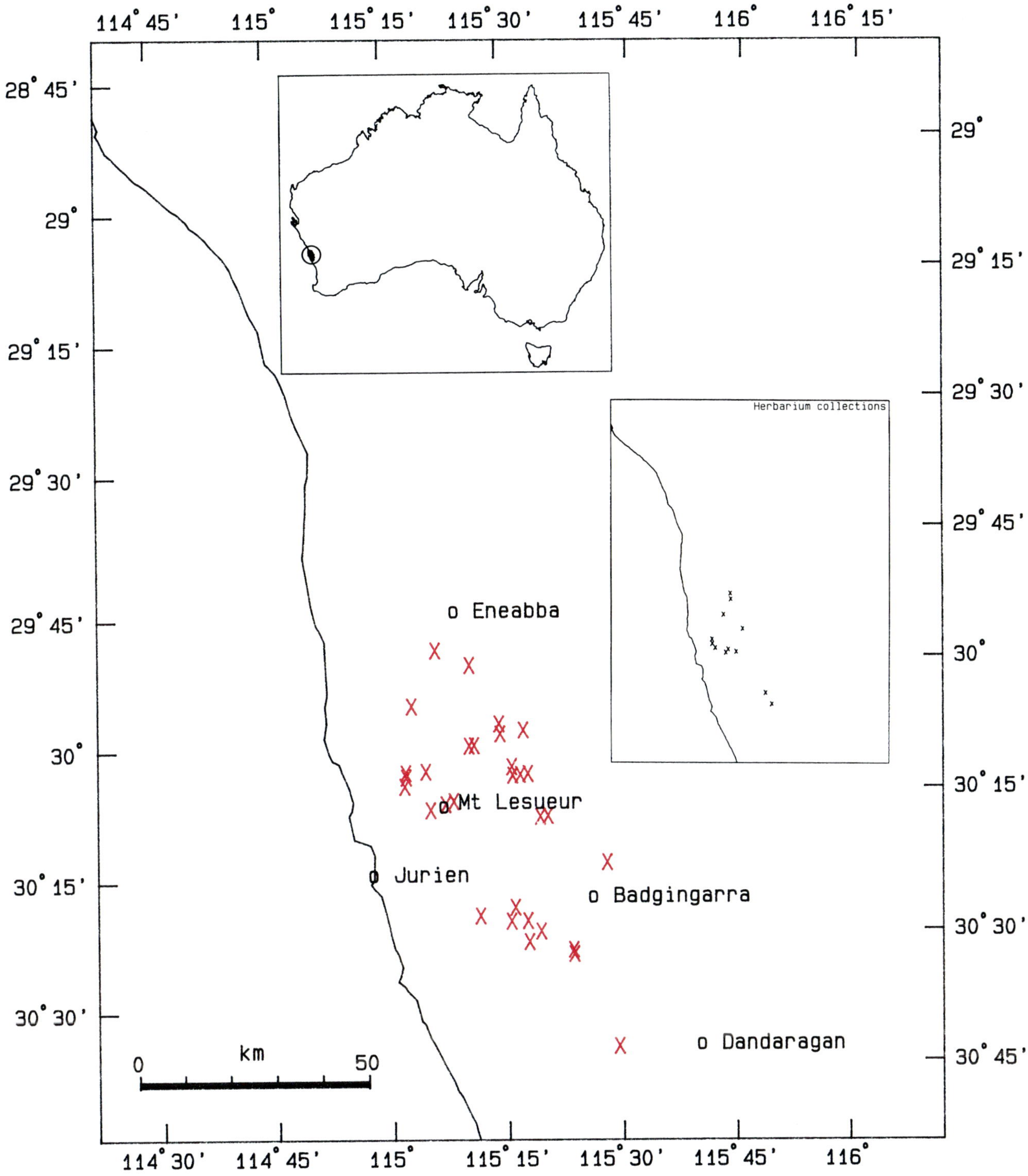

DISCUSSION

Banksia micrantha caused the most identification problems thoughout the Atlas, being frequently confused with B. sphaerocarpa var. sphaerocarpa. The latter, in its northern heathland form, is a low growing sprawling shrub with short pungent leaves and small flower spikes and flowers falling well within the range of B. micrantha described by George (1981). The single main difference lay in the perianth limb of B. sphaerocarpa var. sphaerocarpa being hairy whilst that of B. micrantha is mostly glabrous. Many submitted specimens of B. micrantha were subsequently changed to B. sphaerocarpa var. sphaerocarpa. The distribution map of B. micrantha should therefore be considered within the context of some possible mis-identifications having taken place.

Western Australian Herbarium records indicated a distribution between Eneabba and Badgingarra National Park centred mainly in an area around Mt Lesueur. Atlas contributors recorded several sites from Badgingarra National Park and extended the range to Cataby and to north-east of Badgingarra. Most recorded populations are of less than 100 plants, with almost one-quarter in conservation reserves.

B. micrantha is typically a low sprawling shrub less than a metre in height. It grows in low heath on sands or sand over laterite, generally on the gentle slopes of the low lateritic hills which characterise the area in which it occurs. Flowering was recorded in December and also from March to May. There is a lack of records from the intervening months of January and February. However, George (1981) stated that the flowering period is from January to May. Further data are needed on pollinators and new shoot growth before reliable conclusions can be drawn. The species is lignotuberous and the resprouting shoots may be flowering within 12 months of being burnt.

BANKSIA NUTANS R.Brown (1810)

Nodding Banksia

360 RECORDS: Jan (29) Feb (23) Mar (30) Apr (22) May (14) Jun (18) Jul (17) Aug (41) Sep (51) Oct (50) Nov (41) Dec (18).

Population Size: 1-10 (107) 10-100 (172) >100 (71) Unspecified (10).

Conservation Status: Restricted to road verge (21%) Not (76%) Unspecified (3%). In conservation reserve (50%) Not (46%) Unspecified (4%).

Tree/Shrub: Tree form (0%) Shrub form (98%) Unspecified (2%).

Height (metres)

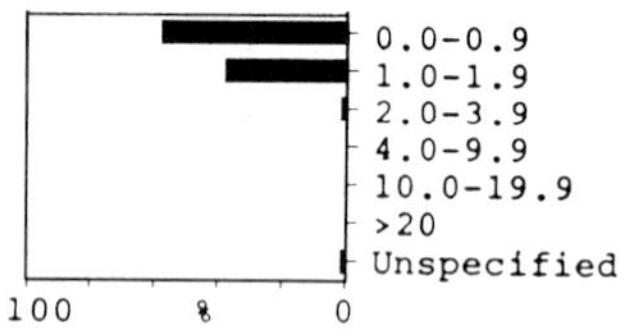

New Shoot Growth

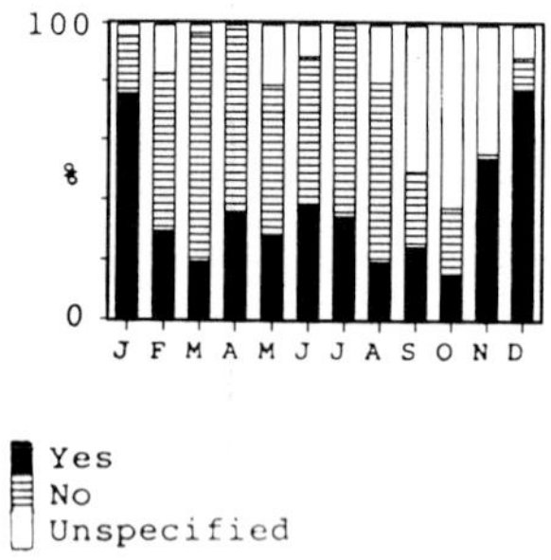

HABITAT

Within 2 km of coast: Yes (15%) No (85%) Unspecified (0%).

Altitude (metres)

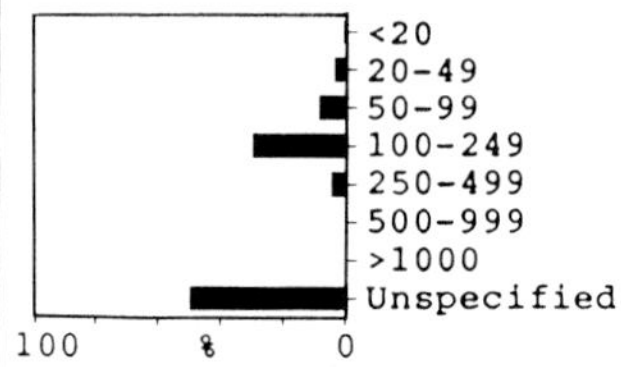

Vegetation

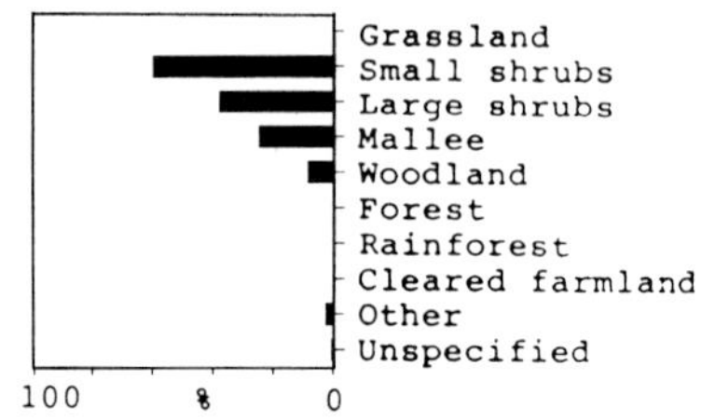

Soil Type

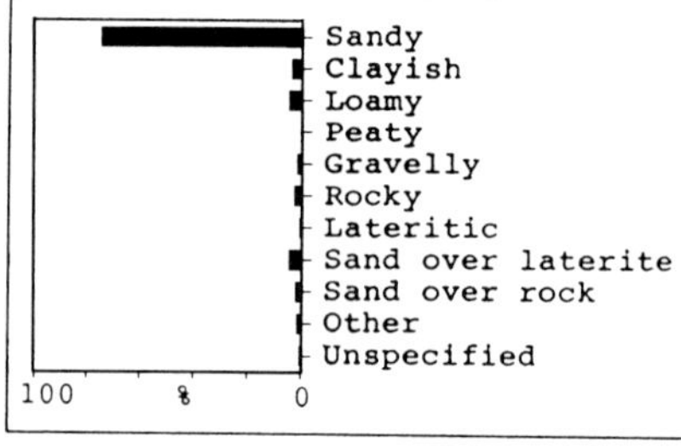

Landform

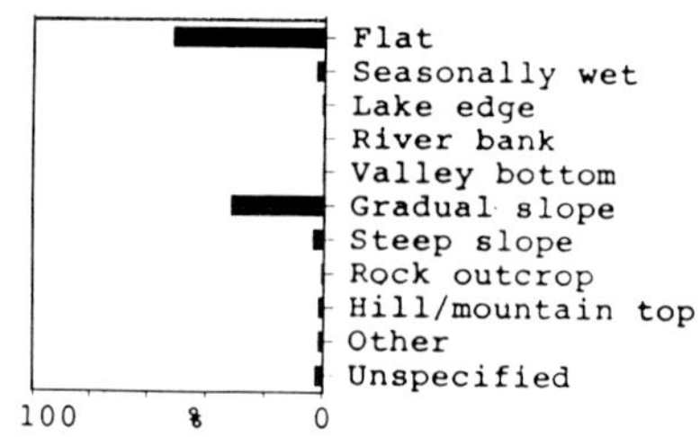

Response to Fire: 26 records

Growth response: Ground resprout (4%) killed, new seedlings (42%) trunk resprout (4%) killed, no seedlings (4%) unspecified (46%).

Flowering response: Number of records where flowering had occurred after fire (15). Median reported time to flowering (13-24 months); minimum reported time to flowering (<1 month). Number of records where flowering had not occurred after fire (9). Median reported time to no flowering (1-12 months); maximum reported time to no flowering (25-36 months).

Flowering

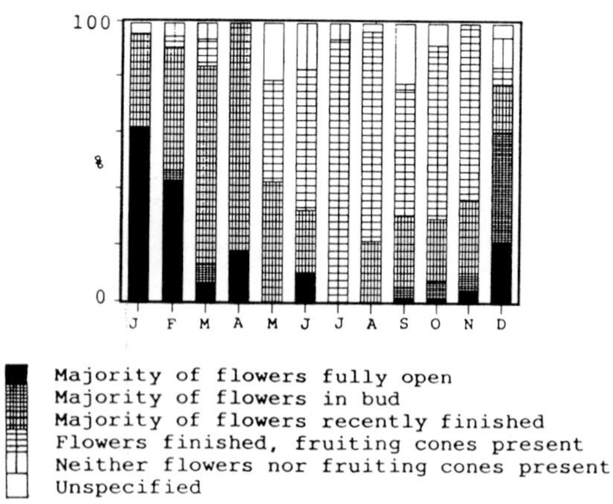

Possible pollinators

Pollinator type 76 records: bird (1%) no pollinator observed (99%).

Specific pollinator 0 records:

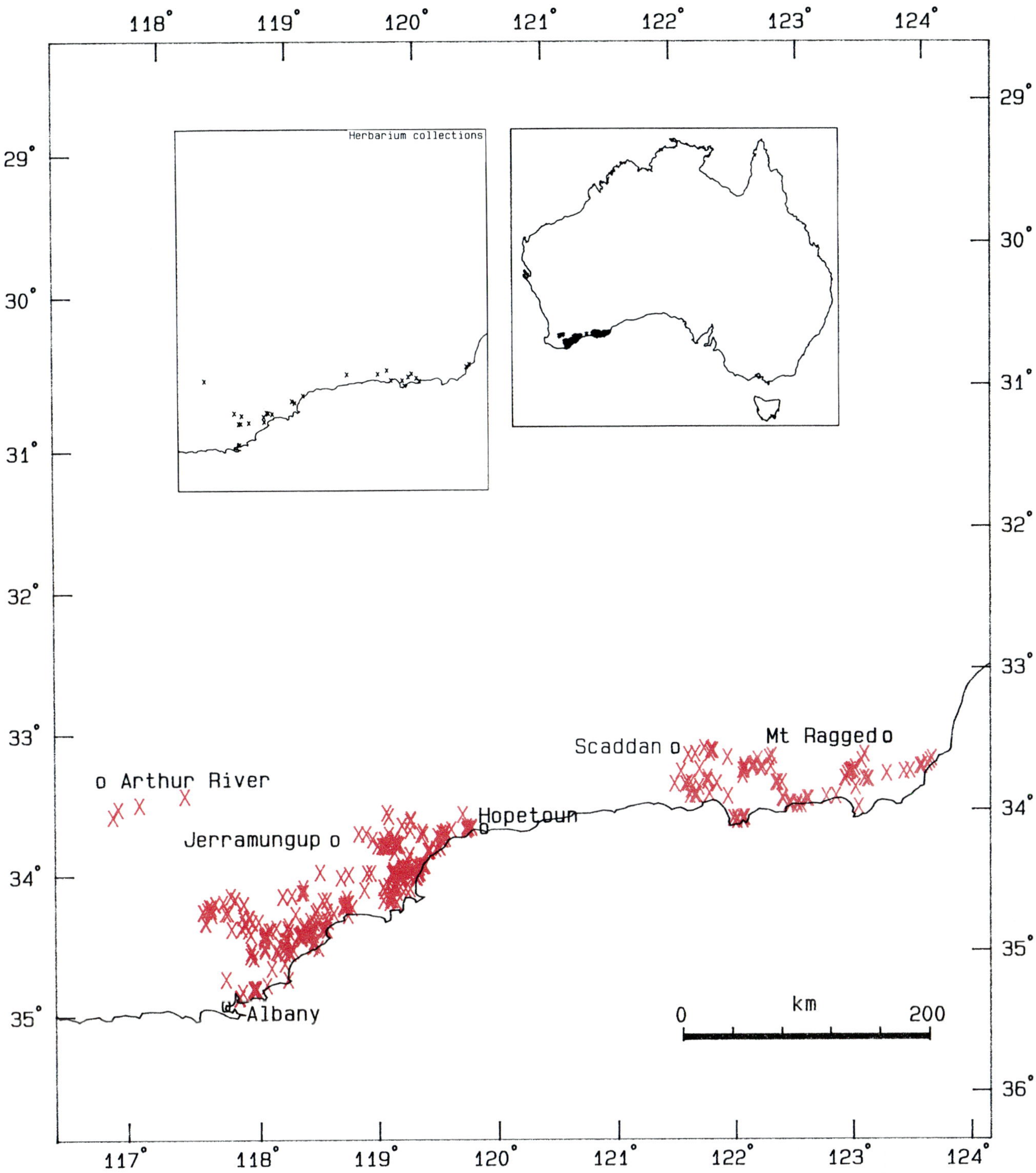

DISCUSSION

B. nutans is a distinctive small shrub with its short blue-green leaves and pendulous flower spikes. There are two varieties distinguished by the size of their follicles and the size of their flowers. Their distributions were also thought to be disjunct, with var. cernuella occurring west of the Pallinup River and var. nutans to the east. The 360 sight records of B. nutans included 302 where a variety was identified and 58 where only the species name was specified. At 16% of the total number of records, the latter figure is high and indicates that this is one of the more difficult of banksias to identify at the varietal level. There were also several specimens submitted which did not conform with the varietal distribution pattern suggested by George (1981). Thus it may be that var. cernuella extends further east and that var. nutans extends further west and that their ranges overlap. This is further discussed under each variety.

B. nutans flowers predominantly in summer though flowers have been recorded as early as September and as late as June. There were very few potential pollinators observed, though the distinctive onion-like scent of the flowers is though to be attractive to small mammals. New shoot growth is mainly in summer. Banksia nutans is generally killed by fire and regenerates from seed. The conservation status of B. nutans seems relatively secure with 50% of its recorded populations being in conservation reserves and only one-fifth being restricted to road verges. However, efforts should be made to protect the outlying populations south of Arthur River which, according to George (1981), have longer and more slender leaves than is typical.

BANKSIA NUTANS R.Brown var. CERNUELLA A.S.George (1981)

100 RECORDS: Jan (3) Feb (10) Mar (9) Apr (6) May (2) Jun (8) Jul (8) Aug (24) Sep (12) Oct (9) Nov (3) Dec (3).

Population Size: 1-10 (26) 10-100 (52) >100 (20) Unspecified (2).

Conservation Status: Restricted to road verge (33%) Not (62%) Unspecified (5%). In conservation reserve (36%) Not (60%) Unspecified (4%).

Tree/Shrub: Tree form (0%) Shrub form (100%) Unspecified (0%).

Height (metres)

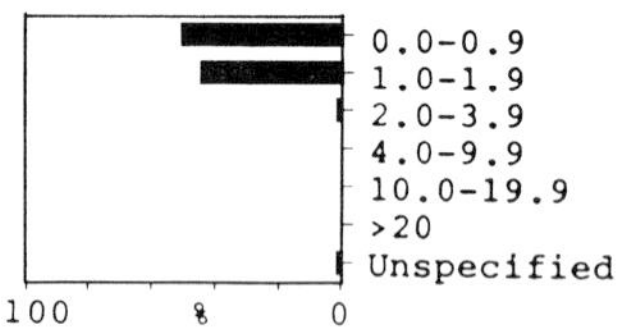

New Shoot Growth

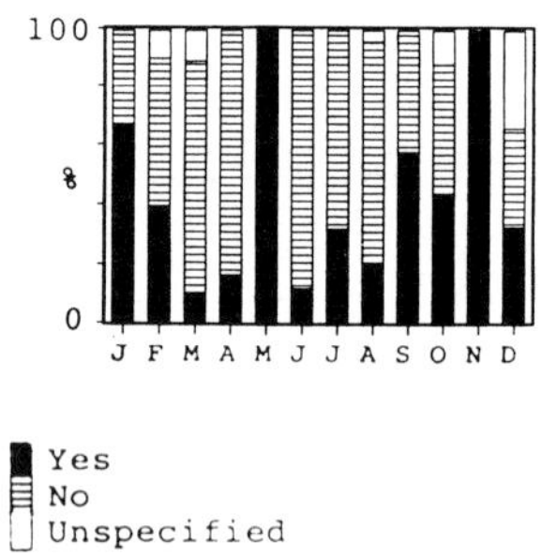

HABITAT

Within 2 km of coast: Yes (6%) No (94%) Unspecified (0%).

Altitude (metres)

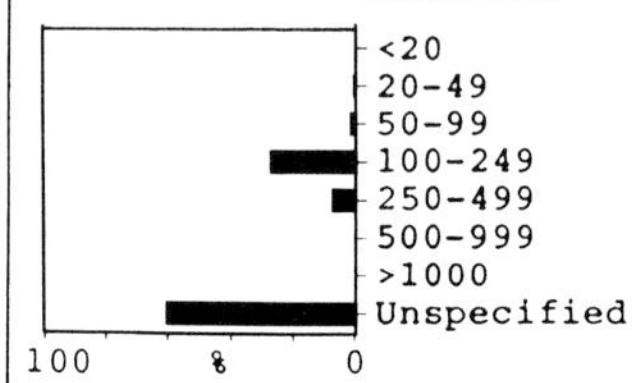

Vegetation

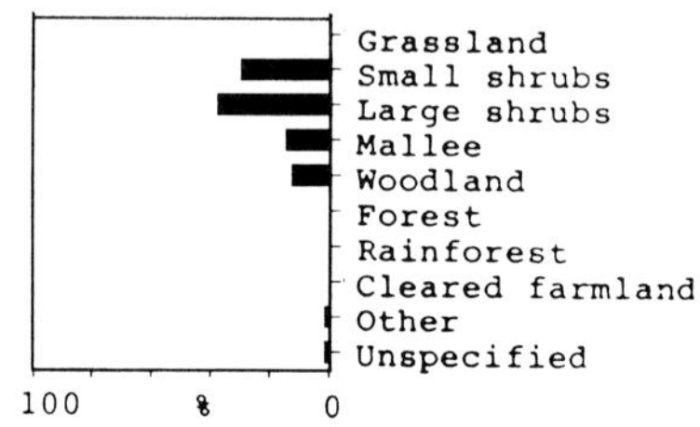

Soil Type

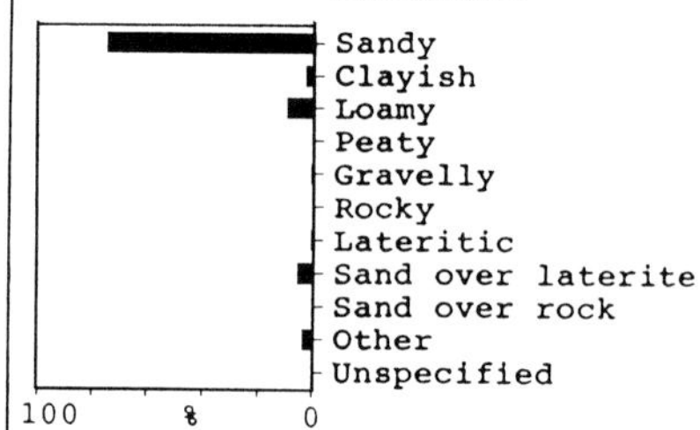

Landform

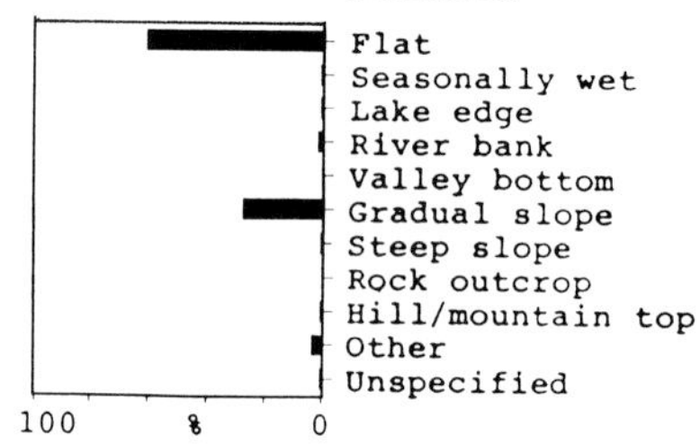

Response to Fire: 6 records

Growth response: Killed, new seedlings (33%) killed, no seedlings (17%) unspecified (50%).

Flowering response: Number of records where flowering had occurred after fire (5). Median reported time to flowering (25-36 months); minimum reported time to flowering (<1 month). Number of records where flowering had not occurred after fire (1). Maximum reported time to no flowering (1-12 months).

Flowering

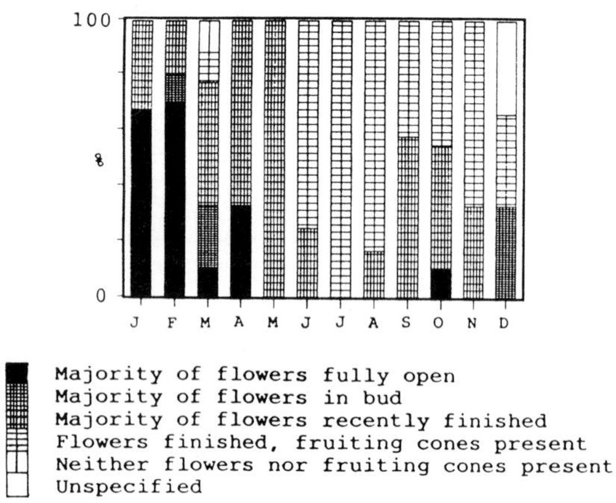

Possible pollinators

Pollinator type 27 records: no pollinator observed (100%).

Specific pollinator 0 records:

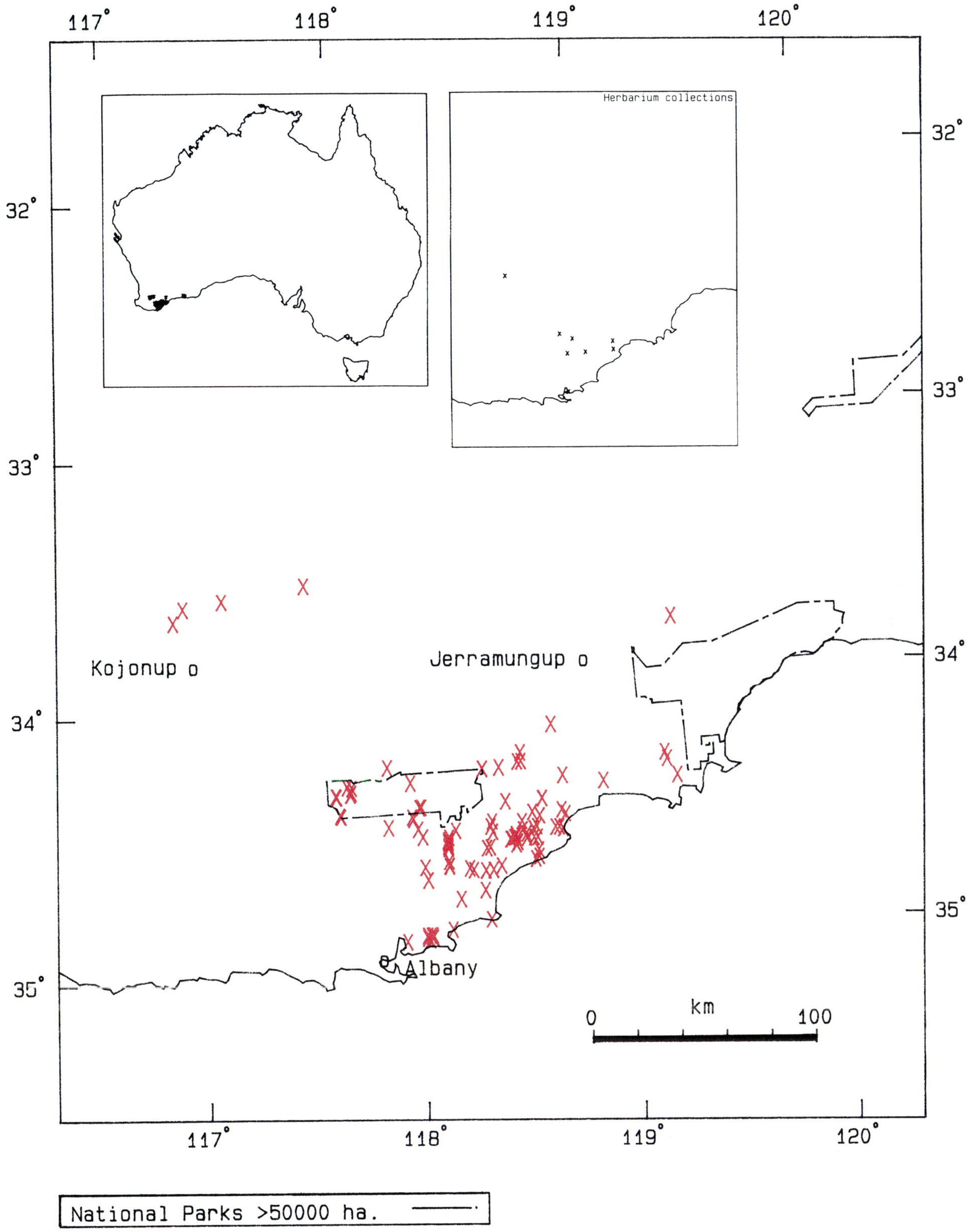

DISCUSSION

Western Australian Herbarium records indicated a distribution for B. nutans var. cernuella from the Stirling Range to Albany and east almost to the Pallinup River, with an outlier near Woodanilling north of Kojonup. Atlas records confirmed this distribution and show the Woodanilling outlier to consist of more populations than was previously thought. The western-most of these records is 3 km south-west of Boscabel which is approximately 40 km west of Woodanilling. Along the south coast the variety may extend further east than was previously thought, beyond the Pallinup River. Some specimens submitted from Bremer Bay area had small flowers well within the dimensions specified for var. cernuella (George 1981). They have therefore been included in the distribution map for this variety. Moreover, a locality just east of the Pallinup River was reported to contain both varieties of B. nutans. It would therefore seem that the geographical separation of the two varieties is not as clear cut as was previously thought.

B. nutans var. cernuella is typicaly a shrub up to 2 m in height growing amongst low or tall shrubland, occasionally in mallee or woodland. Sandy soils are preferred, sometimes overlying laterite. Landforms are usually flat or gently sloping. The main flowering period is January to April, slightly later than that of var. nutans. This is in accordance with the flowering periods suggested by George (1981). A single record of var. cernuella flowering in October is possibly indicative of an incorrect identification. Further data are needed for new shoot growth and pollinators before reliable conclusions can be drawn. Although B. nutans var. cernuella is not as common as var. nutans and is not as well represented in conservation reserves, it seems relatively secure at present. However, the northern outlying populations should be monitored especially since they may be morphologically different from those further south with longer, more slender leaves than is typical (George 1981).

BANKSIA NUTANS R.Brown (1810) var. NUTANS

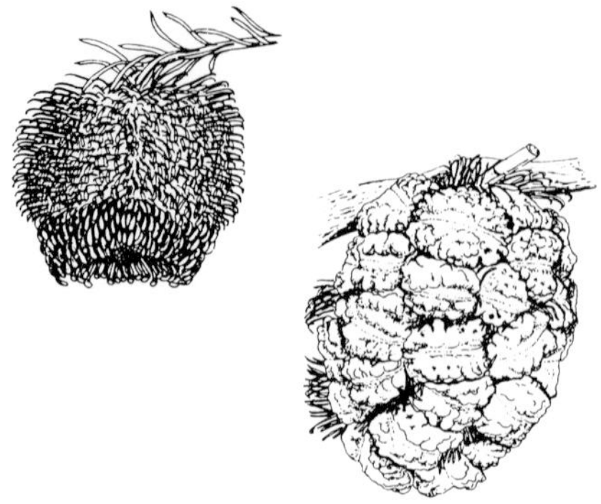

204 RECORDS: Jan (17) Feb (10) Mar (19) Apr (12) May (8) Jun (4) Jul (5) Aug (9) Sep (31) Oct (39) Nov (36) Dec (11).

Population Size: 1-10 (69) 10-100 (95) >100 (34) Unspecified (6).

Conservation Status: Restricted to road verge (14%) Not (83%) Unspecified (3%). In conservation reserve (59%) Not (38%) Unspecified (2%).

Tree/Shrub: Tree form (0%) Shrub form (100%) Unspecified (0%).

Height (metres)

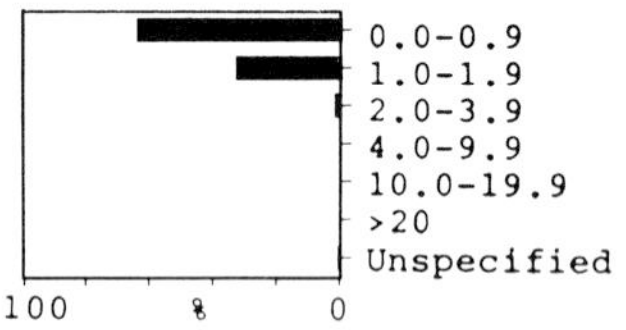

New Shoot Growth

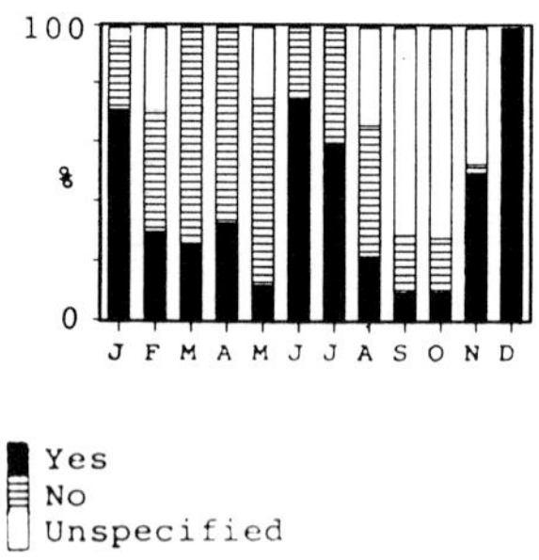

HABITAT

Within 2 km of coast: Yes (21%) No (79%) Unspecified (0%).

Altitude (metres)

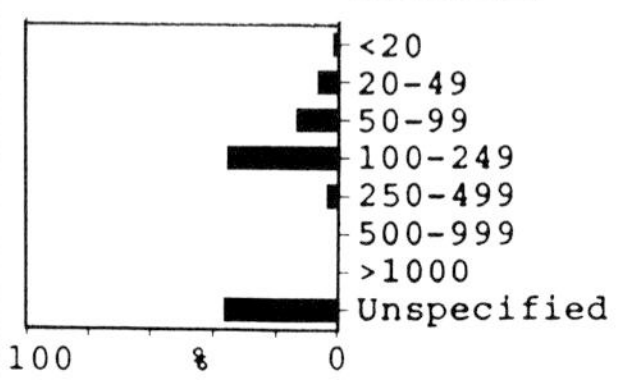

Vegetation

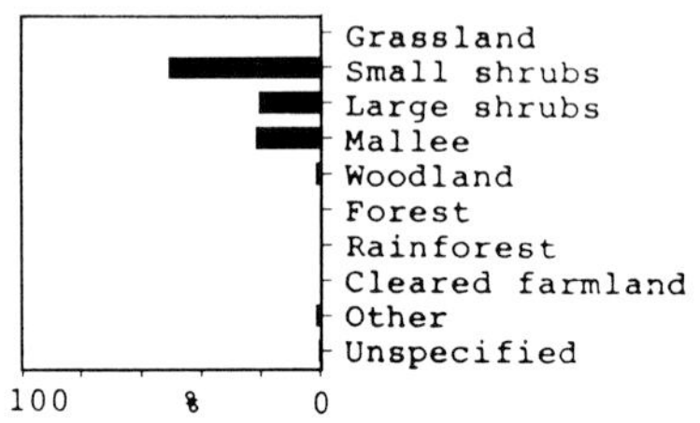

Soil Type

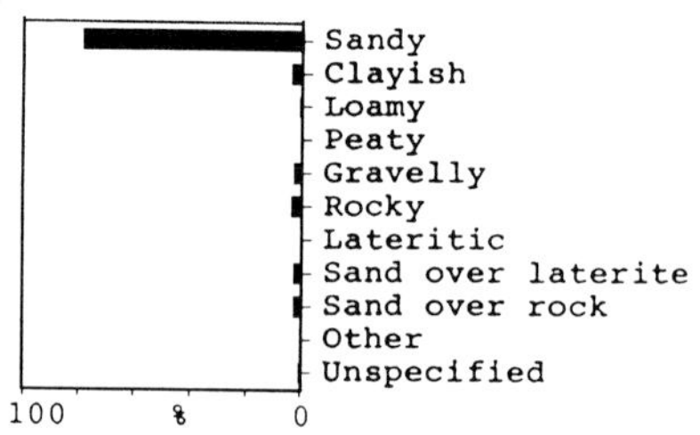

Landform

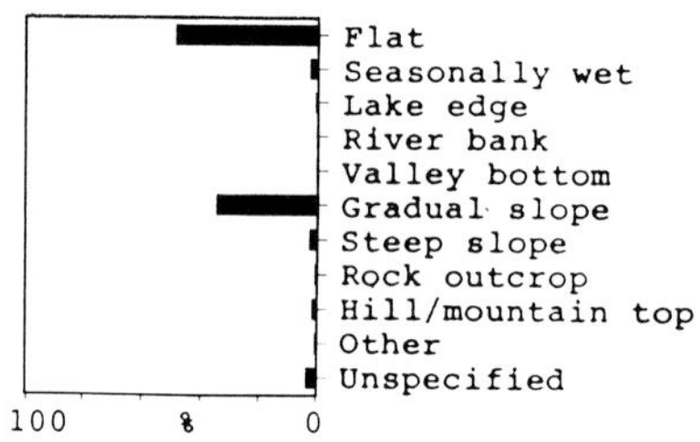

Response to Fire: 13 records

Growth response: Ground resprout (8%) killed, new seedlings (31%) trunk resprout (8%) unspecified (54%).

Flowering response: Number of records where flowering had occurred after fire (8). Median reported time to flowering (13-24 months); minimum reported time to flowering (<1 month). Number of records where flowering had not occurred after fire (4). Median reported time to no flowering (1-12 months); maximum reported time to no flowering (1-12 months).

Flowering

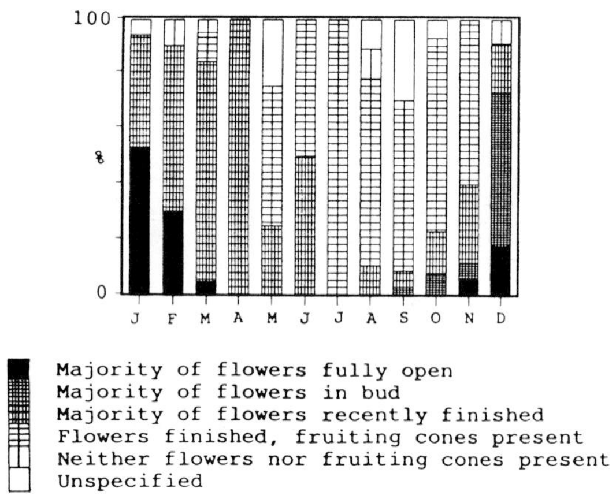

Possible pollinators

Pollinator type 43 records: bird (2%) no pollinator observed (98%).

Specific pollinator 0 records:

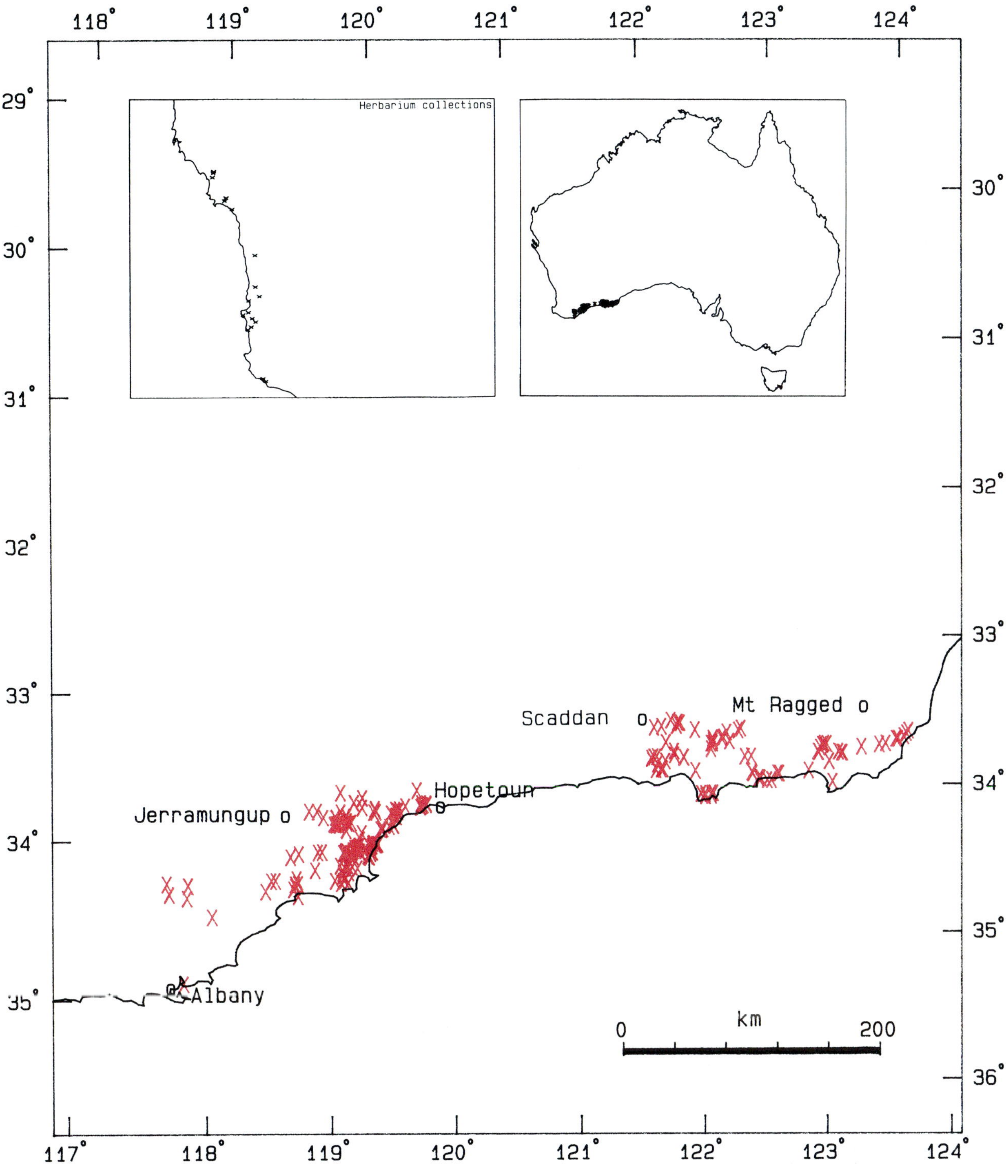

DISCUSSION

Western Australian Herbarium records indicated a distribution for B. nutans var. nutans between the Pallinup River and Israelite Bay. West of the Pallinup River, var. nutans was replaced by var. cernuella (George 1981). Atlas records suggest that var. nutans may extend further west than was previously thought. Specimens submitted from Mt Martin near Albany and from Two Peoples Bay Nature Reserve had large flower and follicle dimensions which placed them within the specified range of var. nutans. Both these specimens are now in the Western Australian Herbarium. Likewise, a few Stirling Range populations were reported to have large flowers and follicles. All the above have been included in the distribution map of var. nutans. It was also reported that just east of the Pallinup River, both varieties appeared to be present. It would therefore seem that the geographical separation of the two varieties is not as clear cut as was previously thought. Aside from the problem of its western boundary, var. nutans occurs in two disjunct populations being absent apparently between Hopetoun and Scaddan.

The variety is typically a shrub generally less than 1 m in height, though sometimes larger. It occurs predominantly in sandy soils within landscapes that are flat or gently sloping. Surrounding vegetation is usually heath, less often shrubland or mallee. The variety is well represented in conservation reserves (notably Fitzgerald River and Cape Arid National Parks) and with correct management should be well protected. The main flowering period is November to March. There appear to be two seasons of new shoot growth (summer and winter), though further studies are needed to confirm this.

BANKSIA OBLONGIFOLIA Cavanilles(1800)

Fern-leaved Banksia

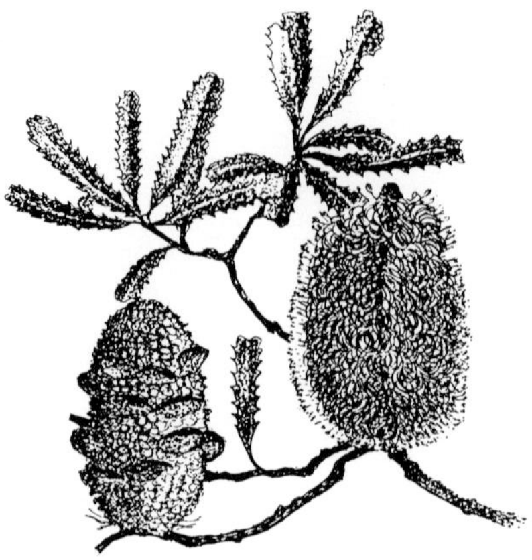

492 RECORDS: Jan (51) Feb (10) Mar (51) Apr (52) May (75) Jun (35) Jul (47) Aug (47) Sep (31) Oct (35) Nov (15) Dec (36).

Population Size: 1-10 (54) 10-100 (138) >100 (299) Unspecified (1).

Conservation Status: Restricted to road verge (1%) Not (97%) Unspecified (2%). In conservation reserve (50%) Not (40%) Unspecified (10%).

Tree/Shrub: Tree form (0.3%) Shrub form (99%) Unspecified (0%).

Height (metres)

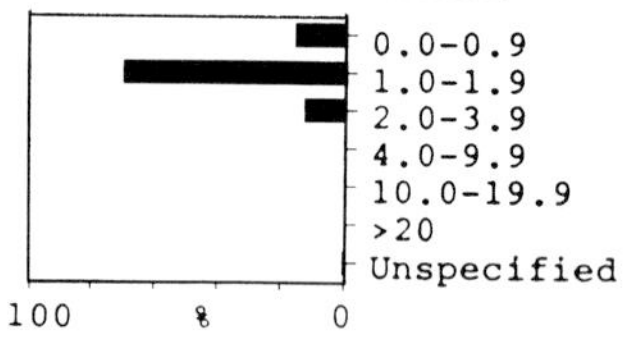

New Shoot Growth

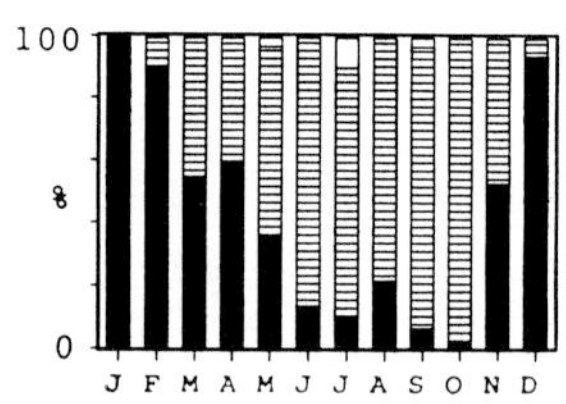

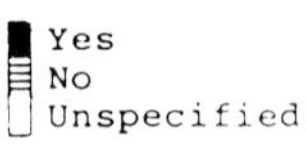

HABITAT

Within 2 km of coast: Yes (38%) No (62%) Unspecified (1%).

Altitude (metres)

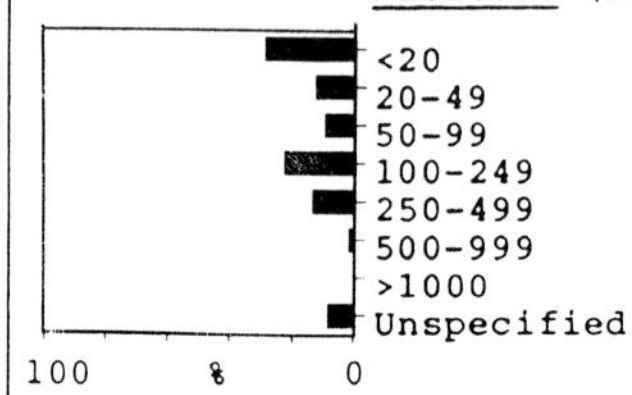

Vegetation

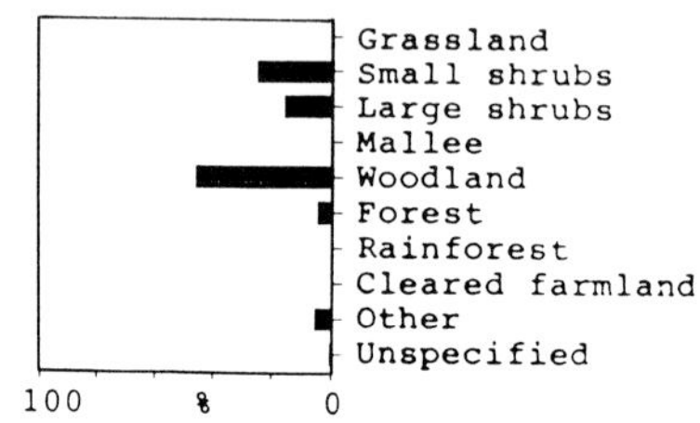

Soil Type

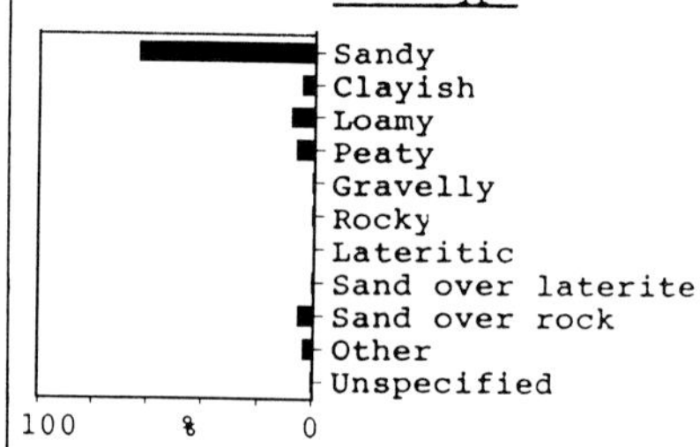

Landform

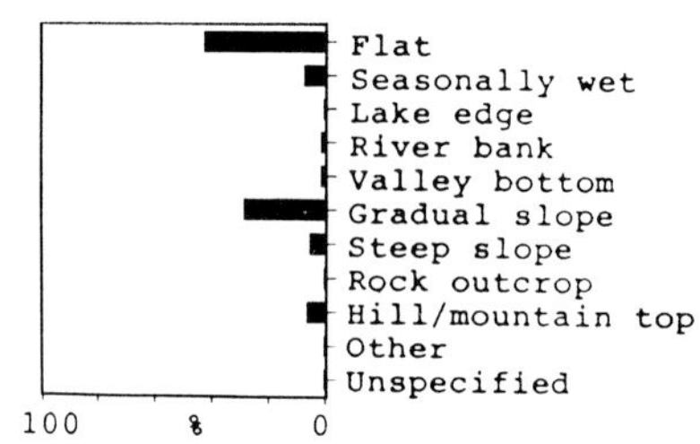

Response to Fire: 98 records

Growth response: Ground resprout (69%) ground resprout and killed, new seedlings (5%) trunk resprout (1%) unspecified (24%).

Flowering response: Number of records where flowering had occurred after fire (47). Median reported time to flowering (13-24 months); minimum reported time to flowering (<1 month). Number of records where flowering had not occurred after fire (23). Median reported time to no flowering (1-12 months); maximum reported time to no flowering (25-36 months).

Flowering

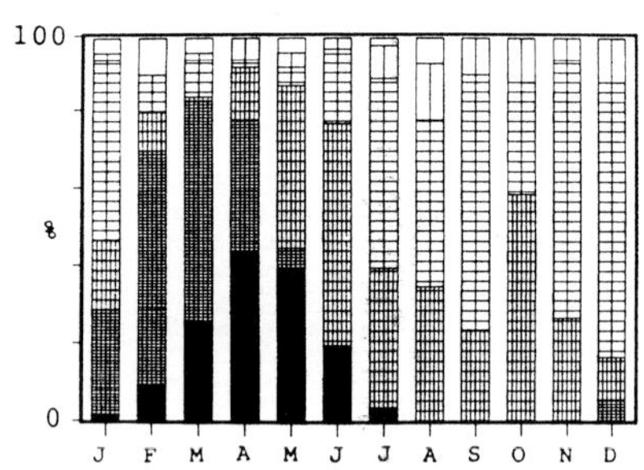

Majority of flowers fully open
Majority of flowers in bud
Majority of flowers recently finished
Flowers finished, fruiting cones present
Neither flowers nor fruiting cones present
Unspecified

Possible pollinators

Pollinator type 243 records: bird (6%) bees, wasps, ants (13%) flies (1%) beetles (0.4%) no pollinator observed (80%).

Specific pollinator 59 records: honeyeater (3) New Holland Honeyeater (2) Brown Honeyeater (6) Tawny-crowned Honeyeater (1) White-plumed Honeyeater (1) Yellow-faced Honeyeater (1) White-cheeked Honeyeater (1) Lewin's Honeyeater (2) Noisy Miner (1) Red Wattlebird (1) Eastern Spinebill (2) friarbird (1) Noisy Friarbird (2) ant (13) bee (13) European honey bee (9)

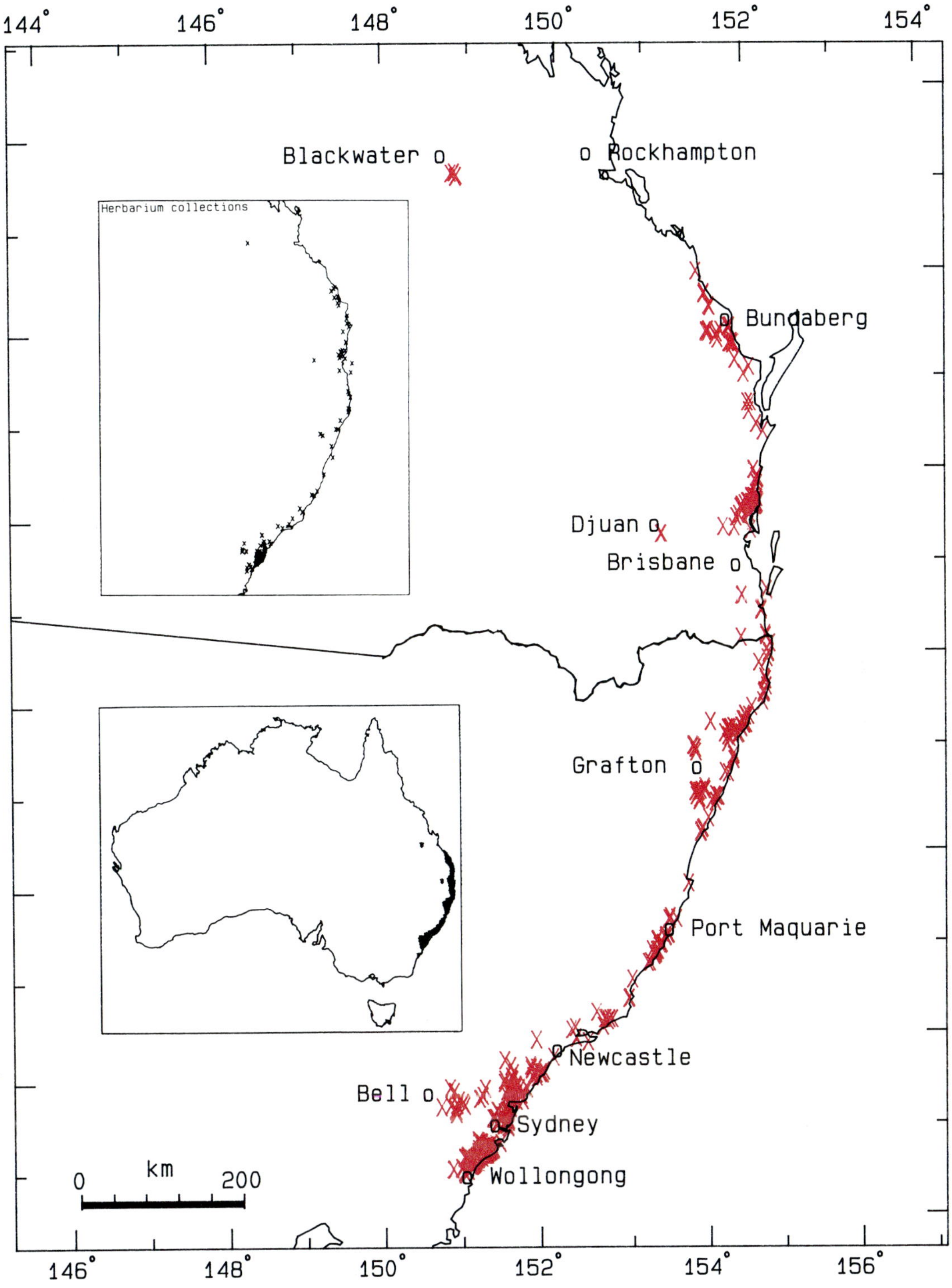

DISCUSSION

Atlas records have confirmed that the main range of B. oblongifolia is from Wollongong to approximately 70 km north of Bundaberg. The latter record extends the previously known range of this species by some 60 km. An isolated population also occurs on the Blackdown Tableland south-east of Blackwater. More than one-third of recorded populations are within 2 km of the coast where the species occupies flat or gently sloping sandy sites often around swamp margins. Here it may grow in woodlands, amongst tall shrubland or heath. Inland, the species occurs on low lying flats as well as the slopes and hilltops of some sandstone outcrops. Surrounding vegetation may again be woodland or heath. In the Blue Mountains west of Sydney, the species extends inland as far as Bilpin and Lawson. In northern New South Wales it occurs inland both north and south of Grafton. In Queensland the species occurs on the lower slopes of the Glasshouse Mountains north of Brisbane. Isolated populations occur at Crows Nest south-east of Djuan and on the Blackdown Tableland.

The species is typically a shrub generally less than 2 m tall, though occasionally larger. It flowers from late January to July when there were several observations of birds, ants and bees visiting flower spikes. New shoot growth occurs in summer. Following fire B. oblongifolia resprouts from its lignotuber. B. oblongifolia hybridises readily with B. robur and 20 such recordings were made by Atlas contributors. A presumed hybrid of B. oblongifolia and B. integrifolia var. integrifolia was found near Caloundra, southern Queensland (see Chapter 3).

Recent research suggests that within B. oblongifolia two varieties can be recognised (J. Conran and H. Clifford 1987). A low growing form with shorter leaves (var. oblongifolia) occurs on the coastal plain and around swamp margins. A taller form with longer leaves (var. minor) generally occurs at higher altitudes and on sandstone outcrops.

BANKSIA OCCIDENTALIS R.Brown (1810)

Red Swamp Banksia, Water Bush Banksi

80 RECORDS: Jan (8) Feb (0) Mar (18) Apr (12) May (7) Jun (0) Jul (2) Aug (8) Sep (5) Oct (5) Nov (13) Dec (1).

Population Size: 1-10 (23) 10-100 (31) >100 (23) Unspecified (3).

Conservation Status: Restricted to road verge (17%) Not (78%) Unspecified (5%). In conservation reserve (41%) Not (49%) Unspecified (10%).

Tree/Shrub: Tree form (25%) Shrub form (72%) Unspecified (3%).

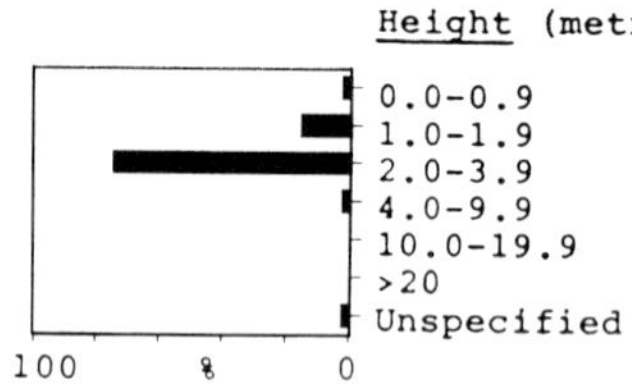

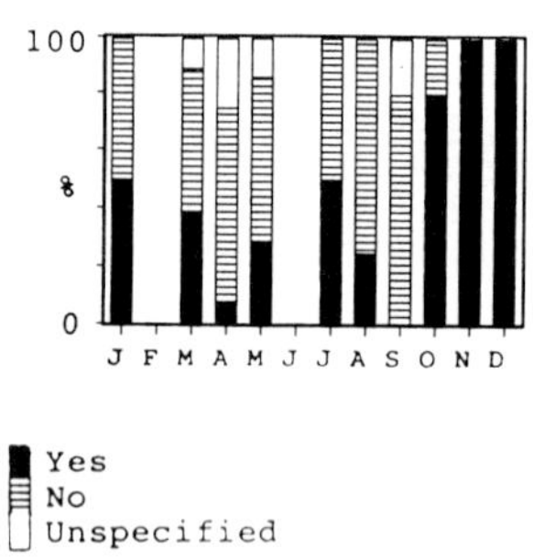

HABITAT

Within 2 km of coast: Yes (30%) No (68%) Unspecified (2%).

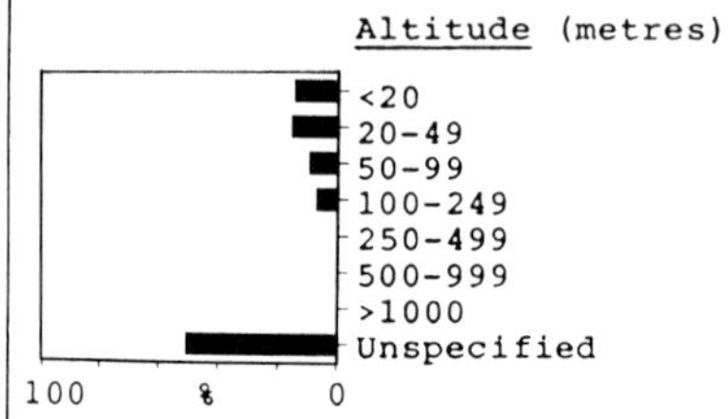

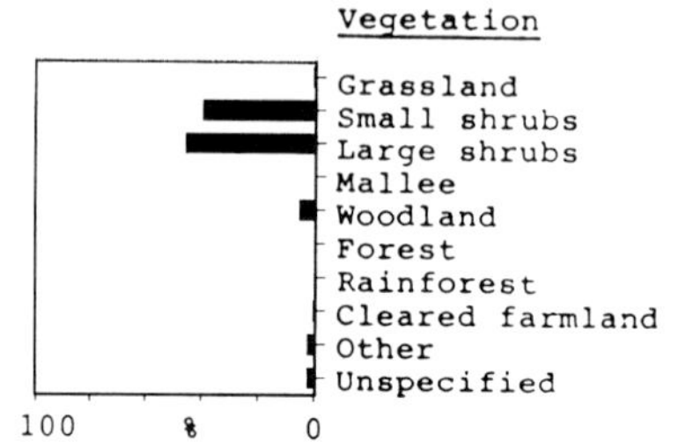

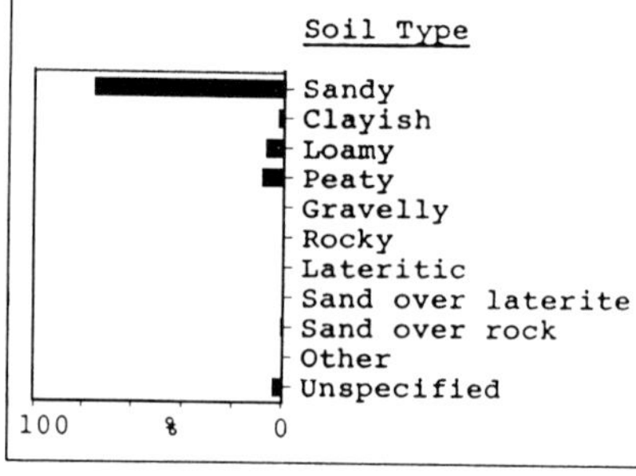

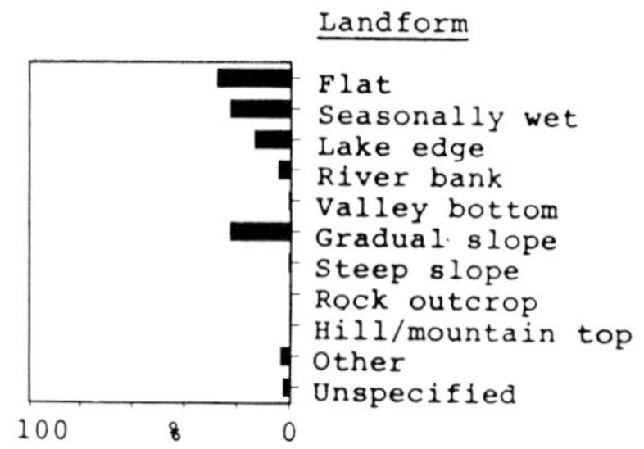

Response to Fire: 9 records

Growth response: Killed, new seedlings (44%) trunk resprout (11%) killed, no seedlings (11%) unspecified (33%).

Flowering response: Number of records where flowering had occurred after fire (8). Median reported time to flowering (25-36 months); minimum reported time to flowering (1-12 months). Number of records where flowering had not occurred after fire (0).

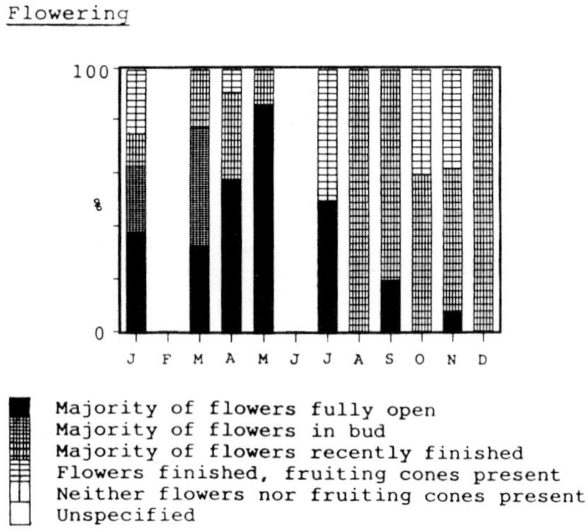

Possible pollinators

Pollinator type 35 records: bird (31%) bees, wasps, ants (29%) no pollinator observed (40%).

Specific pollinator 21 records: honeyeater (1) New Holland Honeyeater (7) Tawny-crowned Honeyeater (1) Little Wattlebird (1) Western Spinebill (1) ant (9) bee (1)

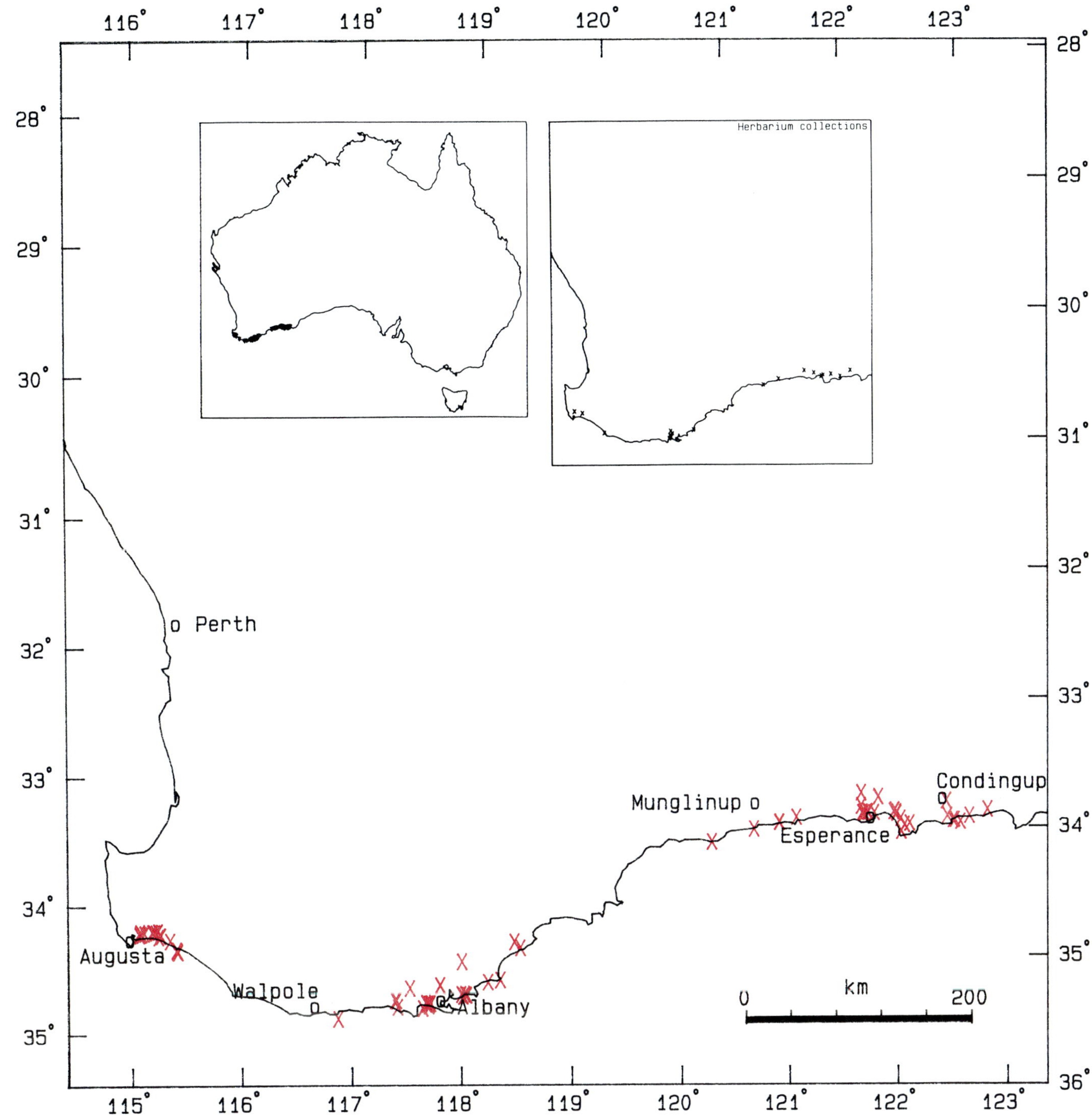

DISCUSSION

B. occidentalis has a coastal distribution with 30% of its recorded populations being within 2 km of the south coast. It appears to occur in three disjunct populations, one between Augusta and Black Point, another between Point Irwin and Cape Riche, and the third between Jerdacuttup Lakes and Cape Arid. A herbarium collection of C. Gardner's from Black Head south-west of Northcliffe was not recorded by Atlas contributors. The species usually occurs in low-lying swampy areas and on lake margins. The interrupted pattern of its distribution is puzzling since in the western part of its range there are several suitable habitats e.g. at Broke Inlet, Lake Maringup and Lake Jasper. The species is killed by fire and regenerates from seed, so it is possible that successive fires over intervals too short to allow seedlings to flower and set seed could have eliminated some populations.

B. occidentalis forms a shrub or small tree up to 4 m in height. It occurs in areas of impeded drainage e.g. low lying flats, swamp margins, seepages and lake edges. Soils are usually sands or peaty sands, sometimes sand over rock e.g. at Cape Arid where the species grows in a winter-wet seepage of shallow sand overlying sandstone and granite. Surrounding vegetation is shrubland, occasionally woodland with similar moisture-loving plants such as Banksia littoralis, Agonis spp., Leptospermum spp. or Melaleuca spp. The species is well represented in conservation reserves, but is, however, very susceptible to dieback disease. Its status in the wild should therefore be monitored. Flowering of B. occidentalis peaks in May, though flowers were recorded from many months of the year. Summer is the main season of new shoot growth. From Denmark there is an interesting record of several plants with cream-coloured flowers growing close to the usual red-flowered form. Hopper (1987) has recently recognised two subspecies, one a small densely floriferous shrub with short leaves confined to coastal basalt (subsp. formosa), and the other the typical subspecies with longer leaves described above (subsp. occidentalis).

BANKSIA OLIGANTHA A.S.George (1987)

1 RECORD: Jan (0) Feb (0) Mar (0) Apr (0) May (0) Jun (0) Jul (0) Aug (0) Sep (0) Oct (0) Nov (1) Dec (0).

Population Size: 1-10 (0) 10-100 (0) >100 (1) Unspecified (0).

Conservation Status: Restricted to road verge (0%) Not (100%) Unspecified (0%). In conservation reserve (100%) Not (0%) Unspecified (0%).

Tree/Shrub: Tree form (0%) Shrub form (100%) Unspecified (0%).

Height (metres)

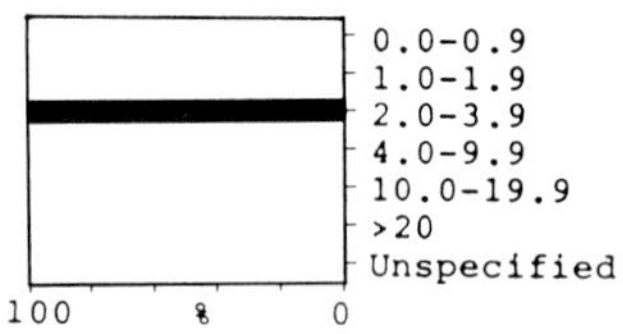

New Shoot Growth

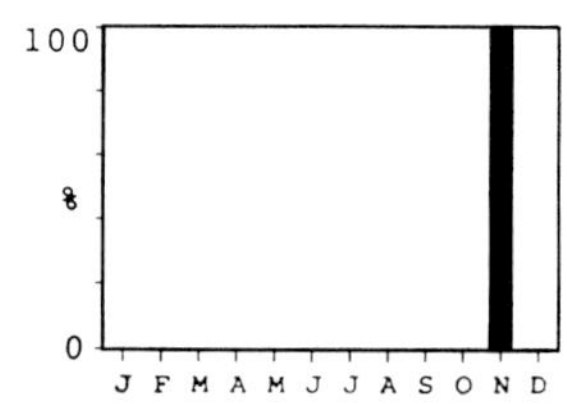

Yes
No
Unspecified

HABITAT

Within 2 km of coast: Yes (0%) No (100%) Unspecified (0%).

Altitude (metres)

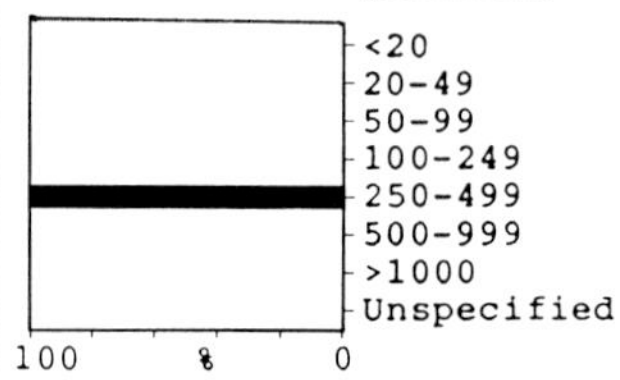

Vegetation

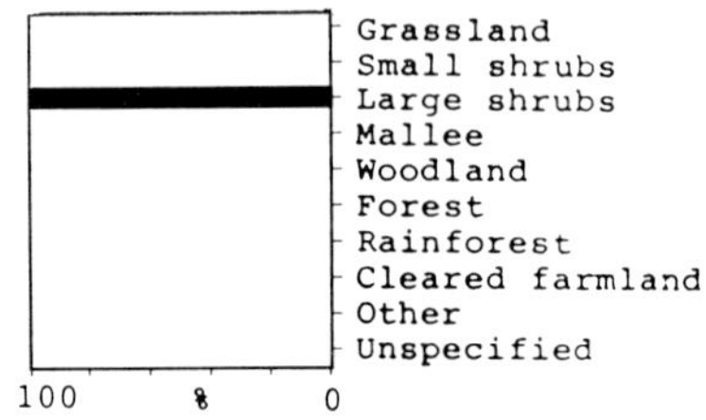

Soil Type

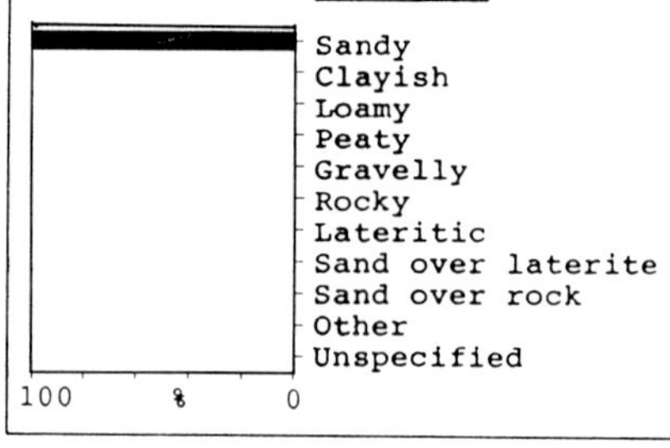

Landform

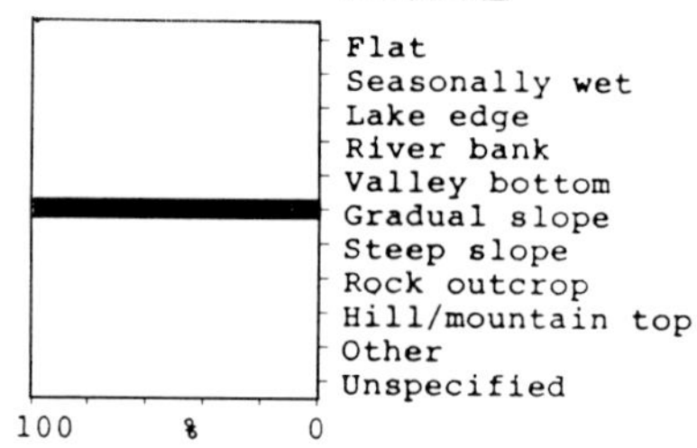

Response to Fire: 0 records

Flowering

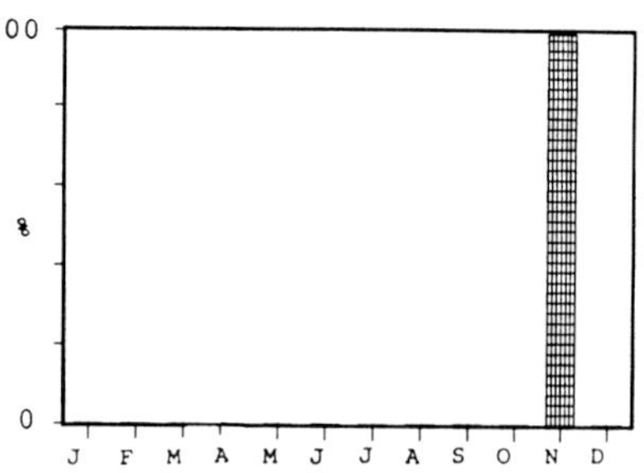

Majority of flowers fully open
Majority of flowers in bud
Majority of flowers recently finished
Flowers finished, fruiting cones present
Neither flowers nor fruiting cones present
Unspecified

Possible pollinators

Pollinator type 1 record: no pollinator observed (100%).

Specific pollinator 0 records.

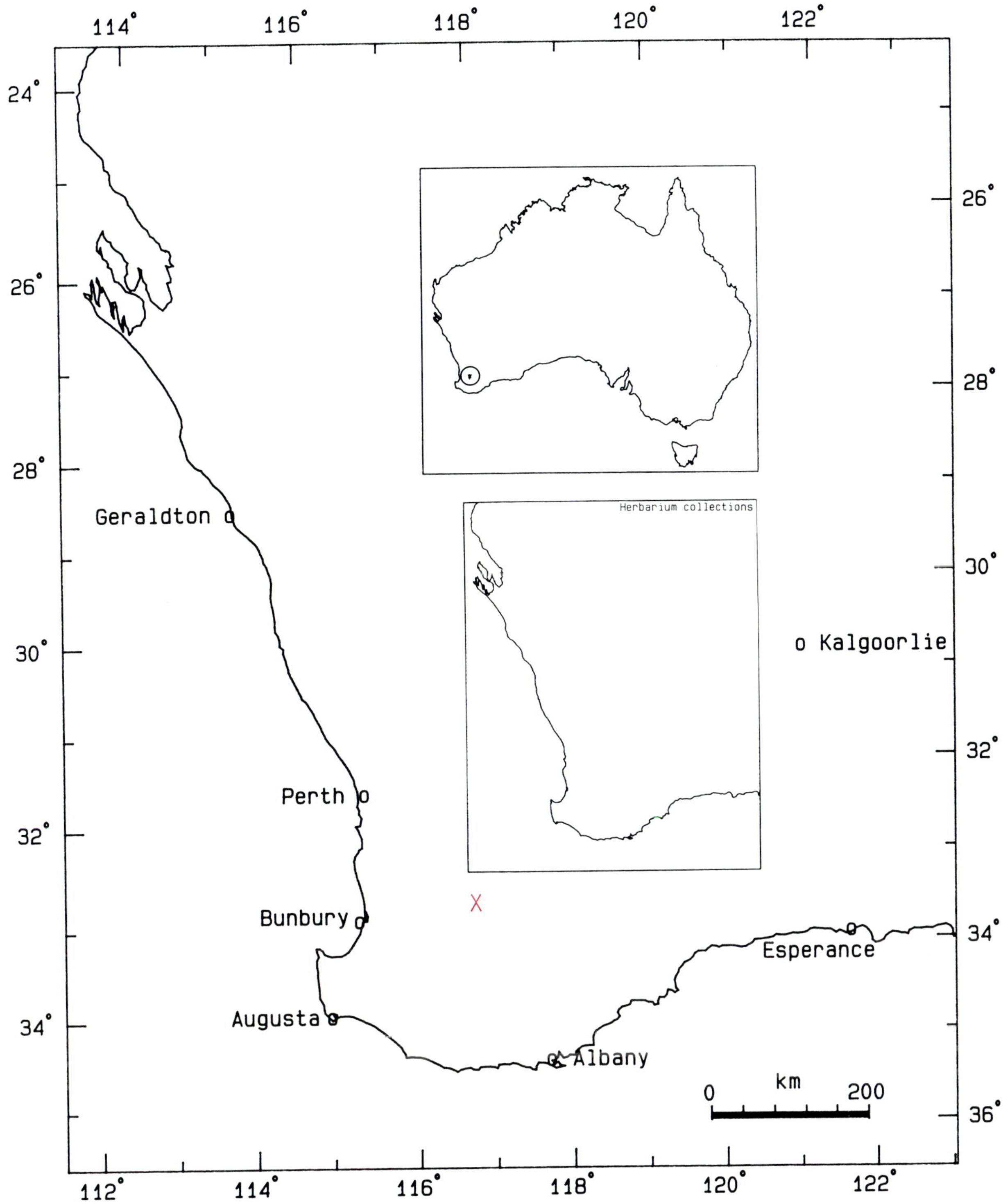

DISCUSSION

Banksia oligantha is a new species that was unknown prior to the Atlas. It is closely related to B. cuneata, differing in its fewer flowered inflorescences, creamy-yellow flowers, roughened basal bark, and its follicles which open readily in the absence of fire. It grows on a small nature reserve amongst otherwise cleared farmland. The species needs close monitoring as it has only been recorded from this single location where there are about 300-400 plants. Although many open seed follicles were observed there was no obvious seedling regeneration. Its susceptibility to dieback disease is unknown. However, its affinity with the highly susceptible B. cuneata suggests that B. oligantha could be under threat.

The species grows as an erect shrub up to 3 m in height on small yellow-brown sand dunes along an ephemeral salt creek system. The surrounding vegetation is shrubland with a few emergent trees (B. attenuata, B. prionotes). When not in flower the species resembles a stand of Dryandra sessilis. The flowering season appears to be very short. In early October, plants were reported to be "mainly in bud", by mid November about 20% were in full flower, and by the end of November most flowers had faded. Growth of greenish brown new shoots was observed in November. Its response to fire is unknown, but the population appears to be of a very even-aged nature, suggesting that plants may have regenerated from seedlings following a previous fire.

BANKSIA OREOPHILA A.S.George (1981)

Western Mountain Banksia

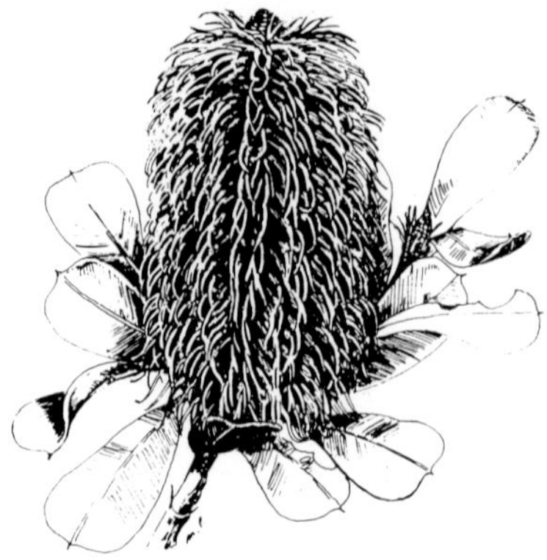

31 RECORDS: Jan (3) Feb (0) Mar (2) Apr (0) May (2) Jun (2) Jul (0) Aug (6) Sep (5) Oct (4) Nov (2) Dec (5).

Population Size: 1-10 (2) 10-100 (17) >100 (9) Unspecified (3).

Conservation Status: Restricted to road verge (0%) Not (90%) Unspecified (10%). In conservation reserve (100%) Not (0%) Unspecified (0%).

Tree/Shrub: Tree form (0%) Shrub form (90%) Unspecified (10%).

Height (metres)

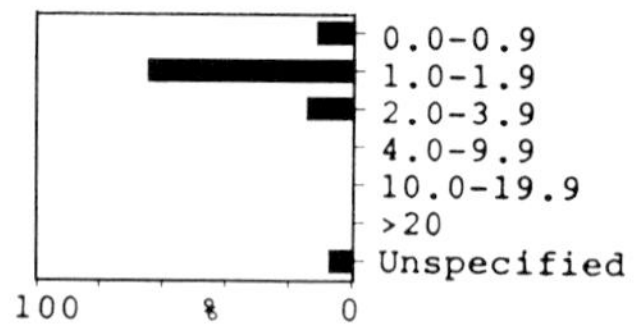

New Shoot Growth

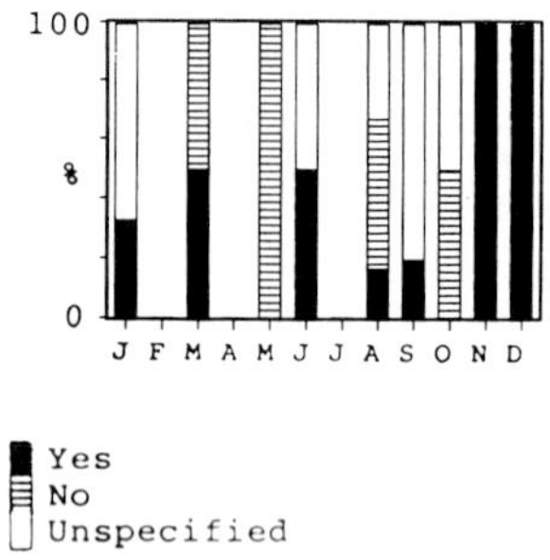

HABITAT

Within 2 km of coast: Yes (32%) No (68%) Unspecified (0%).

Altitude (metres)

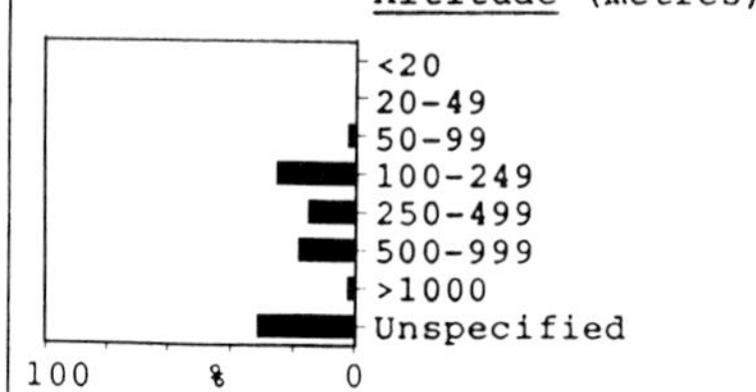

Vegetation

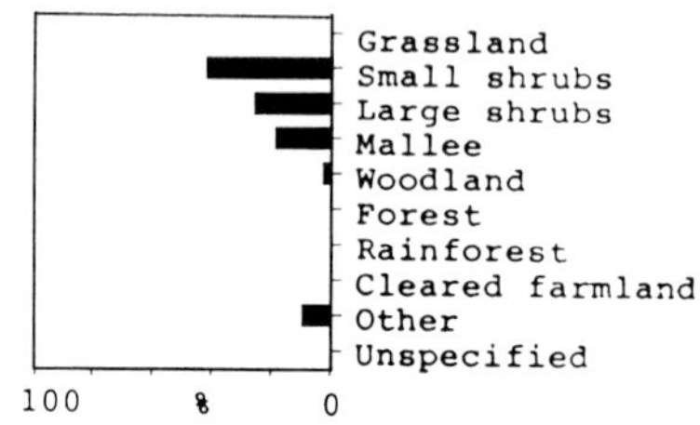

Soil Type

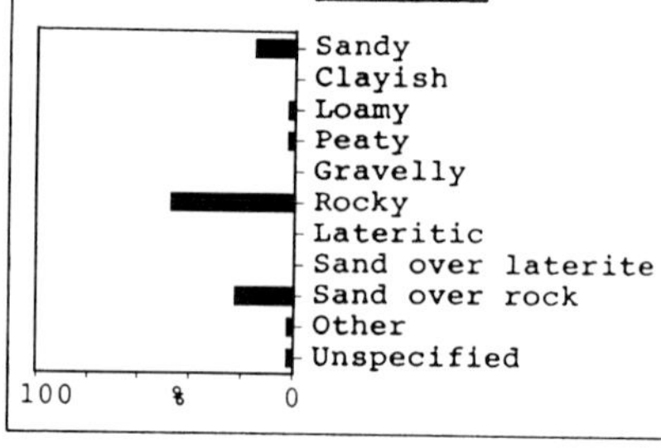

Landform

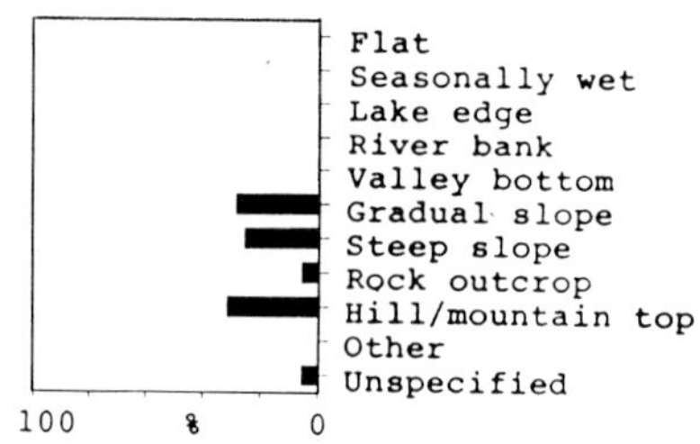

Response to Fire: 6 records

Growth response: Killed, new seedlings (67%) unspecified (33%).

Flowering response: Number of records where flowering had occurred after fire (3). Median reported time to flowering (25-36 months); minimum reported time to flowering (13-24 months). Number of records where flowering had not occurred after fire (1). Maximum reported time to no flowering (25-36 months).

Flowering

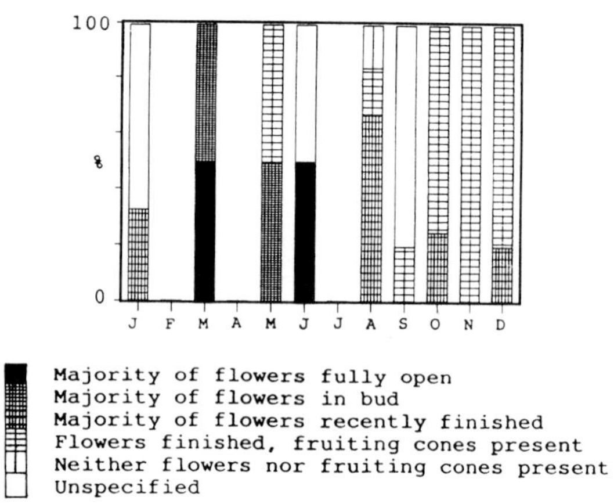

Possible pollinators

Pollinator type 9 records: bees, wasps, ants (22%) no pollinator observed (78%).

Specific pollinator 2 records: ant (2)

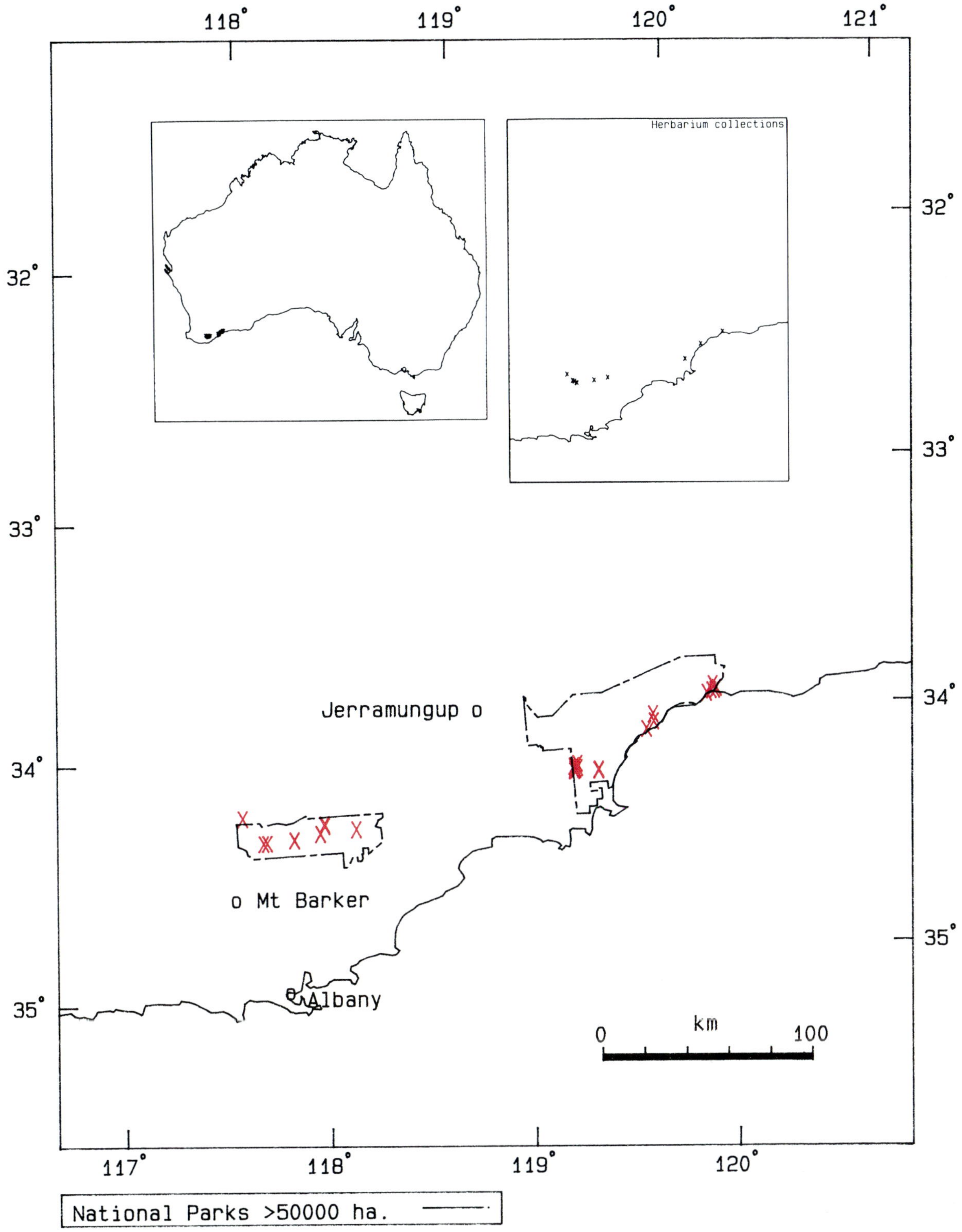

DISCUSSION

Atlas records confirm that B. oreophila is restricted to the slopes and mountain tops of the Stirling Range (e.g. Mt Magog, Bluff Knoll, Mondurup Peak) and the Barrens (e.g. Mt Maxwell, West Mt Barren, East Mt Barren) where it occurs at altitudes of between 100 and more than 1 000 metres. Although there are relatively few populations recorded and most are of less than 100 plants, the entire range of the species is within conservation reserves, and with correct management, therefore, should be well protected. However, the species is susceptible to dieback disease and populations near the more popular Stirling Range walking tracks are severely infected. It is now generally agreed that human activity resulting in the movement of infected soil is one of the major causes of the spread of dieback disease along these tracks.

The species grows as a shrub generally less than 2 m in height, but is occasionally larger. It occurs in rocky soil or shale, sometimes in white sand derived from underlying quartzite (e.g. Mt Maxwell). Surrounding vegetation is a species-rich low heath or tall shrubland including banksias such as B. lemanniana, B. nutans, B. baueri, B. solandri or B. gardneri. Sometimes mallee eucalypts form the dominant vegetation e.g. E. lehmannii, E. coronata. Flowering was recorded in March and June, though George (1981) defined the flowering period as June and July. New shoot growth is predominantly in November and December. The species is killed by fire and regenerates from seed. It may require at least four years for young plants to reach maturity and flower.

BANKSIA ORNATA F.Muell. ex Meissner (1853)

Desert Banksia

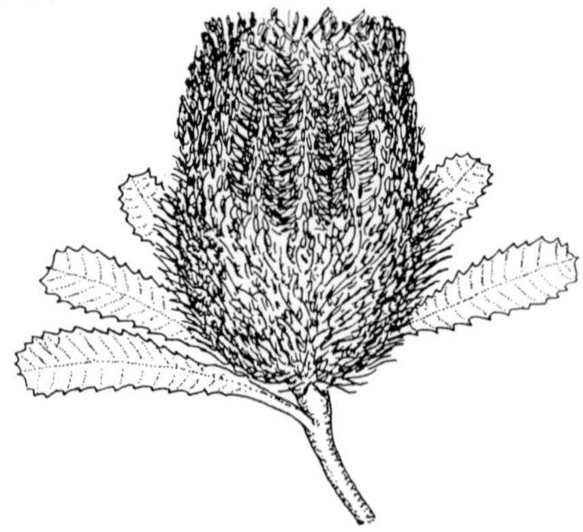

839 RECORDS: Jan (35) Feb (66) Mar (49) Apr (38) May (218) Jun (162) Jul (38) Aug (22) Sep (41) Oct (43) Nov (19) Dec (57).

Population Size: 1-10 (82) 10-100 (126) >100 (547) Unspecified (84).

Conservation Status: Restricted to road verge (27%) Not (61%) Unspecified (12%). In conservation reserve (35%) Not (60%) Unspecified (4%).

Tree/Shrub: Tree form (0.2%) Shrub form (90%) Unspecified (10%).

Height (metres)

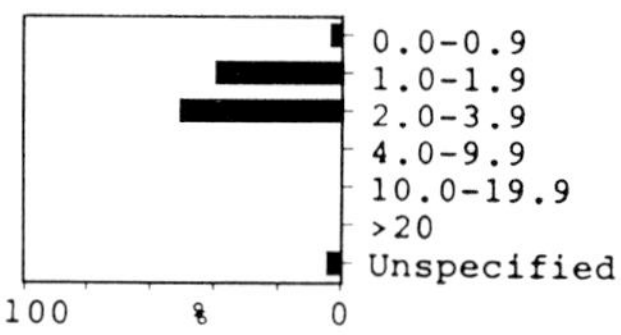

New Shoot Growth

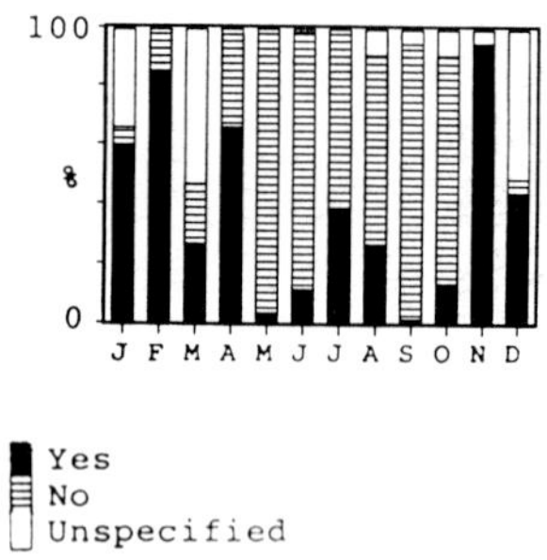

HABITAT

Within 2 km of coast: Yes (2%) No (90%) Unspecified (9%).

Altitude (metres)

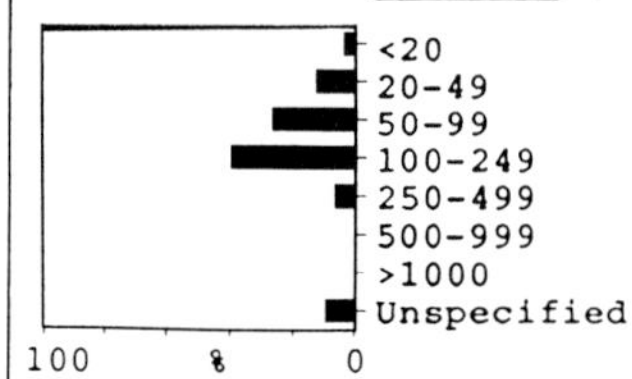

Vegetation

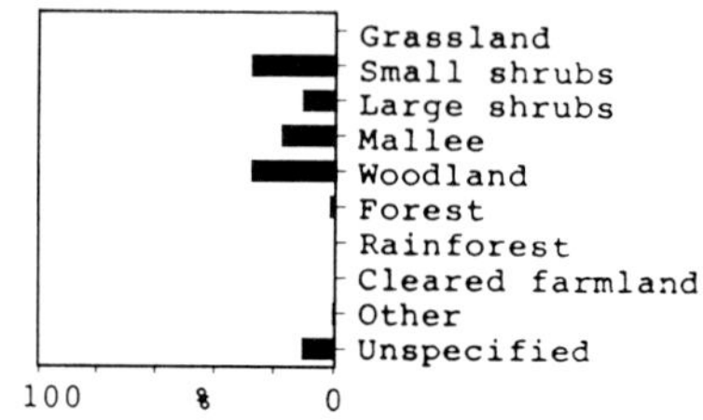

Soil Type

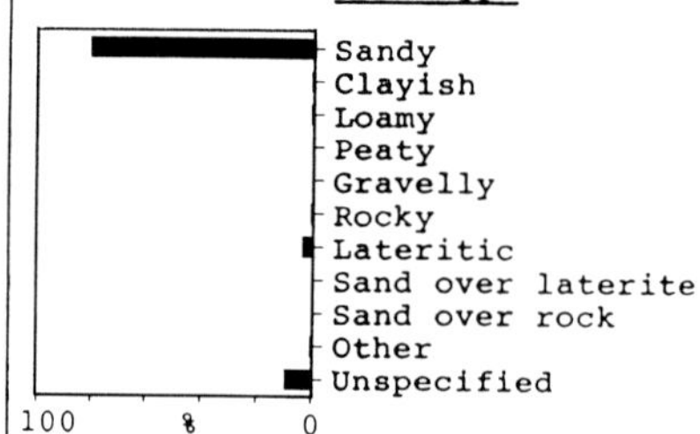

Landform

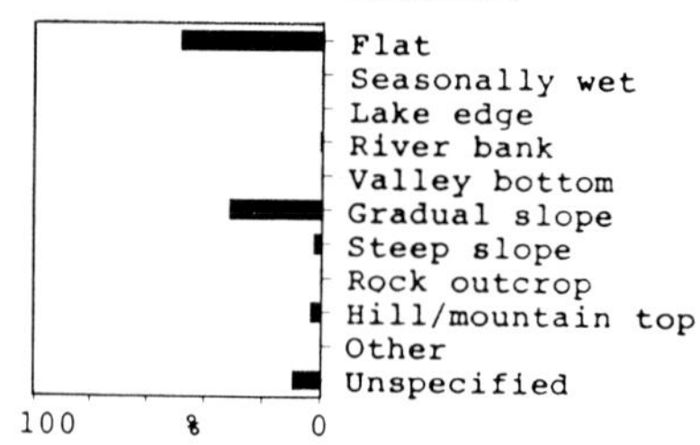

Response to Fire: 93 records

Growth response: Ground resprout (2%) ground resprout and killed, new seedlings (4%) killed, new seedlings (59%) trunk resprout (3%) sucker resprout (1%) killed, no seedlings (2%) unspecified (28%).

Flowering response: Number of records where flowering had occurred after fire (39). Median reported time to flowering (37-48 months); minimum reported time to flowering (<1 month).
Number of records where flowering had not occurred after fire (39). Median reported time to no flowering (13-24 months); maximum reported time to no flowering (49-60 months).

Flowering

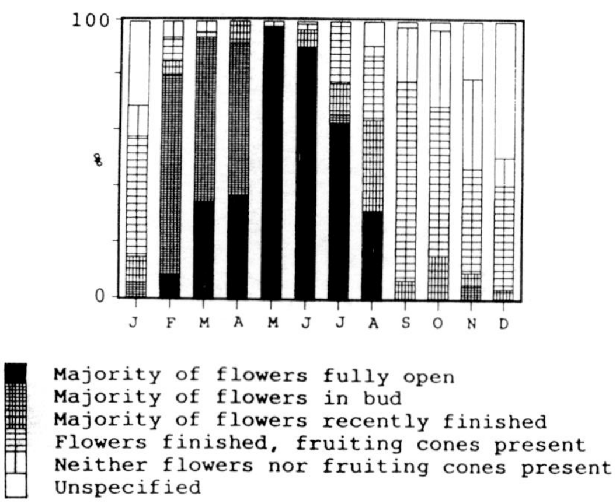

Possible pollinators

Pollinator type 379 records: bird (36%) butterflies/moths (0.3%) bees, wasps, ants (17%) flies (0.3%) mammals (0.3%) no pollinator observed (47%).

Specific pollinator 118 records: honeyeater (12) New Holland Honeyeater (46) Singing Honeyeater (2) Tawny-crowned Honeyeater (1) Noisy Miner (3) Spiny-cheeked Honeyeater (3) White-faced Honeyeater (1) Red Wattlebird (3) Little Wattlebird (3) Little Thornbill (10) White-browed Babbler (1) ant (8) bee (18) European honey bee (7)

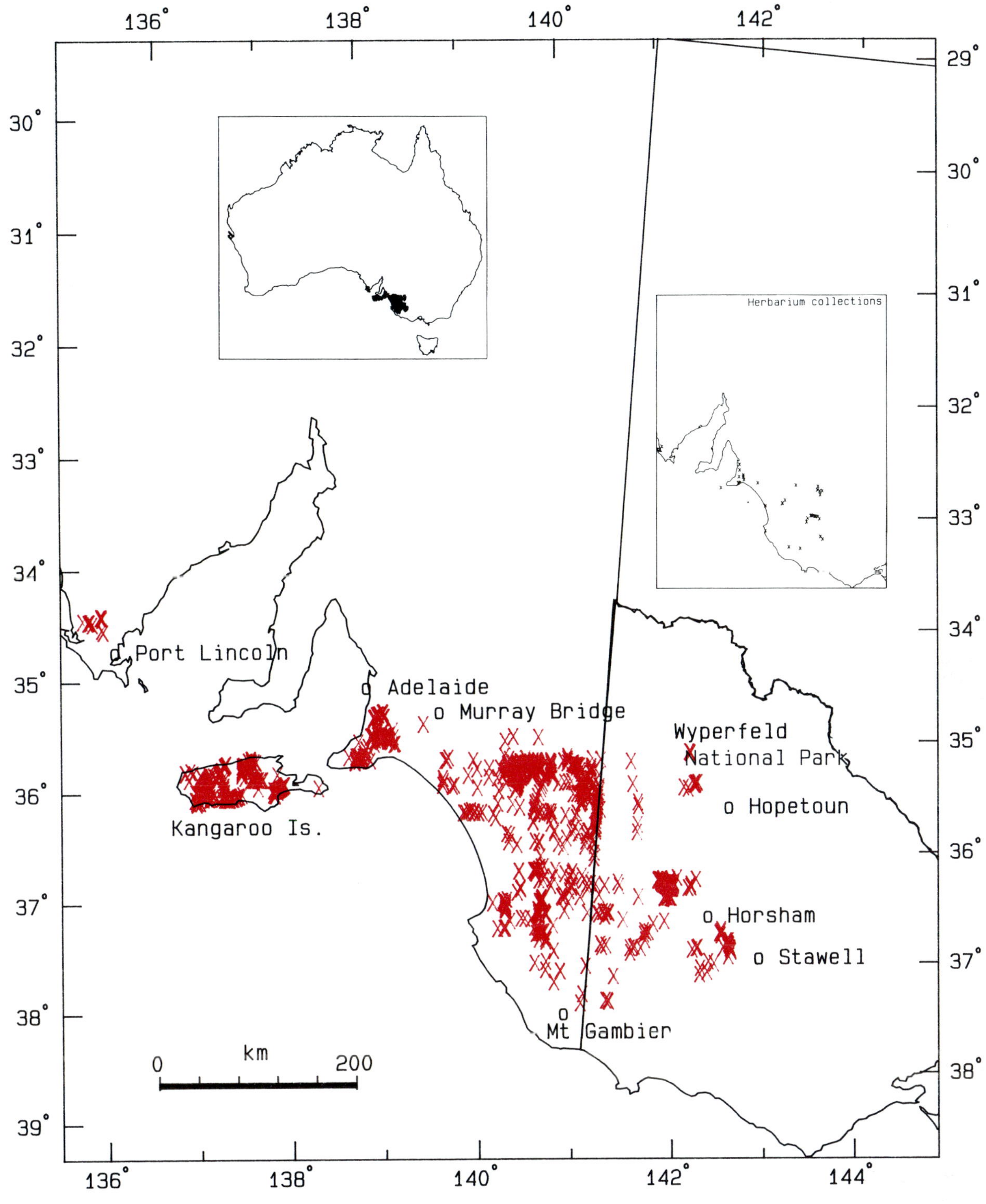

DISCUSSION

Atlas records have confirmed that B. ornata occurs in South Australia and Victoria, from the southern Eyre Peninsula as far east as the Grampians (9 km north of Halls Gap) and Wyperfeld National Park (15 km south of Torrita). The species is apparently absent from Yorke Peninsula but abundant on Kangaroo Island. Southwards, the species extends to Mt Gambier (S.A.) and Casterton (Vic.). The species is locally abundant and most recorded populations are of more than 100 plants. In parts of Ngarkat Conservation Park (S.A.) and Little Desert National Park (Vic.) it is the dominant plant over large areas with record localities of 500 m x 500 m containing thousands of individuals.

The species is typically a shrub up to 3 m in height. A prostrate form, 0.3 m high x 2.5 m wide was recorded in Little Desert National Park. Deep sandy soils are strongly preferred with heavier soils constituting a significant barrier in its distribution. A few Atlas records of it growing in red laterite should be investigated. Landforms are typically flat to gently undulating. Surrounding vegetation may be woodland or heath, less commonly mallee eucalypts or shrubland. In parts of its range the species is liable to severe frosts. Mature plants are generally tolerant but younger plants may be killed. B. ornata flowers in winter when the abundant showy flowers are very attractive to honey-eating birds. Several colour variants exist - the usual cream, to yellow/orange or bronze. The main season of new shoot growth is summer. The species is generally killed by fire and regenerates from seed. Resultant seedlings may take up to 5 years to reach maturity and flower. Individual plants have also been observed resprouting both from below ground and from their trunks. There is also evidence of sucker regrowth (A. Salkin pers. comm.). In very hot fires, both mature plants and seed may be incinerated leaving patches of bare ground with no seedling regrowth (L. Gray pers. comm).

BANKSIA PALUDOSA R.Brown (1810)

Swamp Banksia, Marsh Banksia

291 RECORDS: Jan (21) Feb (28) Mar (30) Apr (19) May (44) Jun (63) Jul (2) Aug (19) Sep (14) Oct (9) Nov (31) Dec (11).

Population Size: 1-10 (32) 10-100 (91) >100 (167) Unspecified (1).

Conservation Status: Restricted to road verge (0%) Not (96%) Unspecified (4%). In conservation reserve (54%) Not (42%) Unspecified (4%).

Tree/Shrub: Tree form (0%) Shrub form (98%) Unspecified (2%).

Height (metres)

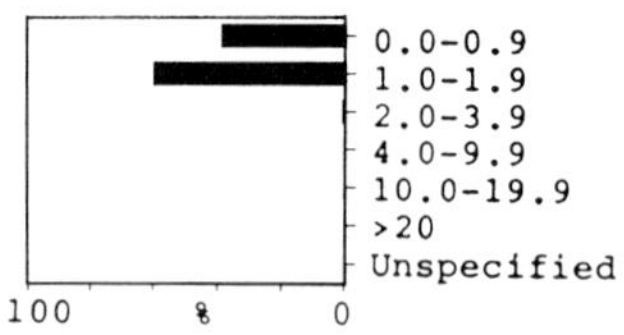

New Shoot Growth

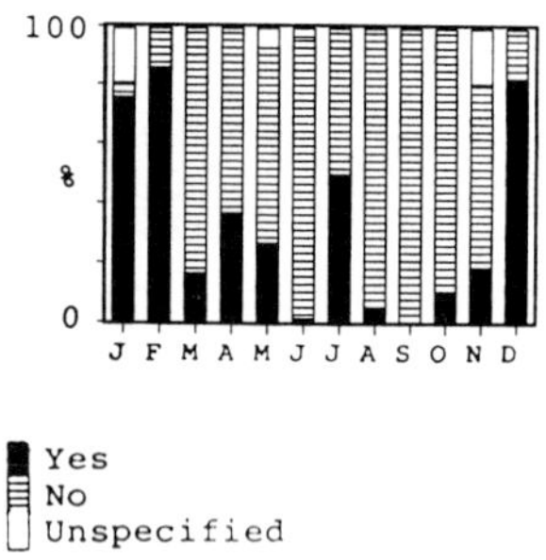

HABITAT

Within 2 km of coast: Yes (15%) No (82%) Unspecified (2%).

Altitude (metres)

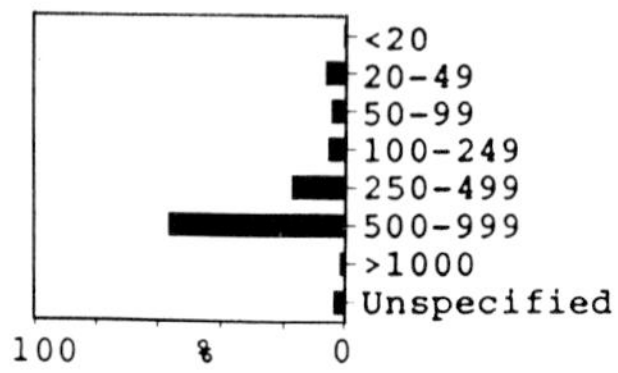

Vegetation

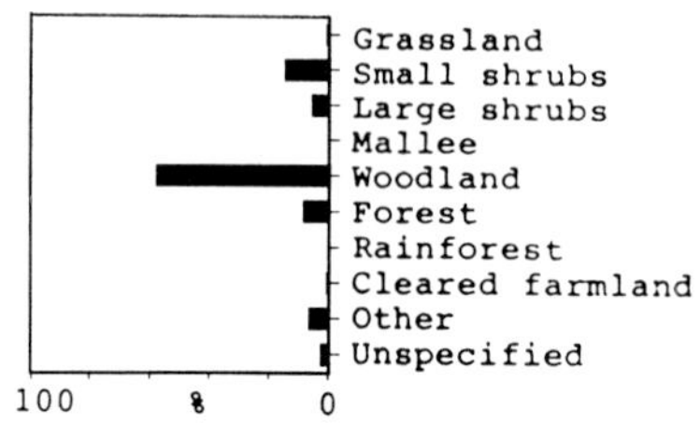

Soil Type

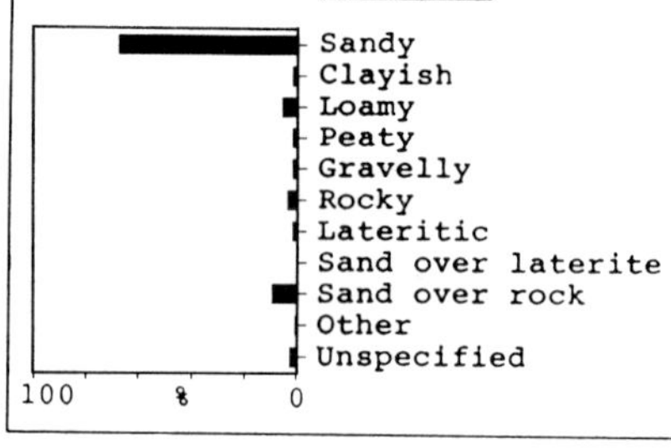

Landform

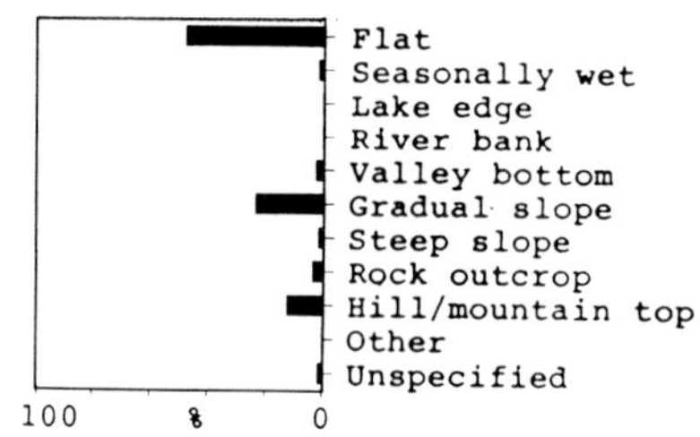

Response to Fire: 51 records

Growth response: Ground resprout (73%) killed, new seedlings (4%) unspecified (24%).

Flowering response: Number of records where flowering had occurred after fire (11). Median reported time to flowering (25-36 months); minimum reported time to flowering (13-24 months). Number of records where flowering had not occurred after fire (21). Median reported time to no flowering (13-24 months); maximum reported time to no flowering (61-72 months).

Flowering

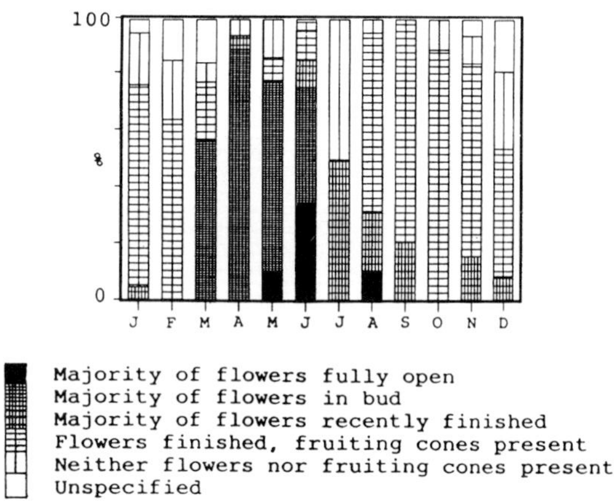

Possible pollinators

Pollinator type 191 records: bird (3%) bees, wasps, ants (2%) mammal (1%) no pollinator observed (95%).

Specific pollinator 18 records: New Holland Honeyeater (4) White-eared Honeyeater (1) Yellow-faced Honeyeater (1) Crescent Honeyeater (2) Red Wattlebird (2) Eastern Spinebill (3) ant (2) European honey bee (2) Brown Antechinus (1)

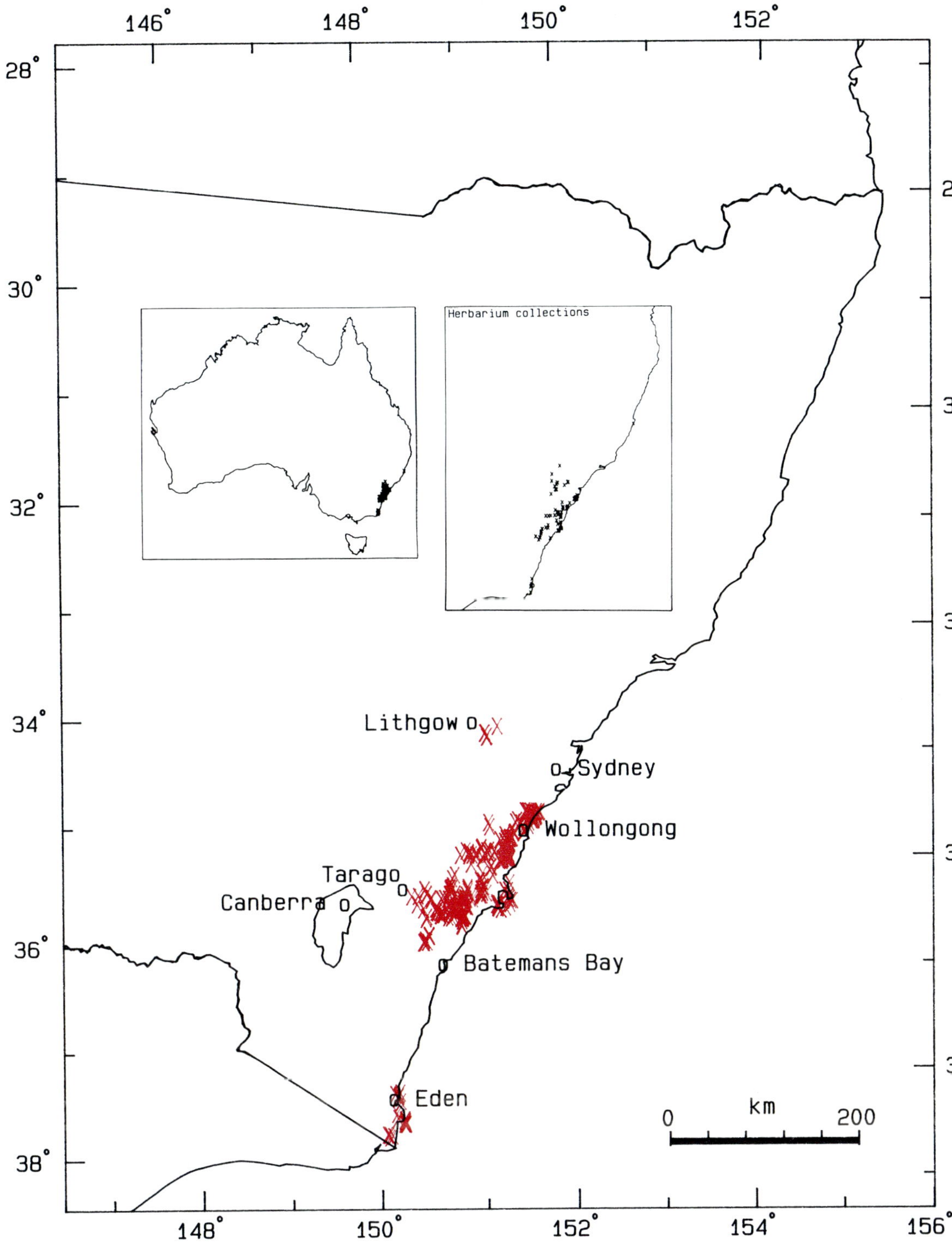

DISCUSSION

From herbarium records it appeared that Banksia paludosa occurred in two disjunct populations in New South Wales, one along the coastal strip and adjoining mountains between Ulladulla and Glen Davis, and the other from Merimbula south almost to the Victorian Border. There was also an isolated 1966 collection from Hat Head, east of Kempsey (L. Johnson). Atlas contributors have confirmed the disjunct nature of the species. The outyling population at Hat Head was not recorded even though the area was extensively surveyed. Compared to herbarium records, the absence of many sight records from the Blue Mountains and Sydney areas is puzzling since both areas were again well covered. B. paludosa can be confused with B. conferta, but in the areas concerned this species was not recorded either. Inland from Ulladulla, B. paludosa extends almost to Tarago. This is approximately 50 km further inland than herbarium records had indicated.

B. paludosa is typically a shrub up to 2 m in height. However, a population from Ben Boyd National Park in the extreme south of its range is completely prostrate with plants 30 cm high and 1.5 m wide. The species occurs predominantly in sandy soil which in mountainous areas can be fairly shallow with rock beneath. Most records were from the coastal plain where landforms are generally flat or gently undulating. Inland, it may occur on hill tops, plateaus, slopes or rocky outcrops. Sites often have impeded drainage making them seasonally wet. Surrounding vegetation is usually eucalypt woodland, but sometimes heath. The species flowers from late autumn through winter when several honey-eating birds were observed visiting flower spikes. On the upland heaths at Barren Grounds Nature Reserve, the Brown Antechinus (A. stuartii) has been frequently observed and is thought to be the main pollinator (R. Whelan and R. Goldingay 1986). Summer appears to be the main season of new shoot growth. Presumed hybrids of B. paludosa x B. marginata and of B. paludosa x B. integrifolia var. integrifolia have been recorded at Nadgee Nature Reserve in southern New South Wales.

BANKSIA PETIOLARIS F.Muell. (1864)

51 RECORDS: Jan (3) Feb (4) Mar (15) Apr (7) May (0) Jun (0) Jul (3) Aug (4) Sep (4) Oct (3) Nov (1) Dec (7).

Population Size: 1-10 (8) 10-100 (9) >100 (33) Unspecified (1).

Conservation Status: Restricted to road verge (37%) Not (63%) Unspecified (0%). In conservation reserve (18%) Not (82%) Unspecified (0%).

Tree/Shrub: Tree form (0%) Shrub form (98%) Unspecified (2%).

Height (metres)

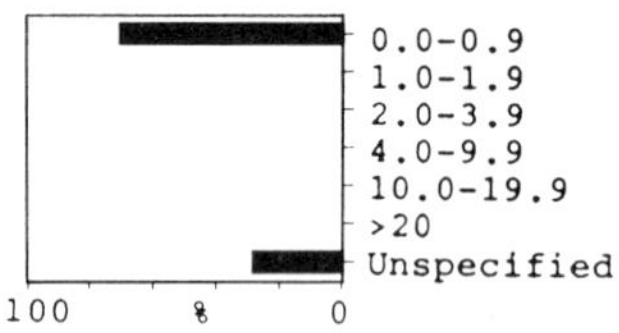

New Shoot Growth

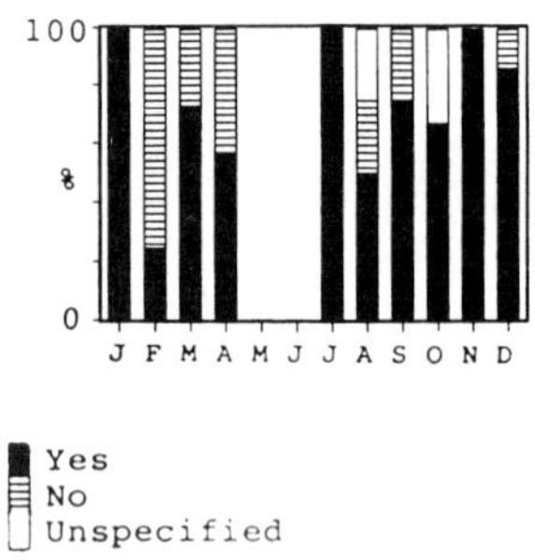

HABITAT

Within 2 km of coast: Yes (0%) No (100%) Unspecified (0%).

Altitude (metres)

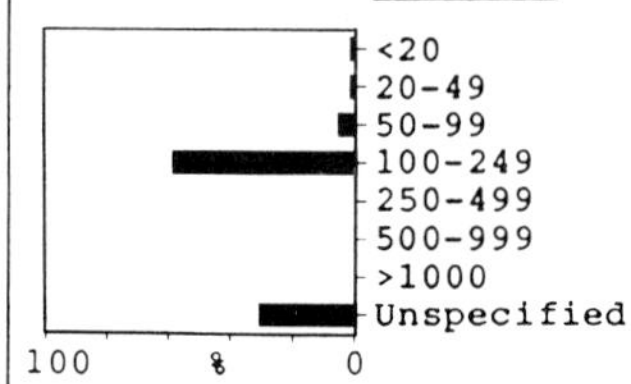

Vegetation

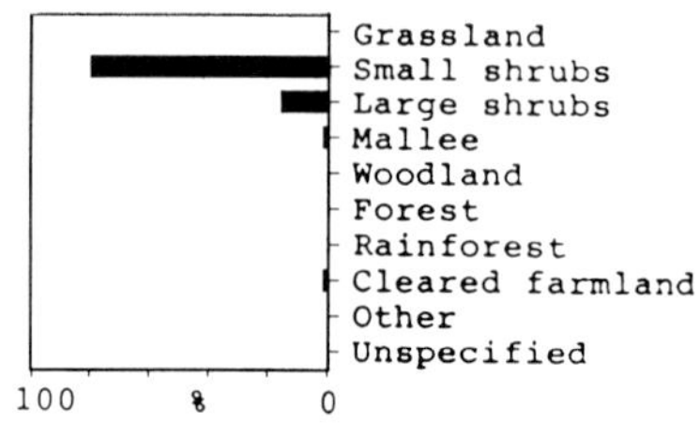

Soil Type

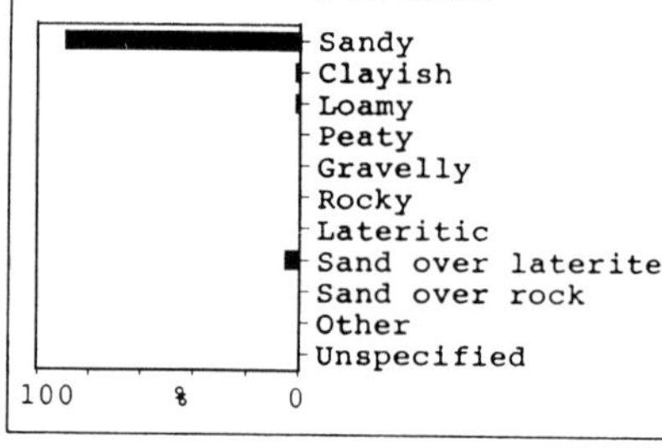

Landform

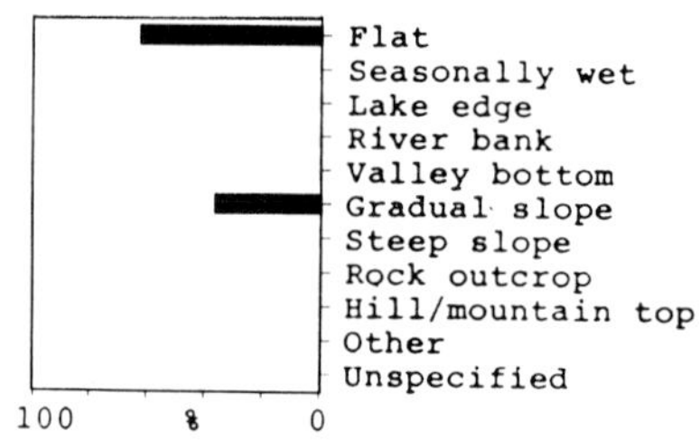

Response to Fire: 7 records

Growth response: Killed, new seedlings (14%) unspecified (86%).

Flowering response: Number of records where flowering had occurred after fire (0). Number of records where flowering had not occurred after fire (7). Median reported time to no flowering (1-12 months); maximum reported time to no flowering (25-36 months).

Flowering

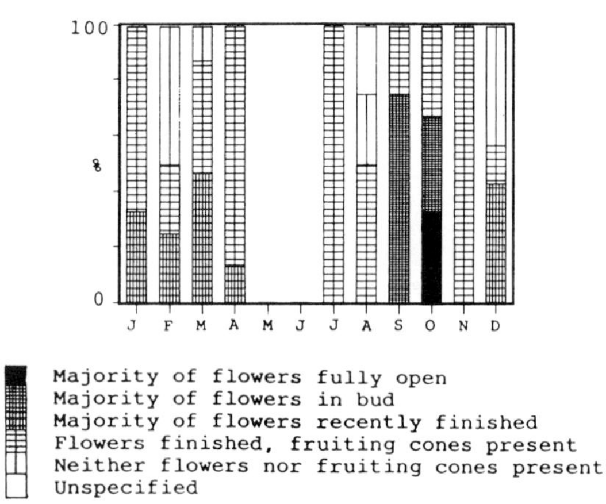

Possible pollinators

Pollinator type 25 records: bees, wasps, ants (4%) no pollinator observed (96%).

Specific pollinator 0 records:

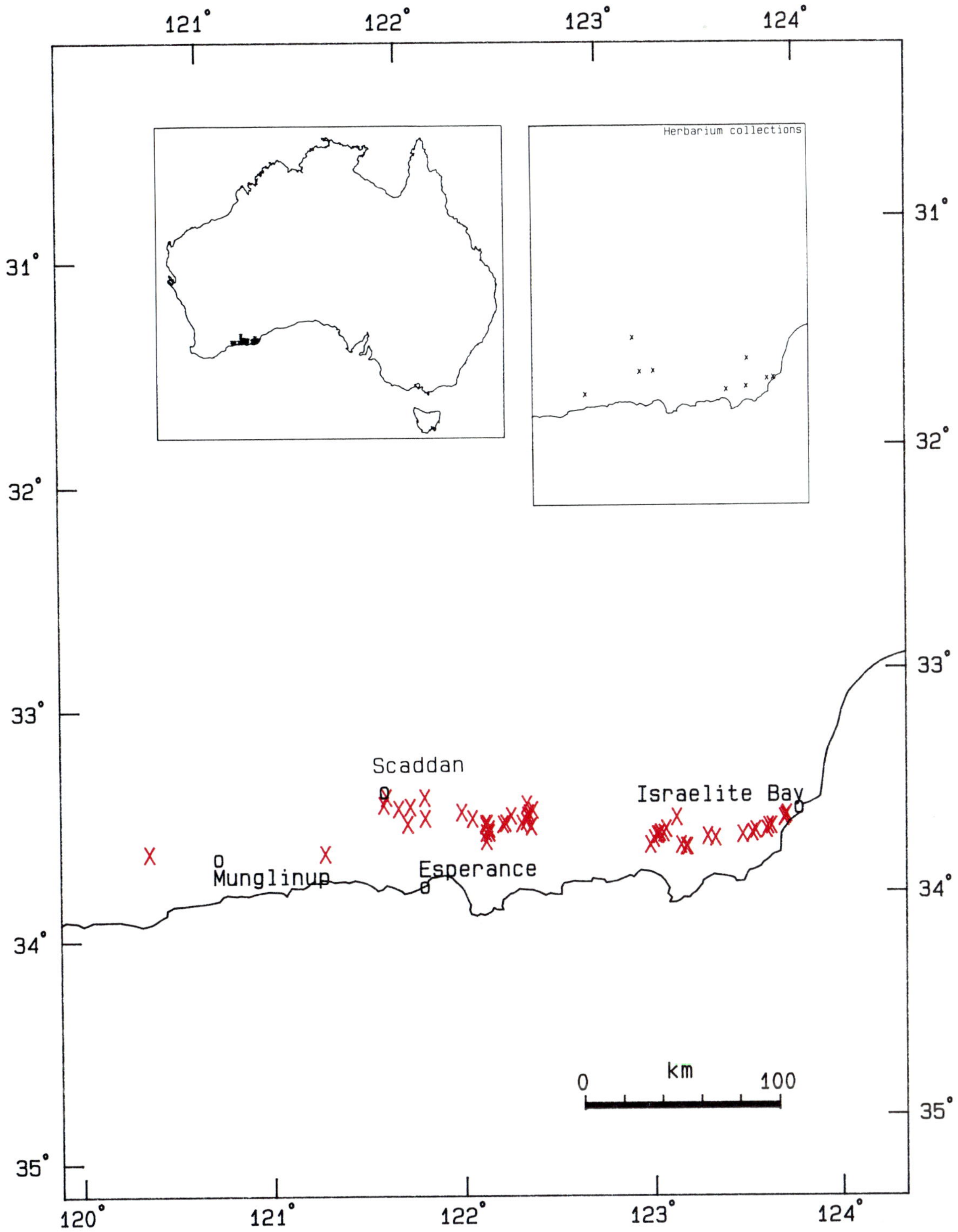

DISCUSSION

Atlas records indicate that, for B. petiolaris, the bulk of its populations are in two disjunct areas, one between Scaddan and Condingup and the other centred on Cape Arid National Park between Boyatup Hill and Israelite Bay. To the west are two isolated records from the Esperance to Ravensthorpe highway. One is 54 km from Esperance and the other is 4 km west of the Rabbit Proof Fence. The latter extends the range of this species by about 50 km. Neither the 1962 collection of J. Beard's from Salmon Gums nor R. Hnatiuk's 1976 collection from northern Cape Arid National Park were recorded by Atlas contributors. The species is locally common with two-thirds of its recorded population having more than 100 plants. It is one of the least susceptible banksias to dieback disease and currently seems under no threat. However, most of its western populations occur on road verges in areas that are largely cleared for agriculture. They could be at risk if road verges become degraded through fertilizer drift and weed invasion.

B. petiolaris is a prostrate shrub with leaves and flower spikes at ground level. It has a strong preference for deep white sands on flat or gently undulating land. Surrounding vegetation is usually low heath, though it is sometimes taller shrubland with species such as B. speciosa. There were few Atlas records of flowering, but George (1981) stated that it is October-December. New shoot growth is apparent in many months of the year, with no particular season being favoured. A single record confirms George's (1981) observations on the species being non-lignotuberous.

BANKSIA PILOSTYLIS C.Gardner (1964)

102 RECORDS: Jan (14) Feb (8) Mar (7) Apr (11) May (1) Jun (2) Jul (12) Aug (16) Sep (2) Oct (11) Nov (12) Dec (6).

Population Size: 1-10 (8) 10-100 (37) >100 (57) Unspecified (0).

Conservation Status: Restricted to road verge (44%) Not (55%) Unspecified (1%). In conservation reserve (14%) Not (79%) Unspecified (7%).

Tree/Shrub: Tree form (2%) Shrub form (97%) Unspecified (1%).

Height (metres)

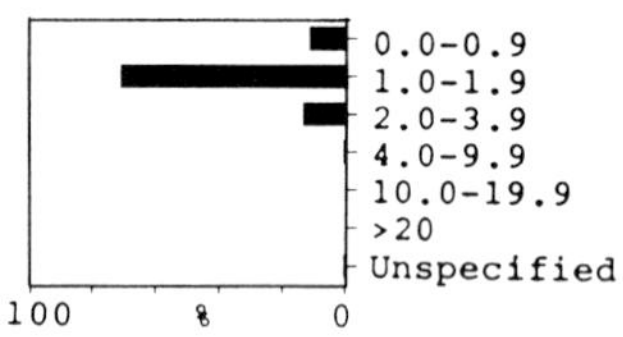

New Shoot Growth

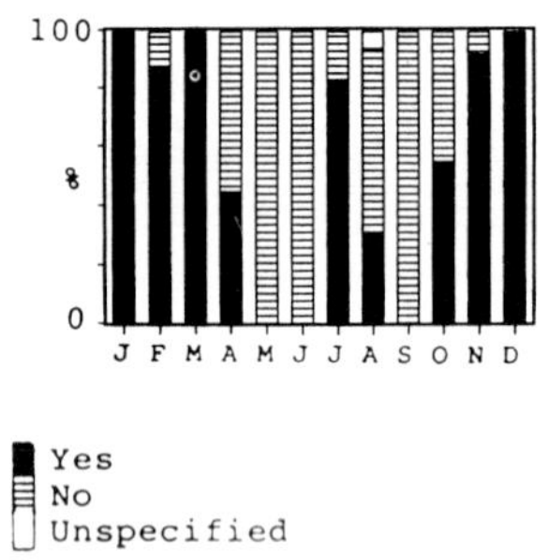

HABITAT

Within 2 km of coast: Yes (1%) No (98%) Unspecified (1%).

Altitude (metres)

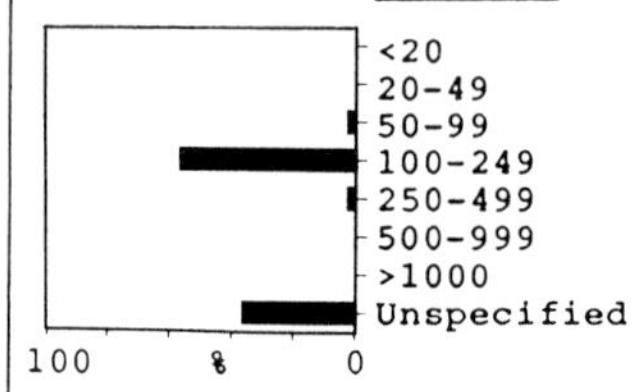

Vegetation

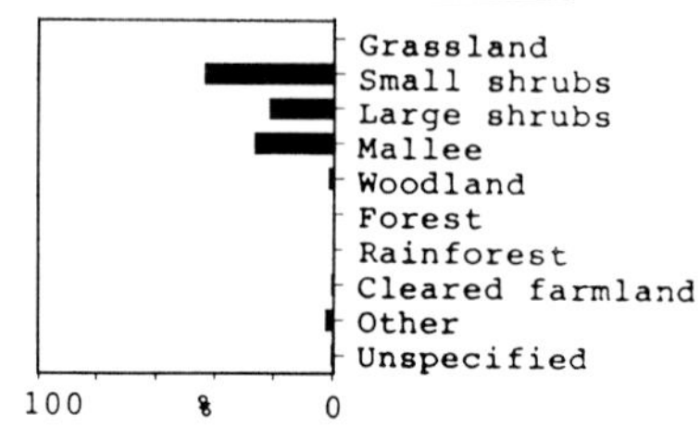

Soil Type

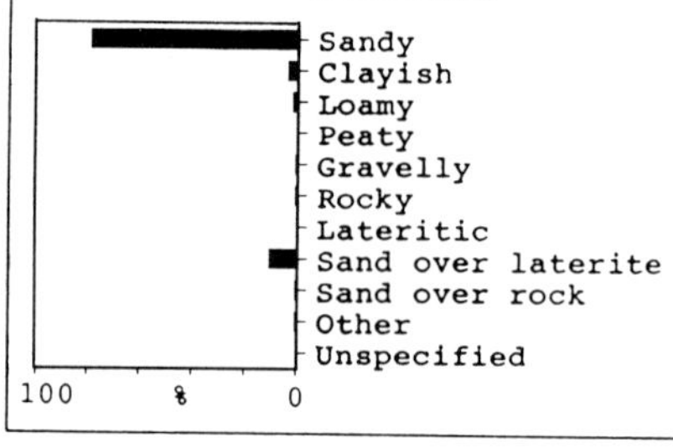

Landform

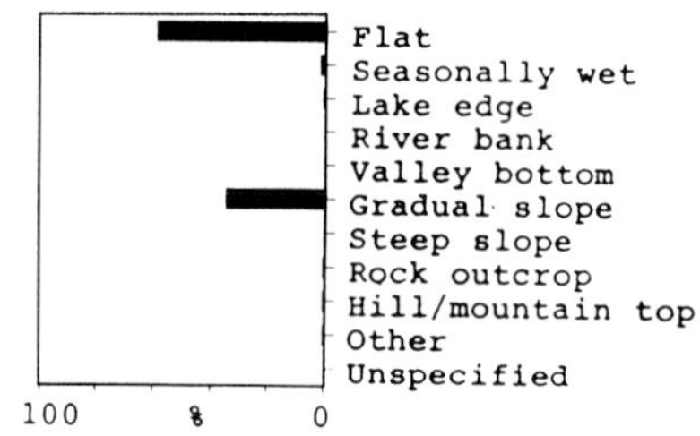

Response to Fire: 6 records

Growth response: Ground resprout (50%) killed, new seedlings (33%) unspecified (17%).

Flowering response: Number of records where flowering had occurred after fire (4). Median reported time to flowering (49-60 months); minimum reported time to flowering (25-36 months). Number of records where flowering had not occurred after fire (0).

Flowering

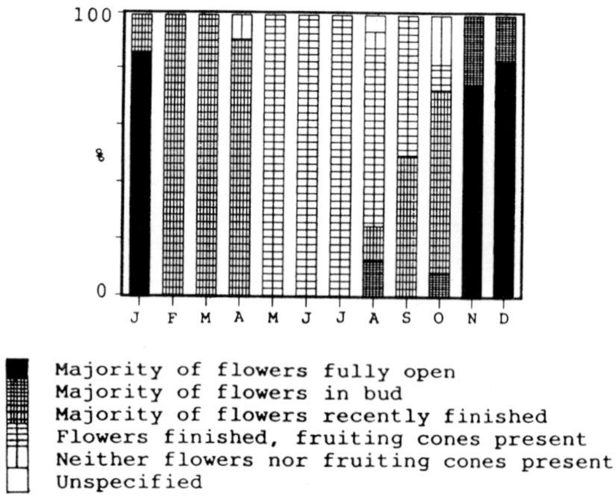

Possible pollinators

Pollinator type 60 records: bees, wasps, ants (25%) flies (3%) beetles (7%) no pollinator observed (65%).

Specific pollinator 14 records: ant (13) bee (1)

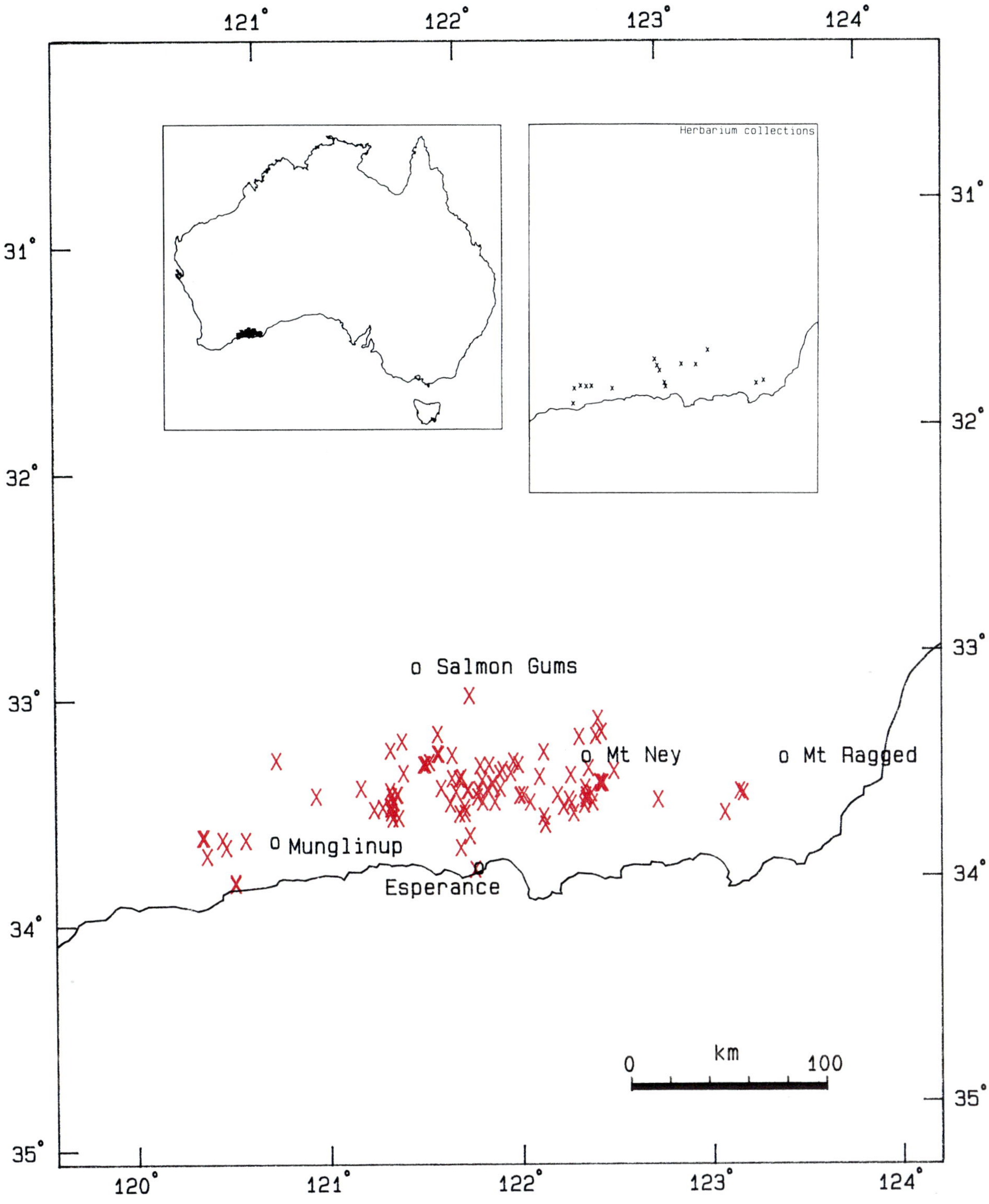

DISCUSSION

Atlas records for B. pilostylis have a similar range to herbarium collections but are much more continuous. The westernmost location is near the Rabbit Proof Fence approximately 50 km east of Ravensthorpe. The easternmost location is in Cape Arid National Park on the track to Israelite Bay. The species is locally common with the majority of its recorded populations being of more than 100 plants. However, most of its range lies within agricultural areas and almost half its recorded populations occur on road verges. With only 14% of its populations in conservation reserves, the species is considered vulnerable and its status, therefore, should be monitored. Fortunately, it is one of the least susceptible of Western Australian banksias to dieback disease. George (1981) stated that B. pilostylis and the similar-looking B. media never occur together although overlapping in their geographic range. However, Atlas contributors recorded several occurrences of the two species within the same record locality e.g. at Truslove Nature Reserve north of Scaddan.

B. pilostylis is typically a shrub less than 2 m in height, occasionally up to 4 m. It occurs in white sand, sandy-loam or sand over laterite, amongst heath, mallee-heath or tall shrubland. Landforms are generally flat or gently undulating. The species flowers from November to January. New shoot growth is predominantly in summer. There were several recordings of ants visiting the flower spikes, but it is not known whether these can be effective pollinators. The fire response of the species requires further investigation. Atlas contributors recorded three occurrences of lignotuberous regrowth and two of plants being killed by fire and regenerating from seed. However, George (1981) stated that the species is always non-lignotuberous.

BANKSIA PLAGIOCARPA A.S.George (1981)

Dallachy's Banksia

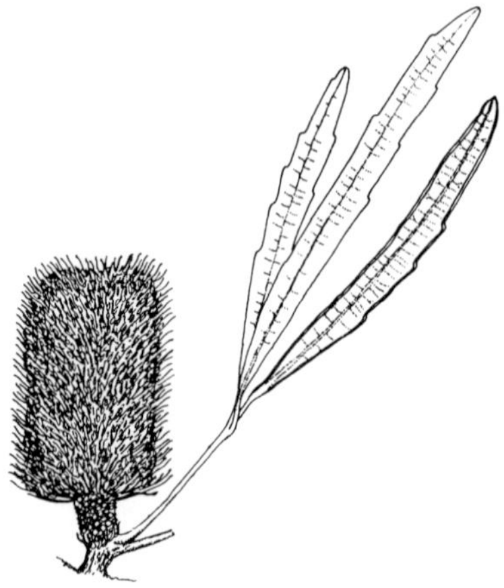

21 RECORDS: Jan (1) Feb (0) Mar (0) Apr (0) May (3) Jun (0) Jul (9) Aug (0) Sep (7) Oct (1) Nov (0) Dec (0).

Population Size: 1-10 (3) 10-100 (7) >100 (11) Unspecified (0).

Conservation Status: Restricted to road verge (0%) Not (100%) Unspecified (0%). In conservation reserve (57%) Not (43%) Unspecified (0%).

Tree/Shrub: Tree form (38%) Shrub form (38%) Unspecified (24%).

Height (metres)

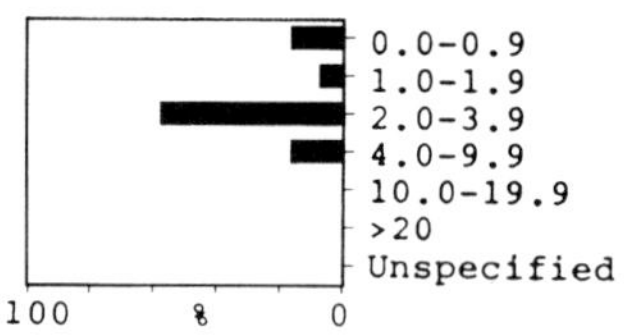

New Shoot Growth

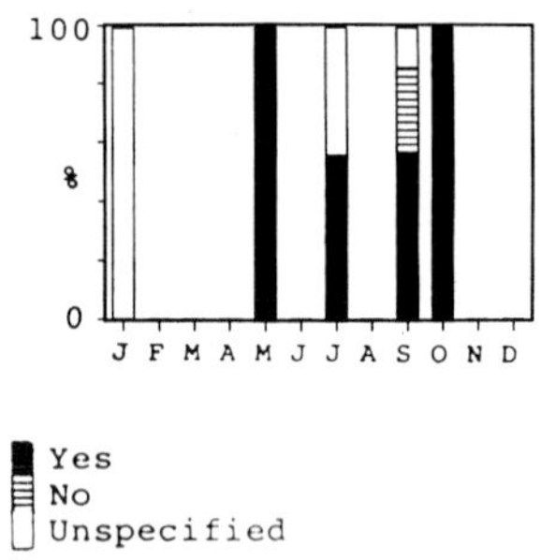

HABITAT

Within 2 km of coast: Yes (29%) No (71%) Unspecified (0%).

Altitude (metres)

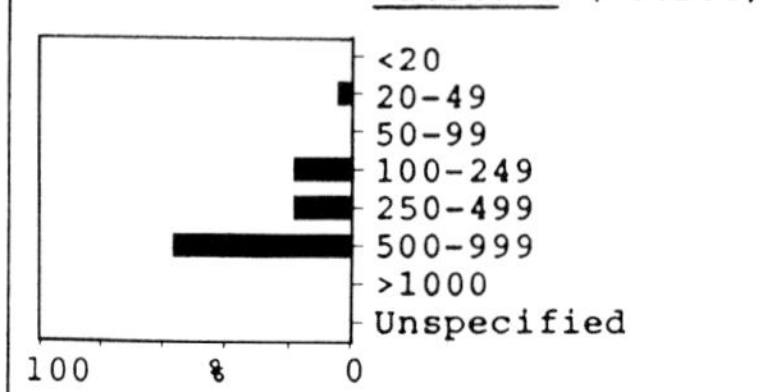

Vegetation

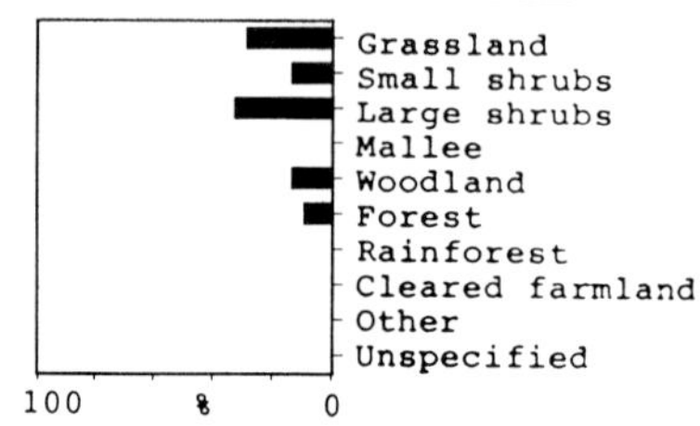

Soil Type

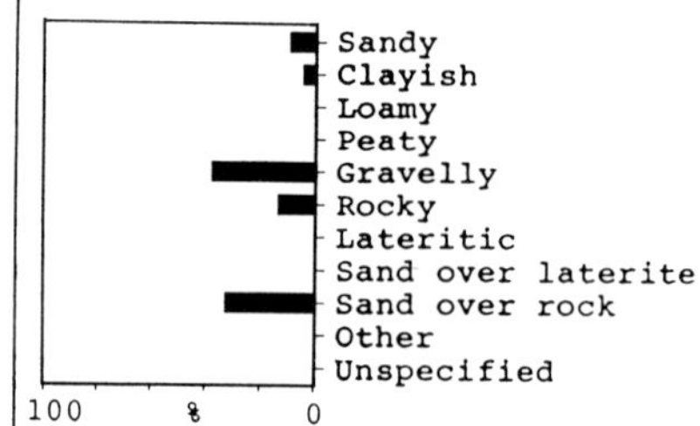

Landform

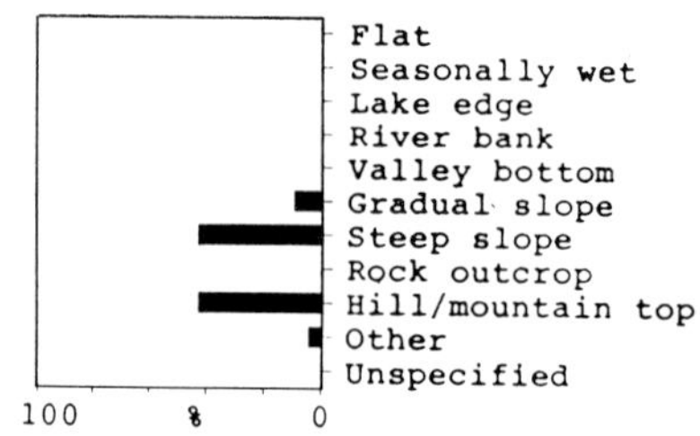

Response to Fire: 9 records

Growth response: Killed, new seedlings (44%) killed, new seedlings and trunk resprout (11%) trunk resprout (11%) unspecified (33%).

Flowering response: Number of records where flowering had occurred after fire (5). Median reported time to flowering (1-12 months); minimum reported time to flowering (1-12 months). Number of records where flowering had not occurred after fire (3). Median reported time to no flowering (1-12 months); maximum reported time to no flowering (1-12 months).

Flowering

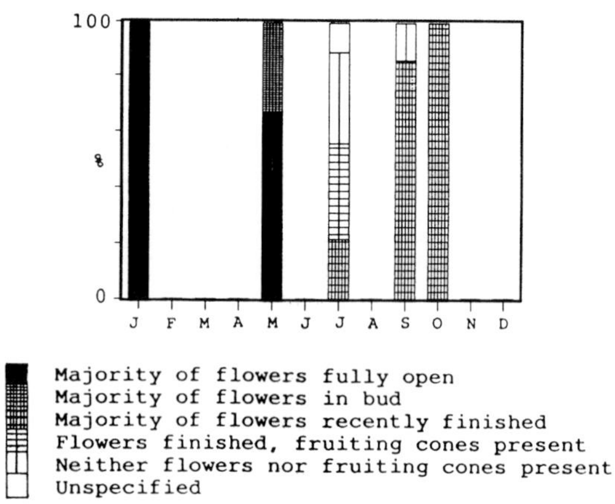

Possible pollinators

Pollinator type 0 records:

Specific pollinator 0 records:

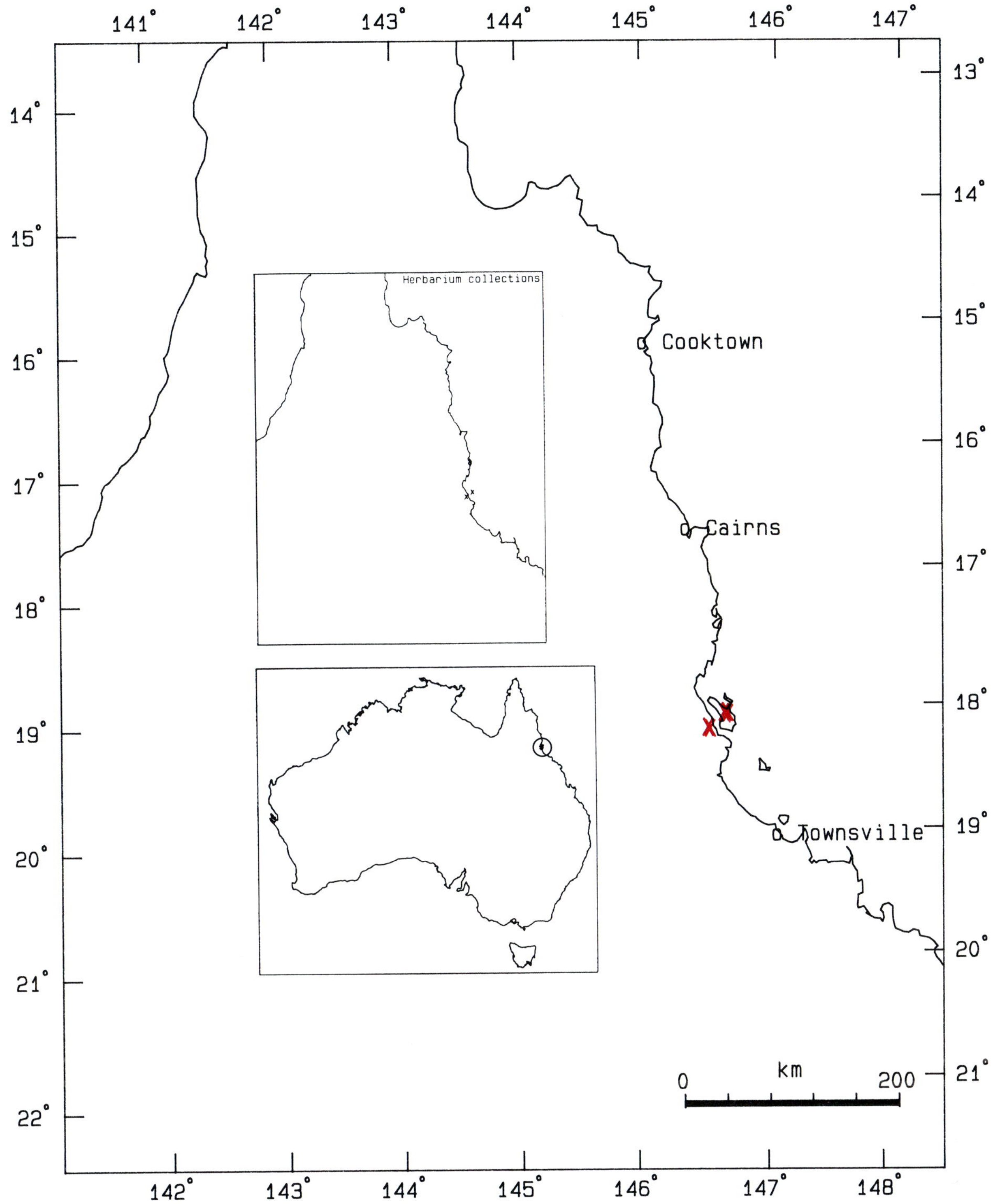

DISCUSSION

Banksia plagiocarpa is a shrub or tree up to 4 m in height. It occurs in Queensland where it is restricted to Hinchinbrook Island and the adjacent mainland. On Hinchinbrook Island, the species had been previously recorded only on the north slope of Mt Bowen. Atlas contributors found it on the slopes and tops of Nina Peak and Mt Bowen, sometimes in such large numbers as to form the dominant plant e.g. on the ridge leading up to Mt Bowen and on the mountain top itself. On Hinchinbrook Island, the species generally grows between 200 and 720 m. At less than 400 m, B. integrifolia var. aquilonia often occurs within the same record locality. B. plagiocarpa grows in gravelly, rocky or clayey soils generally on steep slopes and hilltops, surrounded by tall shrubland e.g. Leptospermum flavescens, Casuarina sp. It also occurs in forest.

On the mainland, the species is found on the slopes and top of Bishops Peak. Above 400 m it is often very abundant and is the dominant plant. Here, the soil type is sand over rock and the species is sometimes in very wet areas e.g. near creeks and soaks. The vegetation is grassland or woodland of B. plagiocarpa, B. integrifolia var. aquilonia and Eucalyptus spp. Much of the population on Bishops Peak was burnt in late 1985 and although some large mature trees resprouted from epicormic buds, the majority of plants were killed. Atlas contributors reported thousands of young seedlings a few centimetres high (July 1986). It would therefore seem that the species is only partly fire tolerant, with some large plants also being killed by fire and regenerating from seed.

B. plagiocarpa is a rare banksia known only from a few locations less than 30 km apart. On Hinchinbrook Island it lies with a conservation reserve, but not so at Bishops Peak. Its apparent susceptibility to fire places the species at great risk. Successive fires at intervals too short to allow sufficient flowering and seed set could greatly reduce the population.

BANKSIA PRAEMORSA Andrews (1802)

Cut-leaf Banksia

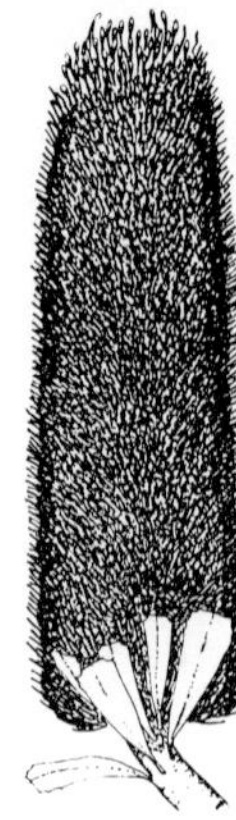

28 RECORDS: Jan (0) Feb (1) Mar (5) Apr (3) May (0) Jun (1) Jul (1) Aug (6) Sep (3) Oct (6) Nov (1) Dec (0).

Population Size: 1-10 (5) 10-100 (10) >100 (12) Unspecified (0).

Conservation Status: Restricted to road verge (0%) Not (100%) Unspecified (0%). In conservation reserve (74%) Not (22%) Unspecified (4%).

Tree/Shrub: Tree form (26%) Shrub form (74%) Unspecified (0%).

Height (metres)

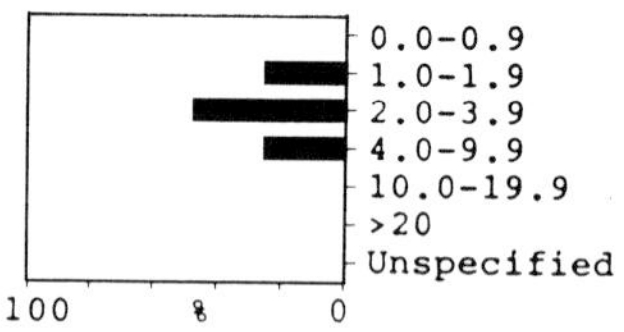

New Shoot Growth

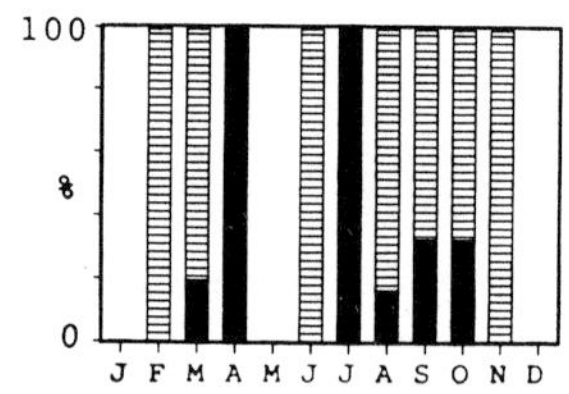

HABITAT

Within 2 km of coast: Yes (100%) No (0%) Unspecified (0%).

Altitude (metres)

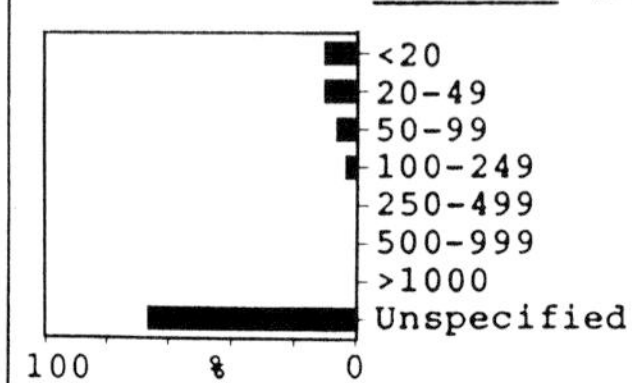

Vegetation

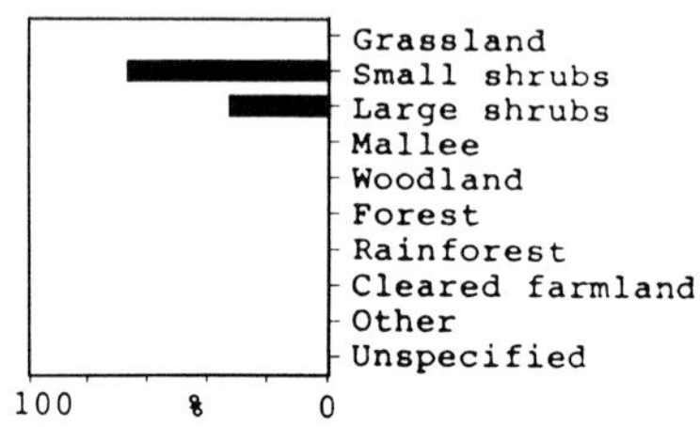

Soil Type

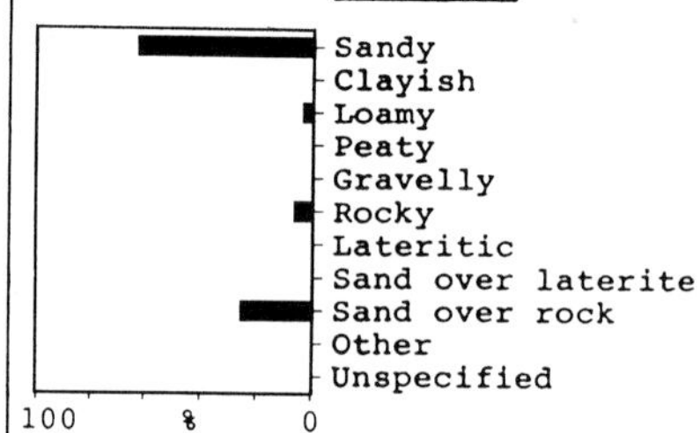

Landform

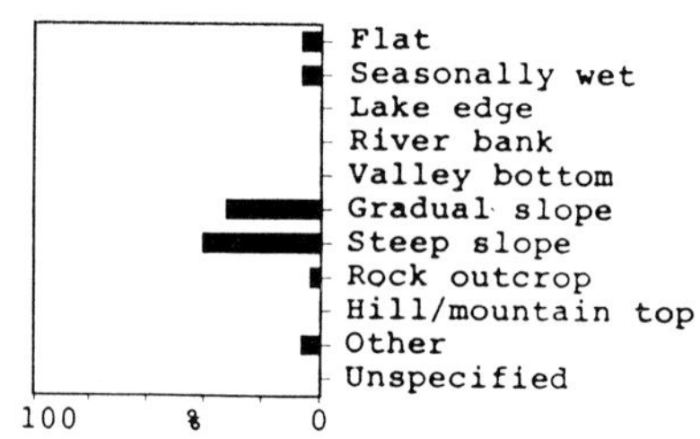

Response to Fire: 3 records

Growth response: Killed, new seedlings (33%) trunk resprout (33%) killed, no seedlings and trunk resprout (33%).

Flowering response: Number of records where flowering had occurred after fire (2). Median reported time to flowering (24-25 months); minimum reported time to flowering (13-24 months). Number of records where flowering had not occurred after fire (1). Maximum reported time to no flowering (13-24 months).

Flowering

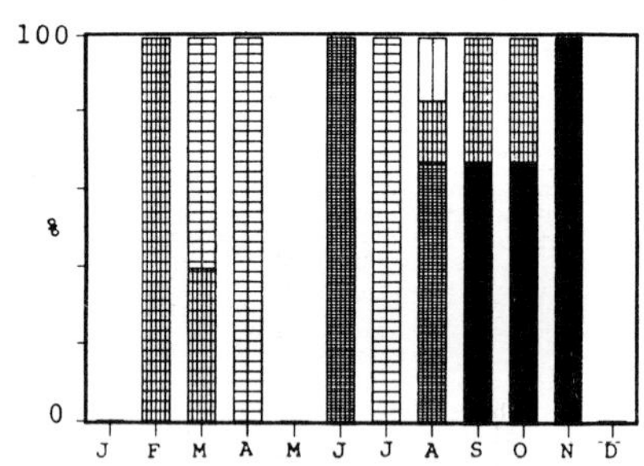

Majority of flowers fully open
Majority of flowers in bud
Majority of flowers recently finished
Flowers finished, fruiting cones present
Neither flowers nor fruiting cones present
Unspecified

Possible pollinators

Pollinator type 9 records: bird (22%) bees, wasps, ants (56%) no pollinator observed (22%).

Specific pollinator 8 records: New Holland Honeyeater (2) ant (6)

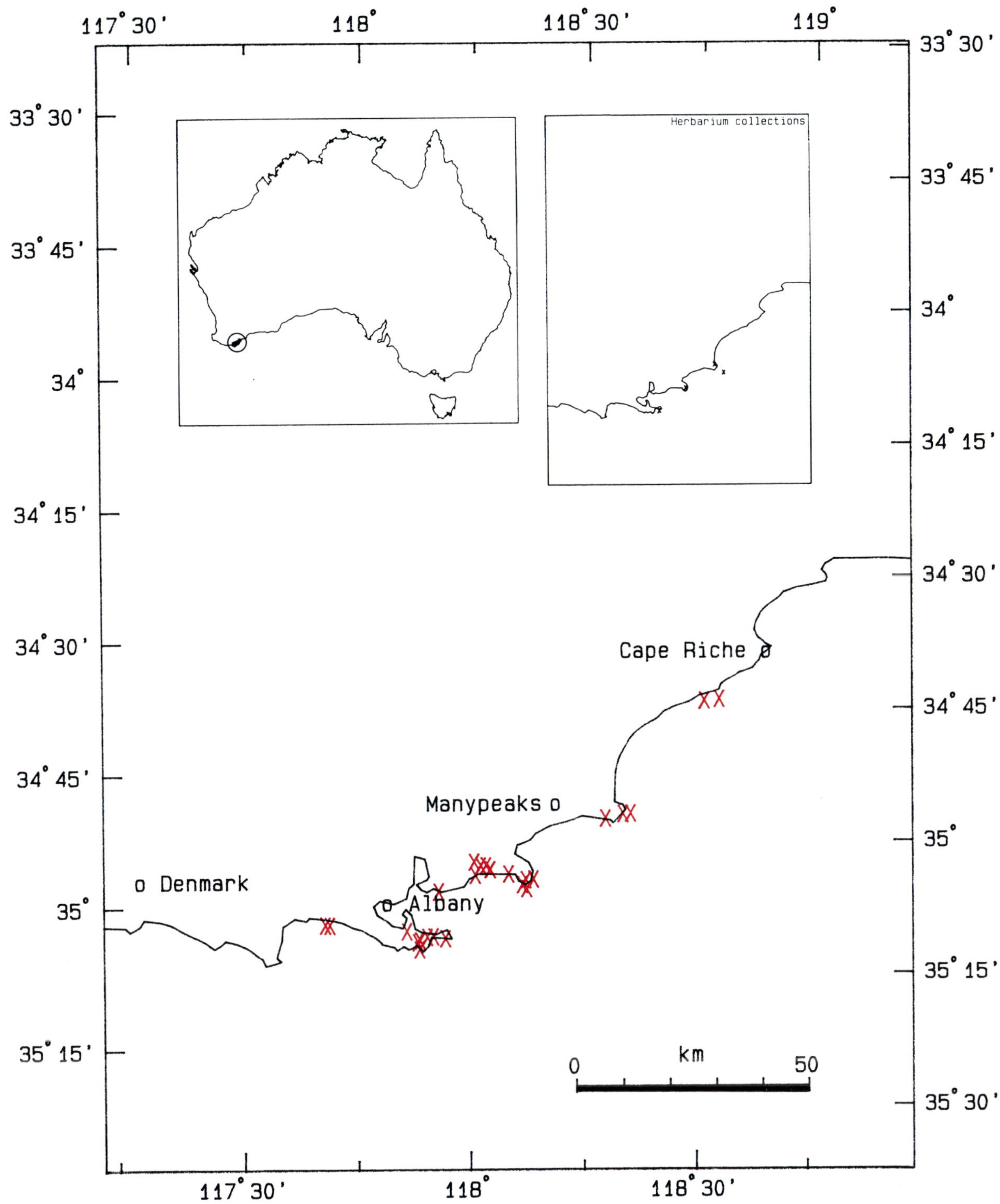

DISCUSSION

B. praemorsa is known from only a few localities along the south coast of Western Australia near Albany. W.A. Herbarium collections were from Torndirrup National Park, Mt Gardner, Bald Island and Cheyne Beach. Apart from Bald Island, all these sites were recorded by Atlas contributors who also showed the populations in these areas to be more extensive than was previously known. Additional records were made at Nannarup Beach, at Torbay, and on the coast towards Cape Riche opposite Haul Off Rock. The latter two records were both range extensions. With relatively few populations known and these being mostly of less than 100 plants, the conservation status of the species should be closely monitored. However, almost three-quarters of recorded populations are in conservation reserves (chiefly Torndirrup National Park and Two Peoples Bay Nature Reserve) so with correct management the species should be secure.

B. praemorsa occurs as both a shrub and tree, generally less than 4 m in height though sometimes larger. Specimens have been observed as large as 6-7 m with single trunks 30-40 cm in diameter. It is always within 2 km of the coast, often growing on the seaward facing slopes of sand dunes or low cliff formations of shallow sand over limestone. In very exposed locations plants are often stunted e.g. on low limestone cliffs at the end of Sinker Reef Road, Two Peoples Bay Nature Reserve, mature plants are only 1 m high. The species flowers from September to November. New shoot growth may be from late autumn to winter though further data are needed to confirm this. Plants with pure yellow flowers have been recorded from a few scattered locations and may come true from seed (D. Davidson, pers. comm.). *B. praemorsa* is thought to be killed by fire and to regenerate from seed. The trunk resprout record was of plants near the edge of a fire where fire intensity was therefore low.

BANKSIA PRIONOTES Lindley (1839)

Acorn Banksia, Orange Banksia

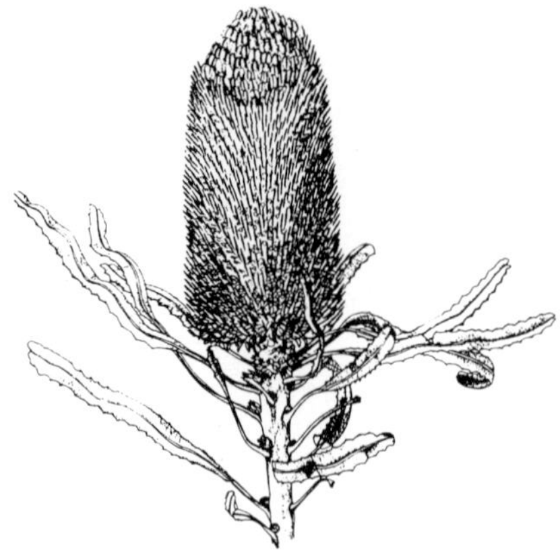

386 RECORDS: Jan (19) Feb (27) Mar (45) Apr (34) May (33) Jun (13) Jul (21) Aug (36) Sep (93) Oct (21) Nov (12) Dec (26).

Population Size: 1-10 (64) 10-100 (150) >100 (167) Unspecified (5).

Conservation Status: Restricted to road verge (26%) Not (69%) Unspecified (5%). In conservation reserve (31%) Not (65%) Unspecified (4%).

Tree/Shrub: Tree form (82%) Shrub form (16%) Unspecified (2%).

Height (metres)

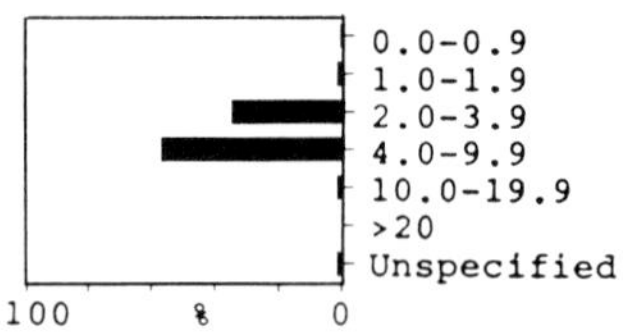

New Shoot Growth

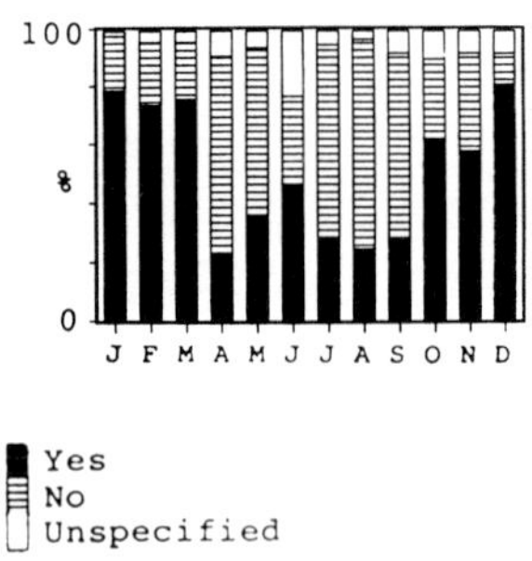

HABITAT

Within 2 km of coast: Yes (3%) No (96%) Unspecified (1%).

Altitude (metres)

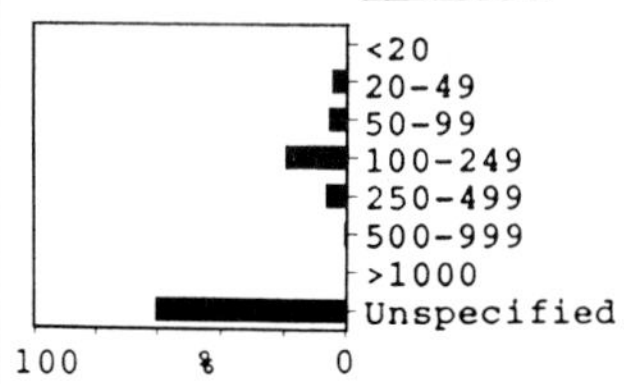

Vegetation

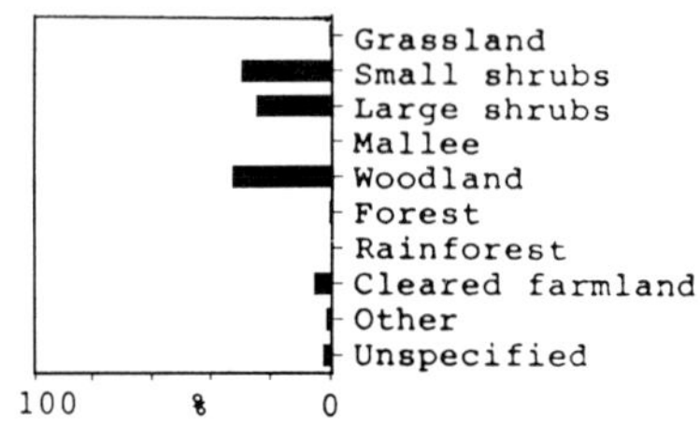

Soil Type

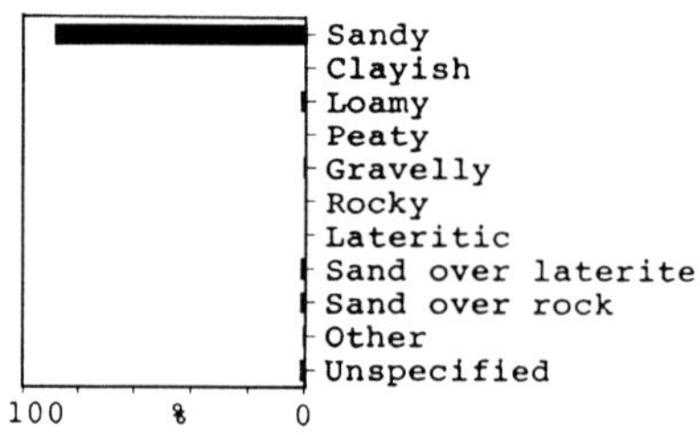

Landform

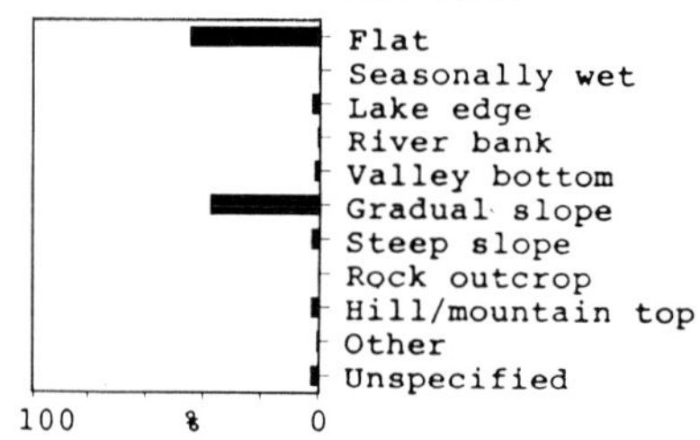

Response to Fire: 35 records

Growth response: Ground resprout (9%) ground resprout and trunk resprout (3%) killed, new seedlings (46%) killed, no seedlings (3%) killed, no seedlings and killed, new seedlings (3%) unspecified (37%).

Flowering response: Number of records where flowering had occurred after fire (15). Median reported time to flowering (25-36 months); minimum reported time to flowering (<1 month).
Number of records where flowering had not occurred after fire (8). Median reported time to no flowering (13-24 months); maximum reported time to no flowering (25-36 months).

Flowering

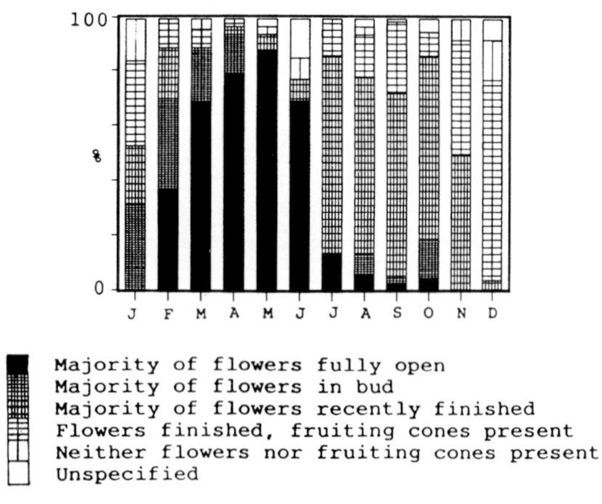

Possible pollinators

Pollinator type 114 records: bird (32%) butterflies, moths (1%) bees, wasps, ants (15%) flies (1%) no pollinator observed (51%).

Specific pollinator 45 records: honeyeater (12) New Holland Honeyeater (2) Brown Honeyeater (2) White-cheeked Honeyeater (3) Singing Honeyeater (4) Tawny-crowned Honeyeater (1) wattlebird (1) Red Wattlebird (2) Eastern Spinebill (1) lorikeets and allies (1) ant (10) bee (5) aphid (1)

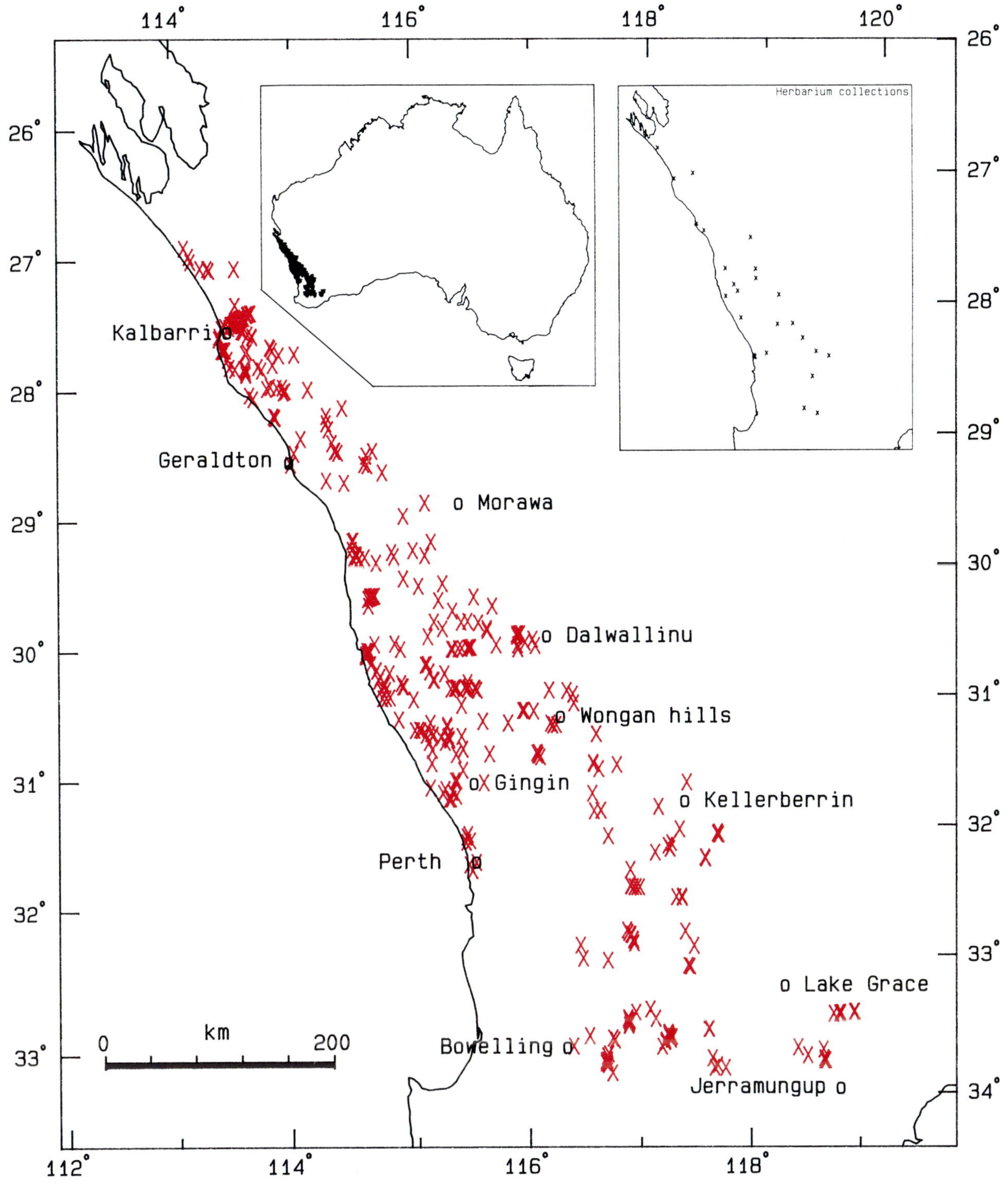

DISCUSSION

B. prionotes was previously recorded from Wagin north almost to Shark Bay and inland as far as Wongan Hills and Quairading. Atlas records have confirmed and extended this distribution. To the south-east of its range, new populations have been recorded from the vicinities of Coyrecup Lake, Holland Rocks, Lake Cairlocup and Lake Lockhart. The latter record is a considerable range extension being some 160 km from Wagin. Further north, the species is recorded from 15 km east of Shackleton and from approximately 10 km north of Kellerberrin. *B. prionotes* shows a strong preference for deep sandy soils, often yellow in colour. In the north of its range where sandy soils are widespread, Atlas records show the species to have a fairly continuous distribution with a well defined eastern boundary which coincides approximately with the 350 mm rainfall isohyet. In the south of its range the heavier soils of the Darling Scarp constitute a significant barrier to its distribution. To the east of the scarp, soils are generally unsuitable and populations of *B. prionotes* are limited to the occasional occurrences of alluvial or aeolian deep yellow sand deposits.

The species may occur as either a tree or shrub up to 10 m in height. It is often locally common forming patches of woodland or tall shrubland where it may constitute the dominant plant. Its showy orange flowers occur mainly between February and June when they are frequently visited by honey-eating birds such as the Singing Honeyeater, White-cheeked Honeyeater, Brown Honeyeater and Red Wattlebird. Flowers were also recorded as late as October. New shoot growth is mainly in summer. The species is generally killed by fire and regenerates from seed. Records of plants resprouting from epicormic buds and at ground level are probably a result of low fire intensities and/or the species occurring as a tall tree. Baird (1977) for example, noted that *B. prionotes* trees were not killed by fire in Kings Park, Perth. Presumed hybrids of *B. prionotes* x *B. menziesii* and of *B. prionotes* x *B. hookeriana* have been recorded and are described in Chapter 3.

BANKSIA PULCHELLA R.Brown (1810)

Teasel Banksia

147 RECORDS: Jan (7) Feb (21) Mar (22) Apr (17) May (12) Jun (2) Jul (9) Aug (17) Sep (5) Oct (10) Nov (18) Dec (6).

Population Size: 1-10 (26) 10-100 (63) >100 (55) Unspecified (3).

Conservation Status: Restricted to road verge (24%) Not (76%) Unspecified (1%). In conservation reserve (24%) Not (66%) Unspecified (10%).

Tree/Shrub: Tree form (0%) Shrub form (98%) Unspecified (2%).

Height (metres)

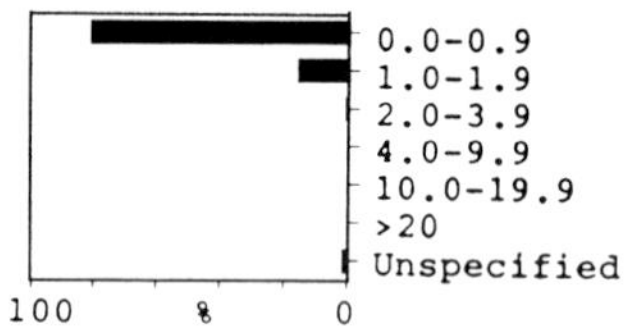

New Shoot Growth

HABITAT

Within 2 km of coast: Yes (23%) No (77%) Unspecified (0%).

Altitude (metres)

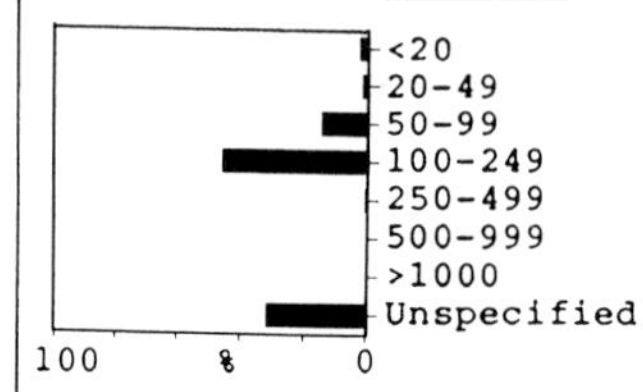

Vegetation

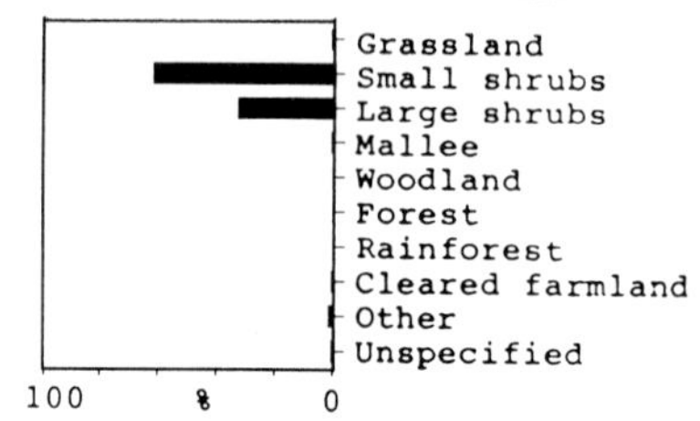

Soil Type

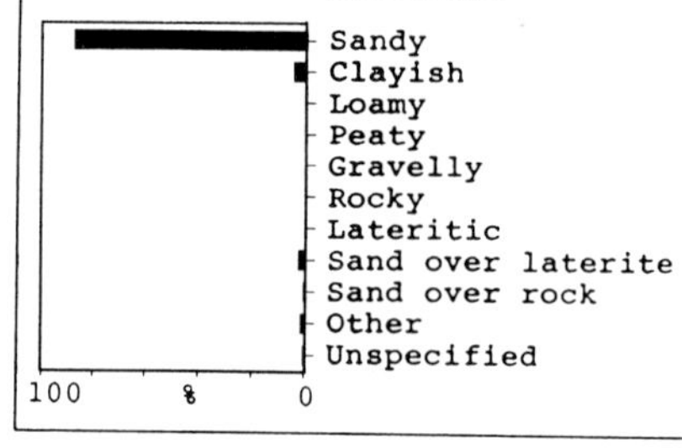

Landform

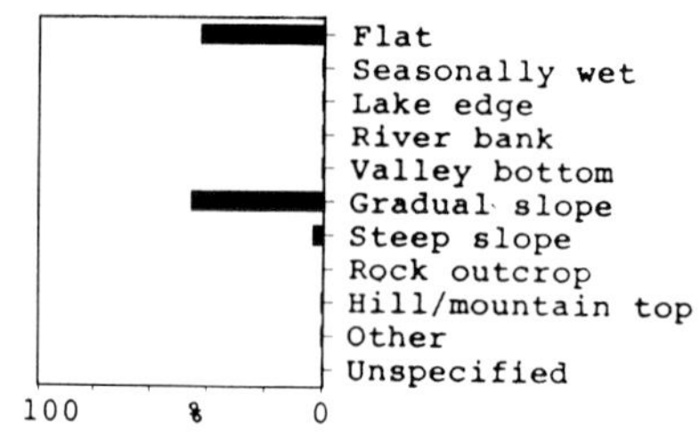

Response to Fire: 11 records

Growth response: Killed, new seedlings (36%) unspecified (64%).

Flowering response: Number of records where flowering had occurred after fire (4). Median reported time to flowering (12-13 months); minimum reported time to flowering (1-12 months). Number of records where flowering had not occurred after fire (3). Median reported time to no flowering (1-12 months); maximum reported time to no flowering (1-12 months).

Flowering

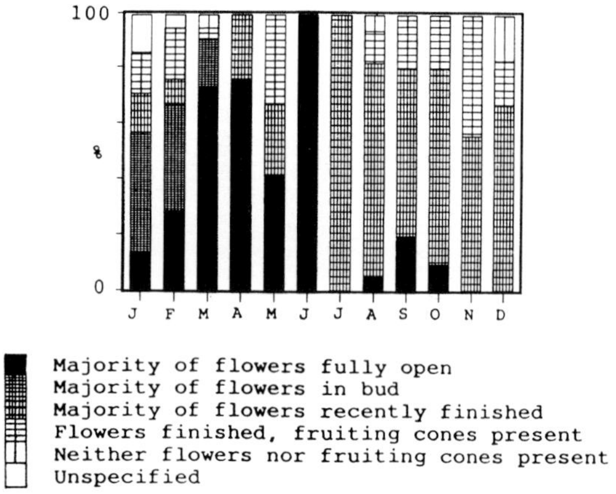

Possible pollinators

Pollinator type 85 records: bees, wasps, ants (20%) flies (4%) beetles (2%) no pollinator observed (74%).

Specific pollinator 14 records: ant (13) bee (1)

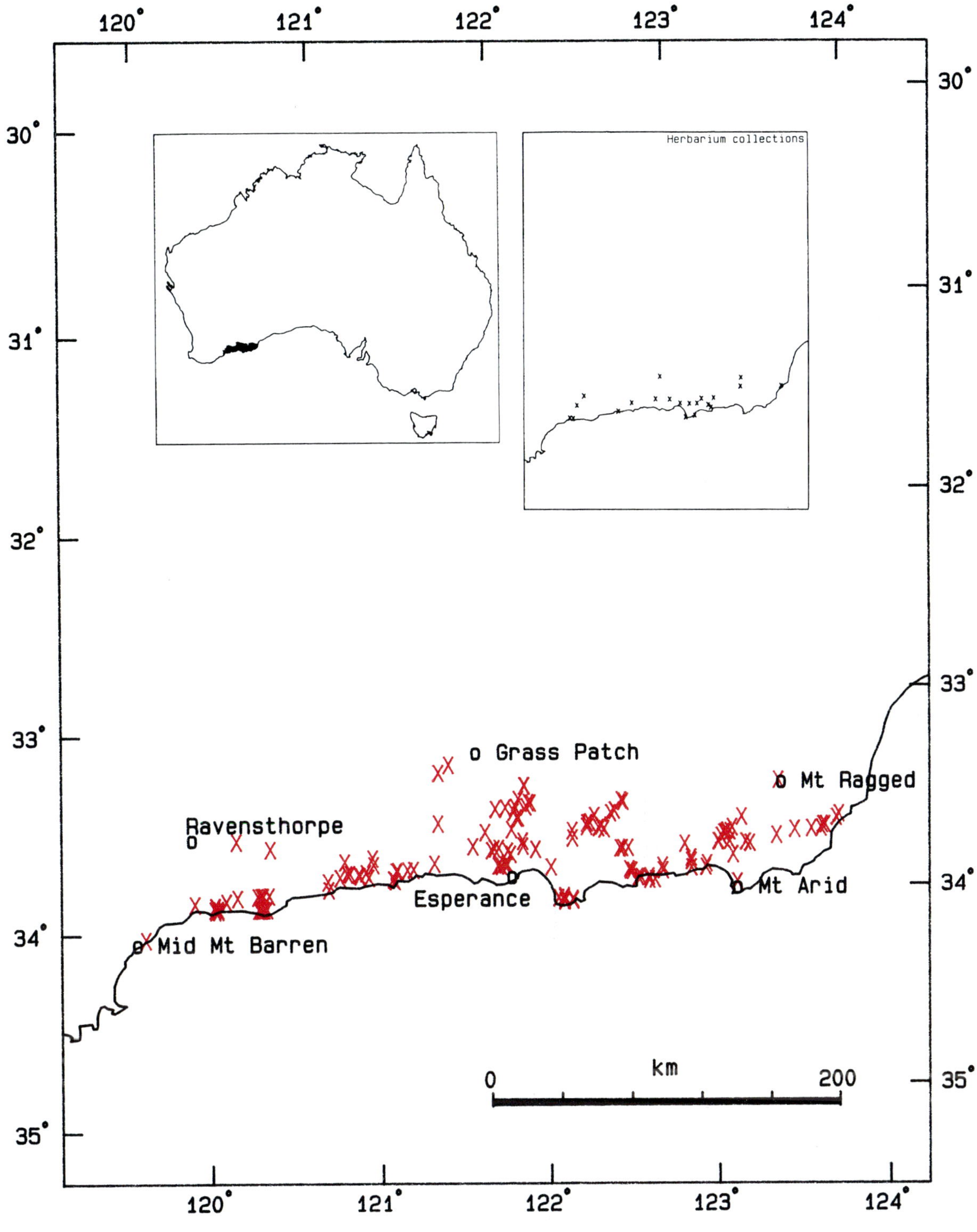

DISCUSSION

B. pulchella extends from Twin Bays in Fitzgerald River National Park eastwards to Israelite Bay, always within 60 km of the coast. The Twin Bay record is a new locality for this species. Previously it had not been recorded west of Culham Inlet, near Hopetoun. Throughout its range B. pulchella is locally common with most populations being of more than 100 plants. Almost 25% of its recorded populations are in conservation reserves. However, the species is susceptible to dieback disease and in national parks such as Cape Arid, Cape Le Grand and Fitzgerald River its numbers are diminishing.

Typically a small shrub less than 1 m in height, B. pulchella grows amongst heath and tall shrubland often in association with B. speciosa and/or B. petiolaris. It occurs in deep white sands on flat or gently sloping landscapes. The main flowering period is summer through to winter, though flowers were also recorded in spring. The flowers were very attractive to ants though the structure of B. pulchella with its very long style means that in visiting the nectaries at the base of the flowers they were almost certainly by-passing the pollen presenter and therefore not acting as pollinators. Atlas data suggest that new shoot growth can occur both in November-December and/or May-July. The species is non-lignotuberous, being killed by fire and regenerating from seed.

BANKSIA QUERCIFOLIA R.Brown (1810)

Oak-leaved Banksia

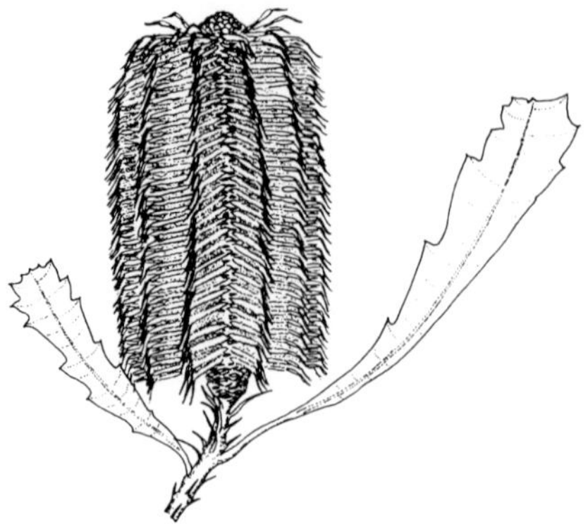

180 RECORDS: Jan (4) Feb (4) Mar (68) Apr (21) May (5) Jun (6) Jul (8) Aug (17) Sep (9) Oct (11) Nov (15) Dec (11).

Population Size: 1-10 (37) 10-100 (81) >100 (61) Unspecified (1).

Conservation Status: Restricted to road verge (4%) Not (91%) Unspecified (6%). In conservation reserve (17%) Not (76%) Unspecified (7%).

Tree/Shrub: Tree form (11%) Shrub form (88%) Unspecified (1%).

Height (metres)

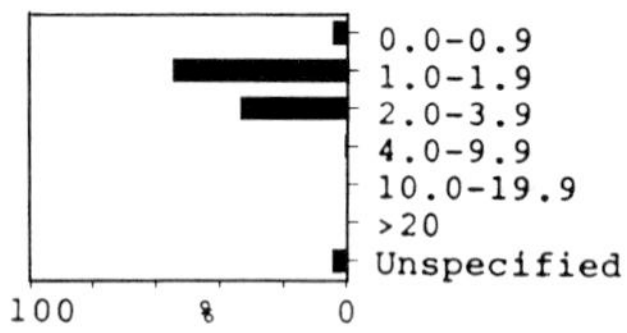

New Shoot Growth

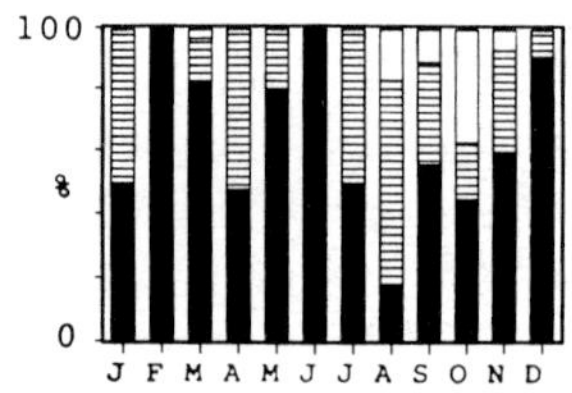

Yes
No
Unspecified

HABITAT

Within 2 km of coast: Yes (24%) No (75%) Unspecified (1%).

Altitude (metres)

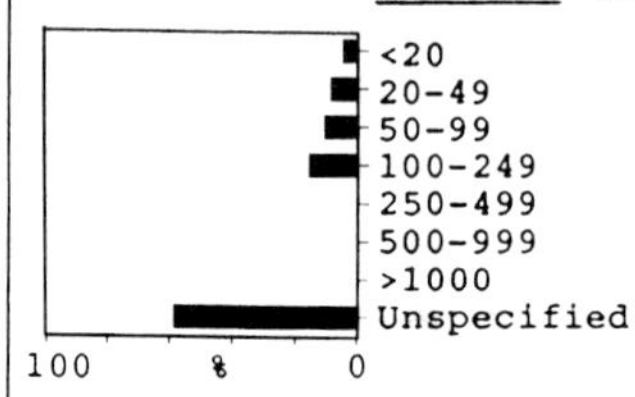

Vegetation

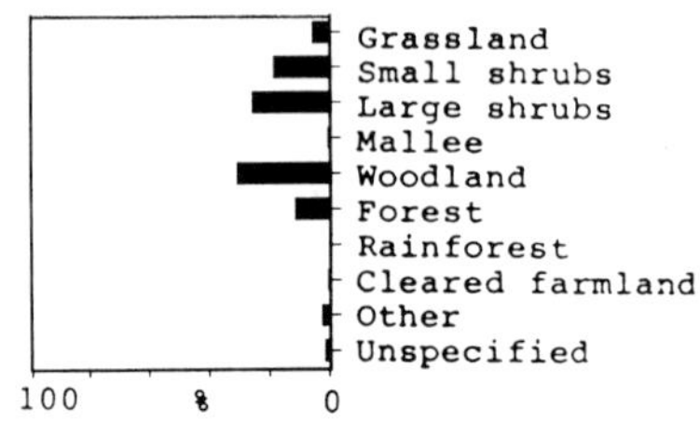

Soil Type

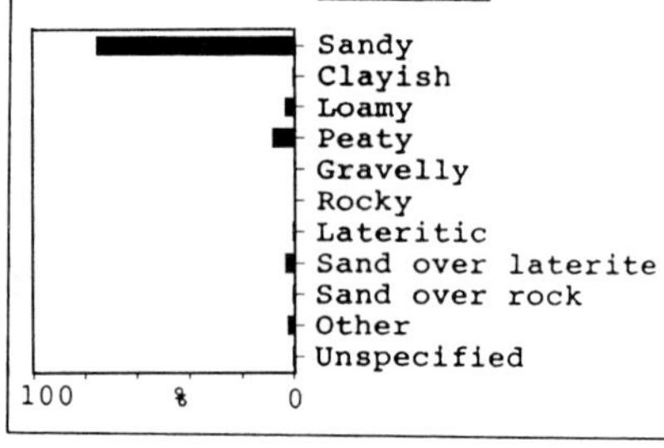

Landform

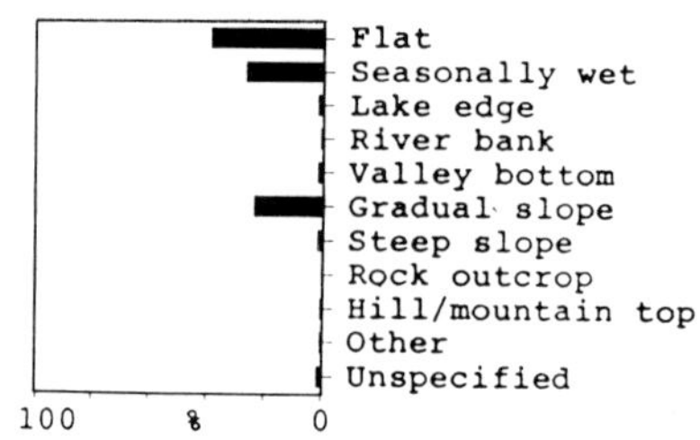

Response to Fire: 49 records

Growth response: Ground resprout and killed, new seedlings (2%) killed, new seedlings (63%) killed, no seedlings (2%) unspecified (33%).

Flowering response: Number of records where flowering had occurred after fire (23). Median reported time to flowering (25-36 months); minimum reported time to flowering (1-12 months). Number of records where flowering had not occurred after fire (14). Median reported time to no flowering (13-24 months); maximum reported time to no flowering (25-36 months).

Flowering

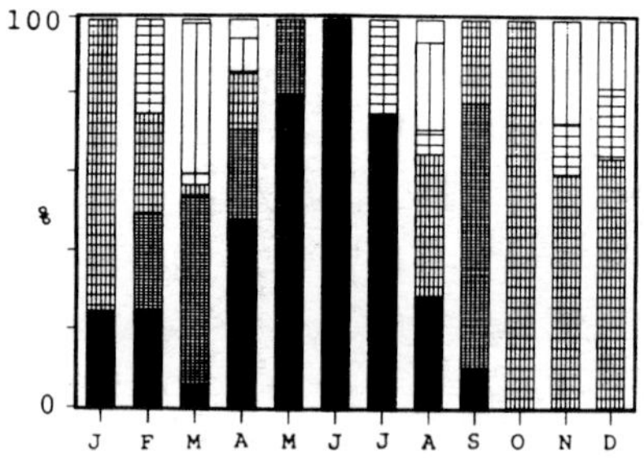

Majority of flowers fully open
Majority of flowers in bud
Majority of flowers recently finished
Flowers finished, fruiting cones present
Neither flowers nor fruiting cones present
Unspecified

Possible pollinators

Pollinator type 43 records: bird (12%) mammals (2%) no pollinator observed (86%).

Specific pollinator 6 records: New Holland Honeyeater (5) Western Spinebill (1)

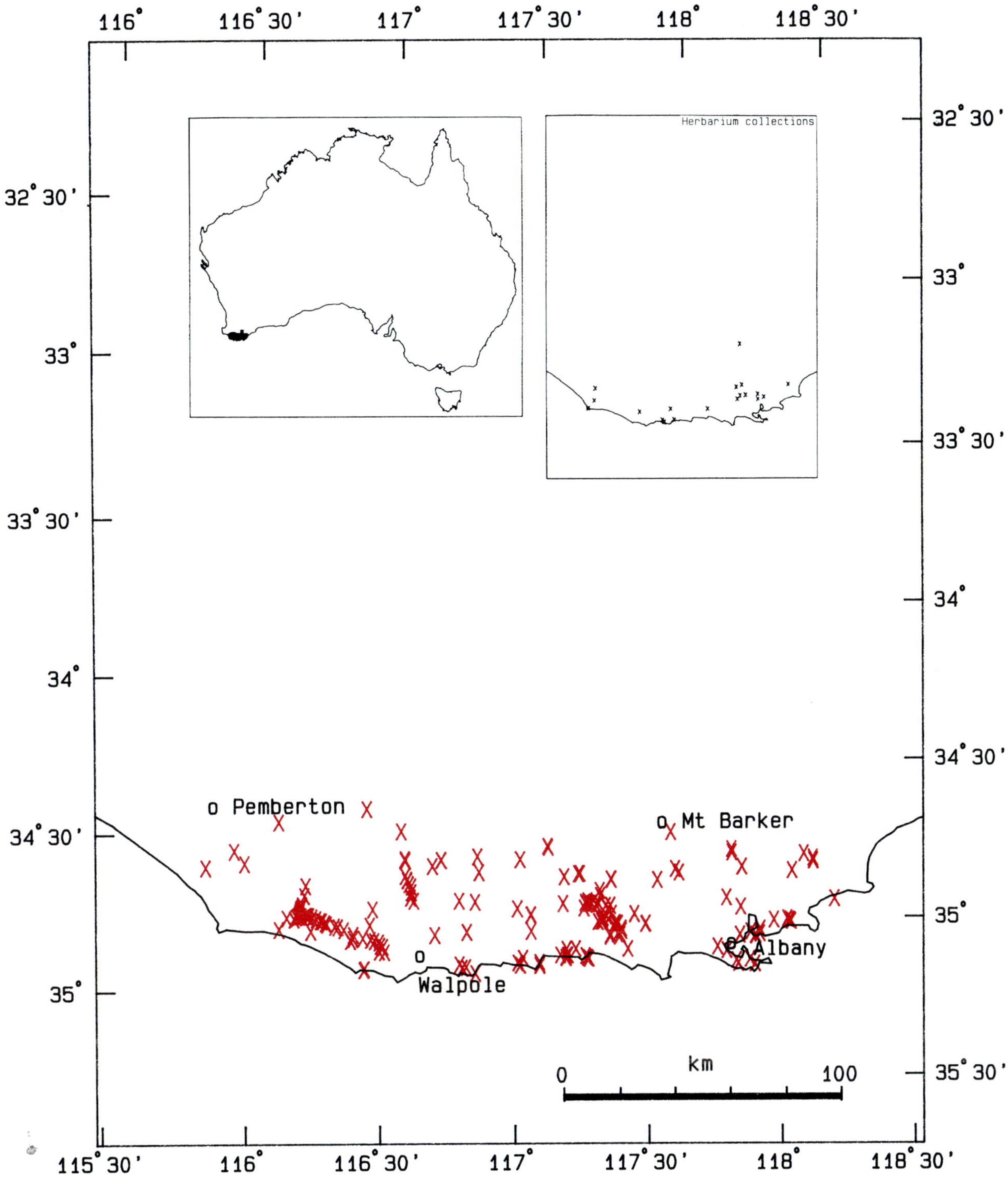

DISCUSSION

Western Australian Herbarium collections indicated that *B. quercifolia* had a near coastal distribution between Windy Harbour and Cheyne Bay, generally within 35 km of the coast. There is also an outlying collection from the Stirling Range (Morrison in 1902). Atlas records have confirmed this general distribution and shown it to extend as far west as D'Entrecasteaux National Park approximately 13 km west of Northcliffe. The most easterly record was at Mt Manypeaks. The most inland record is from west of Lake Muir approximately 50 km inland.

The species is typically a shrub less than 2 m in height, though often larger. Although it was identified as a tree in 11% of Atlas records, this requires rechecking in the field. *B. quercifolia* grows in sand or peaty sand generally in wetter areas, wherever drainage is impeded. Surrounding vegetation may be woodland, shrubland or forest. Compared to herbarium collections, there were many more Atlas records of *B. quercifolia* occurring away from the coast in *E. marginata* woodland or forest where it usually occupies the wetter sites - near rivers, creeks, swamps or small depressions. Almost one-quarter of recorded populations were within 2 km of the coast where the species grows amongst tall shrubland or low heath. Sometimes it is very close to the coast, only metres from high water mark e.g. between Walpole and Denmark. Here, it grows in seepages and soaks in shallow peaty-sand overlying granite. The main flowering season of *B. quercifolia* is from autumn to spring, though some flowers were recorded as early as January. New shoot growth may occur throughout much of the year. In every month except August at least 50% of recorded populations were showing new shoot growth. The species is killed by fire and regenerates from seed. George (1981) stated that flowering may begin in three years when plants may be only 20 cm tall. *B. quercifolia* is susceptible to dieback disease, especially so due to its preference for poorly-drained sites. Although it is relatively widespread at present, its status should be monitored in future years.

BANKSIA REPENS Labillardiere (1800)

Creeping Banksia

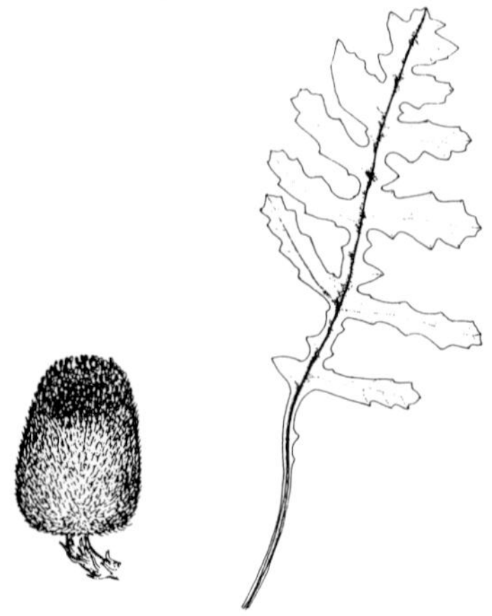

346 RECORDS: Jan (41) Feb (35) Mar (39) Apr (16) May (18) Jun (21) Jul (15) Aug (33) Sep (33) Oct (43) Nov (33) Dec (18).

Population Size: 1-10 (68) 10-100 (134) >100 (137) Unspecified (7).

Conservation Status: Restricted to road verge (21%) Not (77%) Unspecified (3%). In conservation reserve (39%) Not (56%) Unspecified (5%).

Tree/Shrub: Tree form (0%) Shrub form (97%) Unspecified (3%).

Height (metres)

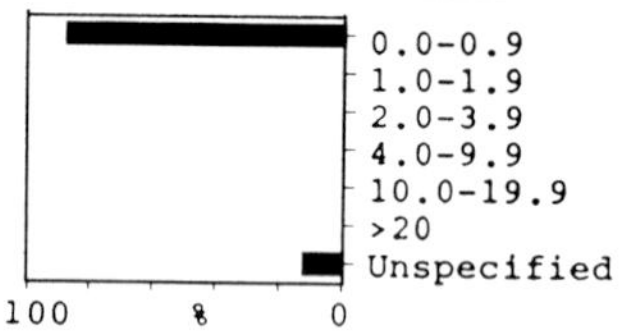

New Shoot Growth

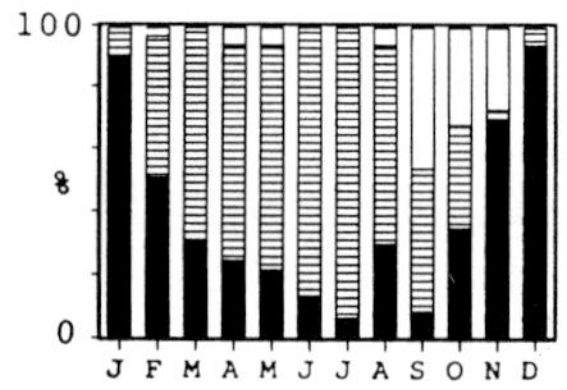

Yes
No
Unspecified

HABITAT

Within 2 km of coast: Yes (20%) No (80%) Unspecified (0%).

Altitude (metres)

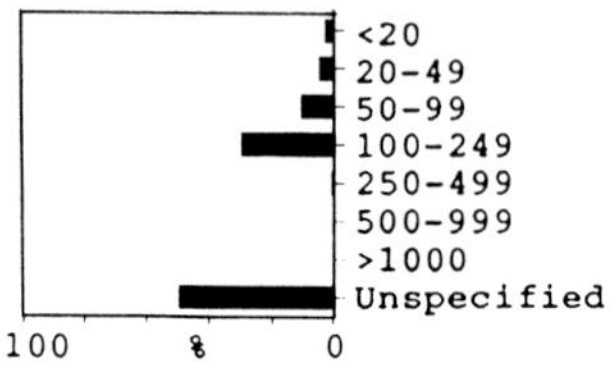

Vegetation

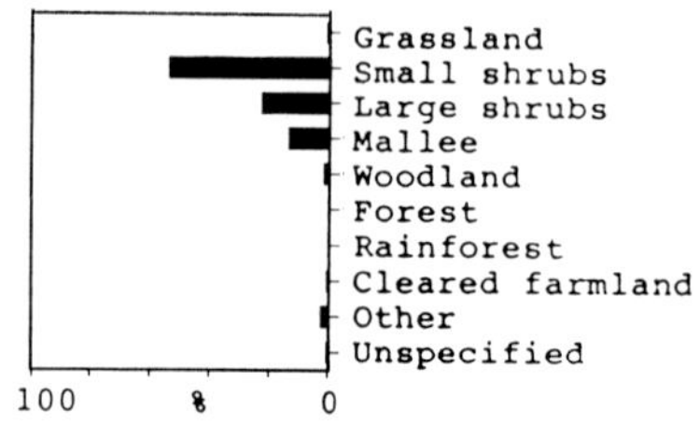

Soil Type

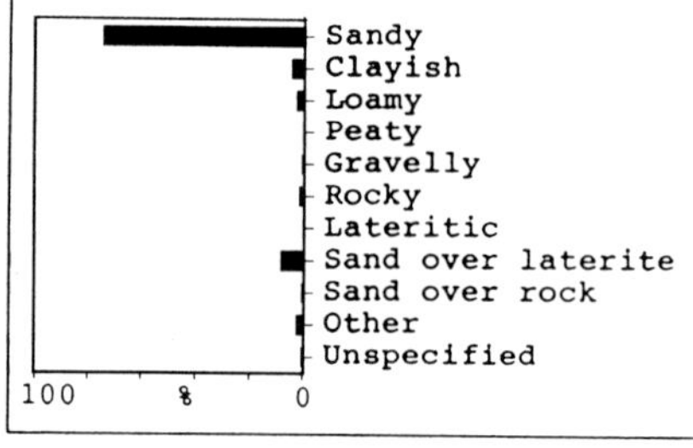

Landform

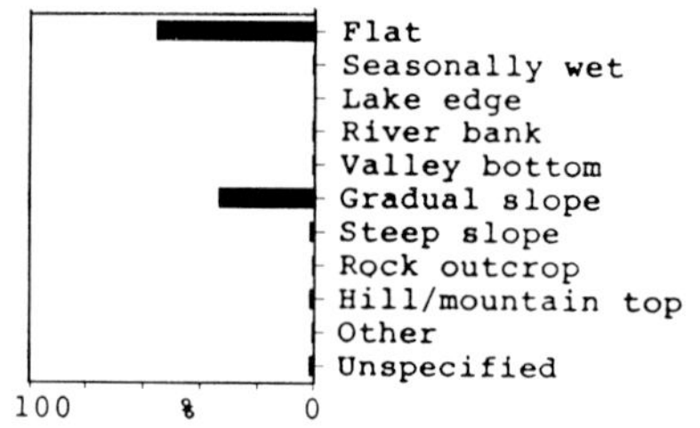

Response to Fire: 38 records

Growth response: Ground resprout (37%) ground resprout and killed, new seedlings (3%) killed, new seedlings (5%) unspecified (55%).

Flowering response: Number of records where flowering had occurred after fire (13). Median reported time to flowering (13-24 months); minimum reported time to flowering (1-12 months). Number of records where flowering had not occurred after fire (16). Median reported time to no flowering (13-24 months); maximum reported time to no flowering (49-60 months).

Flowering

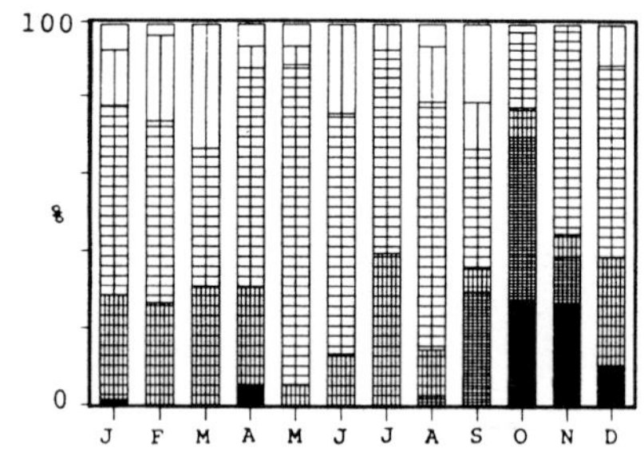

Majority of flowers fully open
Majority of flowers in bud
Majority of flowers recently finished
Flowers finished, fruiting cones present
Neither flowers nor fruiting cones present
Unspecified

Possible pollinators

Pollinator type 117 records: bird (2%) bees, wasps, ants (4%) no pollinator observed (94%).

Specific pollinator 3 records: ant (3)

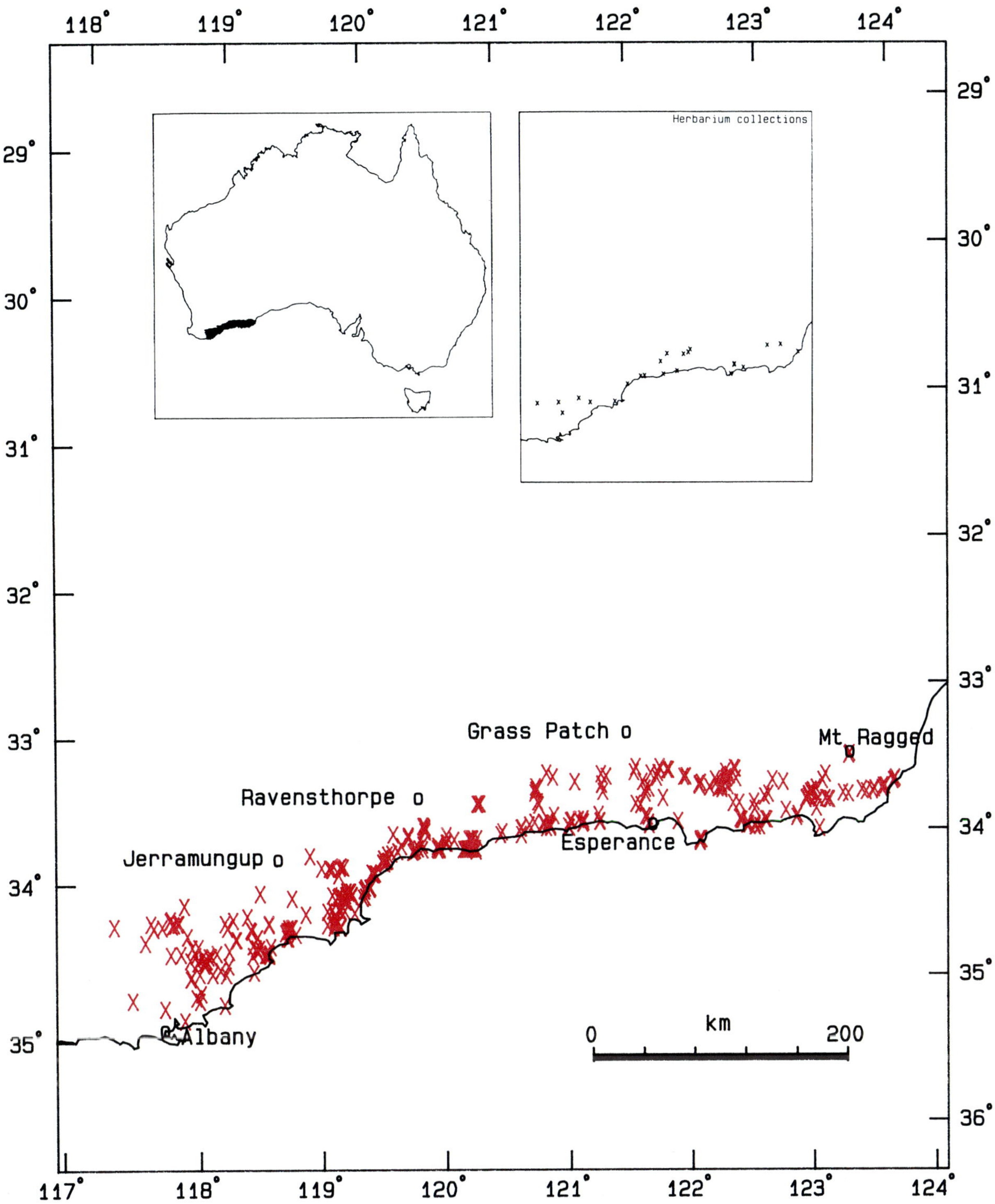

DISCUSSION

The 346 records show B. repens to be relatively common between Cranbrook and Israelite Bay, generally with 80 km of the coast. One-fifth of its recorded populations are within 2 km of the coast. The species is the most widespread of the prostrate shrubs and almost 40% of its recorded populations occur in conservation reserves. It grows predominantly in deep white or grey sandy soils, sometimes in sand over laterite. Surrounding vegetation is usually heath, sometimes tall shrubland or mallee-shrubland. B. repens flowers from October to December. There is also a single record of it flowering in April. Late spring to summer is the main season of new shoot growth.

B. repens is usually easy to identify with its irregularly lobed leaves and cream and pink flowers. Sometimes, however, the new leaves can strongly resemble those of B. gardneri such as those from Camel Lake Nature Reserve (illustration in main text, in W.A. Herbarium, J. Chilvers 1984). Apart from differences in flower colour, the clue to identification is in the lobing of the leaves. If the lobes themselves show even a small degree of irregular lobing, the plant is B. repens.

BANKSIA ROBUR Cavanilles (1800)

Eastern Swamp Banksia, Swamp Banksia, Broad-leaved Banksia

261 RECORDS: Jan (22) Feb (3) Mar (22) Apr (17) May (34) Jun (34) Jul (27) Aug (29) Sep (22) Oct (20) Nov (10) Dec (21).

Population Size: 1-10 (31) 10-100 (65) >100 (163) Unspecified (2).

Conservation Status: Restricted to road verge (5%) Not (91%) Unspecified (4%). In conservation reserve (34%) Not (48%) Unspecified (18%).

Tree/Shrub: Tree form (2%) Shrub form (96%) Unspecified (2%).

Height (metres)

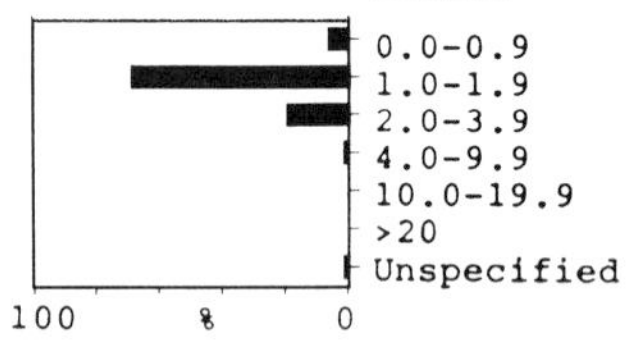

New Shoot Growth

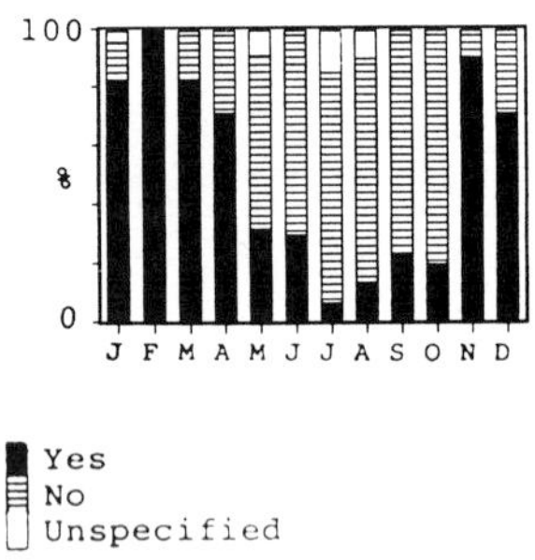

HABITAT

Within 2 km of coast: Yes (38%) No (62%) Unspecified (0%).

Altitude (metres)

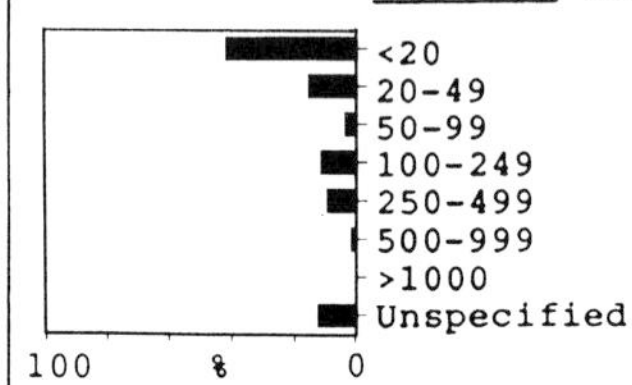

Vegetation

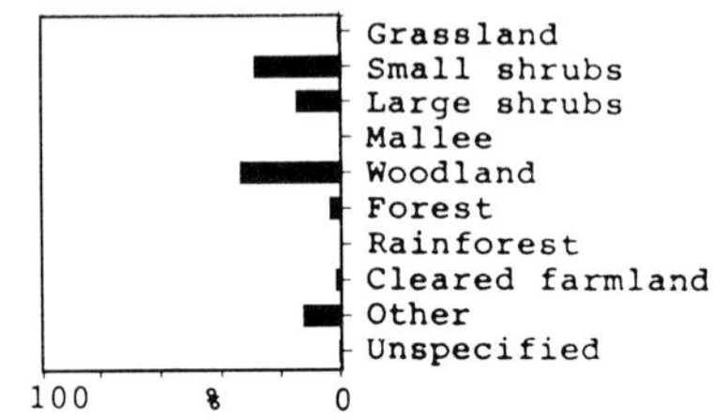

Soil Type

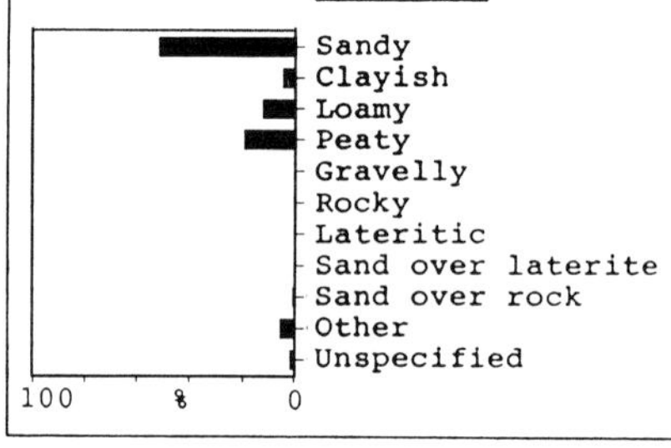

Landform

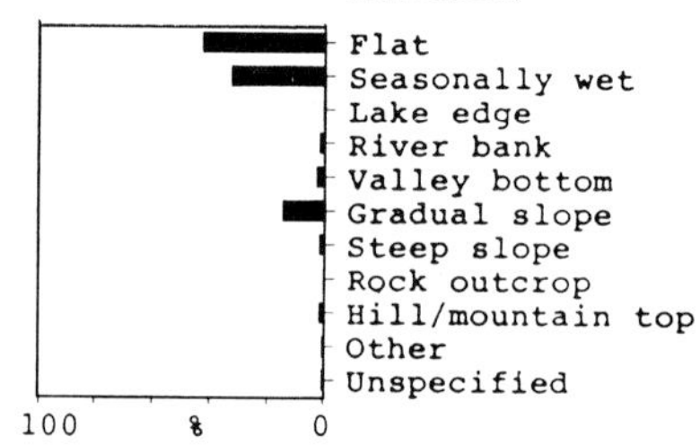

Response to Fire: 40 records

Growth response: Ground resprout (62%) ground resprout and killed, new seedlings (2%) ground resprout and trunk resprout (10%) trunk resprout (2%) unspecified (23%).

Flowering response: Number of records where flowering had occurred after fire (24). Median reported time to flowering (13-24 months); minimum reported time to flowering (<1 month). Number of records where flowering had not occurred after fire (8). Median reported time to no flowering (12-13 months); maximum reported time to no flowering (13-24 months).

Flowering

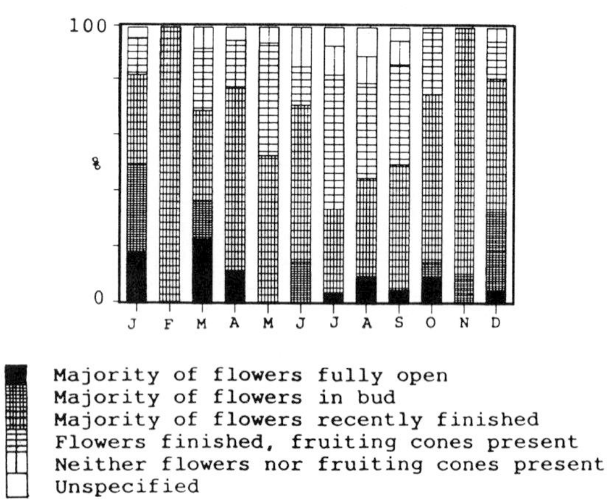

Possible pollinators

Pollinator type 110 records: bird (12%) butterflies, moths (1%) bees, wasps, ants (7%) beetles (2%) no pollinator observed (78%).

Specific pollinator 20 records: honeyeater (2) New Holland Honeyeater (1) Brown Honeyeater (5) Tawny-crowned Honeyeater (1) Lewin's Honeyeater (1) Little Wattlebird (1) friarbird (2) Little Friabird (1) ant (1) bee (1) European honey bee (2) blue butterfly (1) jewel beetle (1)

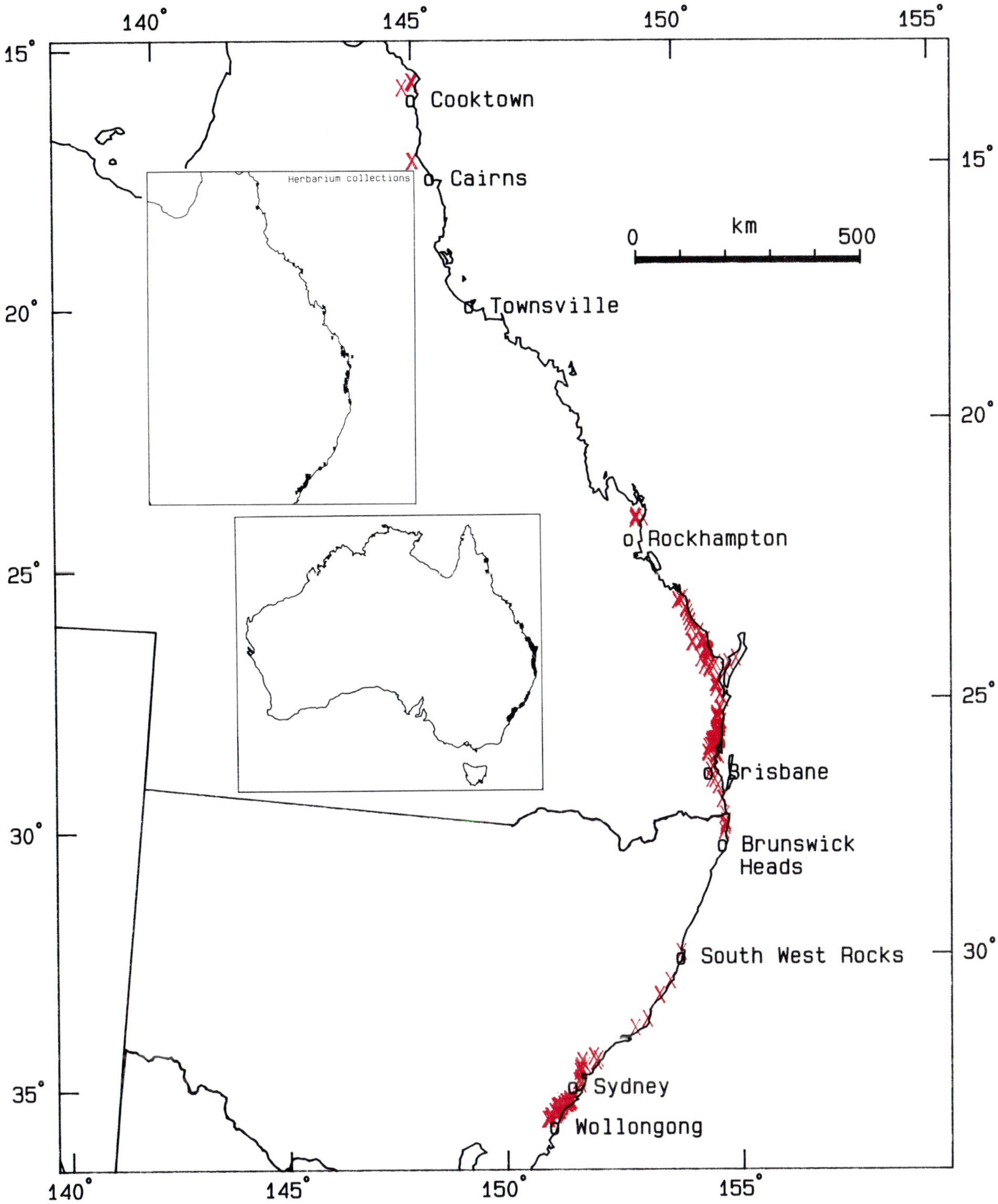

DISCUSSION

Herbarium records indicated that B. robur mainly occurred in two disjunct populations, one in New South Wales between Wollongong and South West Rocks and the other in Queensland between Coolangatta and Shoalwater Bay, north of Rockhampton. Further north, there were outlying populations near Bowen, Julatten and north of Cooktown. Atlas contributors have confirmed this general distribution though the population north of Rockhampton appears to be separated by about 200 km from the main Queensland population further south. The range of this latter population was also extended beyond Coolangatta almost to Brunswick Heads in northern New South Wales. There were new Atlas records from Fraser Island, but B. Kaspiev's 1955 collection from near Bowen was not recorded.

B. robur is typically a shrub less than 2 m in height, occasionally larger. Rarely it can reach tree proportions e.g. in Wooroi State Forest near Caloundra, single-stemmed plants up to 5 m high were recorded. B. robur grows in sandy or peaty soils, sometimes loamy. They are often seasonally or permanently damp. It generally occurs on the low-lying flat to gently undulating coastal plain but can also occur at slightly higher altitudes usually on poorly drained raised plateaus or on slopes close to permanent springs. Surrounding vegetation may be woodland or small shrubs. Sedges are common associates and these largely account for the 13% of "other" vegetation records. There does not appear to be a definite flowering season with flowers recorded in all months of the year except February, May, June and November. George (1981) had stated the main flowering period to be January to July. The main season of new shoot growth is November-April. Following fire, the species resprouts from its lignotuber and may produce its first flowers within only 2 years. B. robur hybridises readily with B. oblongifolia and there were 20 Atlas recordings of hybrid populations between Wollongong (N.S.W.) and Pialba (Qld).

BANKSIA SAXICOLA A.S.George (1981)

Grampians Banksia

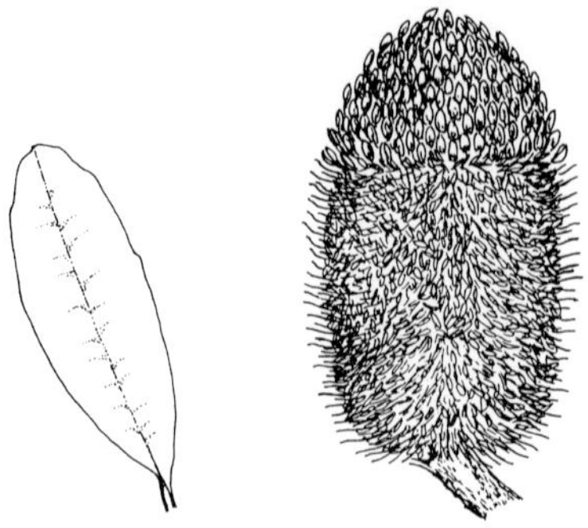

54 RECORDS: Jan (5) Feb (13) Mar (3) Apr (1) May (8) Jun (3) Jul (0) Aug (0) Sep (2) Oct (11) Nov (4) Dec (4).

Population Size: 1-10 (19) 10-100 (20) >100 (15) Unspecified (0).

Conservation Status: Restricted to road verge (0%) Not (98%) Unspecified (2%). In conservation reserve (94%) Not (0%) Unspecified (6%).

Tree/Shrub: Tree form (52%) Shrub form (46%) Unspecified (2%).

Height (metres)

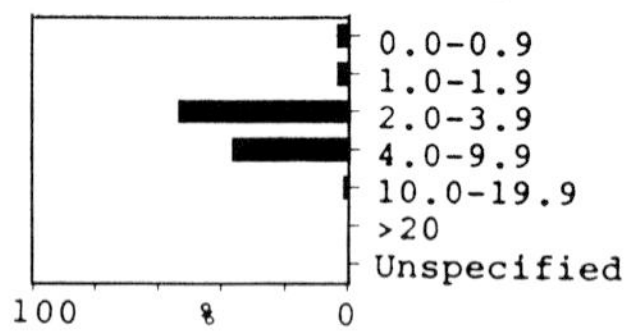

New Shoot Growth

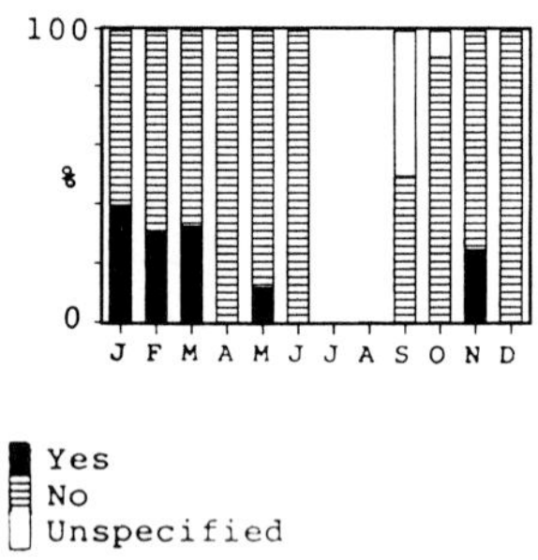

HABITAT

Within 2 km of coast: Yes (4%) No (96%) Unspecified (0%).

Altitude (metres)

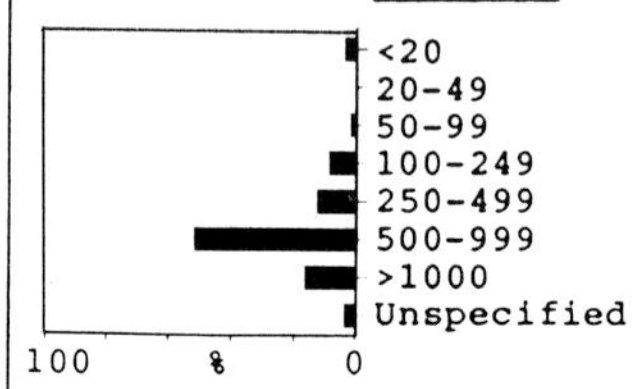

Vegetation

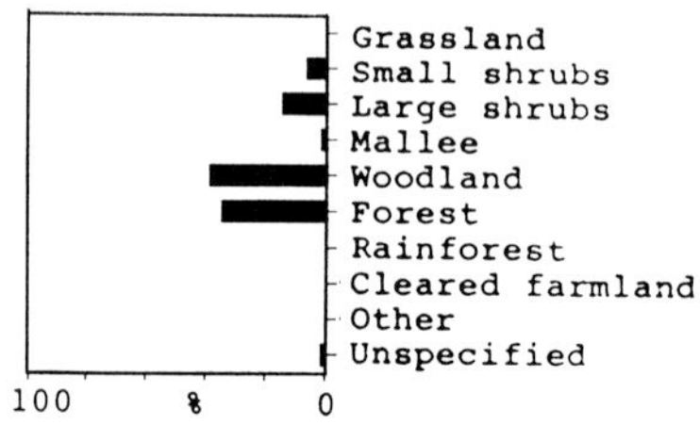

Soil Type

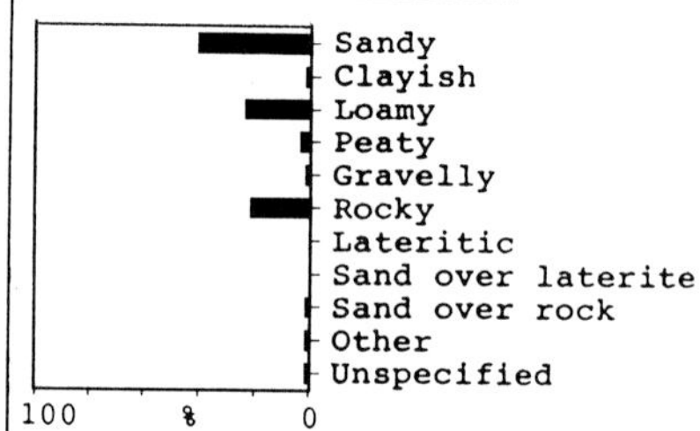

Landform

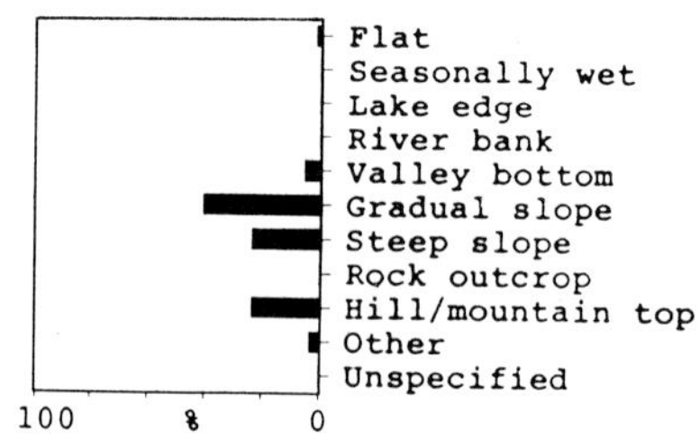

Response to Fire: 3 records

Growth response: Trunk resprout (33%) killed, no seedlings (33%) unspecified (33%).

Flowering response: Number of records where flowering had occurred after fire (1). Minimum reported time to flowering (49-60 months).
Number of records where flowering had not occurred after fire (2). Median reported time to no flowering (48-49 months); maximum reported time to no flowering (85-96 months).

Flowering

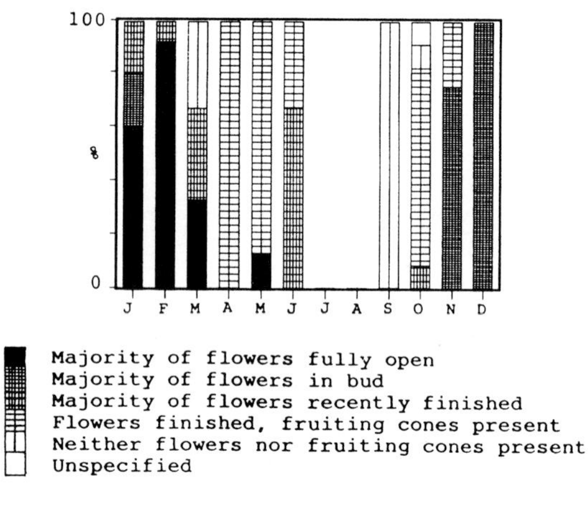

Possible pollinators

Pollinator type 34 records: bird (6%) no pollinator observed (94%).

Specific pollinator 2 records: New Holland Honeyeater (1) Crescent Honeyeater (1)

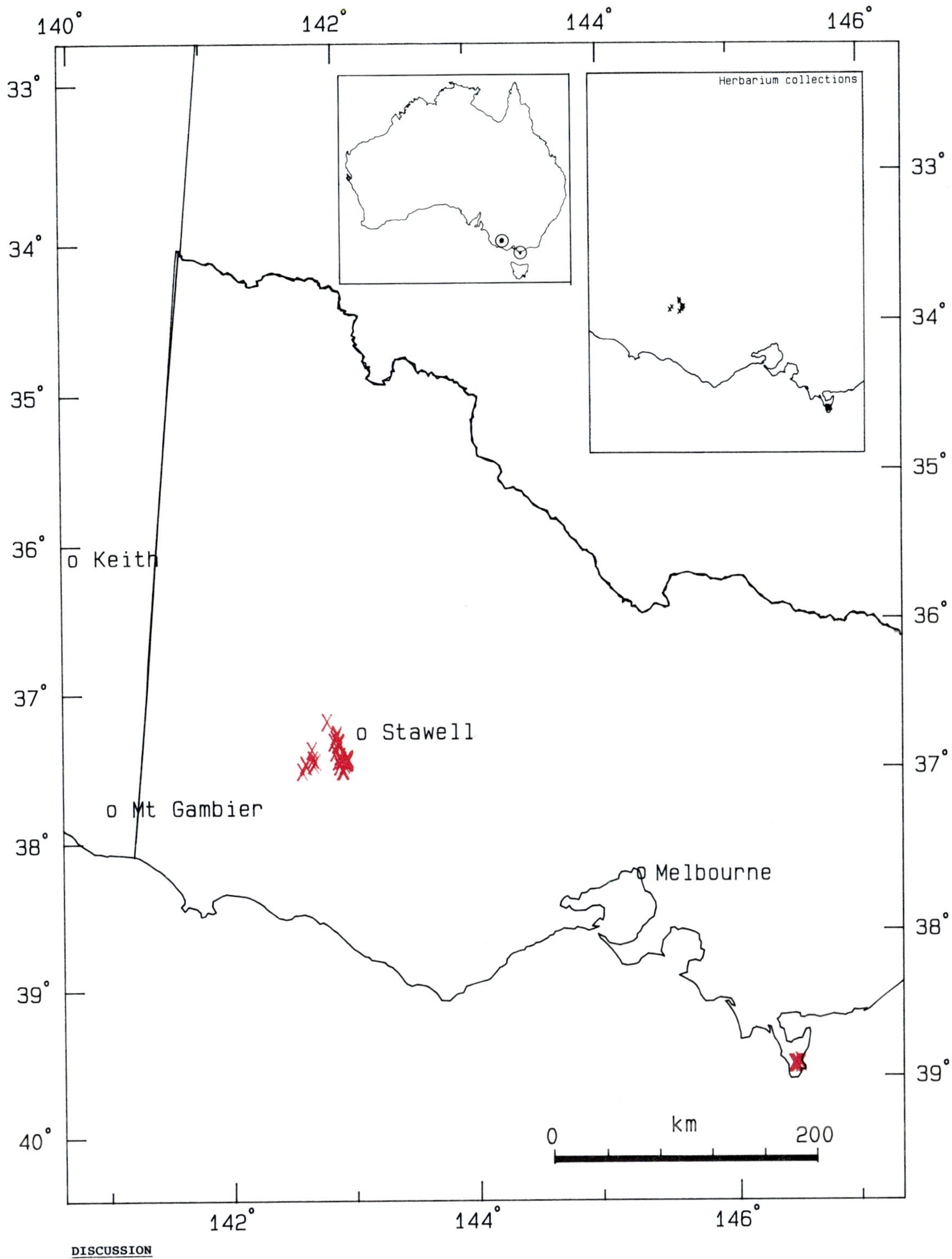

DISCUSSION

Atlas data have confirmed the previously known distribution of two disjunct populations of B. saxicola, one in The Grampians of western Victoria, the other on Wilsons Promontory. Within The Grampians, sightings were made at several locations including Mt Williams, Major Mitchell Plateau, Mt Lubra, Mt Rosea, Mt Difficult, Stony Peak, Mt Thackeray and Chimney Pots Gap. The most northerly record was 9 km south-east of Wartook on the edge of Mt Difficult Plateau. In The Grampians, B. saxicola occurs on slopes and mountain tops above 550 m (1 168 m on the summit of Mt William). The species occurs as a tree or shrub less than 5 m in height. On exposed mountain tops it is usually a bushy shrub of no more than 2 m. It grows in sandy, loamy or rocky soils generally in eucalypt forest or woodland (e.g. E. baxteri, E. obliqua), sometimes amongst tall shrubs.

At Wilsons Promontory the species is a tree 3-8 m tall, growing in grey sand amongst granite rocks in eucalypt forest and woodland. At one location, B. saxicola occurs as 10 m high trees and is the dominant plant. Most of the population is located inland from Sealers Cove up to 300 m above sea level. Thus the species usually occurs inland from B. integrifolia var. integrifolia, which it closely resembles. A single plant of B. saxicola was located at Sealers Cove only 5 m above sea level and co-existing with B. integrifolia var. integrifolia.

The main flowering period of B. saxicola is January to March, though flowers were also recorded in May. New shoot growth may be in summer though further data are needed to confirm this. The species is generally killed by fire and regenerates from seed. A presumed hybrid of B. saxicola x B. marginata was recorded near the summit of Mt William in The Grampians. The plant had intermediate leaves and cones similar to B. saxicola but with old flowers retained. Specimens have been placed in the National Herbarium in Melbourne (A. Salkin).

BANKSIA SCABRELLA A.S.George (1981)

Burma Road Banksia

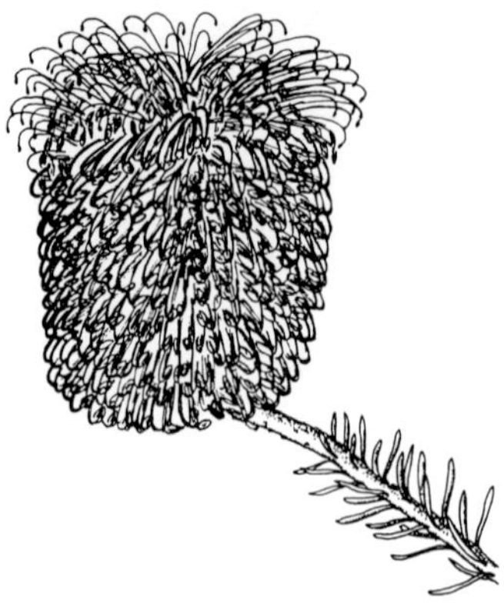

17 RECORDS: Jan (0) Feb (12) Mar (0) Apr (2) May (2) Jun (0) Jul (0) Aug (0) Sep (1) Oct (0) Nov (0) Dec (0).

Population Size: 1-10 (1) 10-100 (6) >100 (10) Unspecified (0).

Conservation Status: Restricted to road verge (53%) Not (47%) Unspecified (0%). In conservation reserve (6%) Not (94%) Unspecified (0%).

Tree/Shrub: Tree form (0%) Shrub form (100%) Unspecified (0%).

Height (metres)

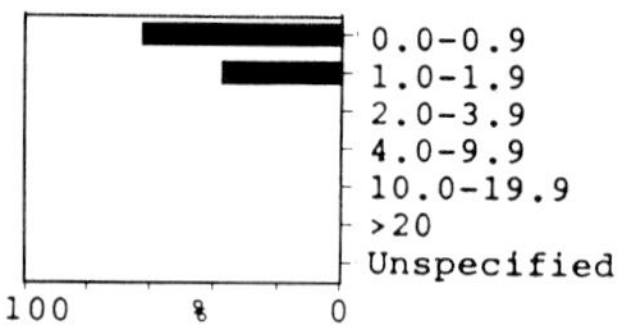

New Shoot Growth

HABITAT

Within 2 km of coast: Yes (0%) No (100%) Unspecified (0%).

Altitude (metres)

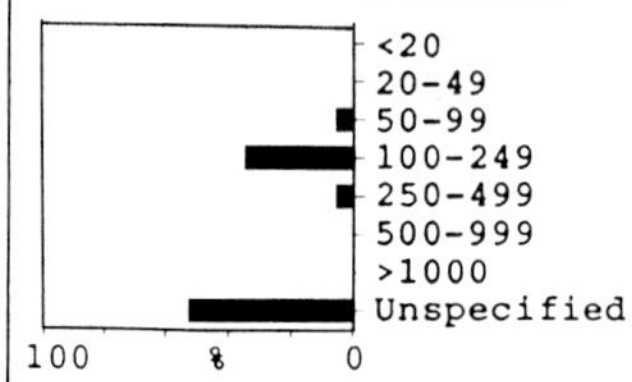

Vegetation

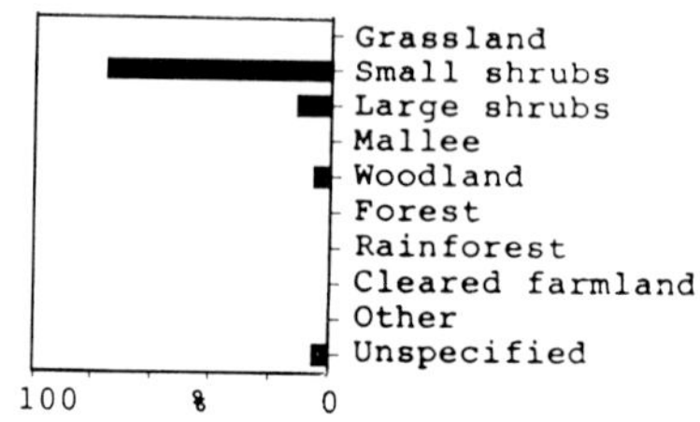

Soil Type

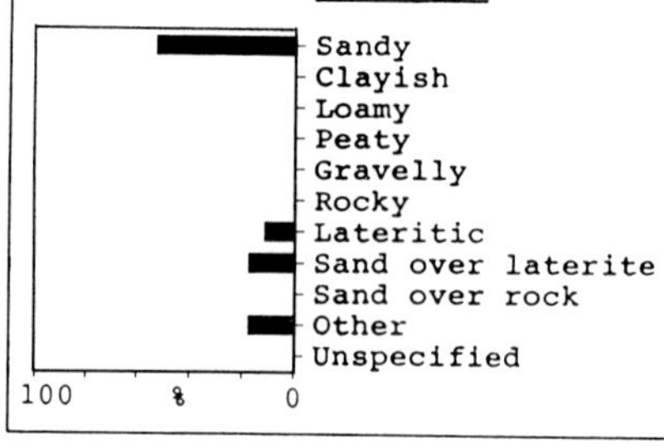

Landform

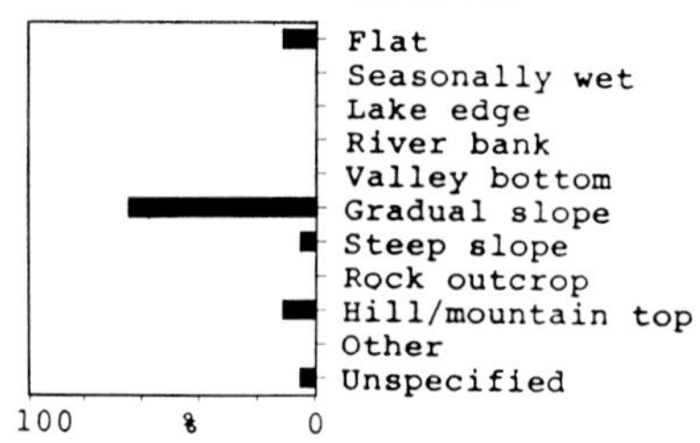

Response to Fire: 1 record

Growth response: Unspecified (100%).

Flowering response: Number of records where flowering had occurred after fire (1). Minimum reported time to flowering (25-36 months).
Number of records where flowering had not occurred after fire (0).

Flowering

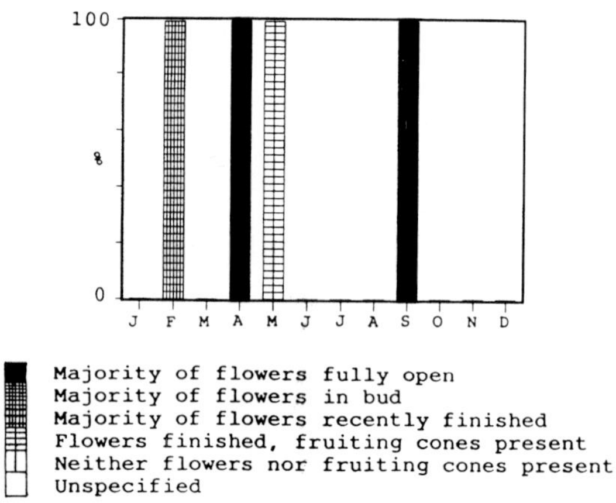

Possible pollinators

Pollinator type 1 record: no pollinator observed (100%).

Specific pollinator 0 records.

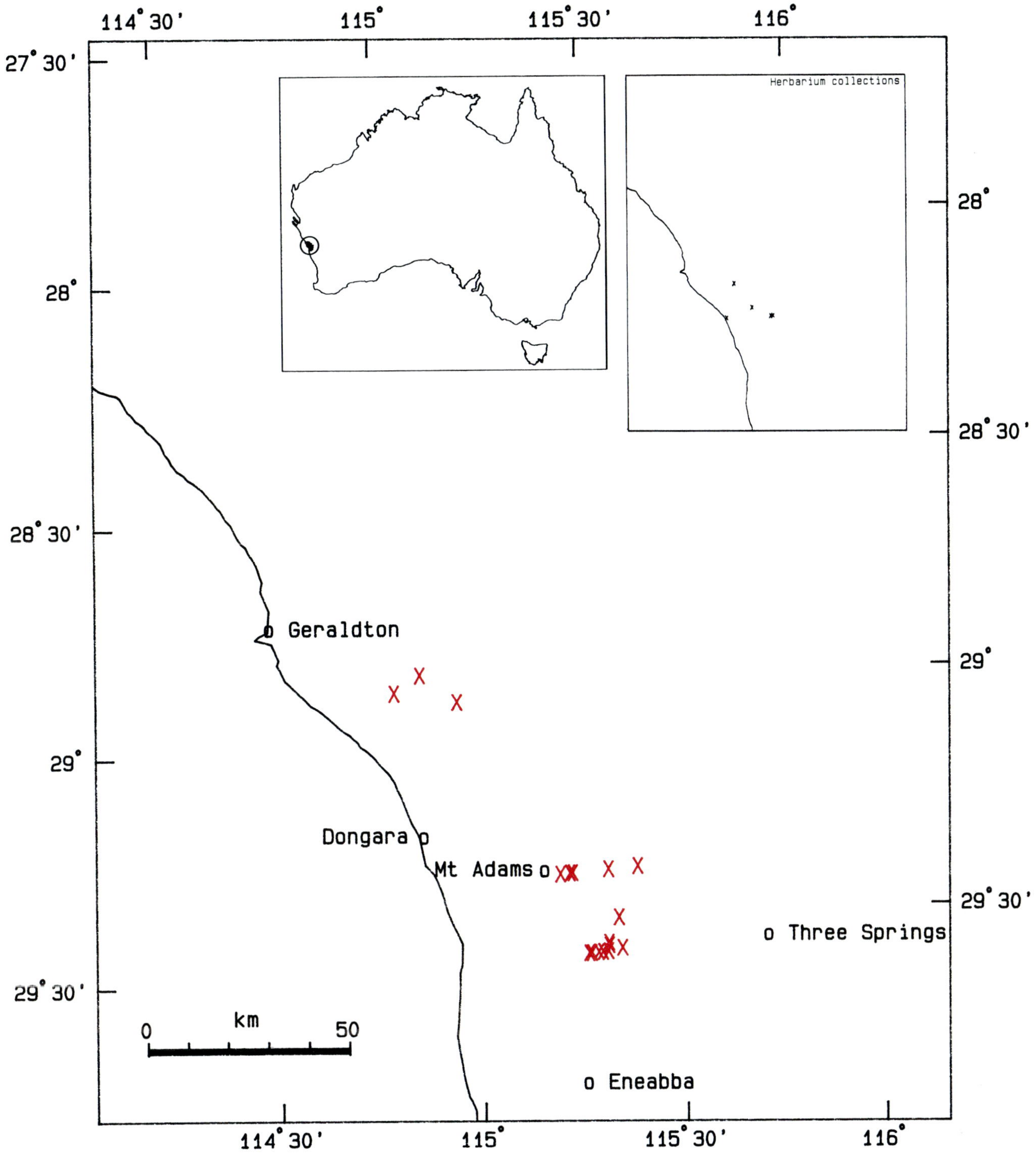

DISCUSSION

Atlas records have added significantly to the known distribution of the relatively rare B. scabrella. Prior to the Atlas, there were only five herbarium records of the species and all were located in a small area to the east and south-east of Walkaway (south of Geraldton). These sites were recorded by Atlas contributors who also recorded several populations in an area to the east and south-east of Mt Adams. Here, the species is locally abundant with most recorded populations being of more than 100 plants. However, almost 50% of this southern population is confined to road verges and as yet this population is totally unprotected as there are no conservation reserves in this area. The proposed nature reserve to the south of Mt Adams would include a population of B. scabrella. Out of a total of 17 records for this species only one occurs on a conservation reserve - that of the Burma Road Nature Reserve south-east of Walkaway. The species has been described as "rare" by Marchant and Keighery (1979) and as "vulnerable, not presently endangered but at risk in the longer term" by George (1984). Atlas records have shown the species to be more widespread than was previously thought. However, it is still considered to be at risk due to the majority of its population occurring only on road verges and to the lack of adequate protection for those that are not.

B. scabrella is typically a shrub up to 2 m in height growing in mixed heath often in association with B. leptophylla and the shrub form of B. attenuata. It favours sandy soils, sometimes overlying laterite and occurs on gentle slopes or flats. Flowering was recorded in September and also in April. However, George (1981) described the flowering period as September to January. Further data are needed for new shoot growth before reliable conclusions can be drawn.

BANKSIA SCEPTRUM Meissner (1855)

Sceptre Banksia

80 RECORDS: Jan (0) Feb (5) Mar (1) Apr (6) May (7) Jun (0) Jul (3) Aug (11) Sep (44) Oct (3) Nov (0) Dec (0).

Population Size: 1-10 (17) 10-100 (32) >100 (31) Unspecified (0).

Conservation Status: Restricted to road verge (10%) Not (84%) Unspecified (6%). In conservation reserve (45%) Not (51%) Unspecified (4%).

Tree/Shrub: Tree form (31%) Shrub form (69%) Unspecified (0%).

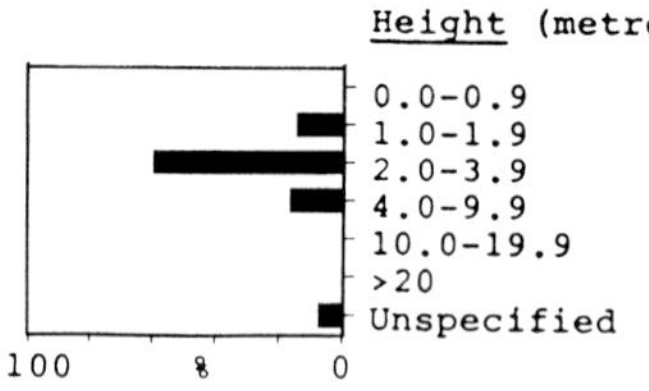

New Shoot Growth

HABITAT

Within 2 km of coast: Yes (1%) No (99%) Unspecified (0%).

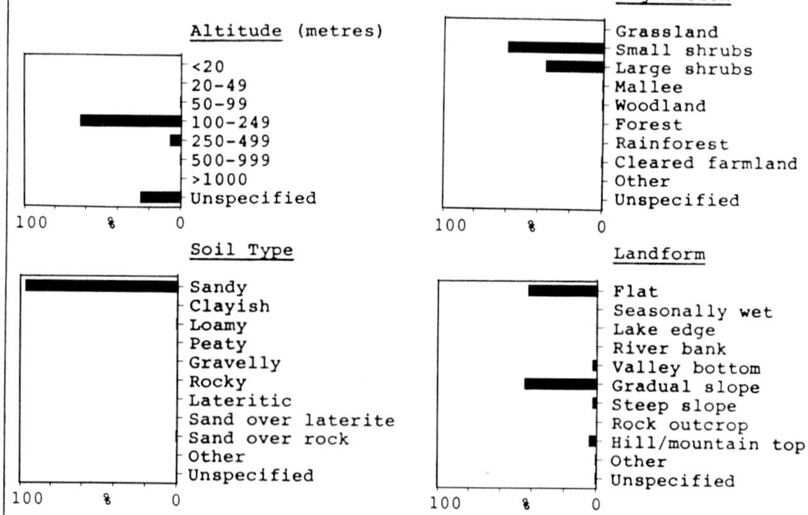

Response to Fire: 8 records

Growth response: Killed, new seedlings (100%).

Flowering response: Number of records where flowering had occurred after fire (4). Median reported time to flowering (36-37 months); minimum reported time to flowering (1-12 months). Number of records where flowering had not occurred after fire (3). Median reported time to no flowering (25-36 months); maximum reported time to no flowering (73-84 months).

Flowering

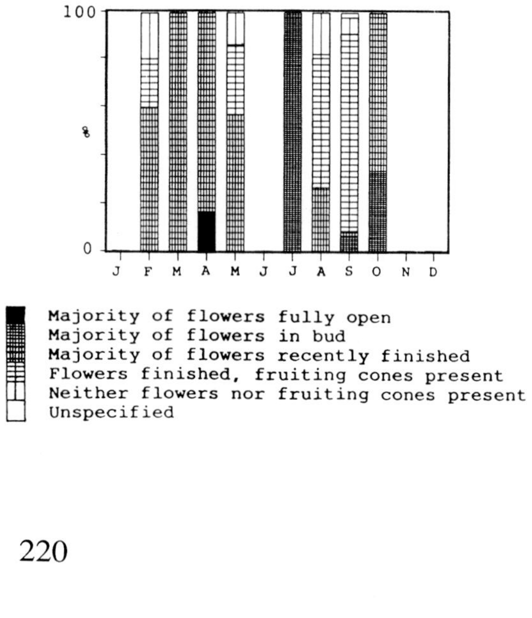

Possible pollinators

Pollinator type 7 records: bird (14%) bees, wasps, ants (29%) no pollinator observed (57%).

Specific pollinator 3 records: honeyeater (1) ant (2)

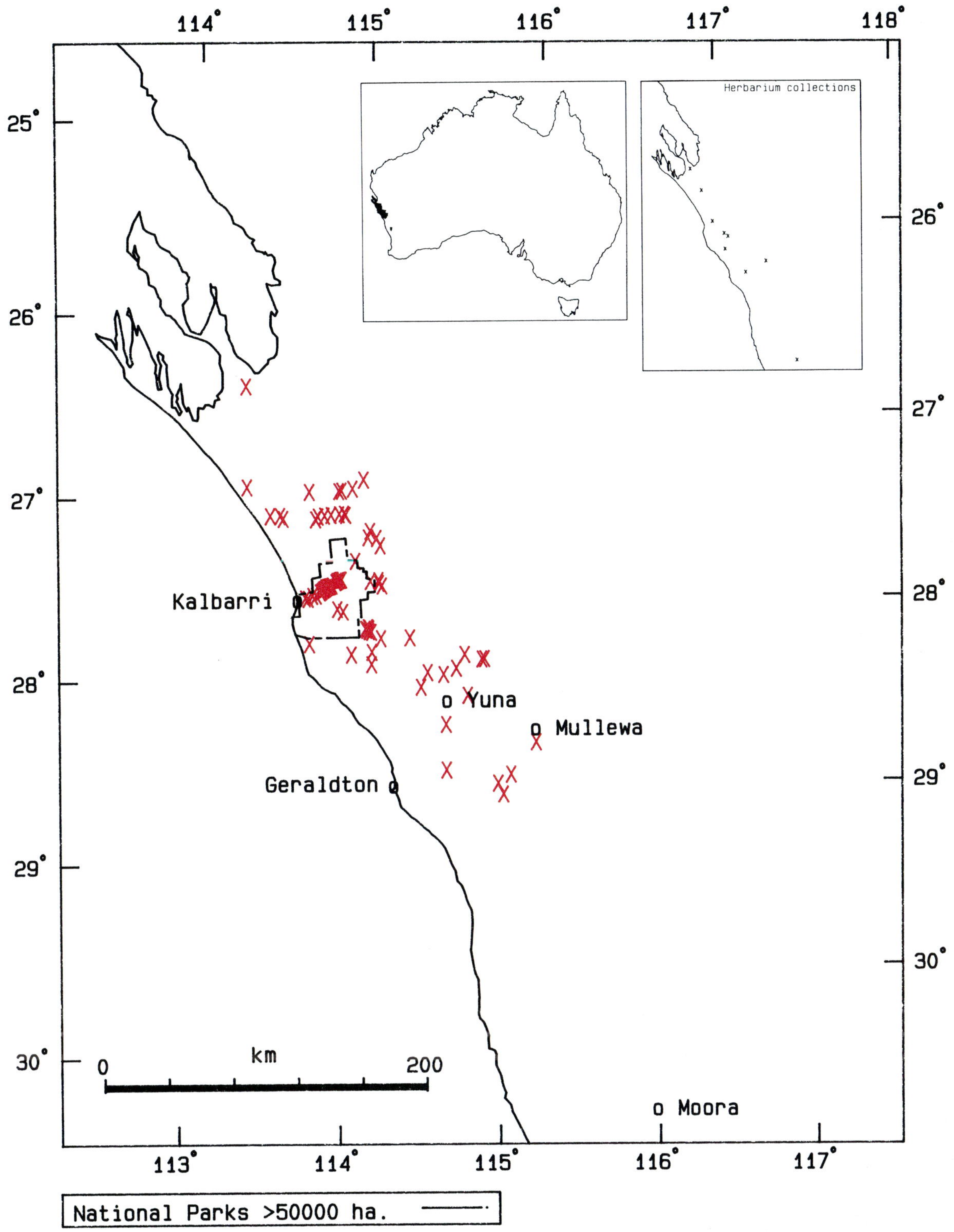

DISCUSSION

B. sceptrum occurs from Hamelin Pool south to an area about 60 km east of Geraldton. Atlas records have shown it to extend further inland than herbarium collections had indicated e.g. to Nerren Nerren Homestead and Wandana Nature Reserve. The sighting just south of Mullewa had been previously recorded. The species is locally common in the lower Murchison River area in Kalbarri National Park but elsewhere occurs in generally scattered populations. The large number of records from this national park would account for most of the 45% of recorded populations being in conservation reserves. The species may be under-recorded as the area in which it occurs was one of the least well covered in W.A. Also the species flowers in summer when temperatures are generally too hot for much field work to be undertaken. The lack of any records for November, December and January and the almost non-existent flowering data, bears witness to this fact. Fieldwork undertaken following the closing date for Atlas observations indicates that the species extends more or less continuously from Shark Bay south to Kalbarri National Park.

B. sceptrum occurs as a shrub or tree generally 2-4 m in height though sometimes larger. It grows in shrubland and deep sandy soils which are yellow or red in colour. Associated plants include Banksia ashbyi, Actinostrobus and mallee eucalypts. Landforms may be flat plains or gently undulating consolidated dunes. New shoot growth occurs in spring and also autumn. Due to lack of records it is not known whether this extends throughout summer. The species is killed by fire and regenerates from seed.

BANKSIA SEMINUDA (A.S.George) B.Rye (1984)

River Banksia

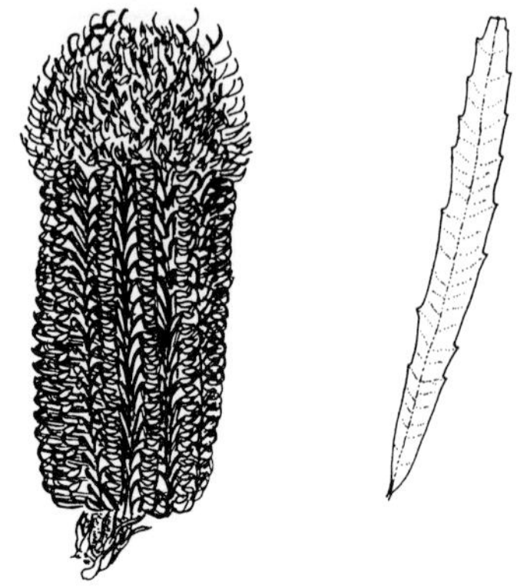

190 RECORDS: Jan (41) Feb (2) Mar (36) Apr (18) May (13) Jun (11) Jul (6) Aug (9) Sep (14) Oct (4) Nov (12) Dec (24).

Population Size: 1-10 (47) 10-100 (91) >100 (50) Unspecified (2).

Conservation Status: Restricted to road verge (4%) Not (88%) Unspecified (7%). In conservation reserve (12%) Not (77%) Unspecified (11%).

Tree/Shrub: Tree form (97%) Shrub form (2%) Unspecified (1%).

Height (metres)

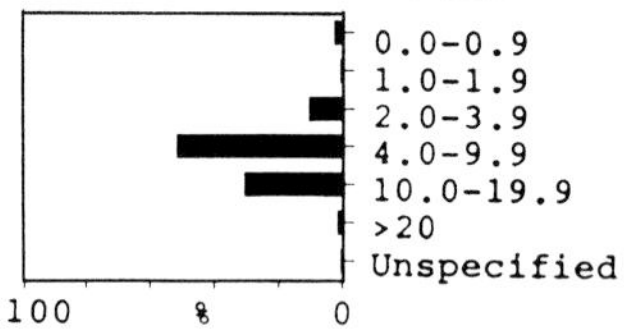

HABITAT

Within 2 km of coast: Yes (7%) No (93%) Unspecified (0%).

Altitude (metres)

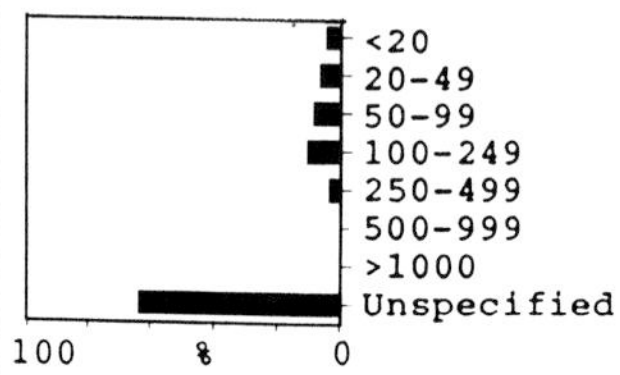

Vegetation

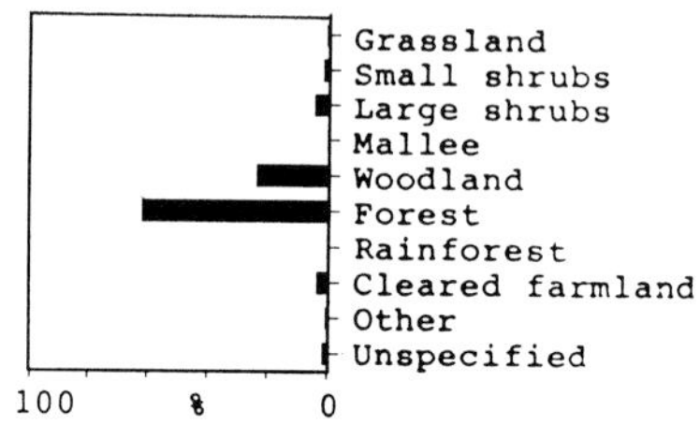

Soil Type

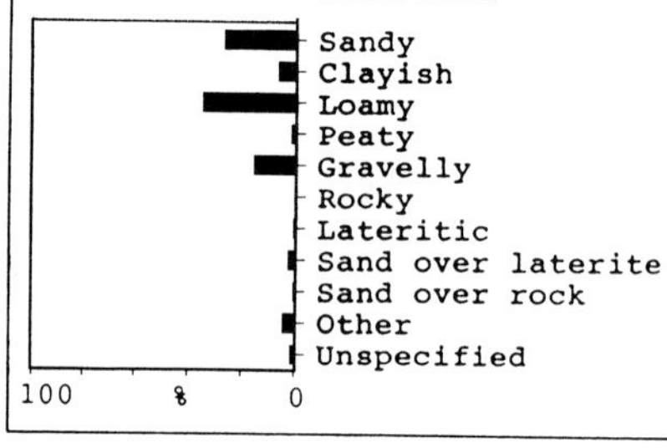

Landform

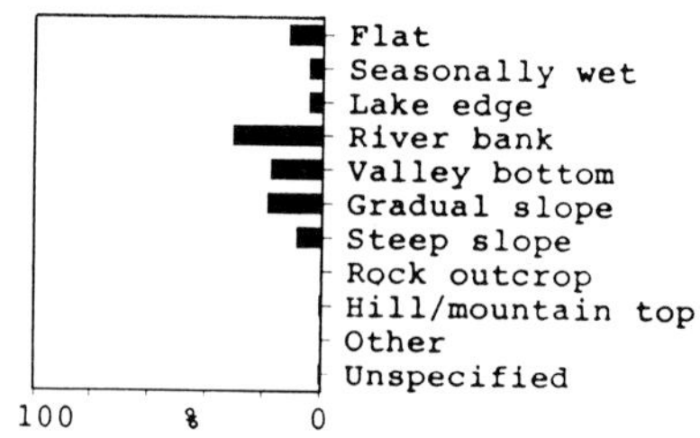

New Shoot Growth

Response to Fire: 22 records

Growth response: Ground resprout (5%) killed, new seedlings (14%) trunk resprout (14%) killed, no seedlings (14%) unspecified (55%).

Flowering response: Number of records where flowering had occurred after fire (14). Median reported time to flowering (24-25 months); minimum reported time to flowering (<1 month). Number of records where flowering had not occurred after fire (4). Median reported time to no flowering (13-24 months); maximum reported time to no flowering (61-72 months).

Flowering

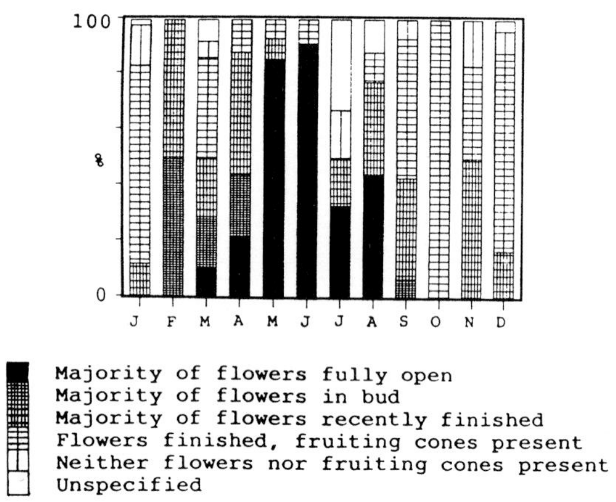

Possible pollinators

Pollinator type 63 records: bird (24%) flies (2%) no pollinator observed (75%).

Specific pollinator 7 records: honeyeater (2) New Holland Honeyeater (2) wattlebird (3)

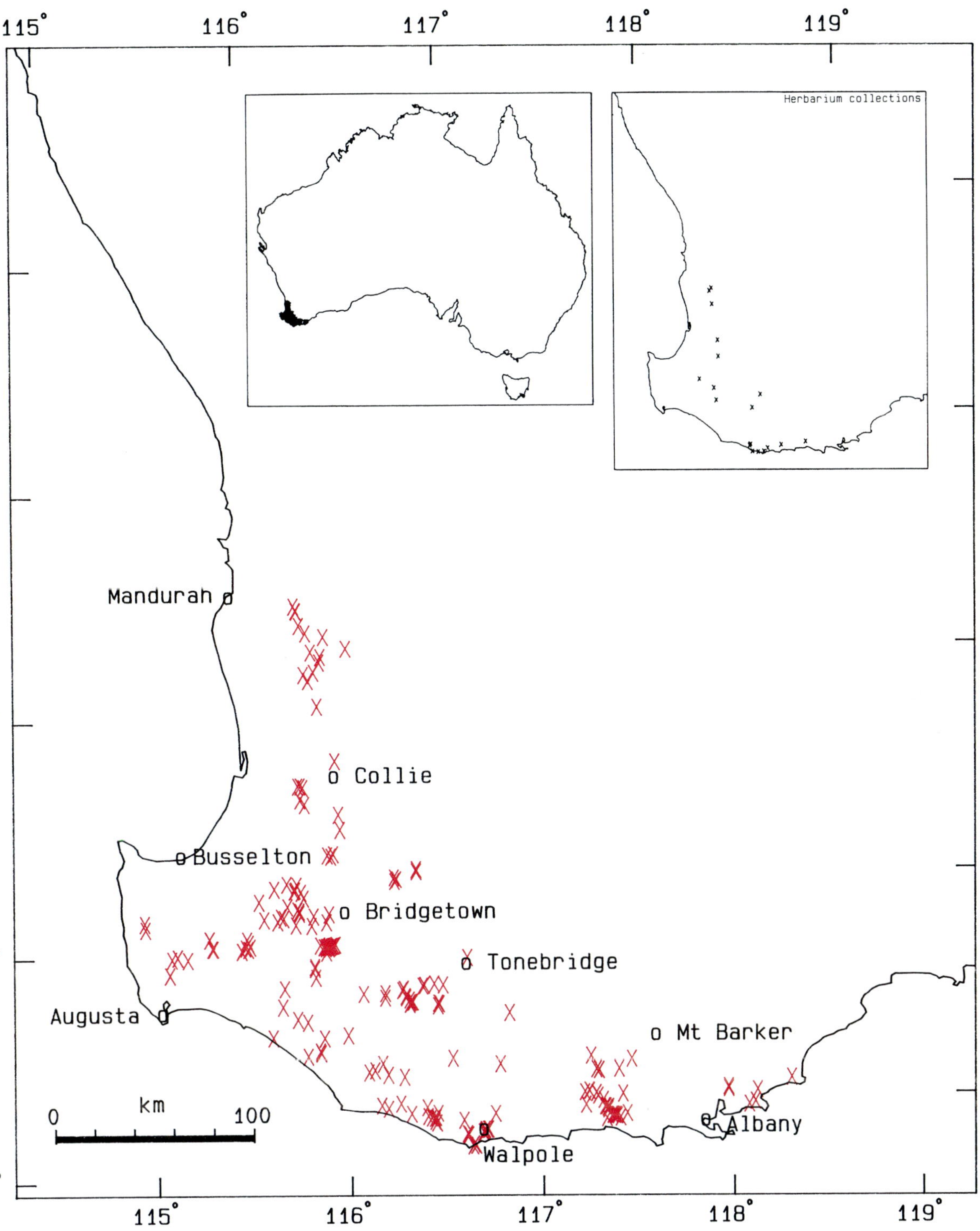

DISCUSSION

B. seminuda is typically a tall forest tree up to 20 m or more in height. It occurs along creeks and river banks in forests of jarrah (*Eucalyptus marginata*), or karri (*E. diversicolor*) from the Murray River (east of Mandurah) to the Blackwood River and along the south coast from the Donnelly River to Waychinicup River near Mt Manypeaks. The most inland records are from Boyup Brook and Tonebridge.

In the northern part of its distribution *B. seminuda* is always on the heavier loams and gravels of the Darling Plateau and never occurs on the sandy coastal plain. In the south, however, it grows in grey sandy soil often within a few kilometres of the coast. Here, the trees are lower being generally less than 8 m in height. They grow on sandy flats, low-lying swamps, around lake edges and on consolidated sand dunes as well as the more typical riverine habitat. Sometimes they grow alongside *B. littoralis* e.g. at Broke Inlet, Lake Maringup, and Two Peoples Bay Nature Reserve. Hopper (1987) recently named subsp. *remanens*, the short-leaved shrubby variant confined to deep coastal sands below granite outcrops in the Walpole-Nornalup National Park.

B. seminuda has a flowering period from March to August. Red style forms have been recorded from several scattered locations along the south coast (and from Nanga Brook, George, pers. comm.). New shoot growth is mainly summer. *B. seminuda* is generally killed by fire and regenerates from seed. However, if only lightly scorched it may resprout from epicormic buds.

BANKSIA SERRATA Linnaeus f. (1782)

Saw Banksia, Red Honeysuckle Banksia

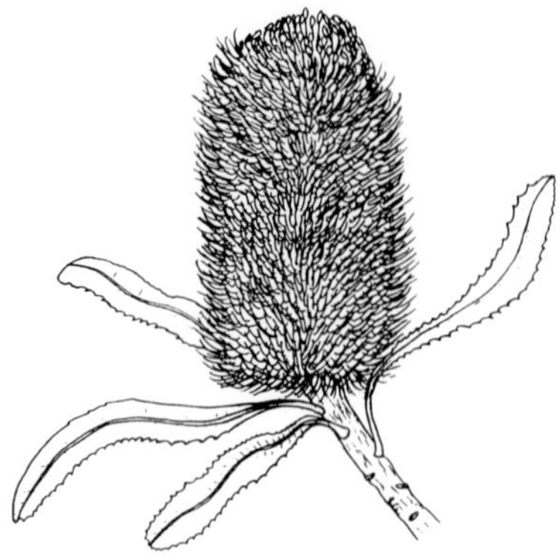

1472 RECORDS: Jan (148) Feb (74) Mar (120) Apr (112) May (334) Jun (322) Jul (69) Aug (66) Sep (45) Oct (31) Nov (58) Dec (88).

Population Size: 1-10 (171) 10-100 (399) >100 (880) Unspecified (22).

Conservation Status: Restricted to road verge (3%) Not (95%) Unspecified (2%). In conservation reserve (51%) Not (43%) Unspecified (6%).

Tree/Shrub: Tree form (82%) Shrub form (16%) Unspecified (2%).

Height (metres)

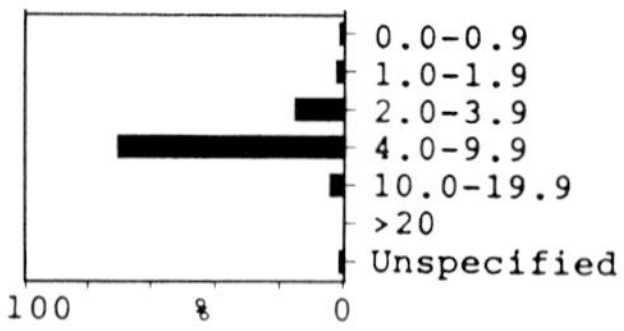

New Shoot Growth

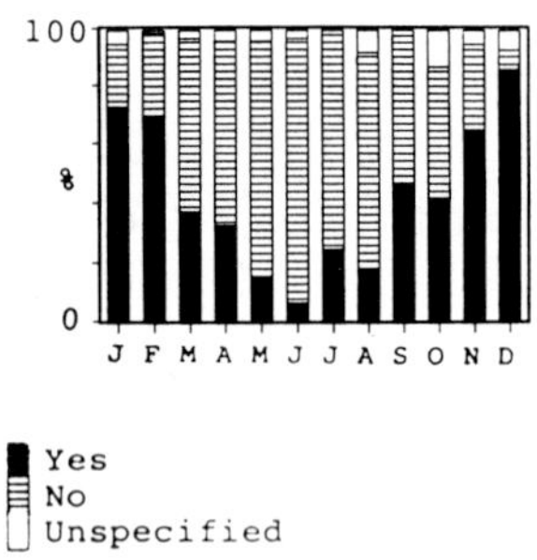

HABITAT

Within 2 km of coast: Yes (35%) No (64%) Unspecified (0%).

Altitude (metres)

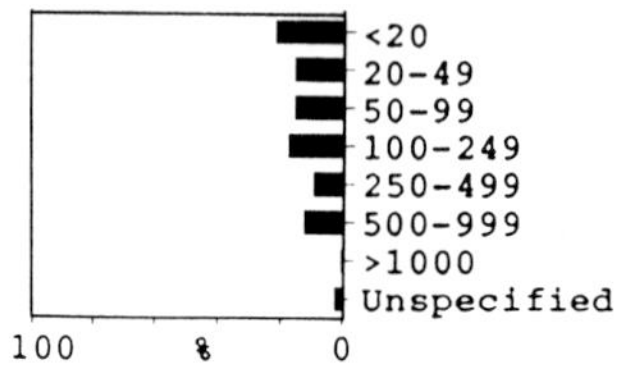

Vegetation

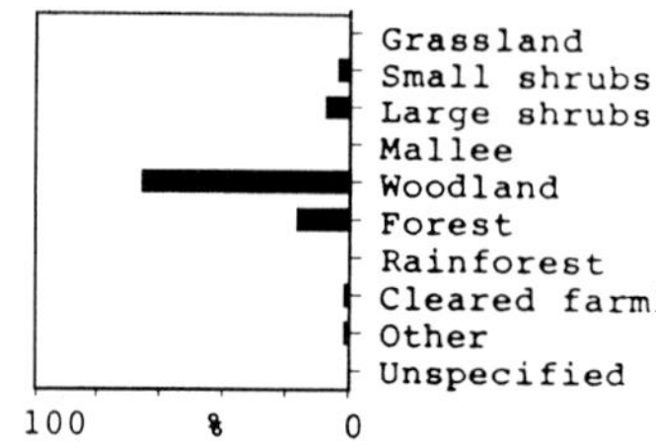

Soil Type

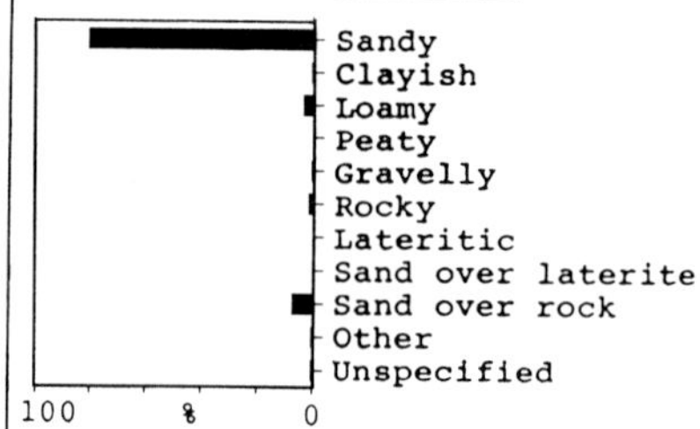

Landform

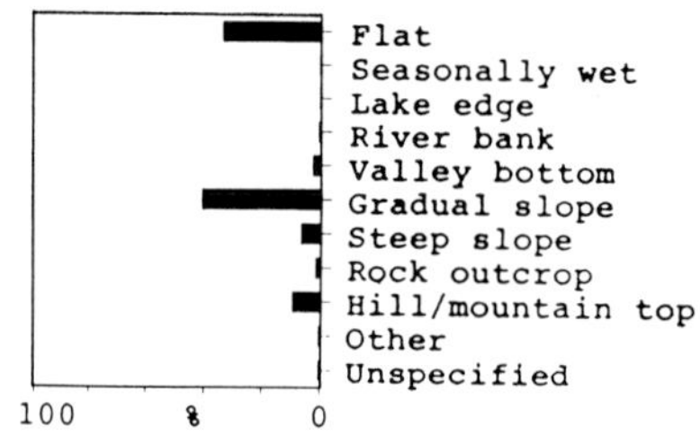

Response to Fire: 285 records

Growth response: Ground resprout (2%) ground resprout and killed, new seedlings (1%) ground resprout and trunk resprout (5%) killed, new seedlings (1%) killed, new seedlings and trunk resprout (3%) trunk resprout (60%) unspecified (28%).

Flowering response: Number of records where flowering had occurred after fire (215). Median reported time to flowering (13-24 months); minimum reported time to flowering (<1 month). Number of records where flowering had not occurred after fire (21). Median reported time to no flowering (25-36 months); maximum reported time to no flowering (85-96 months).

Flowering

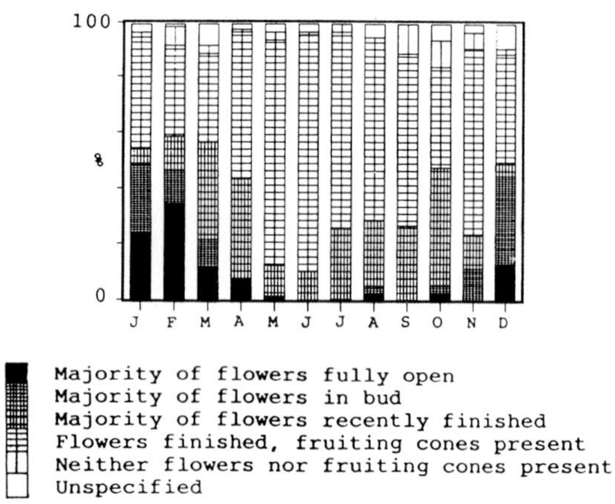

Possible pollinators

Pollinator type 1026 records: bird (5%) bees, wasps, ants (2%) mammals (0.2%) no pollinator observed (93%).

Specific pollinator 100 records: honeyeater (10) New Holland Honeyeater (13) Brown Honeyeater (1) White-cheeked Honeyeater (2) White-eared Honeyeater (1) Tawny-crowned Honeyeater (1) White-naped Honeyeater (2) White-plumed Honeyeater (1) Crescent Honeyeater (1) Noisy Miner (3) Yellow-tufted Honeyeater (1) wattlebird (1) Red Wattlebird (4) Little Wattlebird (13) Eastern Spinebill (11) cockatoos and parrots (1) Yellow-tailed Black-Cockatoo (1) friarbird (1) Noisy Friarbird (4) Drongo (1) Rainbow Lorikeet (2) flycatchers and allies (1) ant (8) bee (11) native bee (1) European honey bee (2)

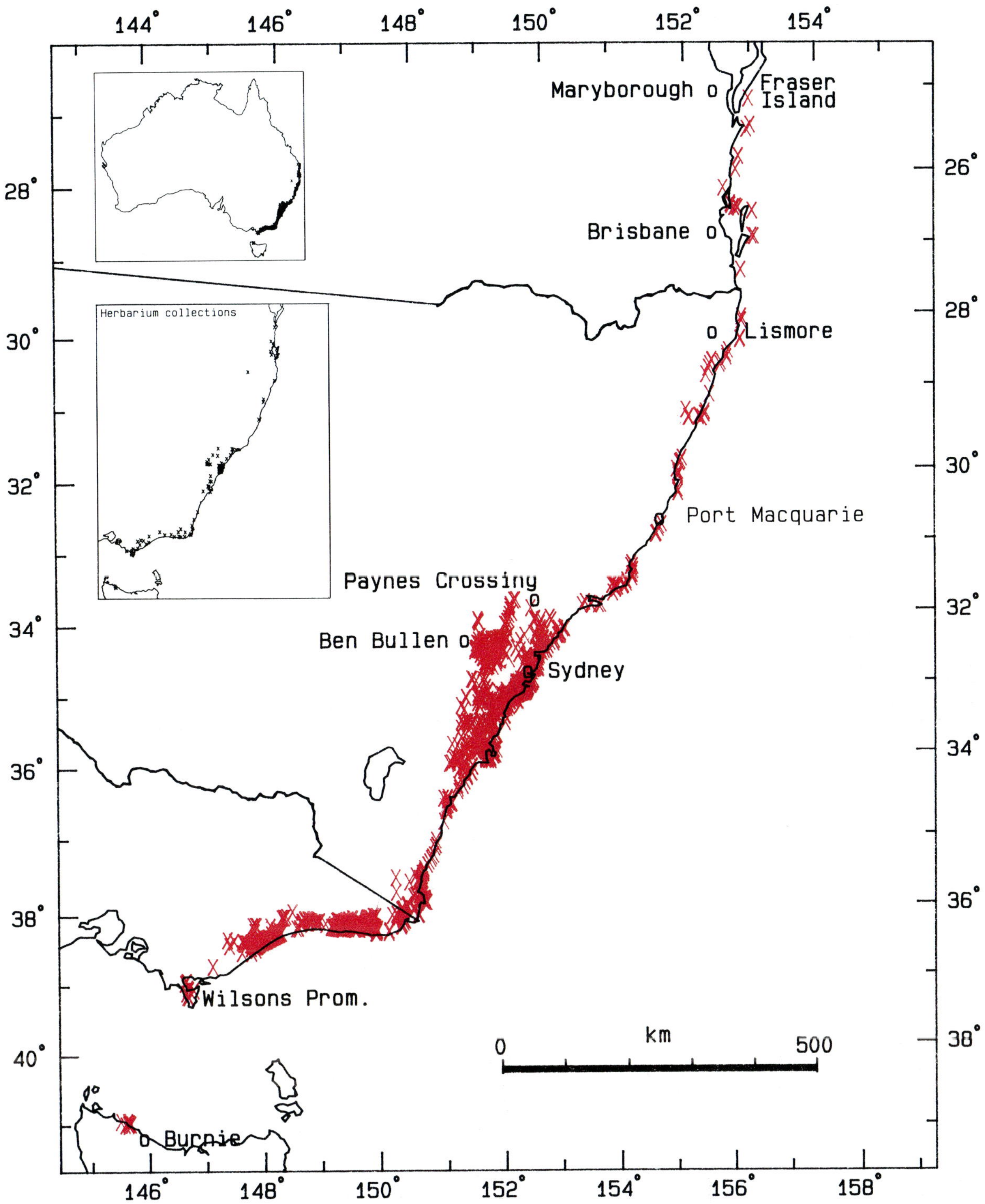

DISCUSSION

Confirmed Atlas records show B. serrata to have a fairly continuous distribution between Wilsons Promontory, (Victoria) and Cooloola (Queensland), mainly on the coastal plain but also extending into the Blue Mountains (New South Wales) where it is found at altitudes in excess of 500 m. There is also a single Atlas record from Fraser Island, Queensland, where, in a few localities, both B. serrata and B. aemula occur together. The species is also recorded at Sisters Creek in north-western Tasmania. An old herbarium collection from near Stanthorpe (Queensland - New South Wales border) was not recorded by Atlas contributors. It seems remarkable that such a significant outlying population has not been re-recorded in more recent years. The exact locality of this record may be in doubt. However, it should be searched for again in the field.

B. serrata usually occurs as a tree, 4-10 m in height, though sometimes it may be a shrub of less than 4 m. On very exposed cliff tops it has been recorded as a prostrate shrub of only 0.5 m in height (e.g. at Point Perpendicular Lighthouse, Beecroft Peninsula and at Green Cape Lighthouse). It generally grows in eucalypt forest or woodland, but sometimes in tall shrubland. Deep sandy soils are preferred though it also occurs in shallower sand over rock (e.g. in the Blue Mountains). The main flowering period is from December to April, though flowers were also recorded in August and October. The large conspicuous flowers are attractive to honey-eating birds with the New Holland Honeyeater, Eastern Spinebill and Little Wattlebird being the most frequently recorded. New shoot growth is mainly in summer. Following fire the species resprouts from both its epicormic buds and lignotuber. Depending on fire intensity, tall trees may be flowering within 1-2 years of being burnt, though smaller shrubs probably require a longer period. A population of B. serrata displaying uniformly large inflorescences and fruiting cones was recorded from near Scotts Head, south of Nambucca Heads (N.S.W.). The inflorescences measure up to 26 cm in length and 9 cm in width. This is 75%-100% longer than usual. Specimens have been sent to the National Herbarium, Sydney (M. Hitchcock).

BANKSIA SOLANDRI R.Brown (1830)

Stirling Range Banksia, Solander's Banksia

19 RECORDS: Jan (2) Feb (2) Mar (0) Apr (0) May (4) Jun (2) Jul (0) Aug (3) Sep (2) Oct (2) Nov (1) Dec (1).

Population Size: 1-10 (6) 10-100 (7) >100 (5) Unspecified (1).

Conservation Status: Restricted to road verge (0%) Not (100%) Unspecified (0%). In conservation reserve (100%) Not (0%) Unspecified (0%).

Tree/Shrub: Tree form (16%) Shrub form (84%) Unspecified (0%).

Height (metres)

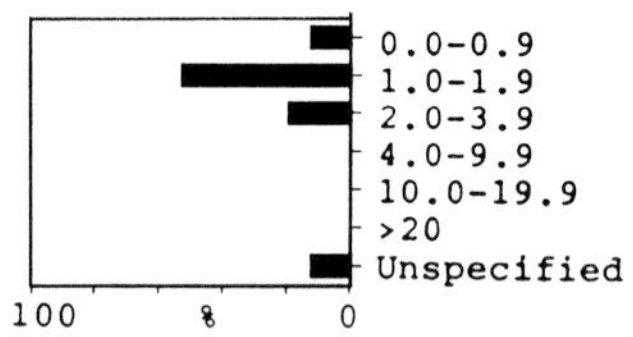

New Shoot Growth

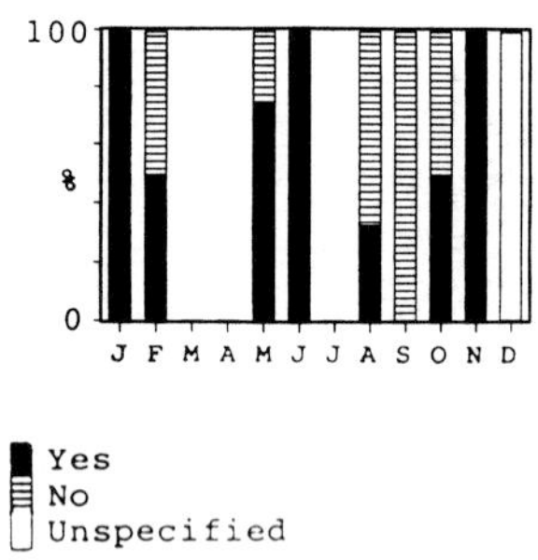

HABITAT

Within 2 km of coast: Yes (0%) No (100%) Unspecified (0%).

Altitude (metres)

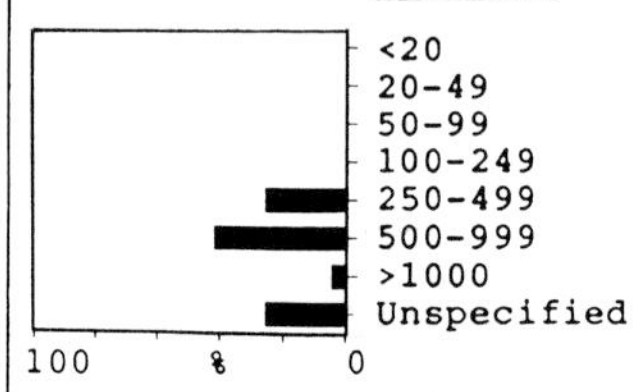

Vegetation

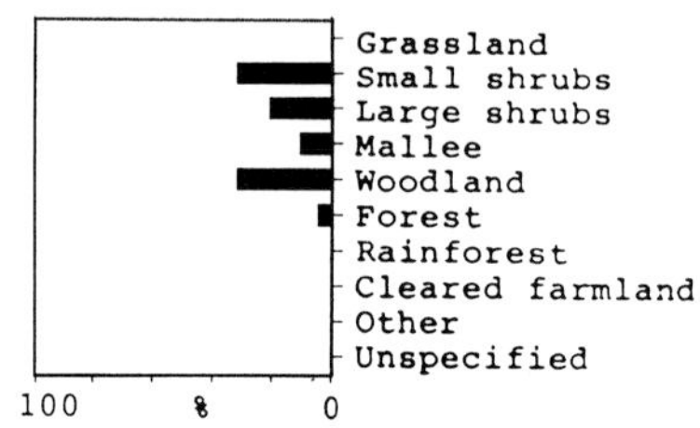

Soil Type

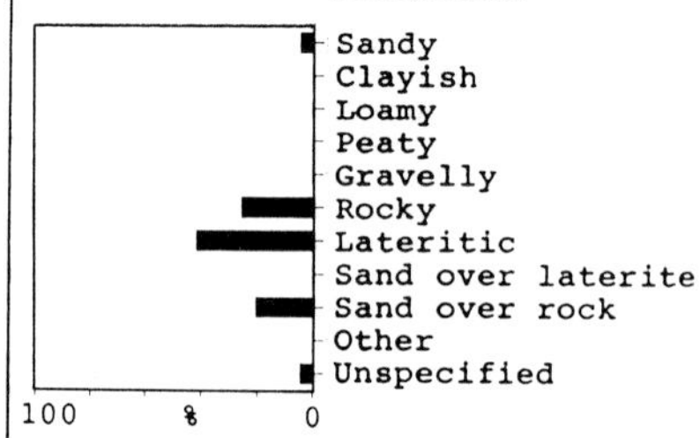

Landform

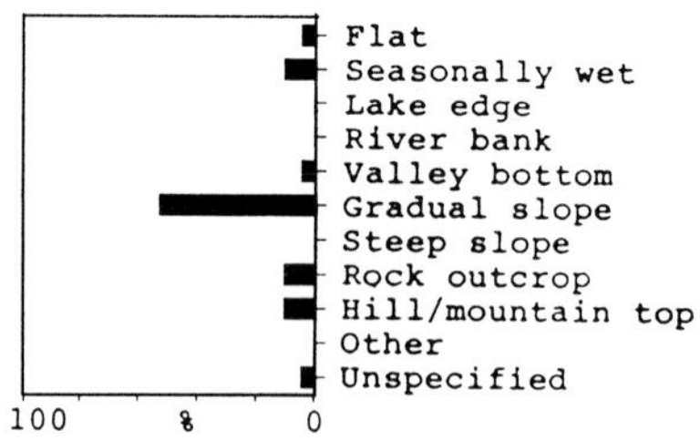

Response to Fire: 3 records

Growth response: Killed, new seedlings (100%).

Flowering response: Number of records where flowering had occurred after fire (1). Minimum reported time to flowering (13-24 months).
Number of records where flowering had not occurred after fire (2). Median reported time to no flowering (25-36 months); maximum reported time to no flowering (25-36 months).

Flowering

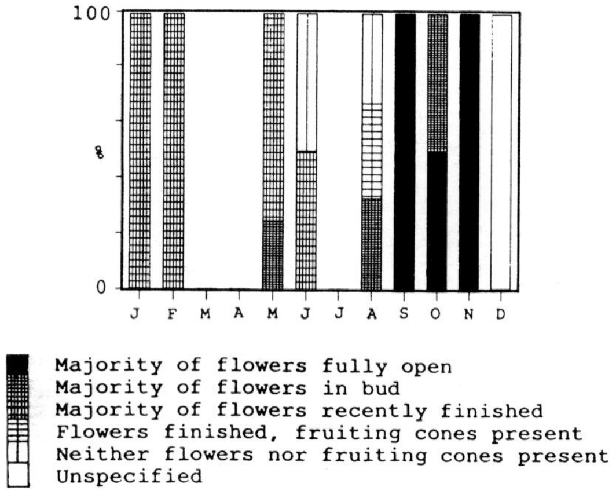

Possible pollinators

Pollinator type 8 records: no pollinator observed (100%).

Specific pollinator 0 records:

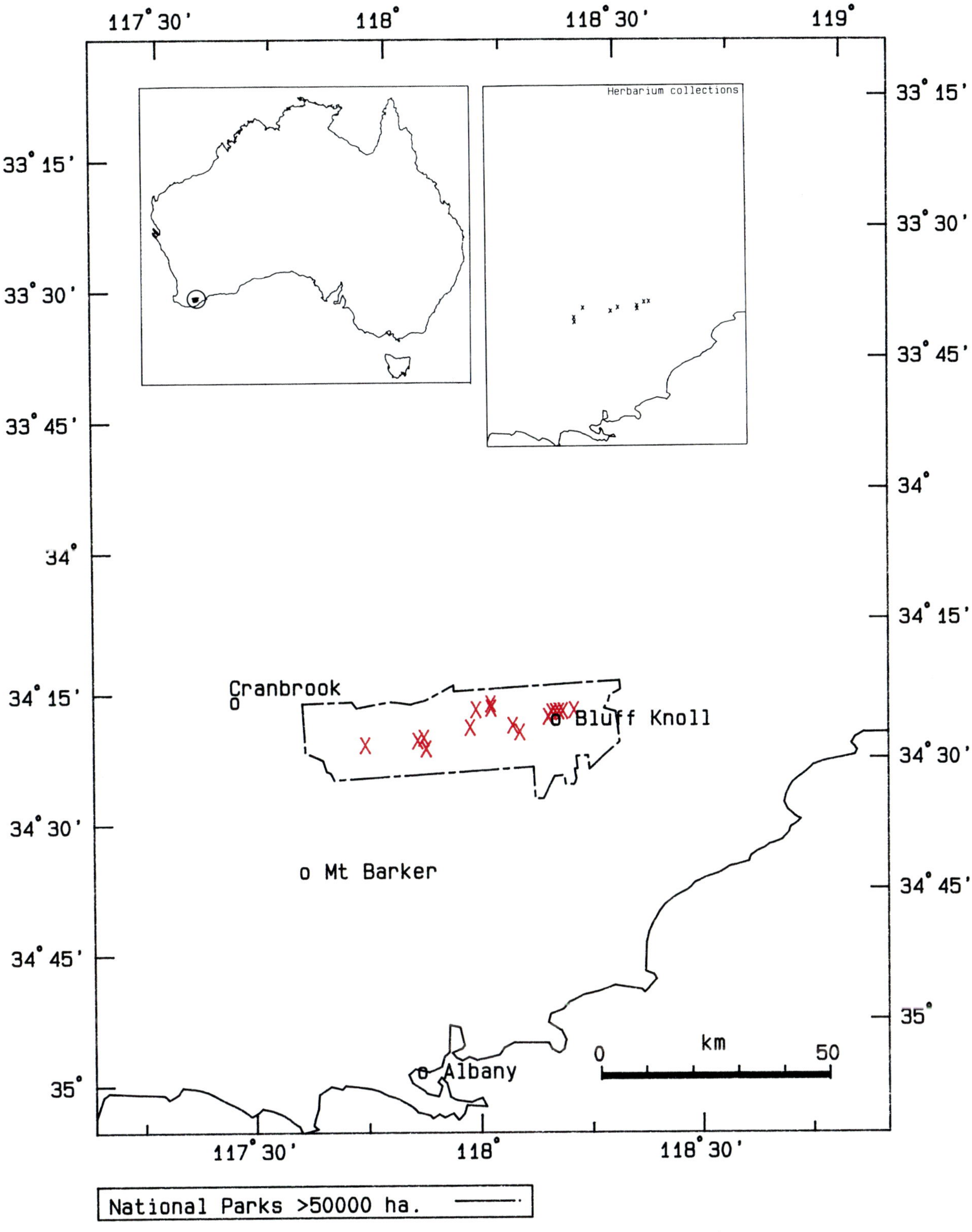

DISCUSSION

Atlas records have confirmed that B. solandri is restricted to the Stirling Range where it occurs on several of the mountain peaks and their steep slopes at altitudes up to and exceeding 1 000 m. New records not previously represented in W.A. Herbarium collections were from Yungermere Peak, Mt Trio and three separate locations along Stirling Range Drive. There was also a record from near the junction of Yungermere Crescent and Mt James Track. B. solandri is totally protected, with all its populations occurring with the Stirling Range National Park. However, its populations are generally small (less than 100 plants) and it is suceptible to dieback disease. Since it often occurs alongside popular walking tracks where dieback can be easily spread, it is particularly at risk. Populations on the summits of Bluff Knoll and Toolbrunup are already dying out. Preservation of the species will depend on management techniques aimed at reducing the spread of this disease. B. solandri is killed by fire and regenerates from seed. Successive fires at intervals too frequent to allow sufficient flowering and seeding could also be a threat.

B. solandri occurs as a large shrub or small tree up to 4 m in height. It grows predominantly in rocky soil, sometimes with a surface layer of sand. Surrounding vegetation is generally shrubland, often including mallee eucalypts. Along Stirling Range Drive, the species occurs on flat or gently sloping land amongst eucalypt woodland. Flowering is from September to November. Further data are needed for pollinators and new shoot growth before reliable conclusions can be drawn.

BANKSIA SPECIOSA R.Brown (1810)

Showy Banksia

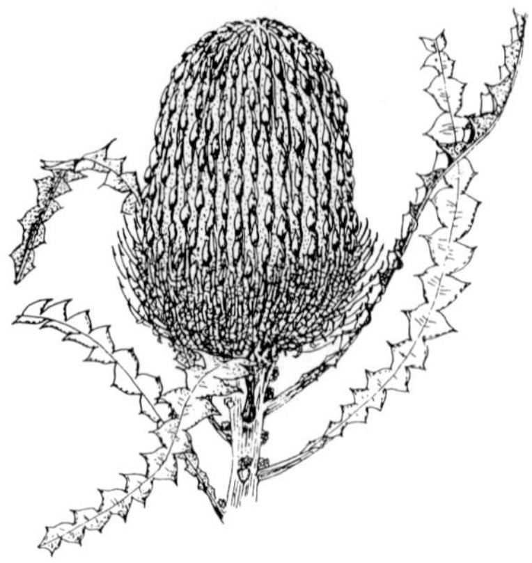

216 RECORDS: Jan (15) Feb (36) Mar (18) Apr (16) May (14) Jun (9) Jul (7) Aug (21) Sep (22) Oct (20) Nov (25) Dec (11).

Population Size: 1-10 (33) 10-100 (58) >100 (122) Unspecified (3).

Conservation Status: Restricted to road verge (29%) Not (69%) Unspecified (2%). In conservation reserve (27%) Not (63%) Unspecified (10%).

Tree/Shrub: Tree form (41%) Shrub form (55%) Unspecified (4%).

Height (metres)

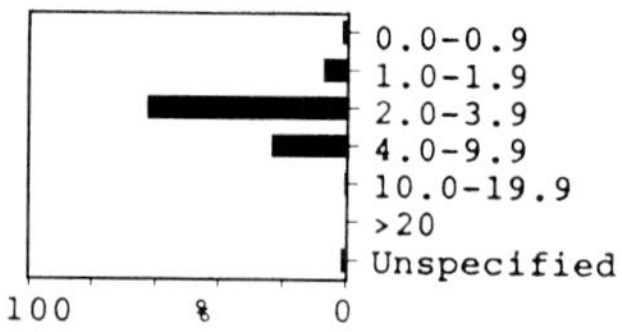

New Shoot Growth

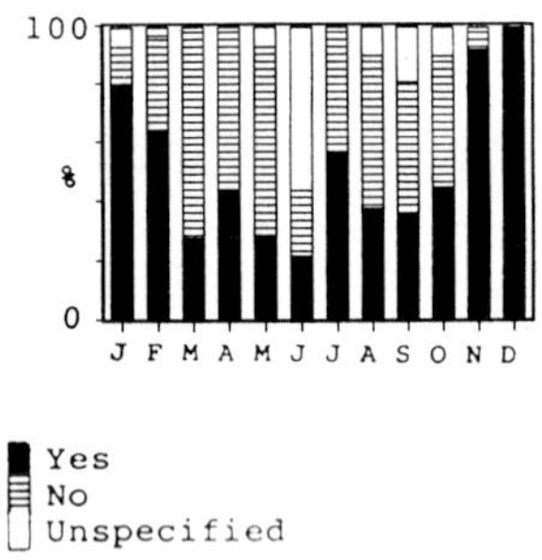

HABITAT

Within 2 km of coast: Yes (29%) No (71%) Unspecified (0%).

Altitude (metres)

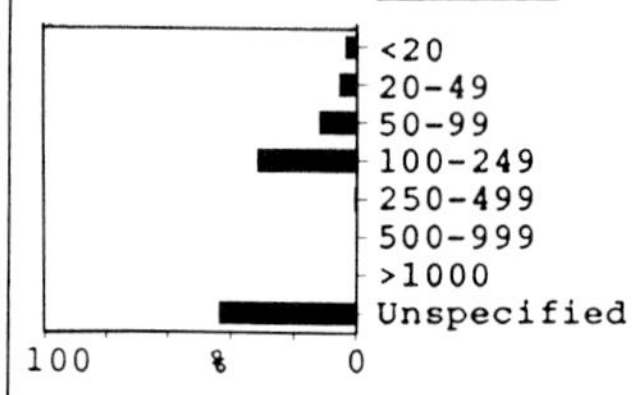

Vegetation

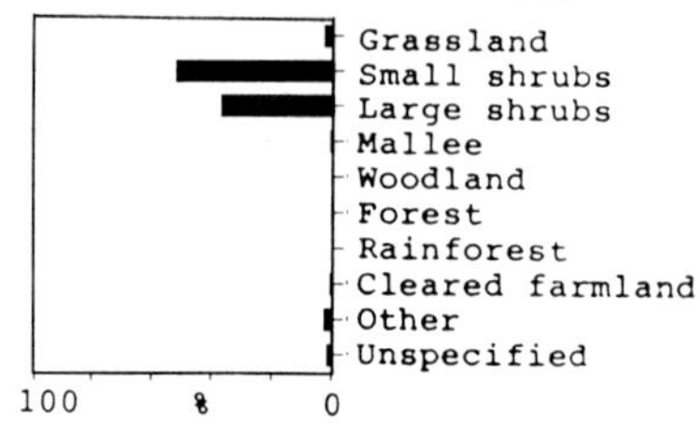

Soil Type

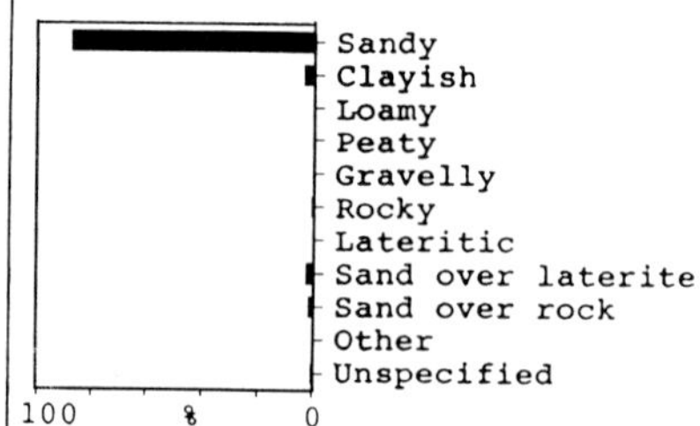

Landform

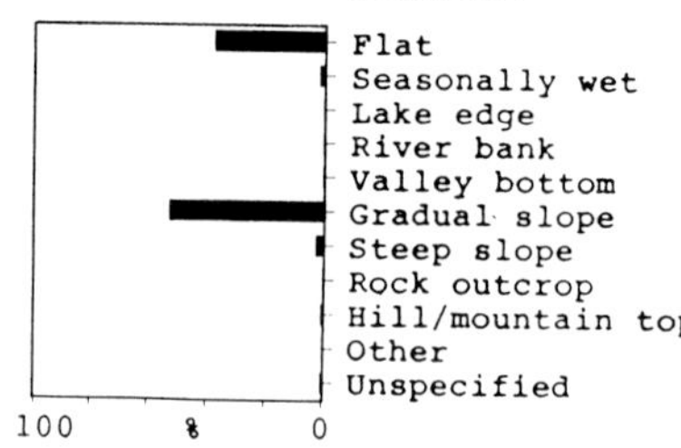

Response to Fire: 24 records

Growth response: Killed, new seedlings (71%) unspecified (29%).

Flowering response: Number of records where flowering had occurred after fire (7). Median reported time to flowering (25-36 months); minimum reported time to flowering (1-12 months). Number of records where flowering had not occurred after fire (12). Median reported time to no flowering (13-24 months); maximum reported time to no flowering (25-36 months).

Flowering

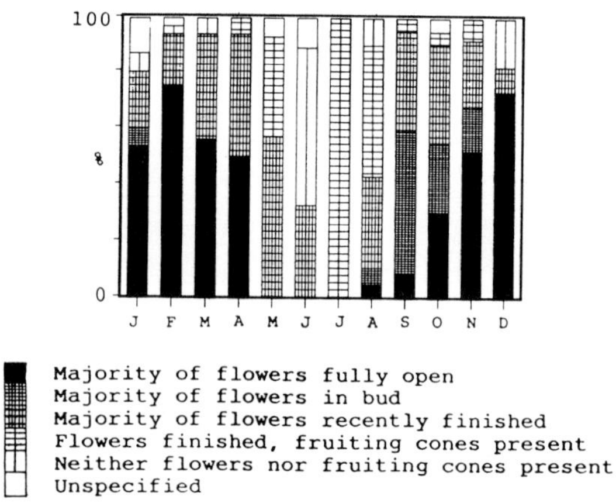

Possible pollinators

Pollinator type 143 records: bird (24%) butterflies, moths (1%) bees, wasps, ants (22%) flies (8%) beetles (4%) no pollinator observed (41%).

Specific pollinator 76 records: honeyeater (11) New Holland Honeyeater (11) Fuscous Honeyeater (4) wattlebird (1) Little Wattlebird (8) Western Spinebill (2) Grey Butcherbird (1) flycatchers and allies (1) thornbill (1) ant (22) bee (11) blowfly (1) aphid (2)

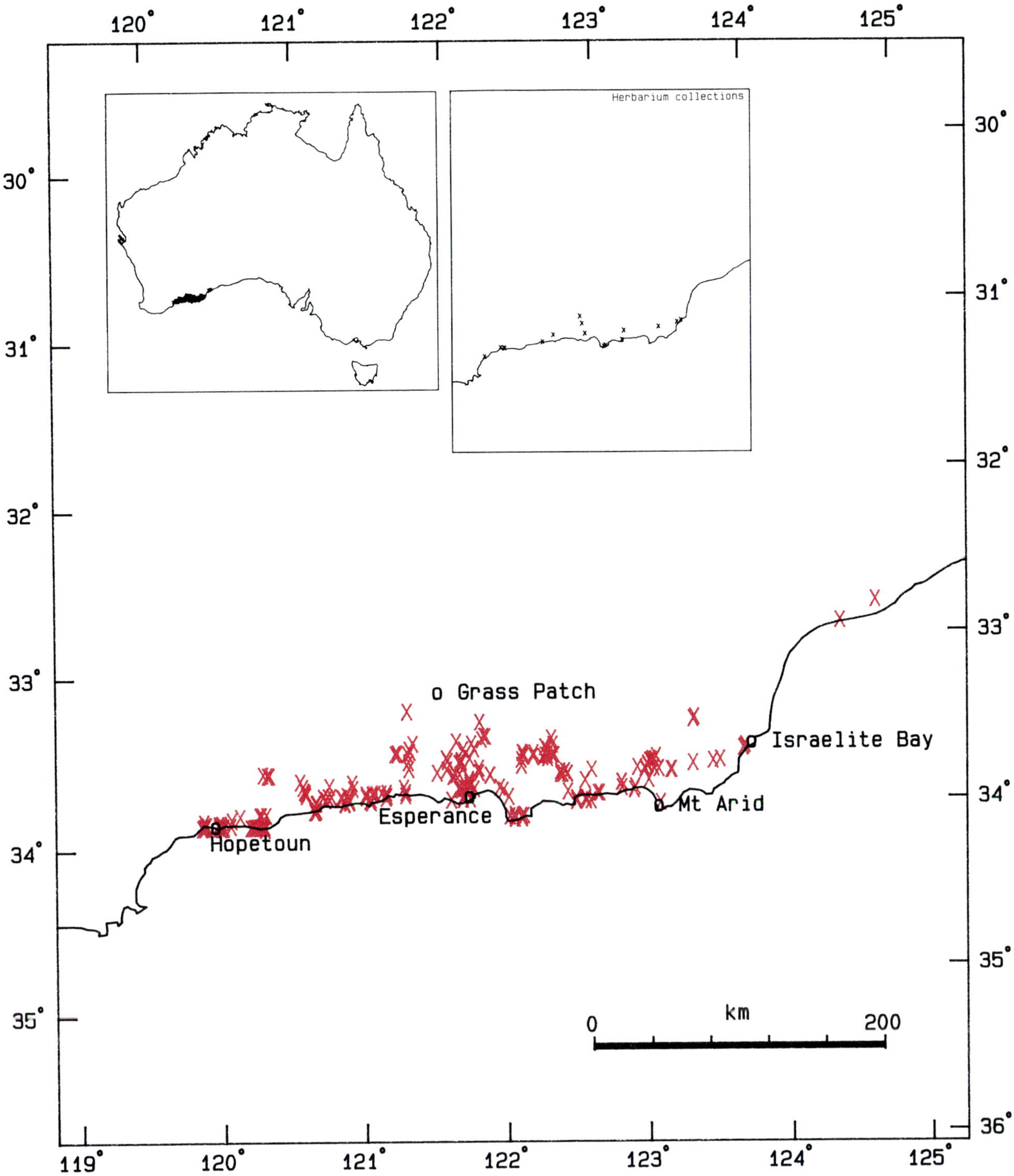

DISCUSSION

Banksia speciosa was recorded between East Mt Barren (Fitzgerald River National Park) and Israelite Bay with outliers near Point Culver in the Great Australian Bight. The latter records were new sightings for this species extending its range by some 150 km eastwards. A 1936 collection by C. Gardner from Middle Mt Barren was not recorded by Atlas contributors. Inland, the species extends to Mt Ragged and approximately 25 km south-west of Grass Patch. Currently, B. speciosa appears relatively widespread and common with most of its populations in excess of 100 plants. It is, however, one of the most susceptible of banksias to dieback disease and in areas such as Cape Le Grand and Cape Arid National Parks populations are already being decimated (G. Keighery, pers. comm.; see also colour plates). It is likely that this species will become increasingly rare in future years.

B. speciosa occurs as a large shrub or small tree generally less than 4 m in height, though sometimes larger. Deep white or grey sandy soils are strongly preferred and surrounding vegetation is usually shrubland. When this is low shrubland, B. speciosa is usually the dominant emergent plant. Other banksia species such as B. petiolaris and B. pulchella are often found in the same locality. The main flowering period is from October to April when the large conspicuous flowers attract many honey-eating birds and insects. The most commonly recorded birds were the New Holland Honeyeater and Little Wattlebird. Summer is the main season for new shoot growth. The species is killed by fire and regenerates from seed. From the Gibson area a large number of strongly upright-growing plants with upright leaves was reported to occur within a population of more typical form. Collected seed is being germinated to see whether the resultant seedlings are similar to their parent.

BANKSIA SPHAEROCARPA R.Brown (1810)

Fox Banksia, Round-fruited Banksia

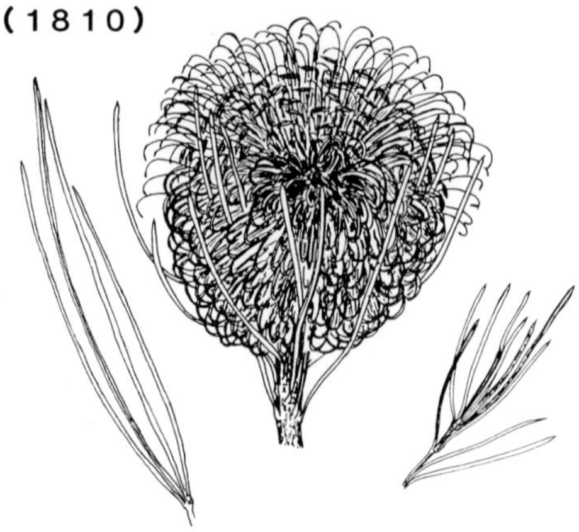

433 RECORDS: Jan (23) Feb (22) Mar (37) Apr (50) May (26) Jun (44) Jul (43) Aug (52) Sep (66) Oct (34) Nov (19) Dec (15).

Population Size: 1-10 (121) 10-100 (206) >100 (103) Unspecified (3).

Conservation Status: Restricted to road verge (33%) Not (65%) Unspecified (2%). In conservation reserve (35%) Not (61%) Unspecified (4%).

Tree/Shrub: Tree form (13%) Shrub form (87%) Unspecified (0%).

Height (metres)

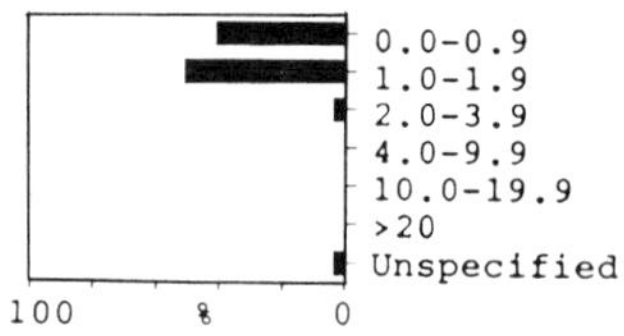

New Shoot Growth

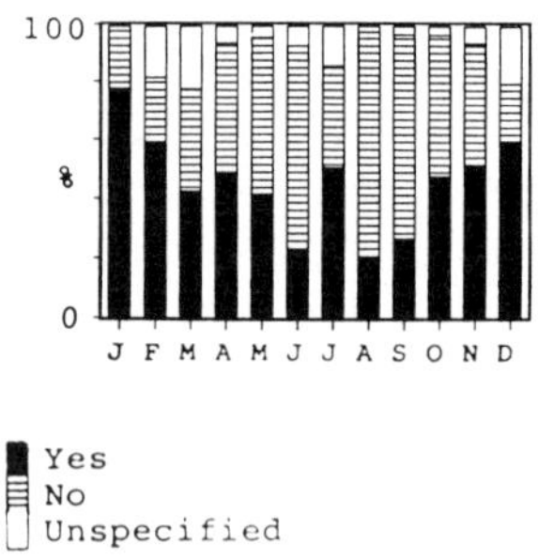

HABITAT

Within 2 km of coast: Yes (2%) No (98%) Unspecified (0%).

Altitude (metres)

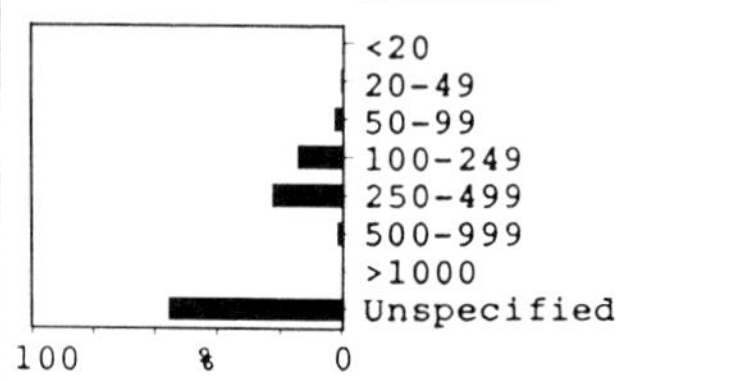

Vegetation

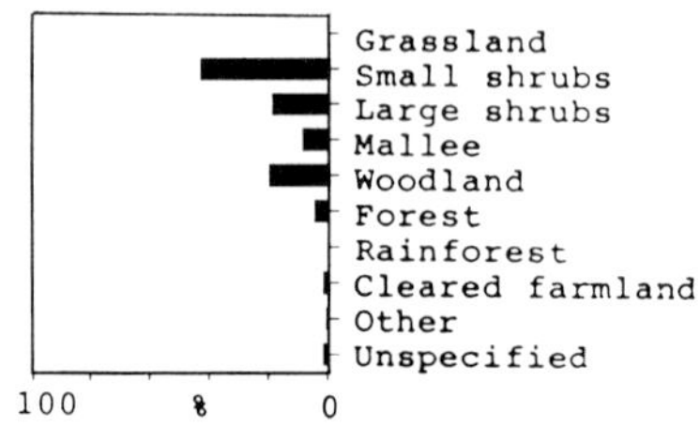

Soil Type

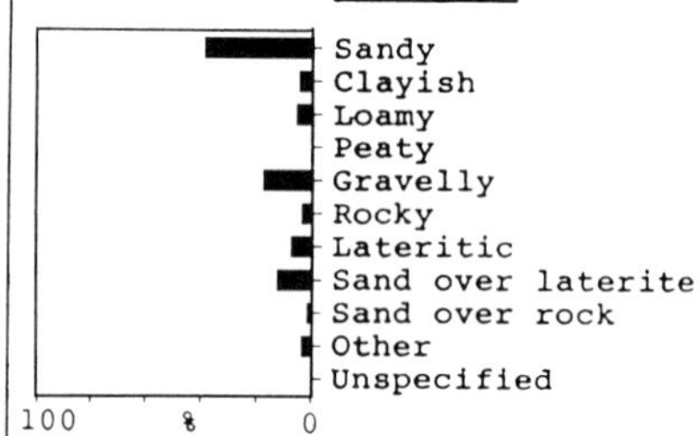

Landform

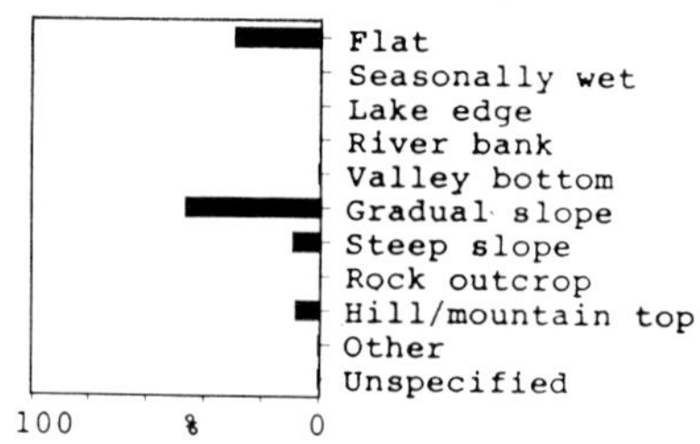

Response to Fire: 59 records

Growth response: Ground resprout (36%) ground resprout and killed, new seedlings (5%) ground resprout and trunk resprout (3%) killed, new seedlings (10%) trunk resprout (3%) unspecified (42%).

Flowering response: Number of records where flowering had occurred after fire (45). Median reported time to flowering (13-24 months); minimum reported time to flowering (<1 month).
Number of records where flowering had not occurred after fire (9). Median reported time to no flowering (1-12 months); maximum reported time to no flowering (25-36 months).

Flowering

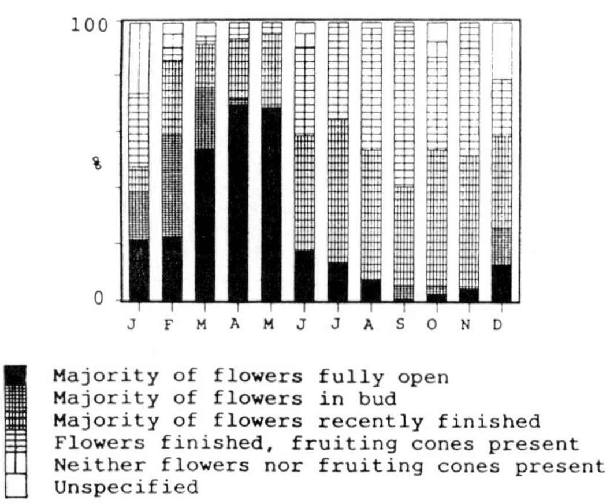

Possible pollinators

Pollinator type 166 records: bird (4%) bees, wasps, ants (19%) flies (3%) beetles (1%) no pollinator observed (73%).

Specific pollinator 34 records: flycatchers and allies (1) ant (11) bee (21) European honey bee (1)

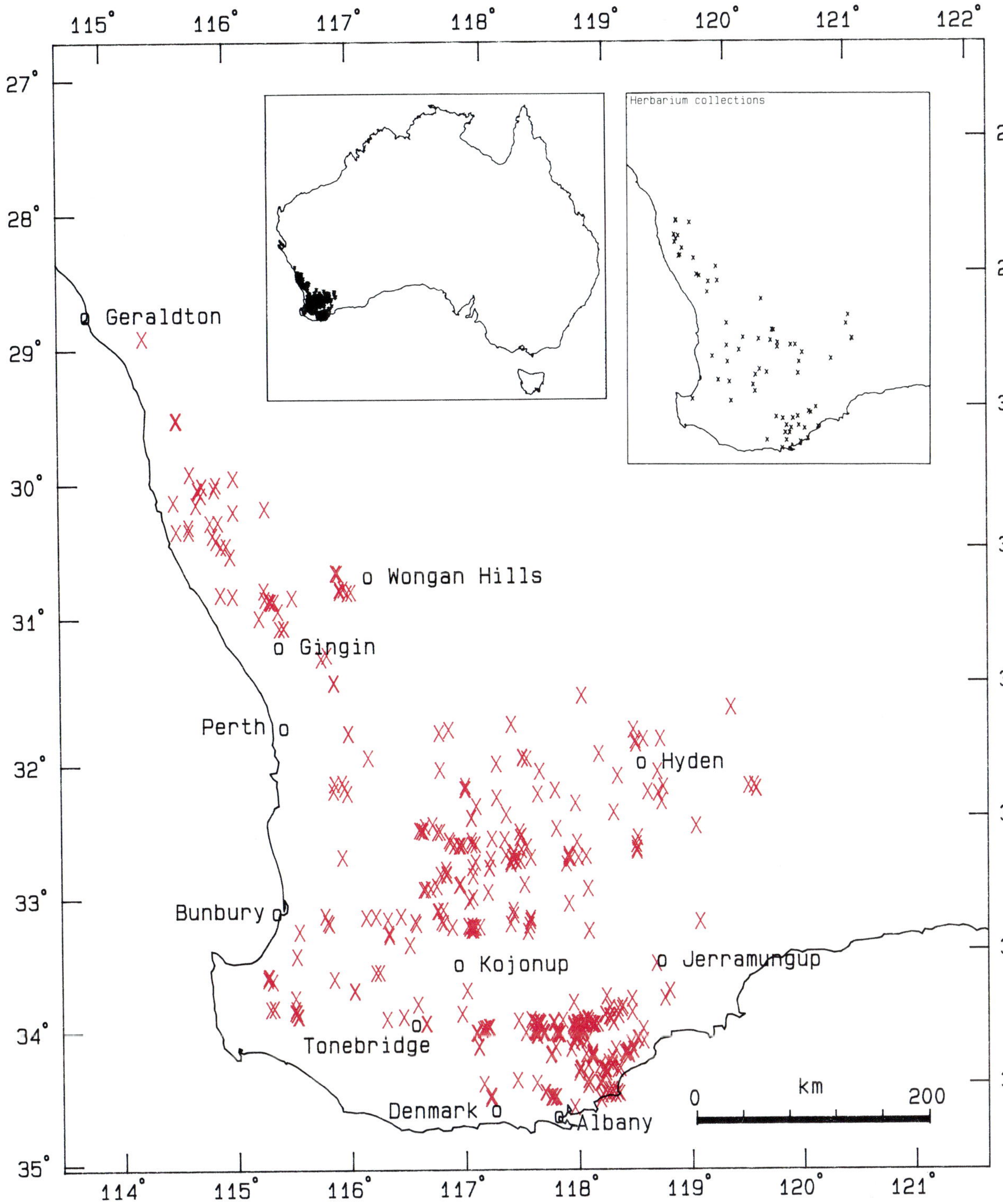

DISCUSSION

B. sphaerocarpa is a common shrub (rarely a small tree) found throughout south-western Australia. It was the first named species of the group with spherical inflorescences and linear leaves, a group that now also includes B. grossa, B. incana, B. leptophylla, B. micrantha, B. lanata, B. telmatiaea and B. scabrella. Until George (1981) named these other seven species, all were called variants of B. sphaerocarpa (e.g. Holliday and Watton (1975). Of the seven new species, only B. micrantha proved troublesome to Atlas contributors in distinguishing it from B. sphaerocarpa.

B. sphaerocarpa ranges from Eneabba south to Nannup, Albany and Jerramungup, and extending through the wheatbelt to east of Hyden. Three varieties were recognised by George (1981). Var. dolichostyla has large flowers (5-7 cm long), green leaves and may grow to 4 m tall, var. caesia is a smaller shrub (usually 1-2 m) with bluish green leaves and smaller flowers, and var. sphaerocarpa is usually smaller again (0.15-1.5 m) with green leaves and small flowers. About 10% of Atlas records for B. sphaerocarpa did not specify the variety observed. Some problems in distinguishing var. sphaerocarpa from var. caesia were noted, particularly near Wagin.

BANKSIA SPHAEROCARPA R.Brown var. CAESIA A.S.George (1981)

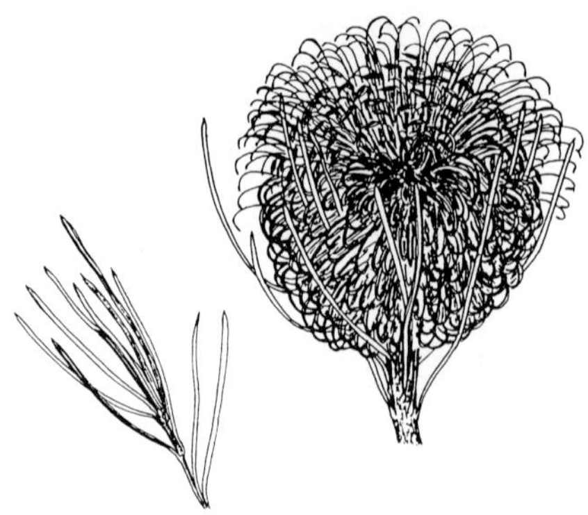

Leaves blue-green

98 RECORDS: Jan (3) Feb (10) Mar (13) Apr (7) May (2) Jun (2) Jul (5) Aug (12) Sep (23) Oct (15) Nov (3) Dec (3).

Population Size: 1-10 (29) 10-100 (38) >100 (31) Unspecified (0).

Conservation Status: Restricted to road verge (41%) Not (59%) Unspecified (0%). In conservation reserve (41%) Not (57%) Unspecified (2%).

Tree/Shrub: Tree form (0%) Shrub form (100%) Unspecified (0%).

Height (metres)

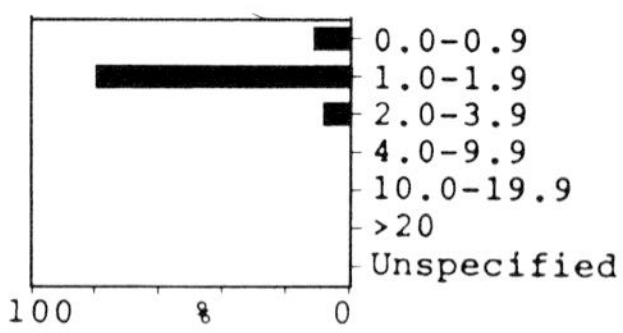

New Shoot Growth

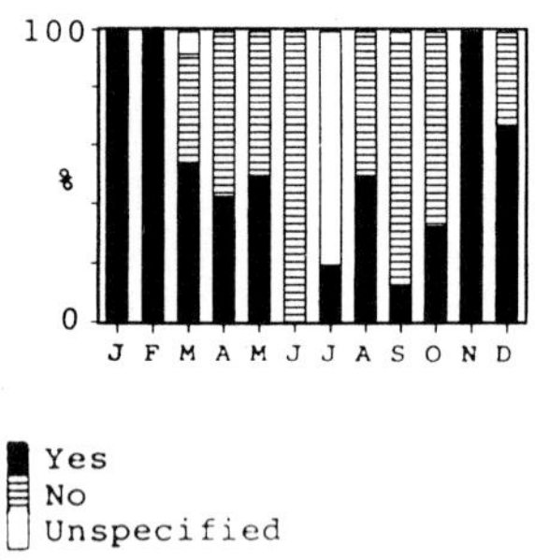

HABITAT

Within 2 km of coast: Yes (0%) No (100%) Unspecified (0%).

Altitude (metres)

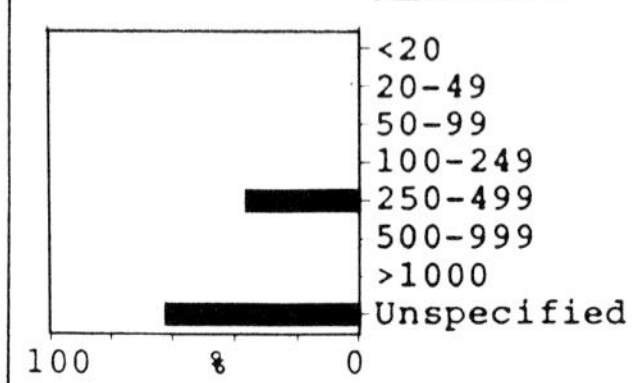

Vegetation

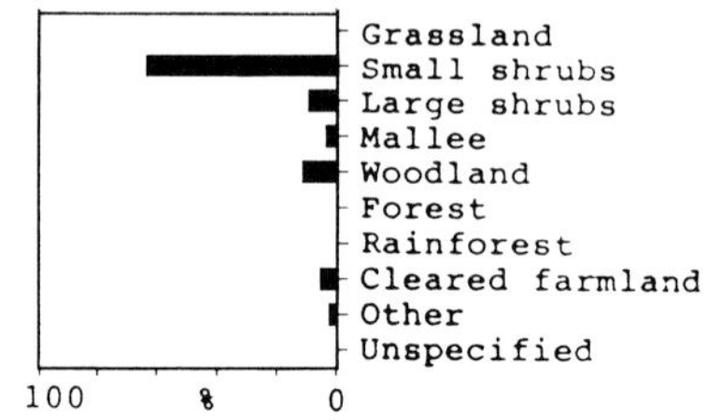

Soil Type

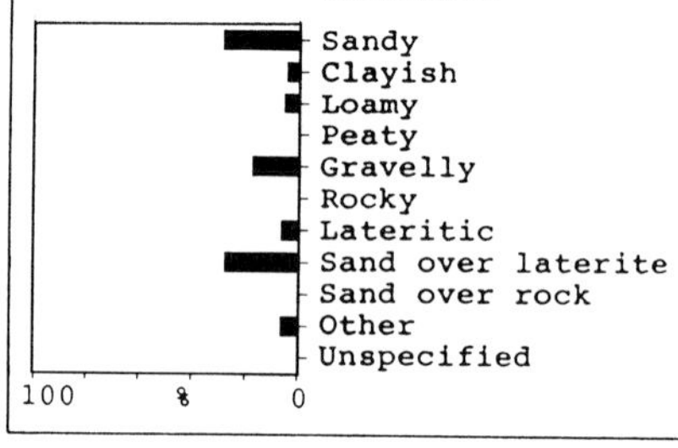

Landform

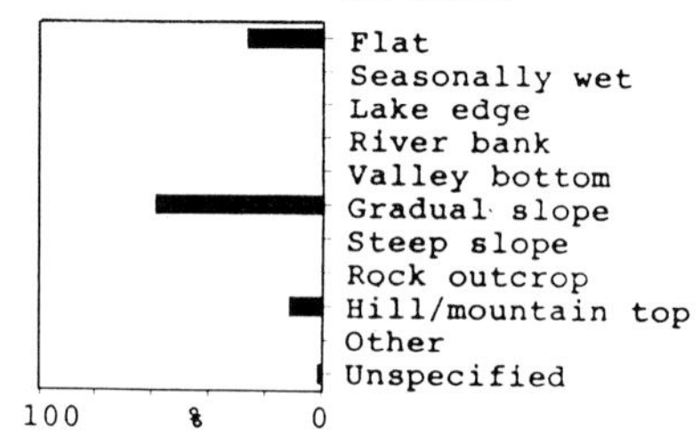

Response to Fire: 3 records

Growth response: Ground resprout (67%) unspecified (33%).

Flowering response: Number of records where flowering had occurred after fire (3). Median reported time to flowering (13-24 months); minimum reported time to flowering (1-12 months). Number of records where flowering had not occurred after fire (0).

Flowering

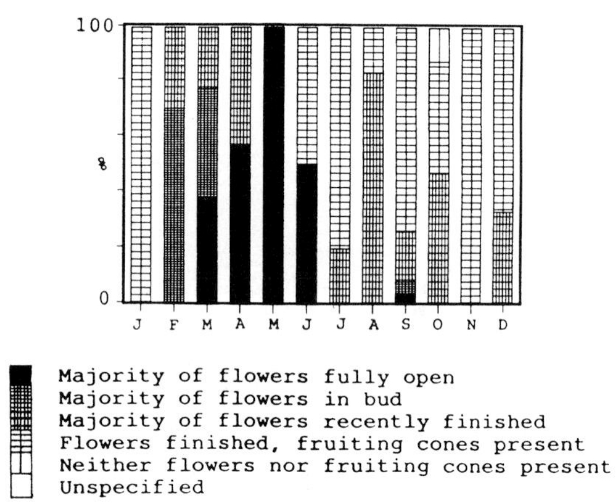

Possible pollinators

Pollinator type 31 records: bird (3%) bees, wasps, ants (13%) no pollinator observed (84%).

Specific pollinator 5 records: flycatchers and allies (1) ant (3) bee (1).

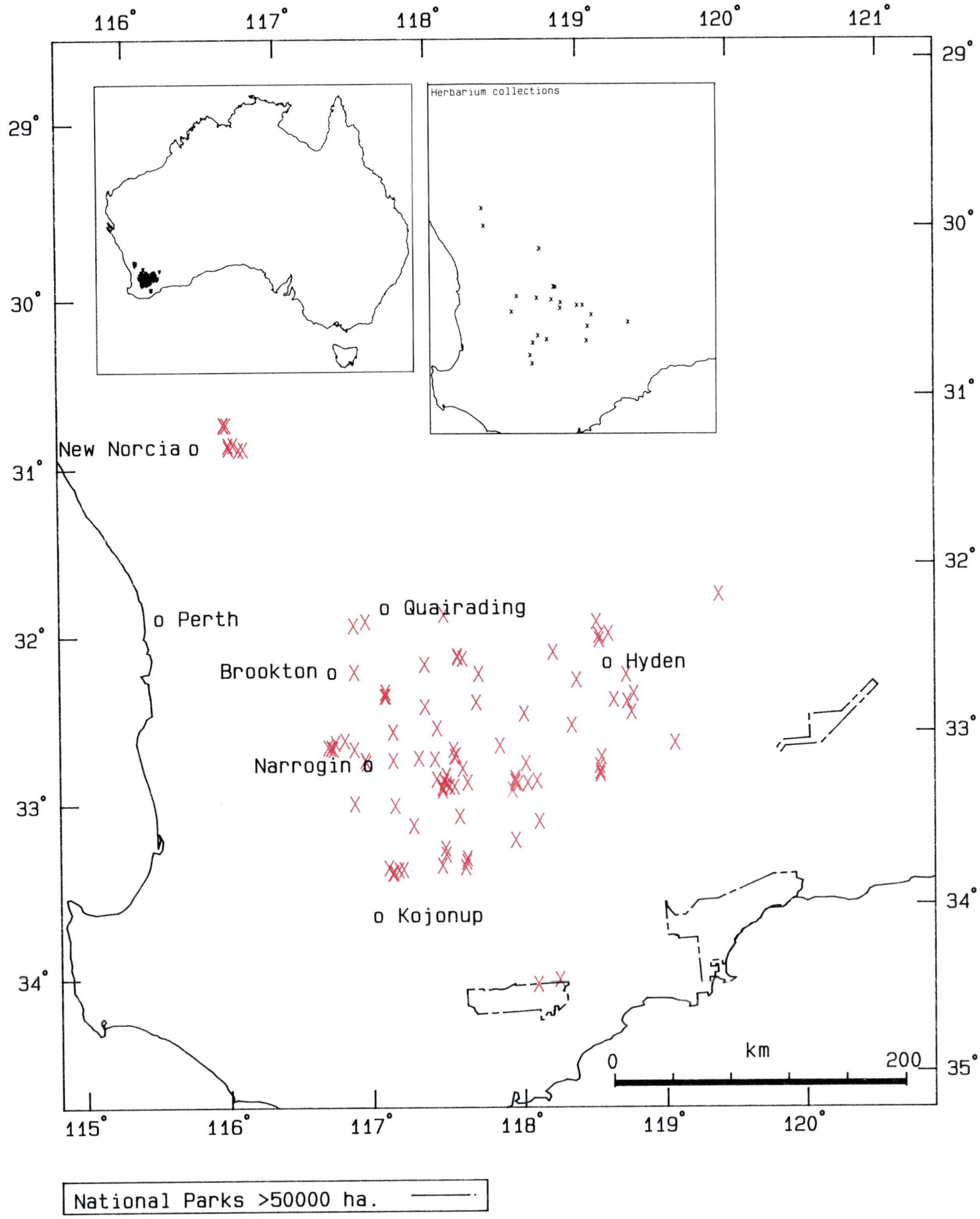

DISCUSSION

Atlas records suggest a distribution for B. sphaerocarpa var. caesia in two disjunct areas, one near Piawanning to the east of New Norcia, and the other extending from Quairading almost to Kojonup and eastwards to beyond Hyden. The 1966 collection of R. Royce from Tammin, which linked these two populations, was not recorded by Atlas contributors. Compared with W.A. Herbarium collections, Atlas contributors have extended the range of this variety further eastwards with several records from both north and south of Hyden, and also an outlying record from the Mt Holland area. The latter is interesting since the rare B. sphaerocarpa var. dolichostyla had been previously recorded in this area. However, from the specimen received and since submitted to the W.A. Herbarium (B. Ballingall) the identification of var. caesia is correct. This population represents a range extension of some 120 km. In the south of its range there are two isolated records from the north-east boundary of Stirling Range National Park (specimens identified by A.S. George) where var. caesia was growing in the same locality as var. sphaerocarpa. Forty-one percent of B. sphaerocarpa var. caesia records were from conservation reserves and the same percentage were restricted to road verges. The northern population near Piawaning is particularly at risk since all records from this area are on road verges.

B. sphaerocarpa var. caesia is a shrub generally less than 2 m in height though sometimes larger. It grows in laterite or sand over laterite on flat or gently sloping land. Surrounding vegetation is usually heath, sometimes tall shrubland or woodland. Atlas records suggest the main flowering period is from March to June which is shorter than the January to July period stated by George (1981). Summer appears to be the main season of new shoot growth. Following fire the variety resprouts from its lignotuber.

BANKSIA SPHAEROCARPA R.Brown var. DOLICHOSTYLA A.S.George (1981)

Ironcaps Banksia

5 RECORDS: Jan (0) Feb (1) Mar (0) Apr (1) May (0) Jun (0) Jul (0) Aug (0) Sep (0) Oct (1) Nov (1) Dec (1).

Population Size: 1-10 (2) 10-100 (2) >100 (1) Unspecified (0).

Conservation Status: Restricted to road verge (20%) Not (80%) Unspecified (0%). In conservation reserve (0%) Not (80%) Unspecified (20%).

Tree/Shrub: Tree form (40%) Shrub form (60%) Unspecified (0%).

Height (metres)

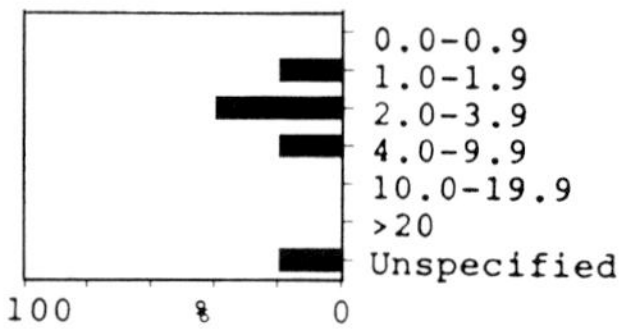

New Shoot Growth

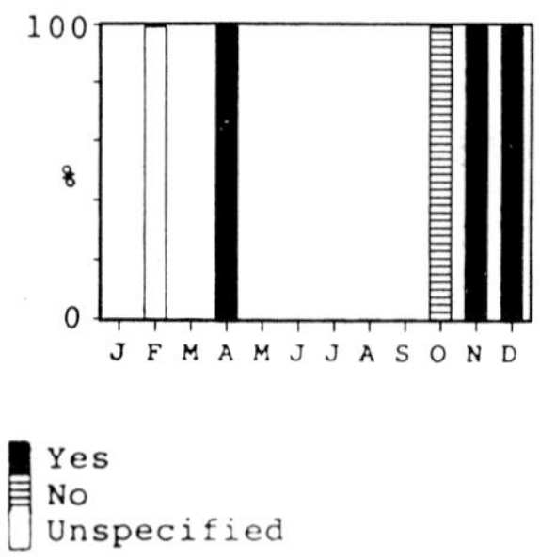

HABITAT

Within 2 km of coast: Yes (0%) No (100%) Unspecified (0%).

Altitude (metres)

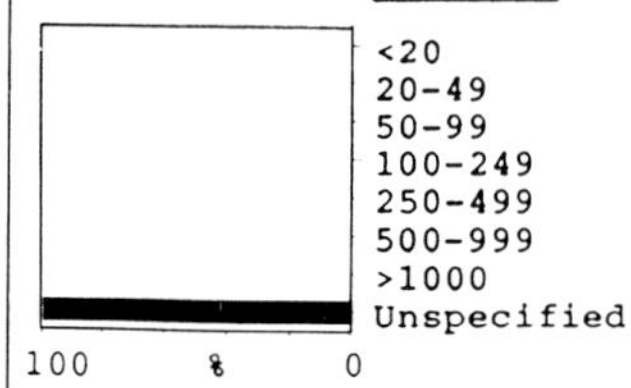

Vegetation

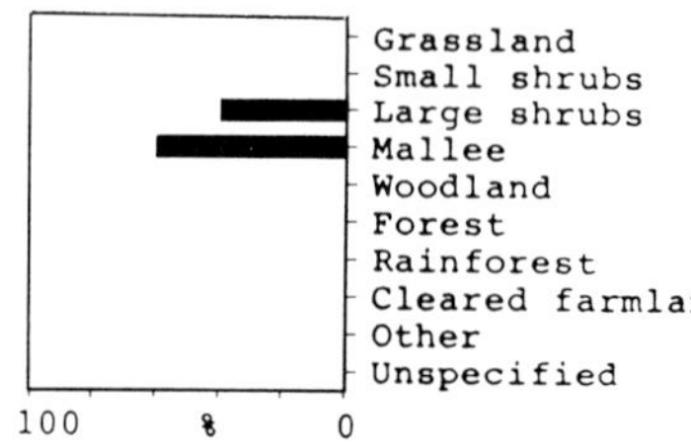

Soil Type

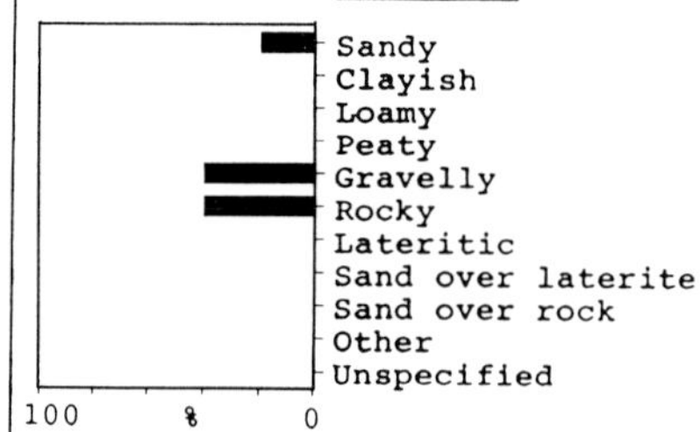

Landform

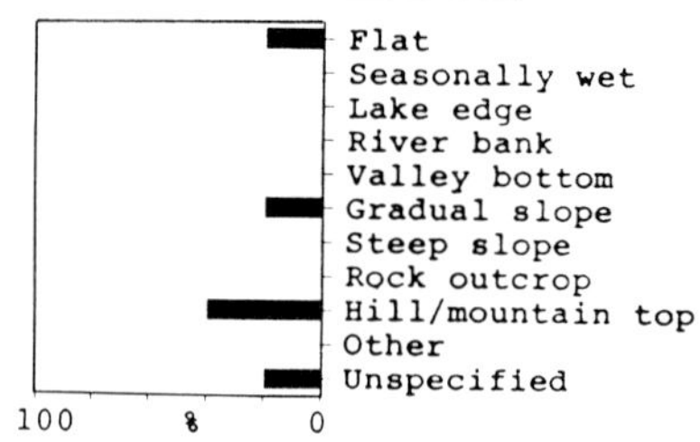

Response to Fire: 0 records

Flowering

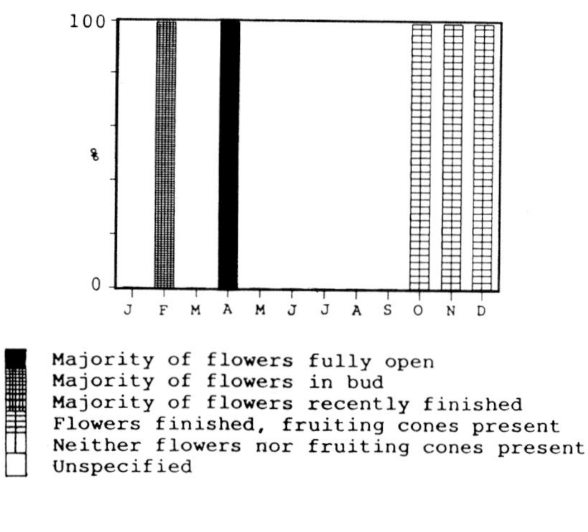

Possible pollinators

Pollinator type 1 record: bees, wasps, ants (100%)

Specific pollinator 1 record: ant (1)

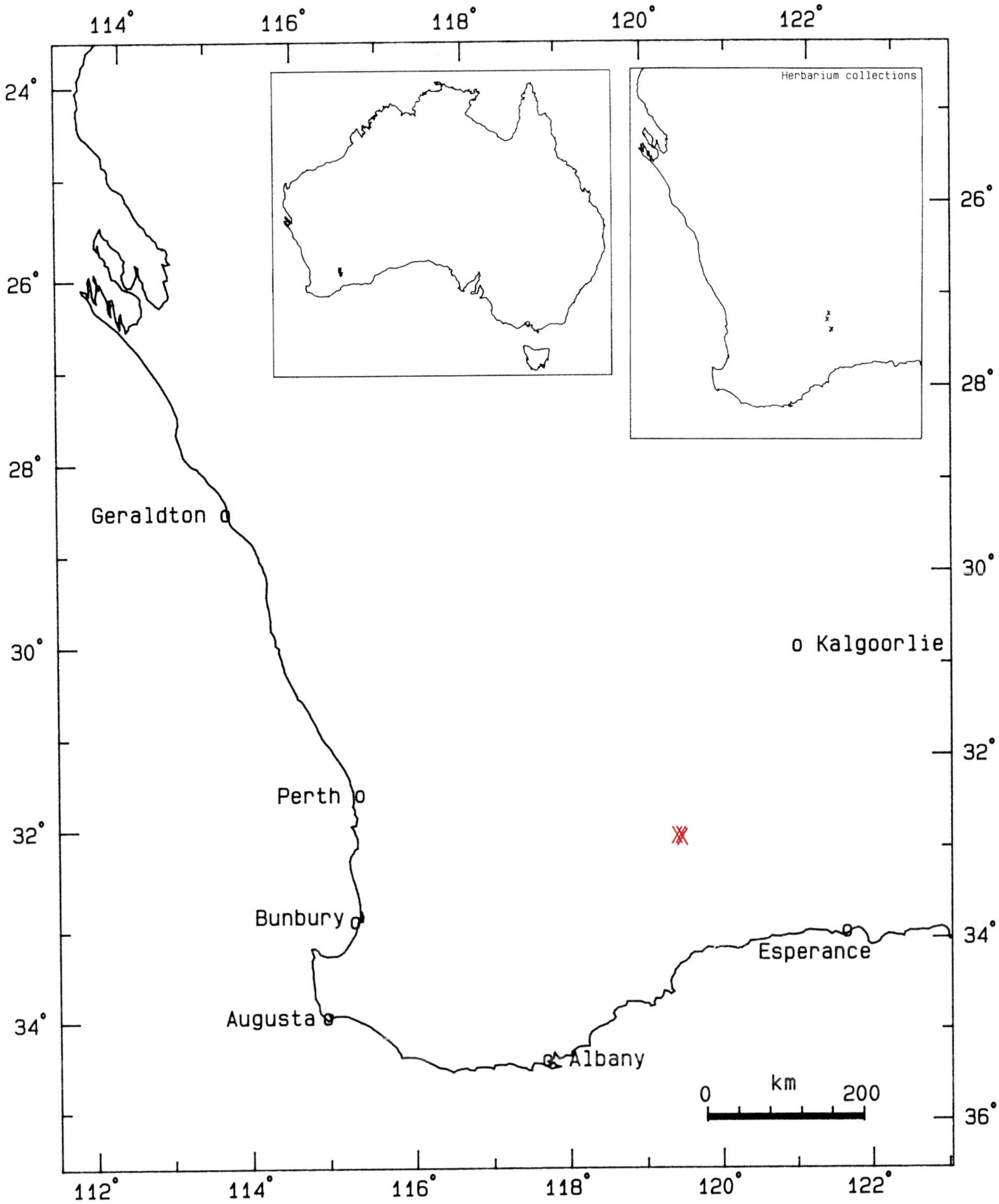

DISCUSSION

B. sphaerocarpa var. dolichostyla is a very rare banksia with only four populations known from herbarium collections and of these, only two were recorded for the Atlas. It is confined to small ironstone hills and ridges in uncleared land to the east of Hyden and Lake King. Although locally common in one or two of its populations, the variety needs to be monitored in the future as all of its habitat is covered by mining leases and exploration has established the presence of significant gold and nickel deposits. Careful management of the variety will be needed should mining proceed.

B. sphaerocarpa var. dolichostyla is a shrub or small tree 1.5 - 4 m in height. It grows on hilltops and flats in lateritic gravel amongst small shrubs and mallee eucalypts. It flowers from March to May (George 1981). Further data are needed on new shoot growth and pollinators before reliable conclusions can be drawn.

This variety was thought to occur further east than var. caesia. However, an Atlas record of var. caesia only 10 km beyond the northernmost location of var. dolichostyla disproves this hypothesis.

BANKSIA SPHAEROCARPA R.Brown var. SPHAEROCARPA

Leaves green

283 RECORDS: Jan (19) Feb (9) Mar (14) Apr (32) May (17) Jun (40) Jul (37) Aug (38) Sep (38) Oct (15) Nov (14) Dec (8).

Population Size: 1-10 (76) 10-100 (141) >100 (64) Unspecified (2).

Conservation Status: Restricted to road verge (29%) Not (68%) Unspecified (2%). In conservation reserve (35%) Not (60%) Unspecified (5%).

Tree/Shrub: Tree form (0%) Shrub form (100%) Unspecified (0%).

Height (metres)

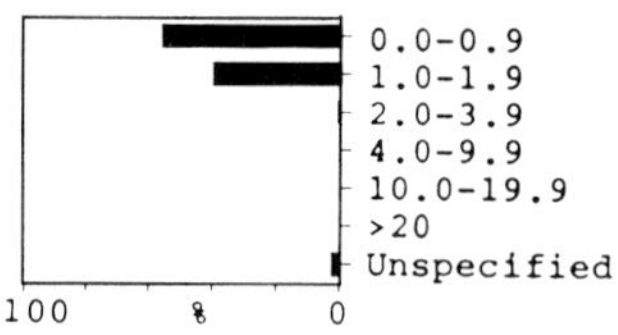

New Shoot Growth

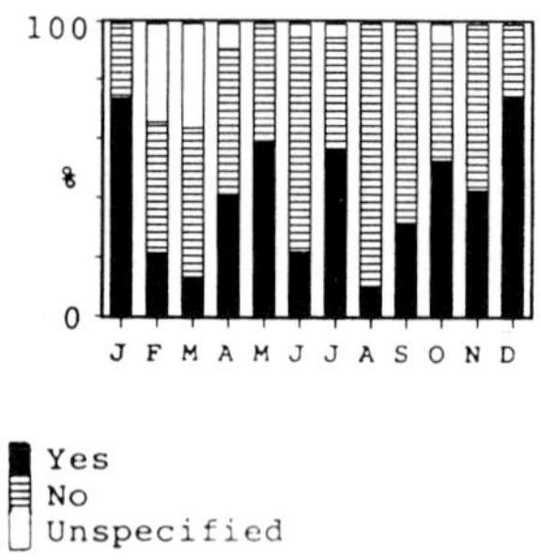

HABITAT

Within 2 km of coast: Yes (3%) No (96%) Unspecified (0%).

Altitude (metres)

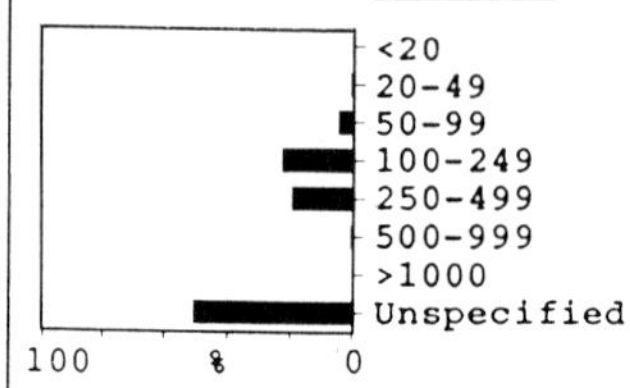

Vegetation

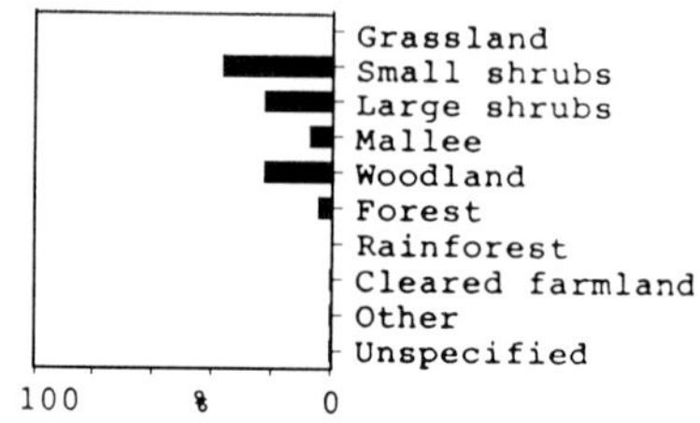

Soil Type

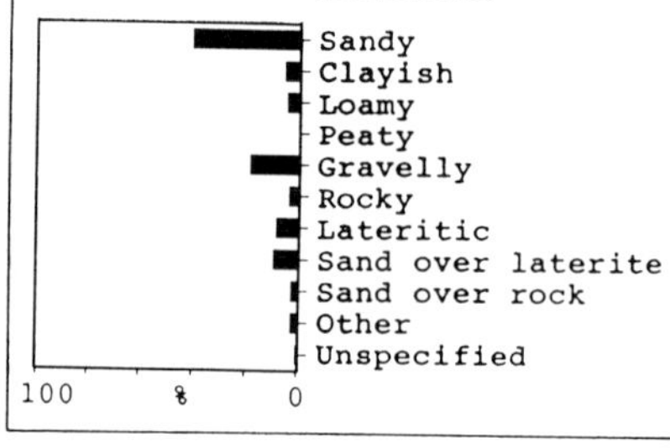

Landform

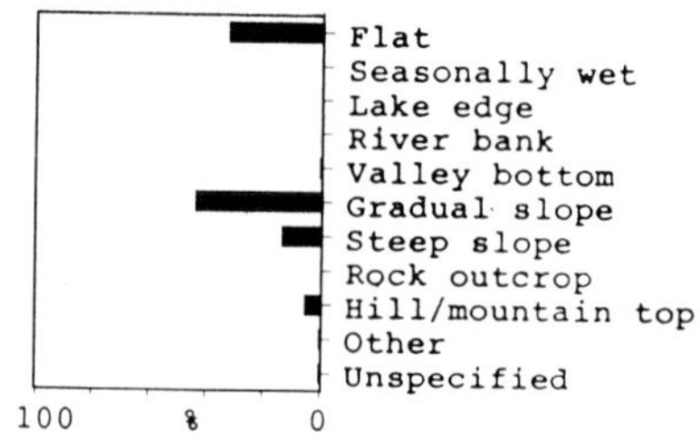

Response to Fire: 48 records

Growth response: Ground resprout (35%) ground resprout and killed, new seedlings (4%) ground resprout and trunk resprout (4%) killed, new seedlings (8%) trunk resprout (4%) unspecified (44%).

Flowering response: Number of records where flowering had occurred after fire (36). Median reported time to flowering (13-24 months); minimum reported time to flowering (<1 month). Number of records where flowering had not occurred after fire (8). Median reported time to no flowering (1-12 months); maximum reported time to no flowering (25-36 months).

Flowering

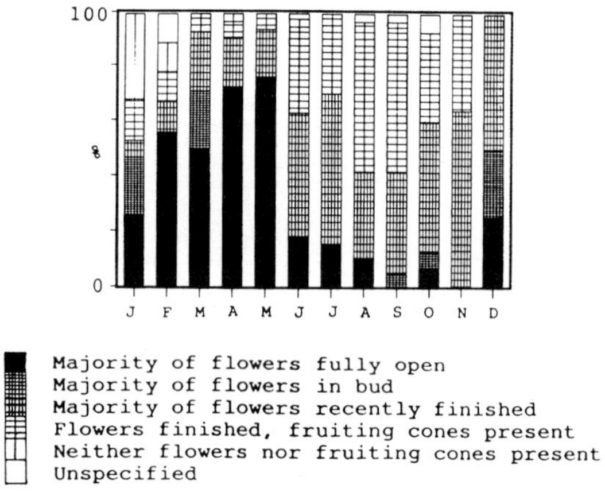

Possible pollinators

Pollinator type 111 records: bird (4%) bees, wasps, ants (15%) flies (2%) beetles (2%) no pollinator observed (77%).

Specific pollinator 19 records: ant (5) bee (13) European honey bee (1)

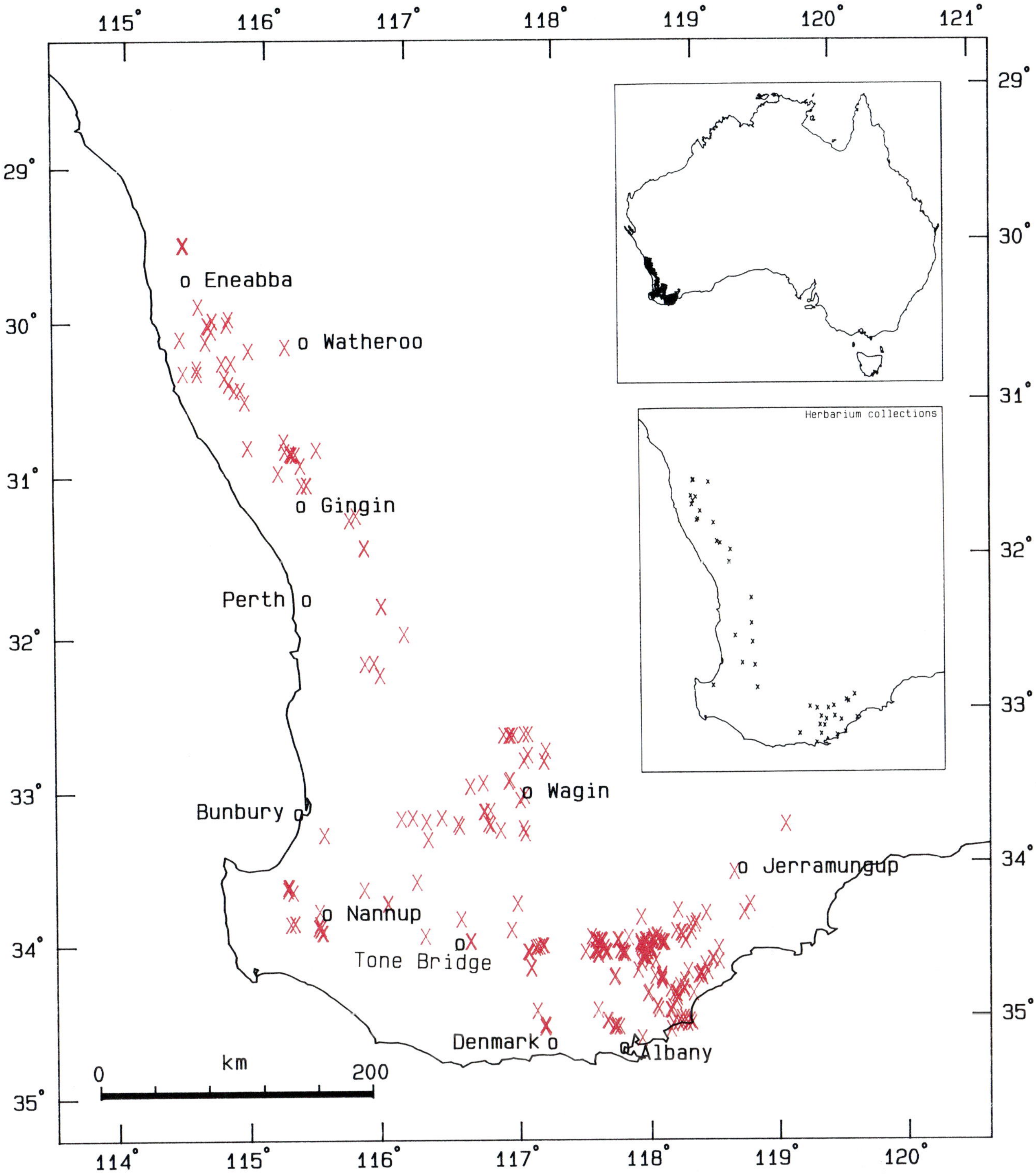

DISCUSSION

Atlas data suggest a more or less continuous distribution for B. sphaerocarpa var. sphaerocarpa from north of Eneabba to west of Nannup and eastwards to beyond Jerramungup. The two apparently disjunct populations suggested by W.A. Herbarium collections are actually connected. The gap in Atlas records north-east of Bunbury is bridged by herbarium collection localities. Atlas contributors have recorded several range extensions. In the south-east there are four new records from west, south and north-east of Jerramungup. The latter extends the known range of this variety by some 100 km. Several populations recorded to the south, west and north-west of Nannup confirm the presence of var. sphaerocarpa in this area. Previously there had been only one herbarium collection from near Busselton. Records from Nannup eastwards to the Stirling Range link the northern and southern populations that herbarium collections indicated. In the northern part of its range, the variety extends both closer to the coast (at Lancelin) and further inland (at Watheroo). In the southern part of its range there may be a large extension inland to Wagin. If plants have been correctly identified, this is a significant range extension bringing var. sphaerocarpa within the same range as var. caesia. George (1981) stated that the two varieties were never sympatric. However, contributors sometimes experienced difficulty in separating the two varieties. The blue-green leaf colour of var. caesia is meant to be distinctive but Atlas contributors made several observations of var. sphaerocarpa with similar coloured leaves. A confirmed recording of the two varieties growing together was made near the north-east boundary of Stirling Range National Park (identifications confirmed by A.S. George).

B. sphaerocarpa var. sphaerocarpa is a highly variable taxon that warrants further taxonomic studies. It is always a shrub. Typical heights vary from 1.5-2 m in the southern part of its range to 50 cm in the north where it is easily confused with B. micrantha (see discussion under that species). In the north it grows in sands or sand over laterite amongst low heath. On the Darling Plateau and east to Wagin it is on lateritic soils generally in woodland or forest of Eucalyptus marginata, E. calophylla or E. wandoo. East of Albany it occurs on sands or sand over laterite amongst tall shrubland, mallee eucalypts or low woodland. Flowers have been recorded in all months of the year except September and November though the main flowering period is from February to May. There does not appear to be a dominant season for new shoot growth.

BANKSIA SPINULOSA Smith (1793)

Hairpin Banksia

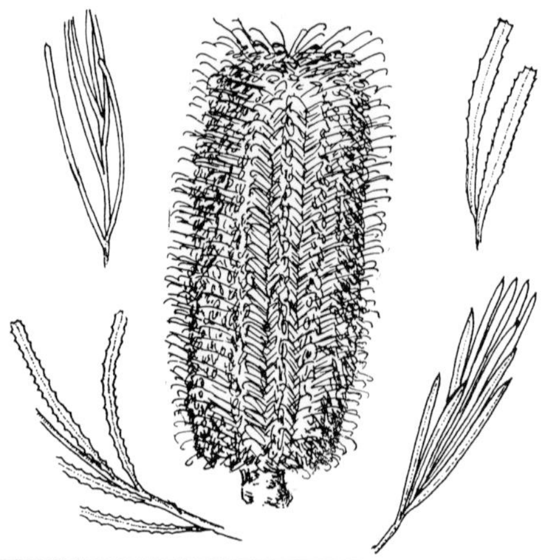

1386 RECORDS: Jan (133) Feb (122) Mar (143) Apr (159) May (176) Jun (192) Jul (92) Aug (85) Sep (65) Oct (47) Nov (76) Dec (86).

Population Size: 1-10 (243) 10-100 (422) >100 (704) Unspecified (17).

Conservation Status: Restricted to road verge (3%) Not (95%) Unspecified (2%). In conservation reserve (42%) Not (52%) Unspecified (6%).

Tree/Shrub: Tree form (9%) Shrub form (89%) Unspecified (2%).

Height (metres)

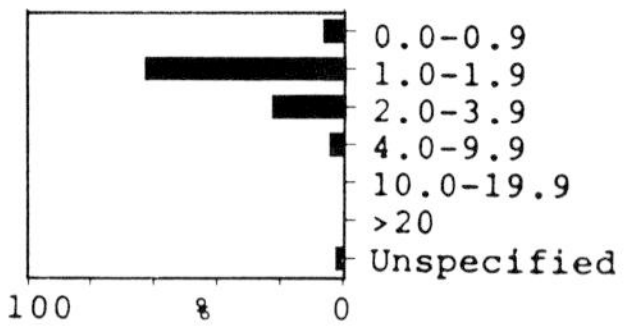

New Shoot Growth

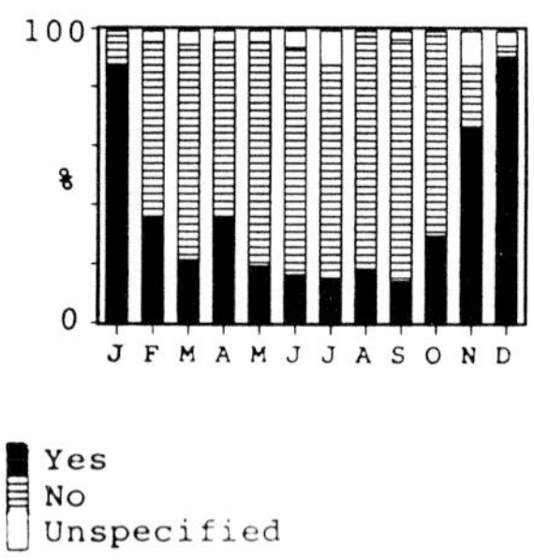

HABITAT

Within 2 km of coast: Yes (10%) No (90%) Unspecified (0%).

Altitude (metres)

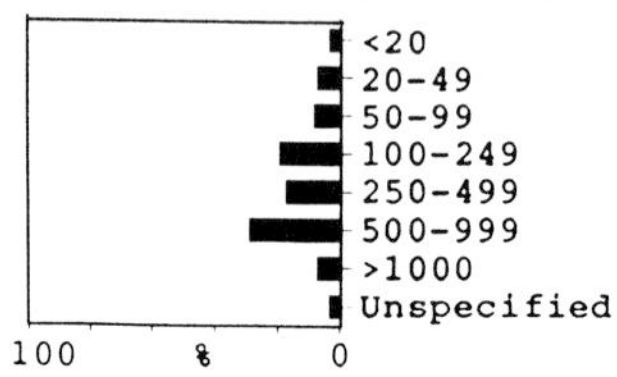

Vegetation

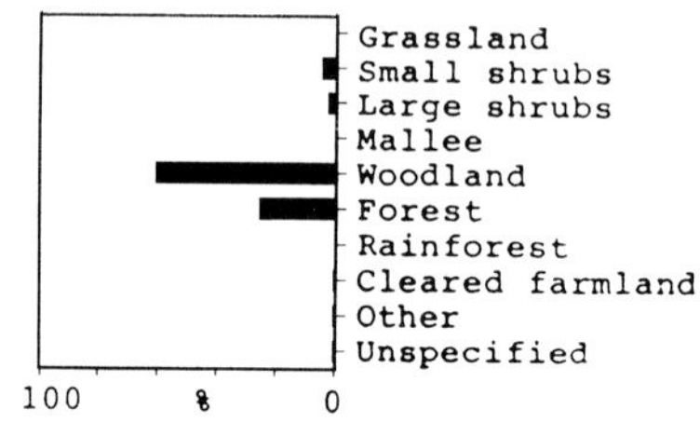

Soil Type

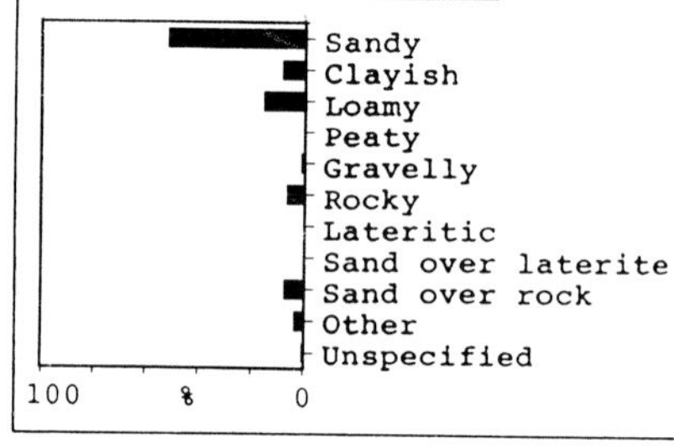

Landform

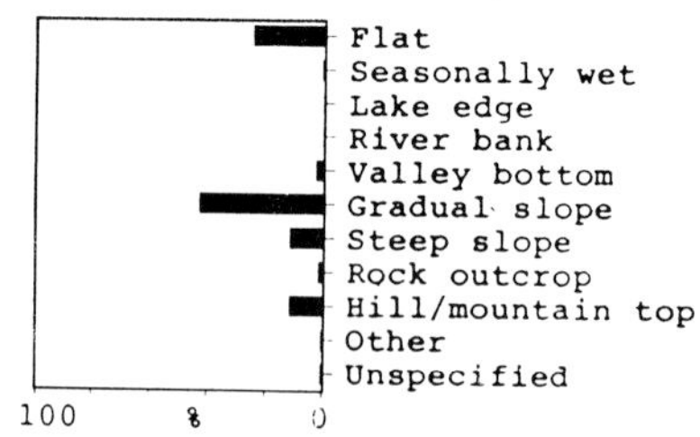

Response to Fire: 235 records

Growth response: Ground resprout (49%) ground resprout and killed, new seedlings (1%) killed, new seedlings (12%) trunk resprout (3%) killed, no seedlings (4%) unspecified (31%).

Flowering response: Number of records where flowering had occurred after fire (125). Median reported time to flowering (25-36 months); minimum reported time to flowering (<1 month). Number of records where flowering had not occurred after fire (35). Median reported time to no flowering (13-24 months); maximum reported time to no flowering (49-60 months).

Flowering

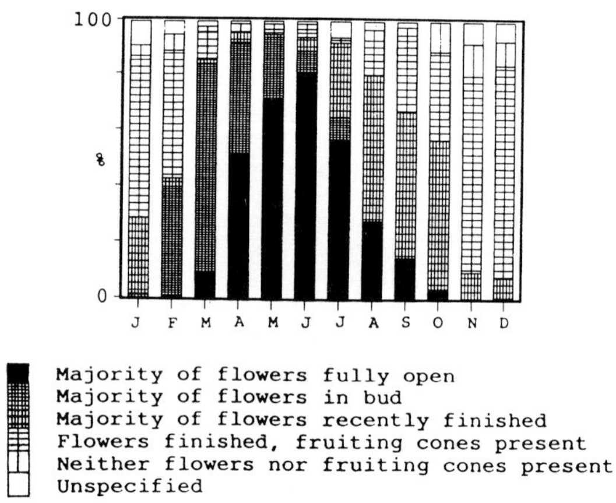

Possible pollinators

Pollinator type 859 records: bird (7%) bees, wasps, ants (5%) flies (0.1%) mammal (0.5%) no pollinator observed (87%).

Specific pollinator 120 records: honeyeater (10) New Holland Honeyeater (8) White-eared Honeyeater (4) Yellow-faced Honeyeater (3) White-plumed Honeyeater (1) Crescent Honeyeater (3) Lewin's Honeyeater (1) White-cheeked Honeyeater (1) Brown Honeyeater (2) Eastern Spinebill (2) Noisy Miner (1) wattlebird (3) Red Wattlebird (7) Noisy Friarbird (3) Eastern yellow Robin (1) ant (15) bee (19) native bee (2) European honey bee (9) European wasp (1) Brown Antechinus (4) Pygmy-possum (1)

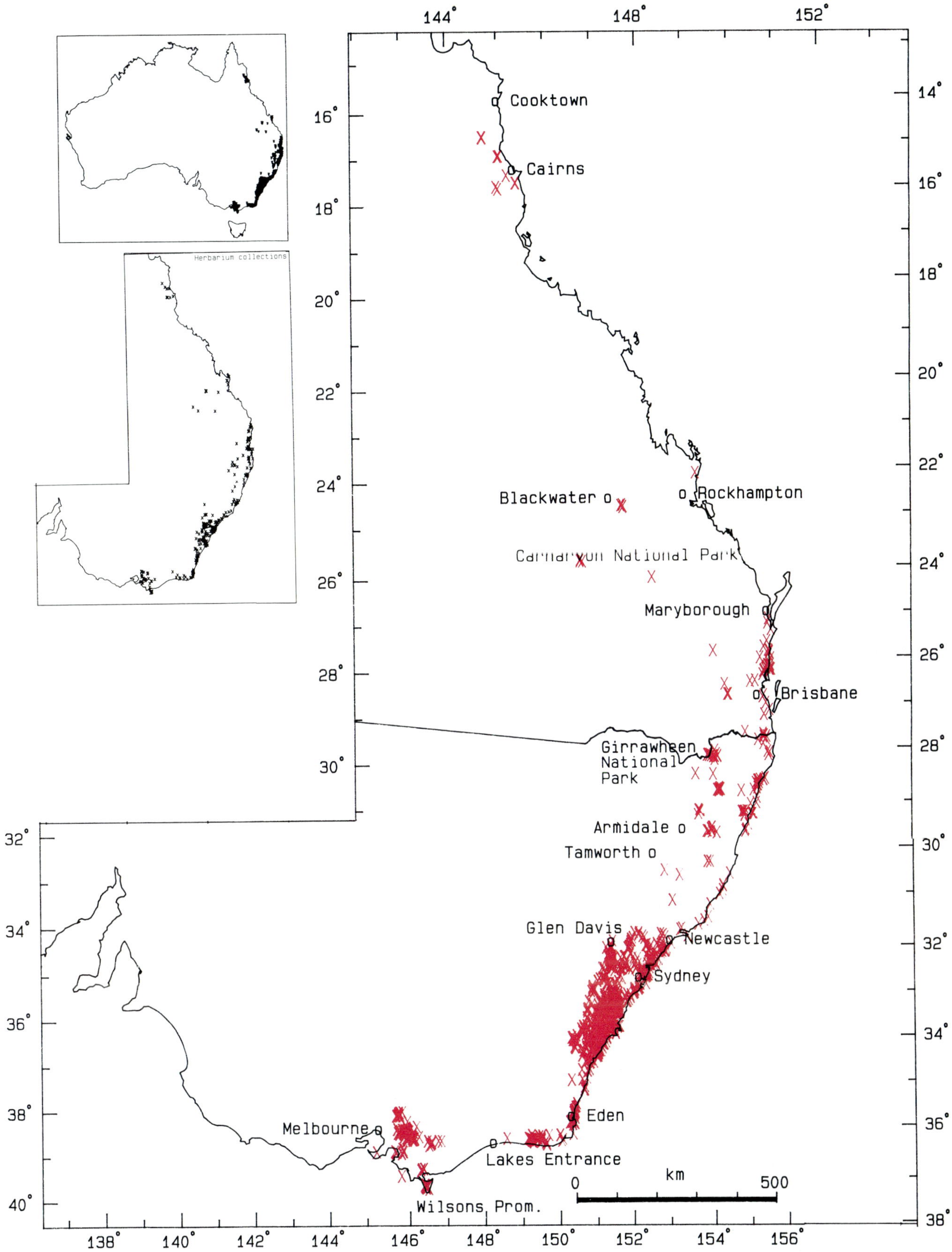

DISCUSSION

B. spinulosa is a common variable shrub usually 1-2 m tall. It has a wide range from the Dandenongs in Victoria east and north through New South Wales into Queensland, where it occurs in disjunct populations as far north as the Mt Windsor and Atherton Tablelands. Concomitant with its extensive distribution, a diversity of habitats is occupied. The species may grow in coastal communities or in inland montane woodlands and forests. Sand is the usual soil type, although the species also occurs in rocky clays and loams. Flowering peaks in autumn, but was recorded from January to October.

Four varieties are recognised (George 1981, 1985). Var. cunninghamii lacks a lignotuber and is killed by fire. The three other varieties all have lignotubers and, thus, may resprout from ground level after fire. Var. neoanglica has entire leaves with a pointed tip and no visible venation on the undersurface. Var. spinulosa similarly lacks such venation, and has revolute leaves generally with several small teeth towards the tip. Occasionally however, its leaves may be serrated throughout. In its typical form, var. collina has leaves that are evenly serrated along their whole length, and with conspicuous venation on the undersurface. When varieties existed in their typical forms, volunteers had little trouble with identification. However, numerous atypical forms do occur in the spinulosa complex, including some which are intermediate between specified varieties (see below). Where these occurred, volunteers had difficulty in identifying them correctly.

BANKSIA SPINULOSA Smith var. COLLINA (R.Brown) A.S.George (1981)

Hill Banksia

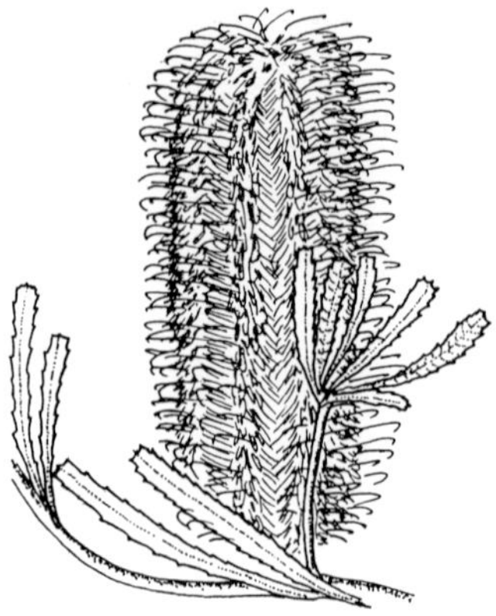

191 RECORDS: Jan (22) Feb (1) Mar (18) Apr (27) May (24) Jun (12) Jul (22) Aug (21) Sep (5) Oct (13) Nov (7) Dec (14).

Population Size: 1-10 (39) 10-100 (79) >100 (71) Unspecified (2).

Conservation Status: Restricted to road verge (2%) Not (93%) Unspecified (5%). In conservation reserve (46%) Not (43%) Unspecified (12%).

Tree/Shrub: Tree form (2%) Shrub form (98%) Unspecified (0%).

Height (metres)

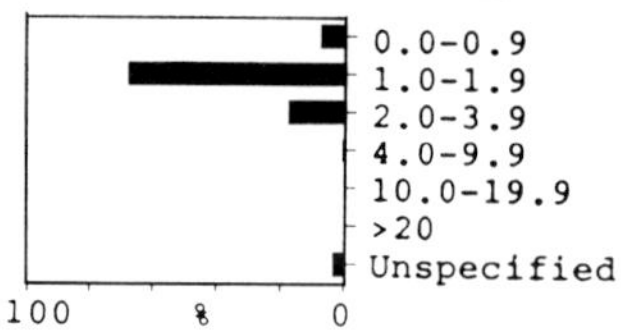

New Shoot Growth

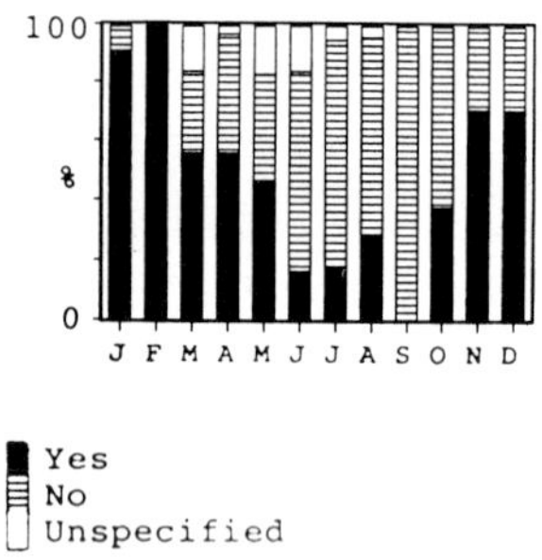

HABITAT

Within 2 km of coast: Yes (21%) No (78%) Unspecified (1%).

Altitude (metres)

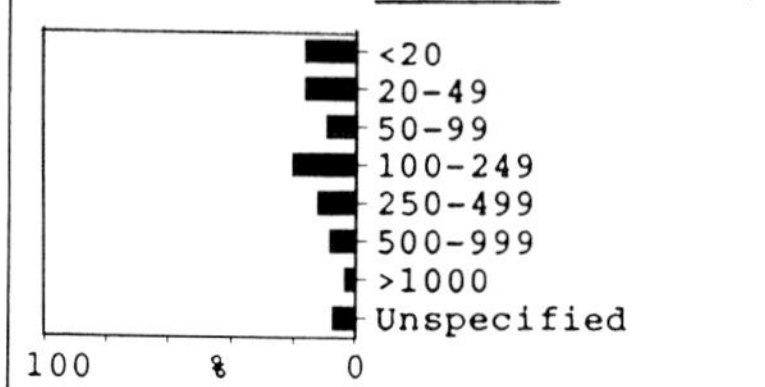

Vegetation

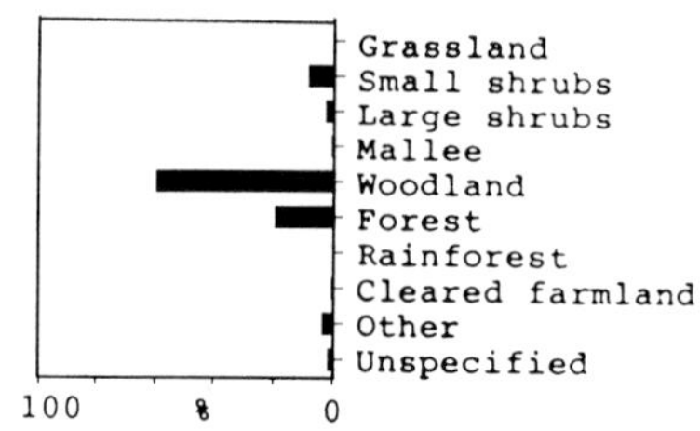

Soil Type

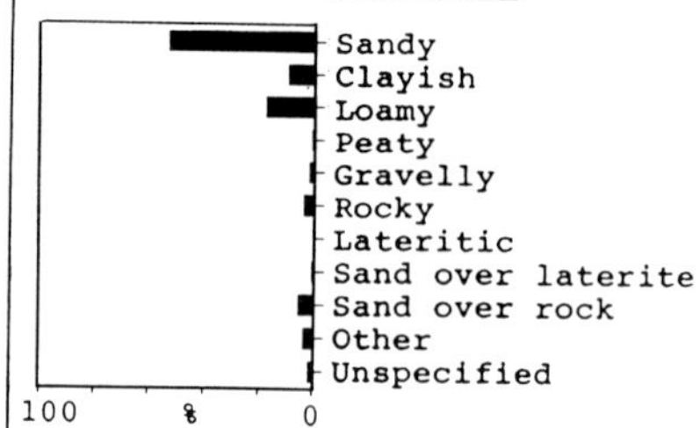

Landform

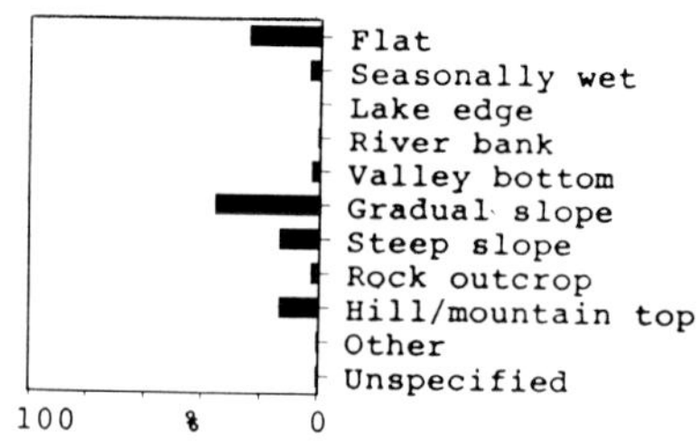

Response to Fire: 45 records

Growth response: Ground resprout (69%) ground resprout and killed, new seedlings (2%) unspecified (29%).

Flowering response: Number of records where flowering had occurred after fire (23). Median reported time to flowering (13-24 months); minimum reported time to flowering (<1 month).
Number of records where flowering had not occurred after fire (6). Median reported time to no flowering (12-13 months); maximum reported time to no flowering (13-24 months).

Flowering

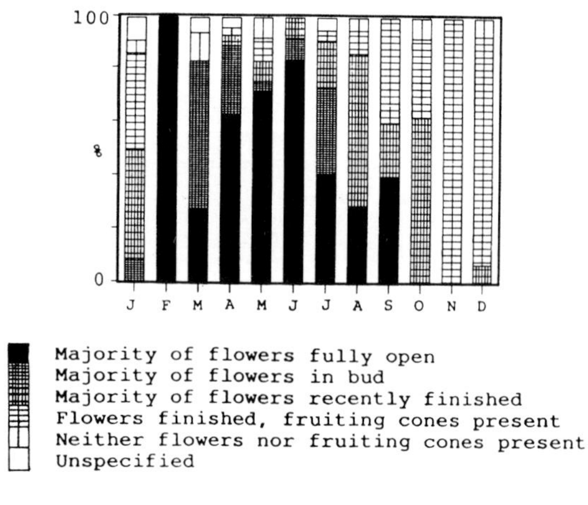

Possible pollinators

Pollinator type 87 records: bird (10%) bees, wasps, ants (9%) no pollinator observed (80%).

Specific pollinator 17 records: Brown Honeyeater (2) White-cheeked Honeyeater (1) White-eared Honeyeater (1) Lewin's Honeyeater (1) Eastern Spinebill (2) Noisy Friarbird (2) ant (2) bee (4) European honey bee (2)

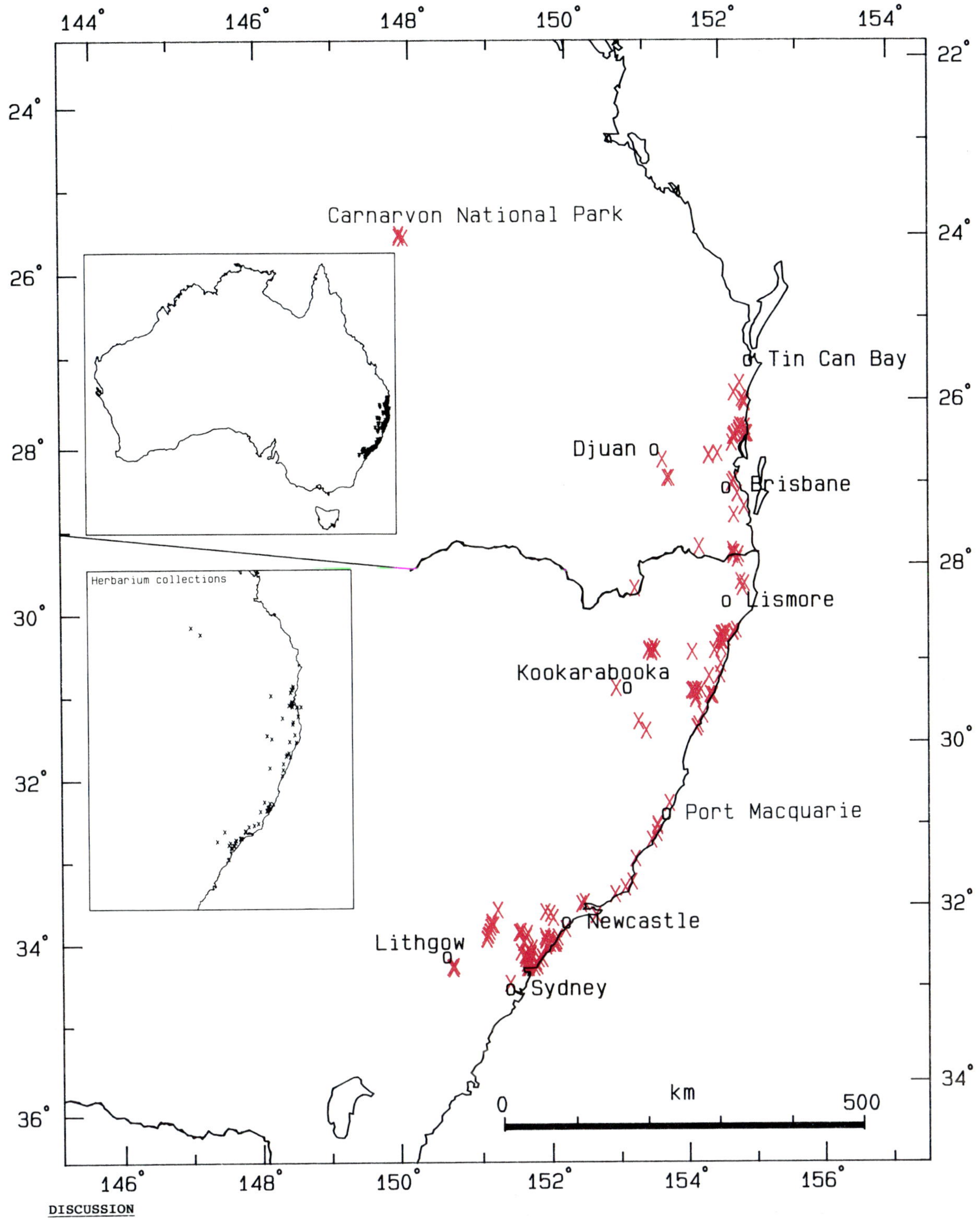

DISCUSSION

The main distribution of B. spinulosa var. collina is from Gympie (Queensland) southwards to Sydney. In Queensland there is an outlying population at Carnarvon National Park, ca. 500 km from the next population. Outliers also occur at Crows Nest (south of Djuan), at Boonoo Boonoo and at Mt Barney National Park. In northern New South Wales there are inland populations at Gibraltar Range and near Mt Mitchell, west of Kookarabooka. However, these populations and the ones from Boonoo Boonoo should possibly be reclassified as the new variety - B. spinulosa var. neoanglica (see below). North of Sydney the range of var. collina overlaps with that of var. spinulosa and intermediate forms were reported from such areas as Bouddi National Park. Inland from Sydney, the variety is found from the Culloul Range to beyond Putty and in the area of Medlow Bath south-east of Lithgow. Herbarium collections show a generally similar pattern to Atlas data, though, as mentioned previously, inland populations from northern New South Wales and the Queensland border should possibly be reclassified as B. spinulosa var. neoanglica.

B. spinulosa var. collina is typically a shrub up to 2 m in height, although sometimes it may be larger. Over 20% of its recorded populations are within 2 km of the coast where it is often a component of coastal heath. Inland, it may reach altitudes of 1 000 metres or more, though several of the highest occurring populations are possibly var. neoanglica. Var. collina grows mainly in sandy soil, sometimes over sandstone. It also occurs in heavier loams and clays. It is generally an understorey shrub of eucalypt woodlands and forests. The main flowering period is from late summer to early spring. Summer appears to be the main season of new shoot growth. Following fire, var. collina resprouts from its lignotuber.

BANKSIA SPINULOSA Smith var. CUNNINGHAMII (Sieber ex Reichenbach) A.S.George (1981)

331 RECORDS: Jan (9) Feb (61) Mar (29) Apr (55) May (52) Jun (28) Jul (12) Aug (6) Sep (17) Oct (9) Nov (35) Dec (16).

Population Size: 1-10 (69) 10-100 (94) >100 (160) Unspecified (8).

Conservation Status: Restricted to road verge (6%) Not (91%) Unspecified (3%). In conservation reserve (43%) Not (48%) Unspecified (9%).

Tree/Shrub: Tree form (27%) Shrub form (70%) Unspecified (2%).

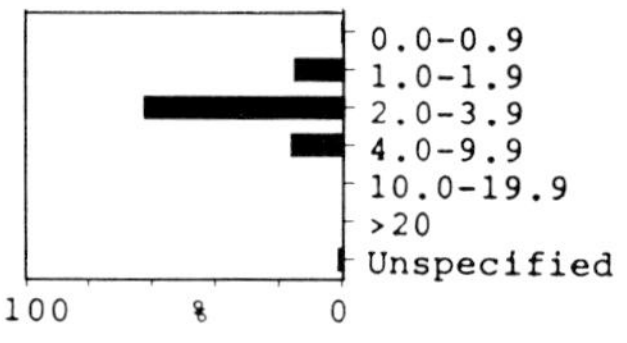

New Shoot Growth

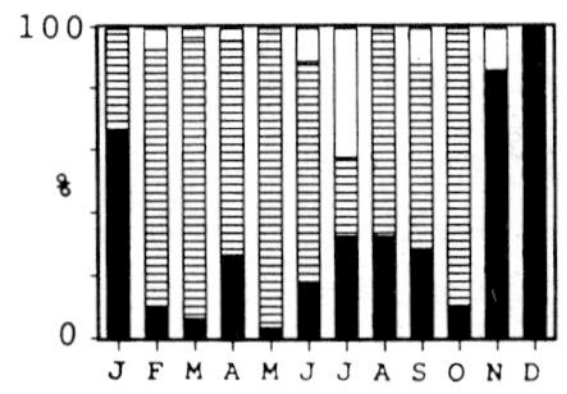

HABITAT

Within 2 km of coast: Yes (9%) No (91%) Unspecified (0%).

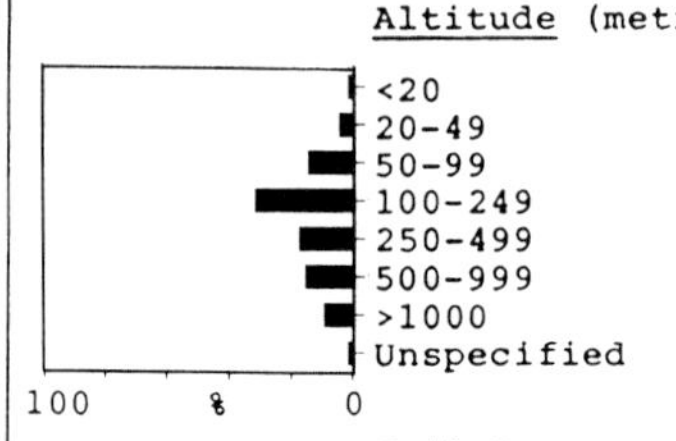

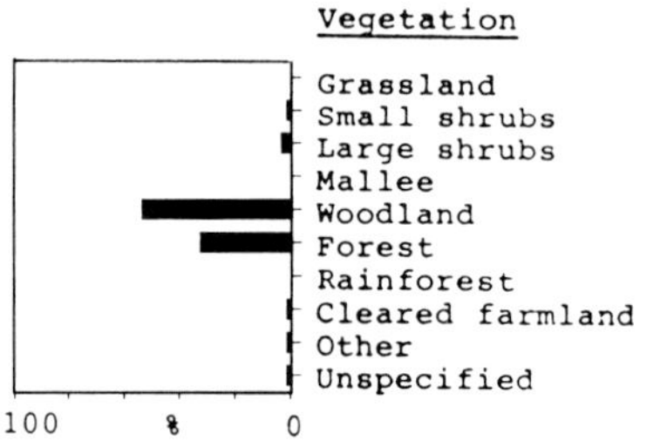

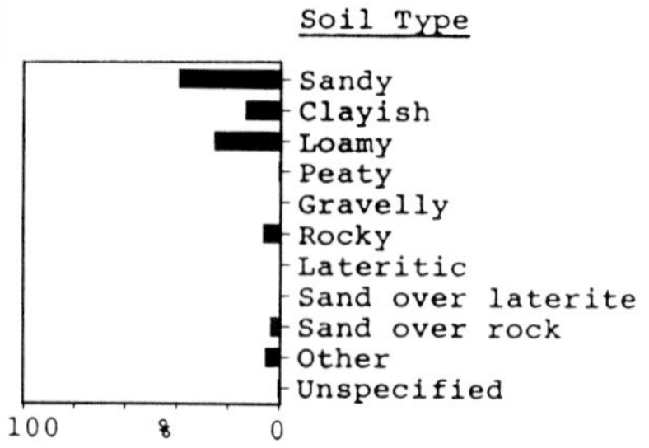

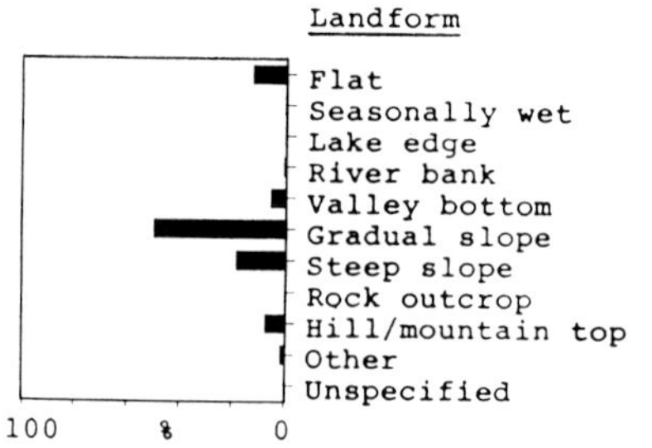

Response to Fire: 50 records

Growth response: Ground resprout (2%) killed, new seedlings (42%) killed, no seedlings (16%) unspecified (40%).

Flowering response: Number of records where flowering had occurred after fire (27). Median reported time to flowering (13-24 months); minimum reported time to flowering (<1 month). Number of records where flowering had not occurred after fire (10). Median reported time to no flowering (13-24 months); maximum reported time to no flowering (25-36 months).

Flowering

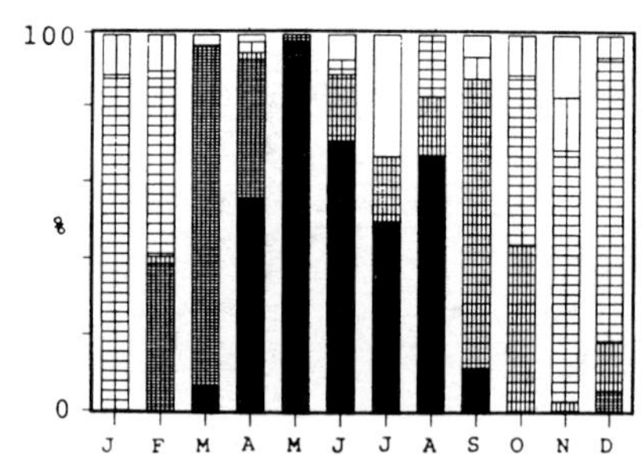

Possible pollinators

Pollinator type 223 records: bird (5%) bees, wasps, ants (2%) no pollinator observed (92%).

Specific pollinator 23 records: honeyeater (6) Crescent Honeyeater (2) wattlebird (2) Red Wattlebird (2) Eastern Spinebill (5) ant (1) bee (2) European honey bee (1) European wasp (1) Brown Antechinus (1)

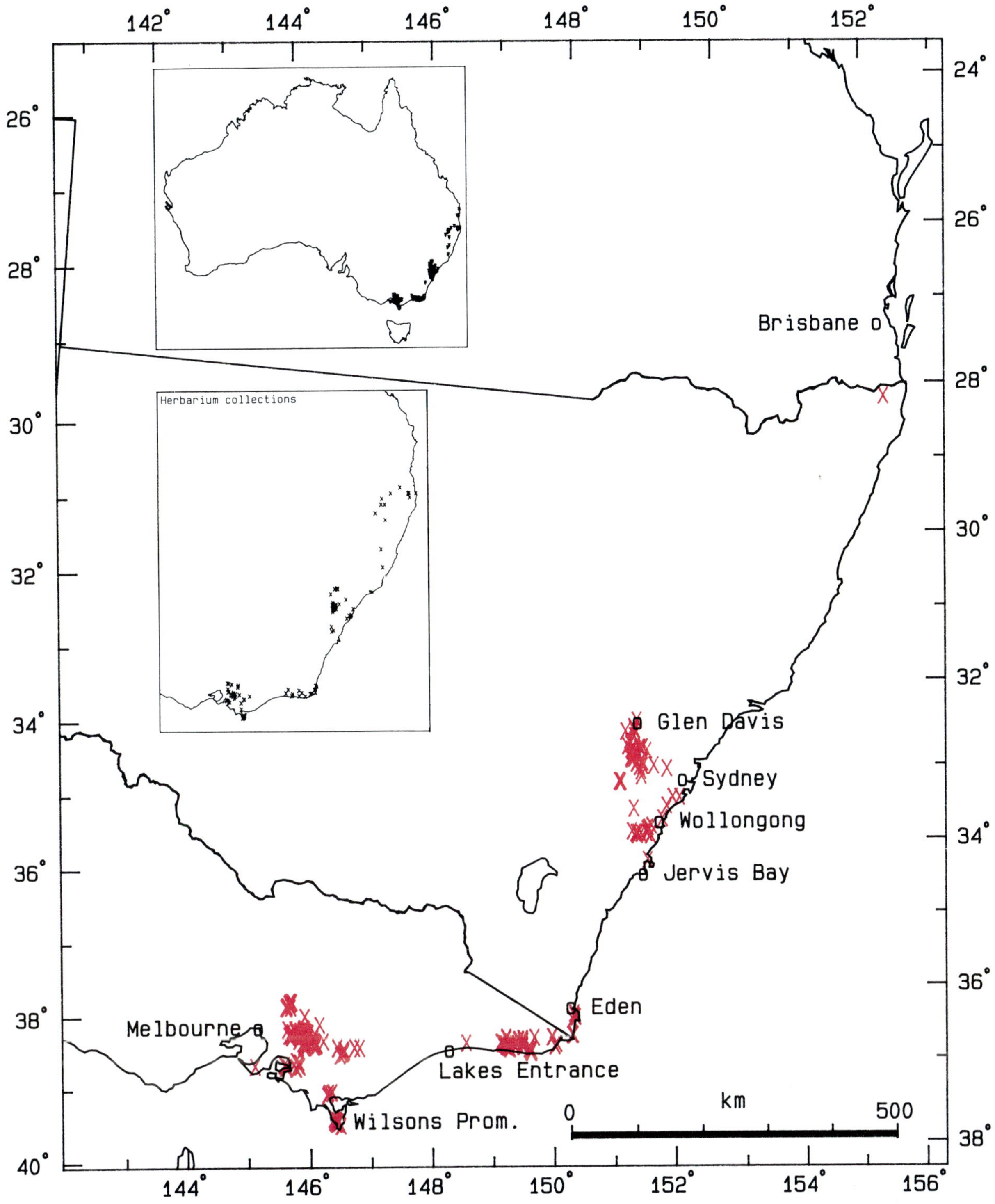

DISCUSSION

Herbarium collections of B. spinulosa var. cunninghamii indicated a scattered distribution from close to the Queensland - New South Wales border to the Dandenongs east of Melbourne. Herbarium collections from mountain areas of northern New South Wales and southern Queensland should most probably be re-classified as B. spinulosa var. neoanglica, a new lignotuberous variety recognised by George (1987). Atlas records from similar locations were identified as var. neoanglica and therefore appear under that variety. A single Atlas record from Mt Warning remains as var. cunninghamii as the observer found no trace of a lignotuber. According to this new classification var. cunninghamii may be largely restricted to three disjunct areas. From Sydney a population extends inland to Glen Davis and south to Jervis Bay. A second population ranges from Eden across the Victorian border almost as far as Lakes Entrance. The third population occurs in the Dandenongs (north-east of Melbourne) eastwards almost to Heyfield and south to Foster and Wilsons Promontory. The variety also occurs on French Island. The Mornington Peninsula record refers to a single old plant on a roadside verge. It is apparently dying with no sign of any seedling recruitment.

B. spinulosa var. cunninghamii occurs as a shrub or tree up to 4 m in height, though sometimes larger. Unlike the other varieties of B. spinulosa, it is non-lignotuberous being killed by fire and regenerating from seedling regrowth. It may take at least 3 years for young plants to reach maturity and flower, although at Bastion Point (Victoria) a population was flowering after a fire only 18 months previously. Var. cunninghamii grows in sandy soils sometimes overlying rock and also in heavier loams and clays. It occurs as an understorey plant of eucalypt woodland and forest often on hillsides at a wide range of altitudes from close to sea level to over 1 000 metres. The main flowering period is from April to August and new shoot growth is largely in summer. Hybrids of B. spinulosa var. cunninghamii and B. ericifolia var. ericifolia were recorded e.g. at Budanoon Creek dam inland from Wollongong.

BANKSIA SPINULOSA Smith var. NEOANGLICA A.S.George (1987)

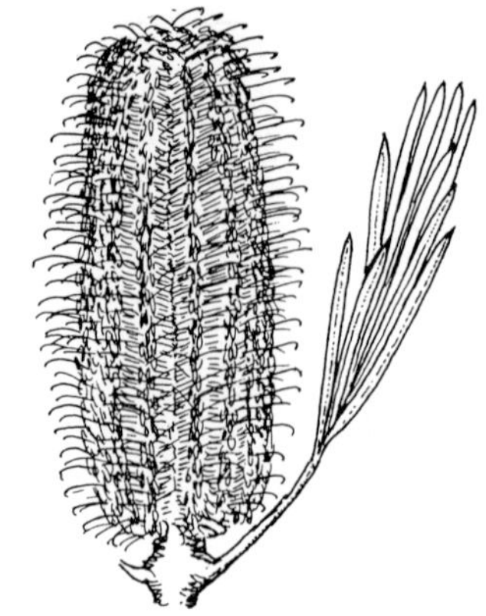

35 RECORDS: Jan (2) Feb (0) Mar (3) Apr (6) May (5) Jun (1) Jul (7) Aug (4) Sep (0) Oct (0) Nov (0) Dec (7).

Population Size: 1-10 (6) 10-100 (14) >100 (15) Unspecified (0).

Conservation Status: Restricted to road verge (3%) Not (94%) Unspecified (3%). In conservation reserve (63%) Not (34%) Unspecified (3%).

Tree/Shrub: Tree form (0%) Shrub form (100%) Unspecified (0%).

Height (metres)

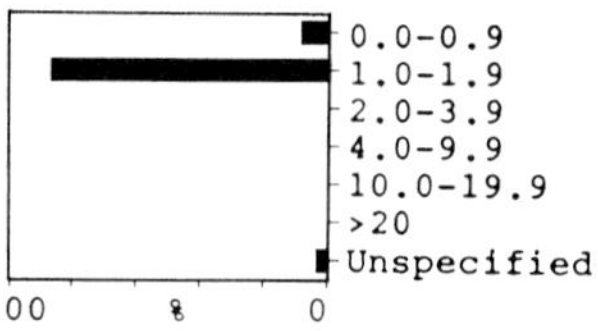

New Shoot Growth

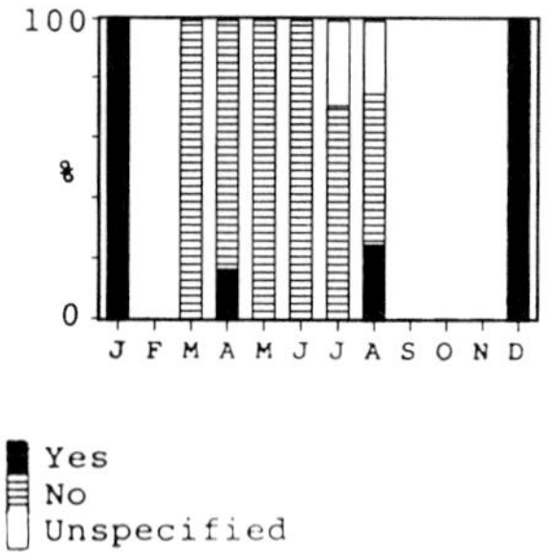

HABITAT

Within 2 km of coast: Yes (0%) No (97%) Unspecified (3%).

Altitude (metres)

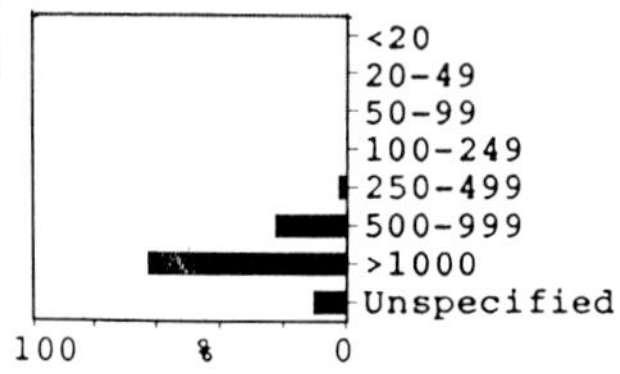

Vegetation

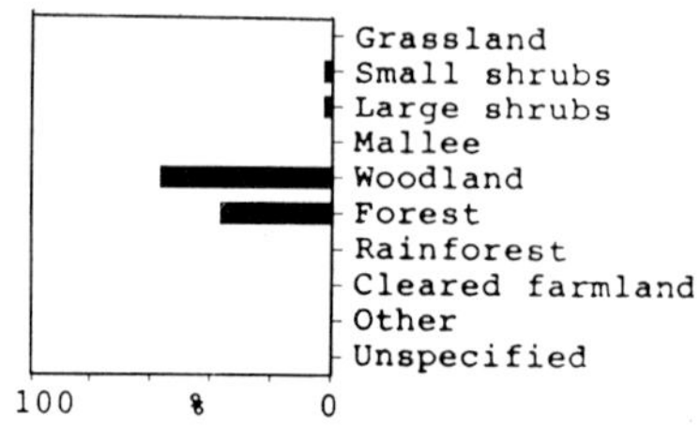

Soil Type

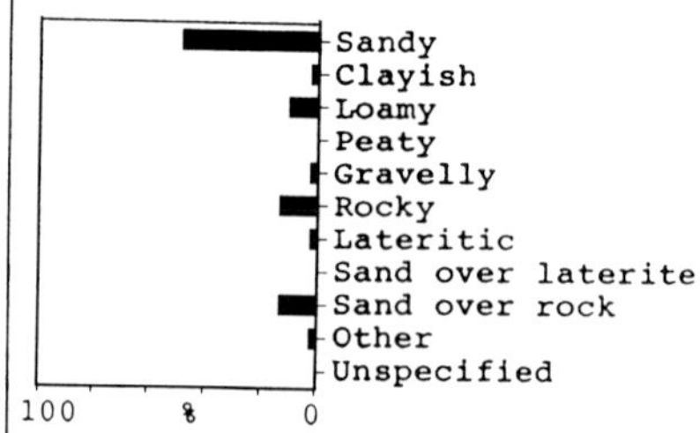

Landform

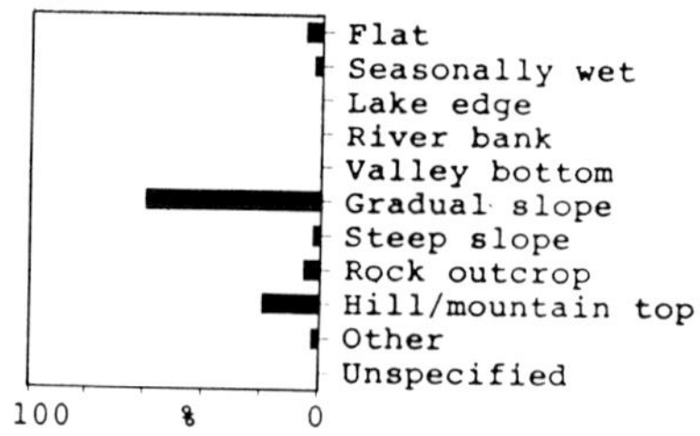

Response to Fire: 14 records

Growth response: Ground resprout (57%) killed, new seedlings (7%) unspecified (36%).

Flowering response: Number of records where flowering had occurred after fire (9). Median reported time to flowering (49-60 months); minimum reported time to flowering (<1 month). Number of records where flowering had not occurred after fire (1). Maximum reported time to no flowering (49-60 months).

Flowering

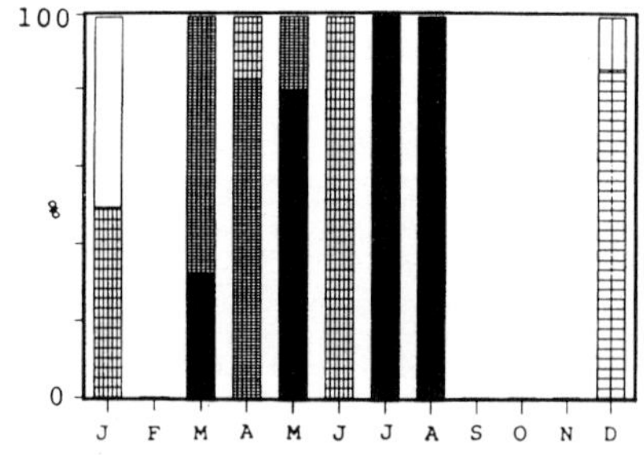

- Majority of flowers fully open
- Majority of flowers in bud
- Majority of flowers recently finished
- Flowers finished, fruiting cones present
- Neither flowers nor fruiting cones present
- Unspecified

Possible pollinators

Pollinator type 10 records: bird (40%) bees, wasps, ants (10%) no pollinator observed (50%).

Specific pollinator 2 records: Eastern Spinebill (2)

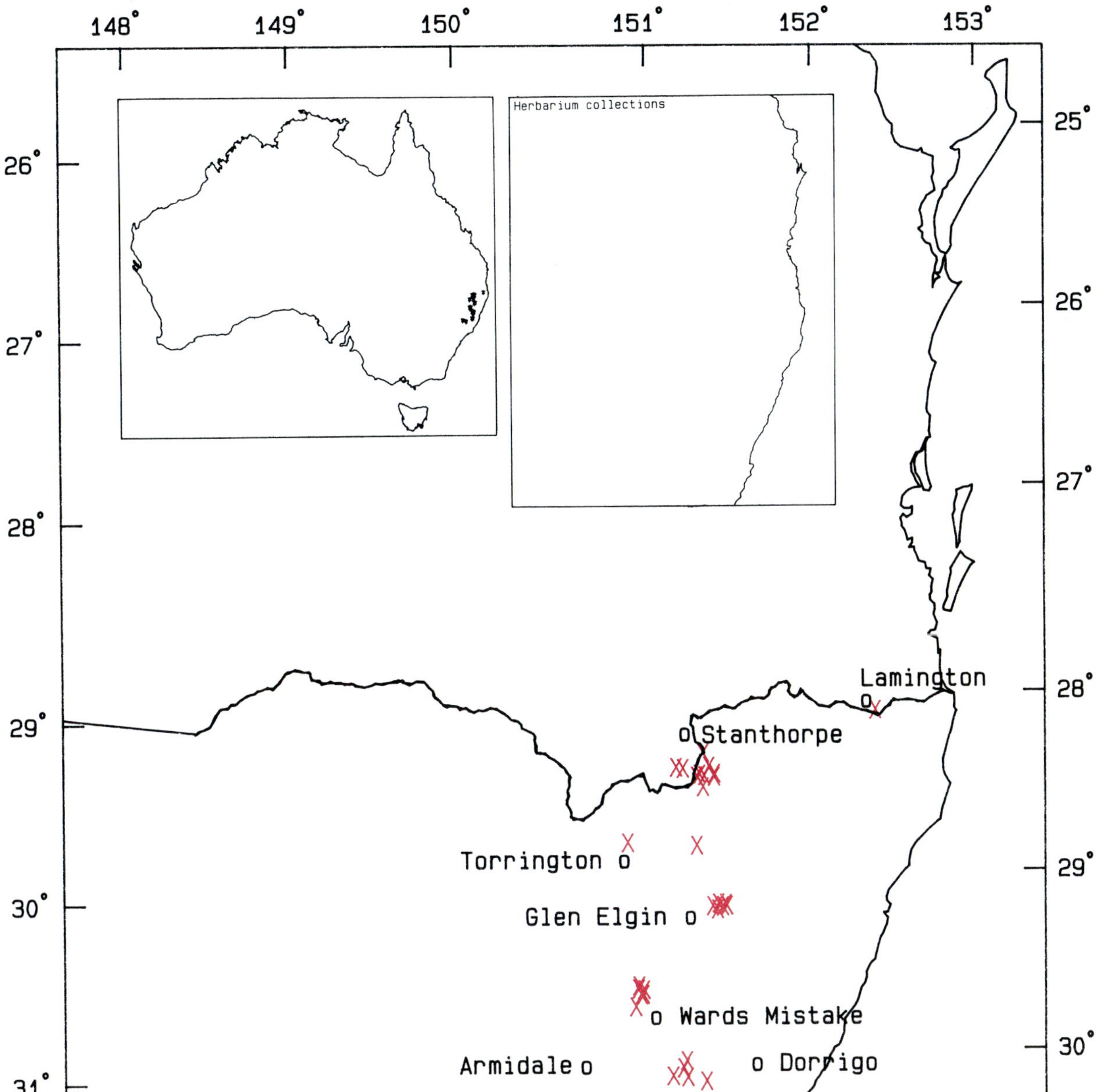

DISCUSSION

B. spinulosa var. neoanglica is a newly recognised variety named by George (1987). It occurs in mountainous regions of northern New South Wales and the Queensland border at altitudes usually in excess of 1 000 metres. Sixty-three percent of recorded populations are from conservation reserves, notably Boonoo-Boonoo, Girraween, Bald Rock, Lamington, Gibralter Range, Cathedral Rocks and Werrikimbe National Parks. The southern-most records are from Nowendoc and Hanging Rock. Prior to the Atlas, specimens of this banksia had usually been included with var. cunninghamii, or sometimes as var. collina. However, close inspection of plants in their natural habitat revealed a low growing, strongly lignotuberous shrub (therefore not var. cunninghamii) with totally entire leaves having a mucronate tip and no visible venation on the undersurface (therefore not var. collina). A description of B. spinulosa var. neoanglica is provided by George (1988).

Atlas records of var. neoanglica suggest that it is always a low shrub less than 2 m in height which is somewhat smaller than other spinulosa varieties. It grows in sands or loams often overlying rock, generally on the slopes or near the tops of mountains. The relatively small amount of flowering data available suggests that it flowers from May to August with new shoot growth predominantly in summer. Approximately one-quarter of all observations made were of populations which had been recently burnt and eight out of these nine records were of plants resprouting from lignotubers. Resultant juvenile leaves are generally serrated over their entire length and strongly resemble mature leaves of var. collina though always lacking that variety's obvious venation on the under-surfaces of the leaves.

BANKSIA SPINULOSA Smith (1793) var. SPINULOSA

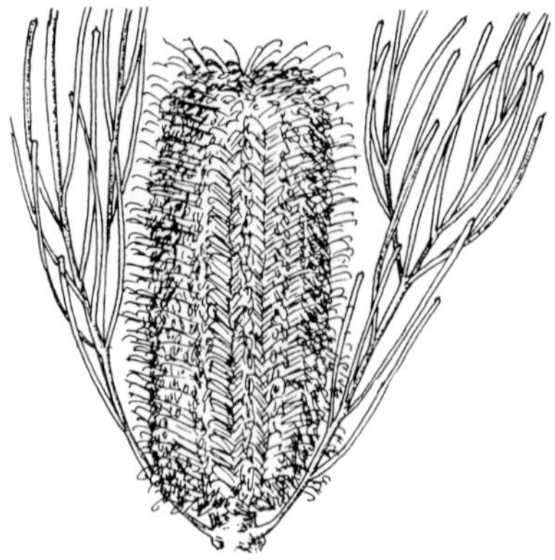

765 RECORDS: Jan (88) Feb (57) Mar (88) Apr (63) May (84) Jun (147) Jul (50) Aug (50) Sep (42) Oct (22) Nov (33) Dec (40).

Population Size: 1-10 (112) 10-100 (210) >100 (440) Unspecified (3).

Conservation Status: Restricted to road verge (2%) Not (97%) Unspecified (1%). In conservation reserve (39%) Not (59%) Unspecified (2%).

Tree/Shrub: Tree form (0.1%) Shrub form (100%) Unspecified (0%).

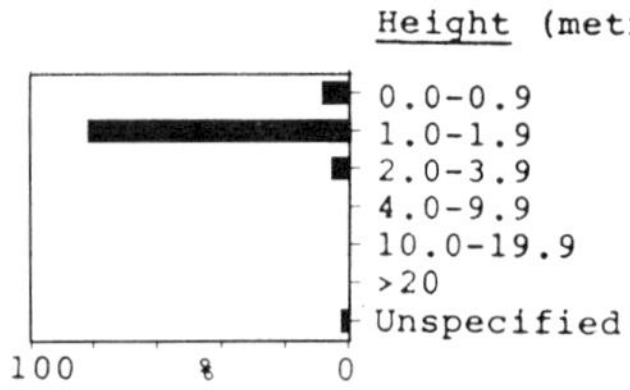

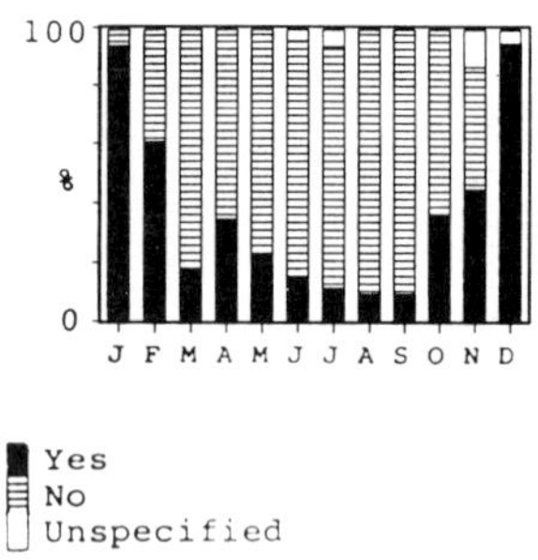

HABITAT

Within 2 km of coast: Yes (7%) No (92%) Unspecified (0%).

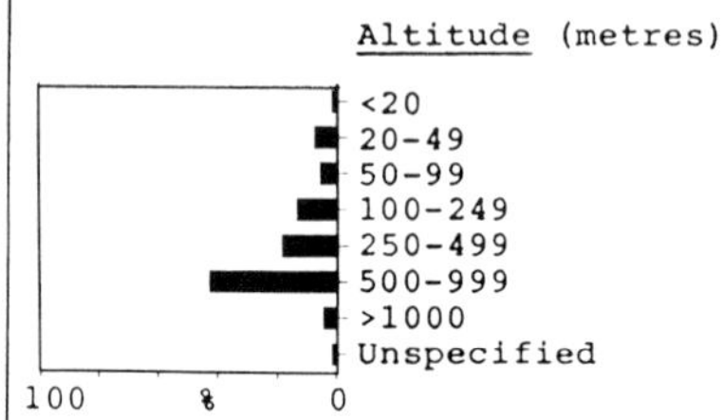

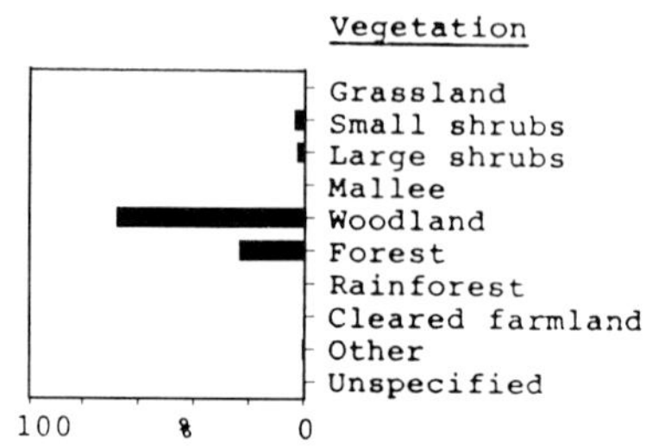

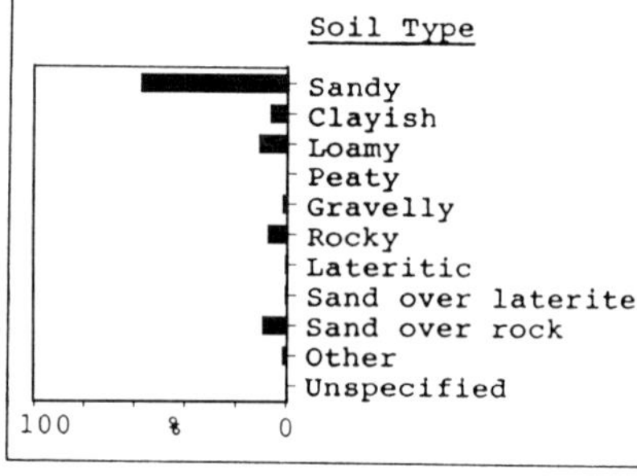

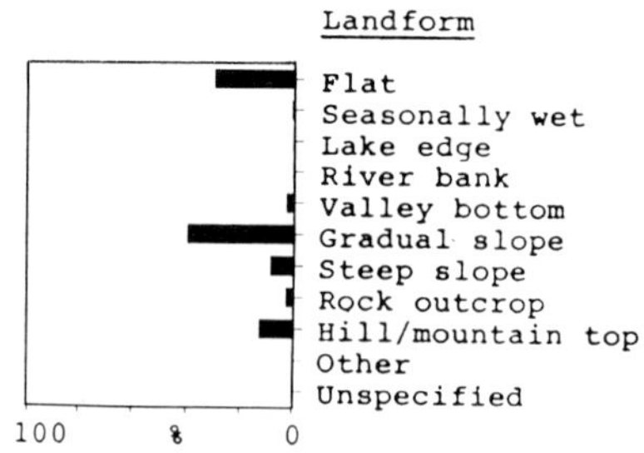

Response to Fire: 103 records

Growth response: Ground resprout (59%) ground resprout and killed, new seedlings (1%) ground resprout and trunk resprout (1%) killed, new seedlings (4%) trunk resprout (7%) unspecified (28%).

Flowering response: Number of records where flowering had occurred after fire (59). Median reported time to flowering (25-36 months); minimum reported time to flowering (<1 month). Number of records where flowering had not occurred after fire (13). Median reported time to no flowering (13-24 months); maximum reported time to no flowering (25-36 months).

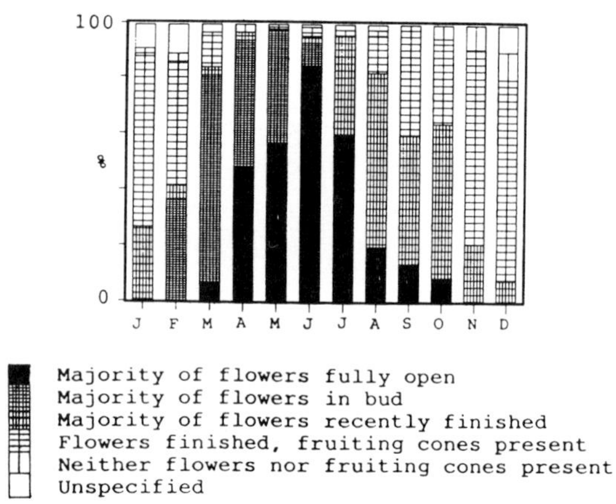

Possible pollinators

Pollinator type 498 records: bird (5%) bees, wasps, ants (5%) flies (1%) no pollinator observed (89%).

Specific pollinator 59 records: honeyeater (3) New Holland Honeyeater (5) White-eared Honeyeater (1) White-plumed Honeyeater (1) Yellow-faced Honeyeater (2) Crescent Honeyeater (1) Noisy Miner (1) wattlebird (1) Red Wattlebird (5) Eastern Spinebill (11) Noisy Friarbird (1) Eastern Yellow Robin (1) ant (12) bee (8) native bee (1) European honey bee (4) Brown Antechinus (1)

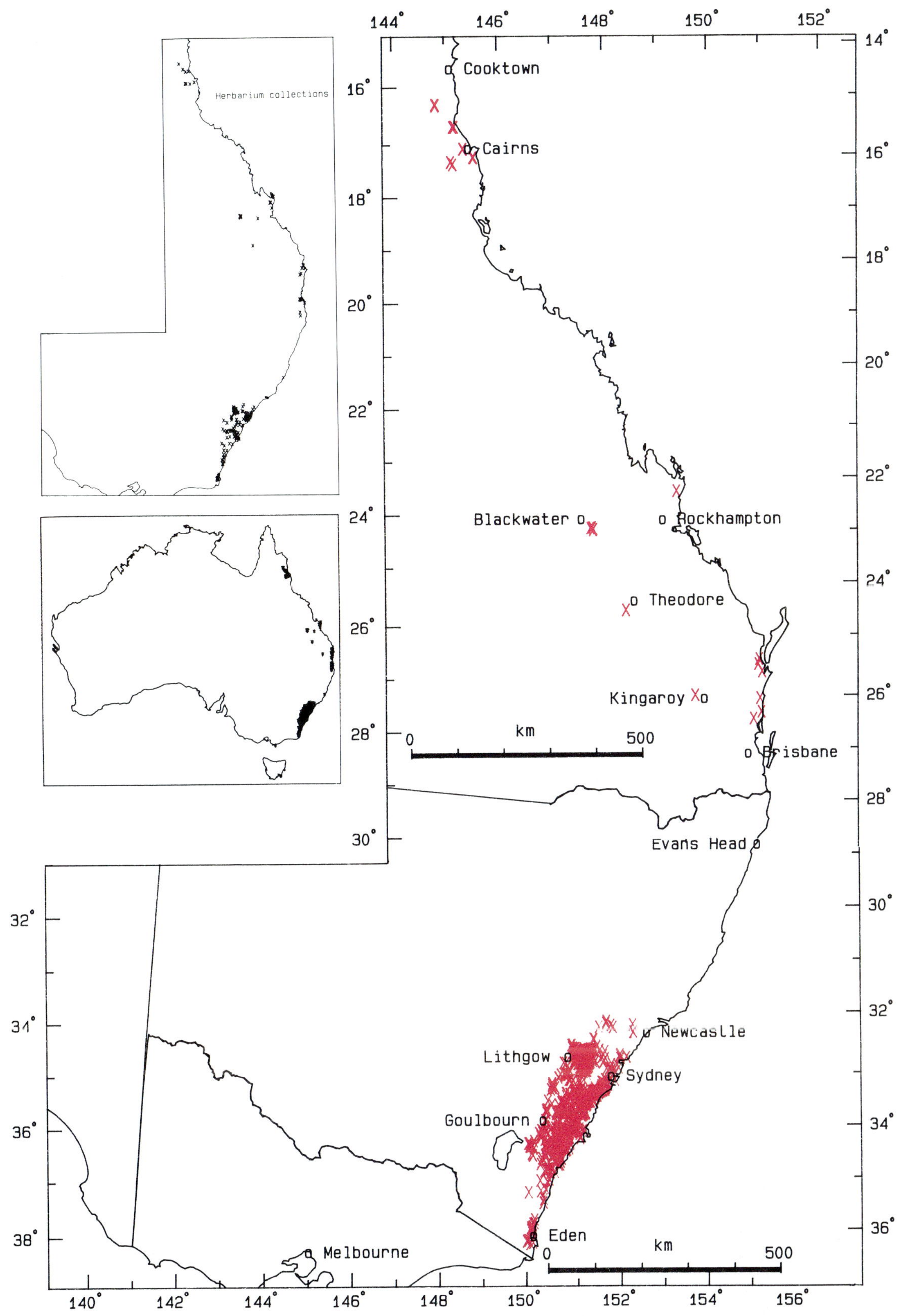

DISCUSSION

Banksia spinulosa var. _spinulosa_ occurs in Queensland and New South Wales, although it is far more common in the latter State. In Queensland it occurs in several disjunct areas. In northern Queensland scattered populations are found from the Mt Windsor Tableland inland from Daintree to the Atherton Tableland and Walsh's Pyramid just south of Cairns. In central east Queensland it occurs at Byfield north of Rockhampton, inland on the Blackdown Tableland and at Isla Gorge National Park south of Theodore. Herbarium collections from Gladstone and south of Biloela were not recorded by Atlas contributors. In south-eastern Queensland, it was recorded between Maryborough and Coloundra. There are two herbarium collections from south of Brisbane but these were not recorded. However, plants intermediate between var. _spinulosa_ and var. _collina_ (confirmed by A.S. George) were recorded from Burbank, a southern suburb of Brisbane. The plants from south of Maryborough and Tin Can Bay had been identified by local botanists and Forestry Department staff as var. _collina_ due to the leaves being serrated over the entire length. However, specimens were found to be var. _spinulosa_ on the basis of their revolute leaf margins. In New South Wales, herbarium collections had indicated a distribution from the Colo and Hawkesbury Rivers south almost to the Victorian border. However, _Atlas_ contributors recorded several populations of var. _spinulosa_ inland from Newcastle. According to George (1981) var. _collina_ is found in this area and not var. _spinulosa._ In southern N.S.W., the inland range of var. _spinulosa_ was extended to Goulbourn and almost to the north-eastern tip of the Australian Capital Territory.

B. spinulosa var. _spinulosa_ is typically a shrub less than 2 m in height, but occasionally it is larger. It occurs on both the coastal plain and inland mountains generally on sandy soil though sometimes they may be loamy or clayish. In mountain areas it may occur on steep rocky hillsides or hilltops, often growing in sand over rock. Typically, it is an understorey shrub of woodlands or forests. Flowering is from March to October, although mainly from April to July. New shoot growth is in summer. Following fire the variety generally resprouts from its lignotuber. However, there are a few reports of it being killed by fire and regenerating from seed.

BANKSIA TELMATIAEA A.S.George (1981)

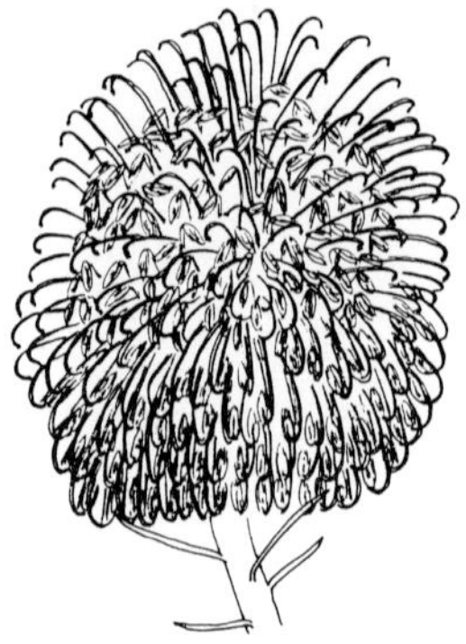

43 RECORDS: Jan (4) Feb (0) Mar (4) Apr (5) May (15) Jun (1) Jul (2) Aug (4) Sep (2) Oct (1) Nov (5) Dec (0).

Population Size: 1-10 (8) 10-100 (9) >100 (23) Unspecified (3).

Conservation Status: Restricted to road verge (23%) Not (74%) Unspecified (2%). In conservation reserve (26%) Not (74%) Unspecified (0%).

Tree/Shrub: Tree form (0%) Shrub form (93%) Unspecified (7%).

Height (metres)

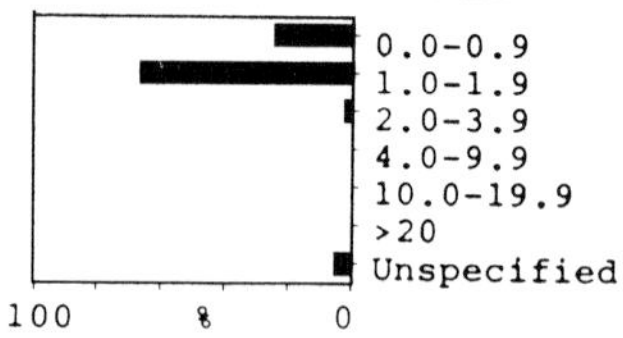

New Shoot Growth

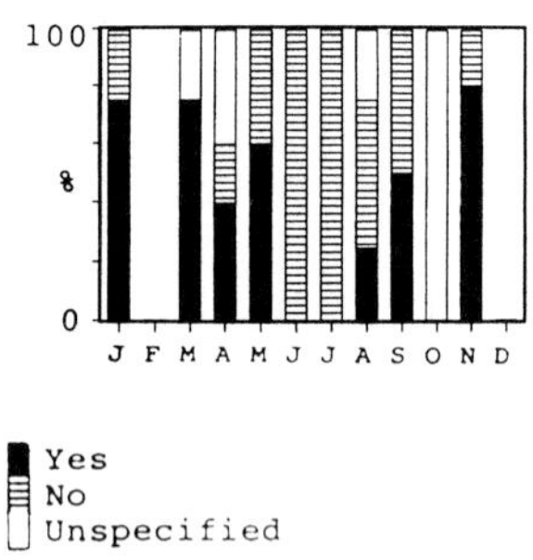

HABITAT

Within 2 km of coast: Yes (0%) No (98%) Unspecified (2%).

Altitude (metres)

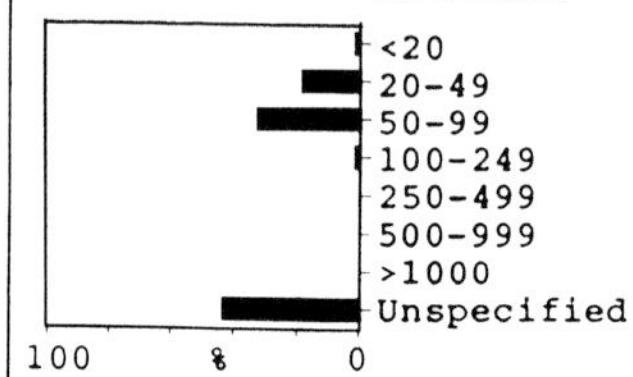

Vegetation

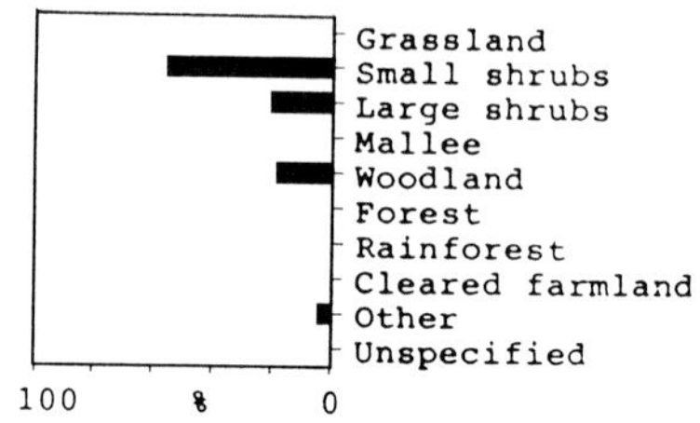

Soil Type

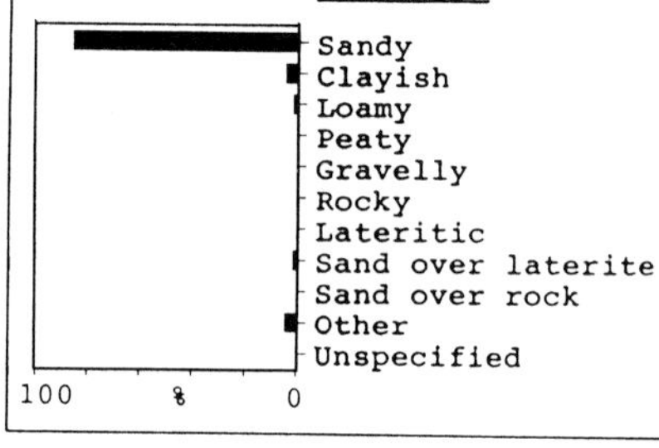

Landform

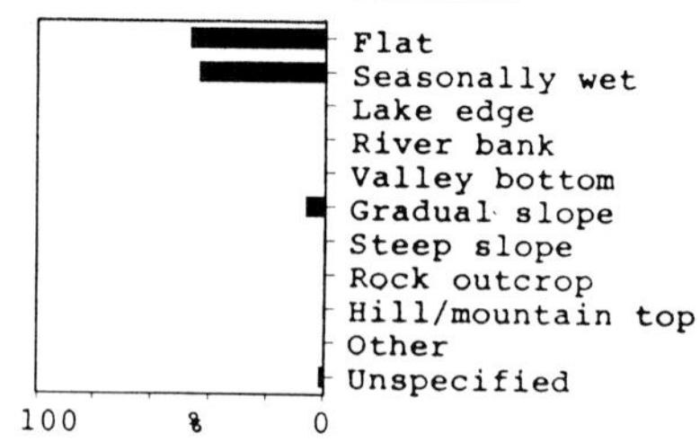

Response to Fire: 4 records

Growth response: Killed, new seedlings (25%) killed, no seedlings (50%) unspecified (25%).

Flowering response: Number of records where flowering had occurred after fire (1). Minimum reported time to flowering (13-24 months).
Number of records where flowering had not occurred after fire (0).

Flowering

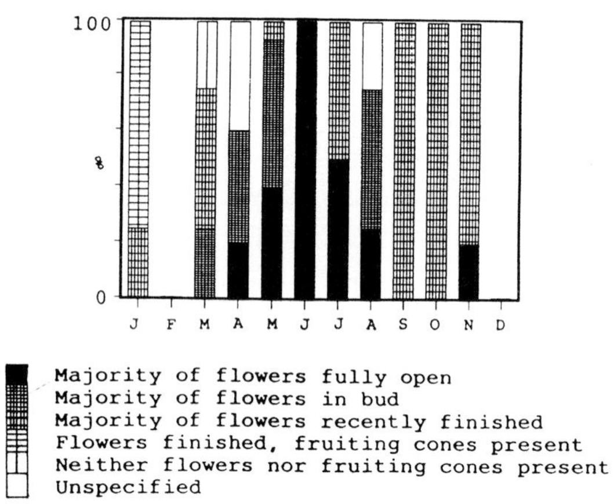

Possible pollinators

Pollinator type 9 records: bird (11%) bees, wasps, ants (11%) no pollinator observed (78%).

Specific pollinator 2 records: Red Wattlebird (1) bee (1)

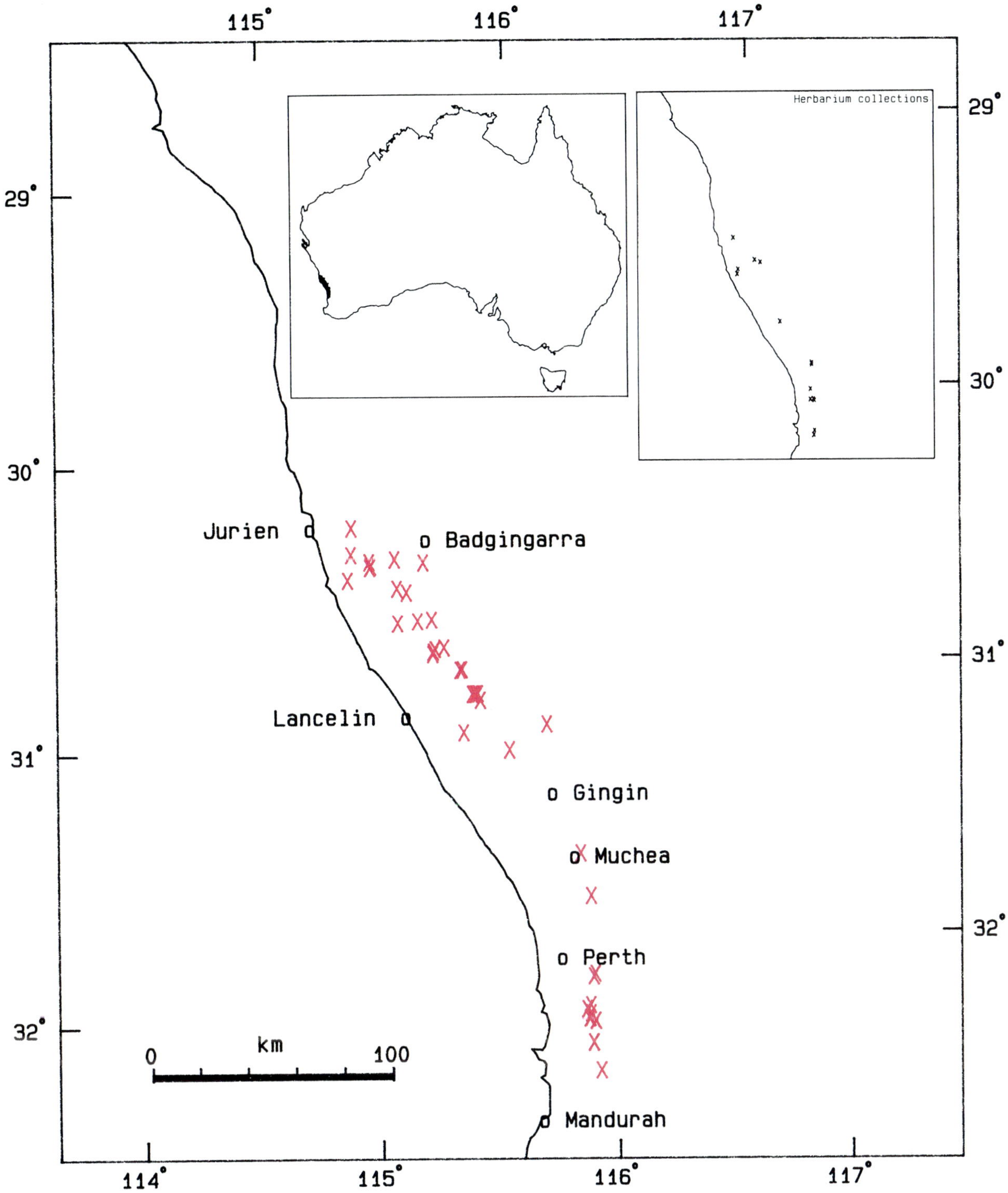

DISCUSSION

B. telmatiaea was recorded in scattered populations between the Hill River and Serpentine, always west of the Darling Scarp. Most of the populations are in the north and south of its range, from Moore River north to Hill River and from Cannington (a Perth suburb) south to Serpentine. The general distribution is therefore similar to that indicated by herbarium records. Most populations are of more than 100 plants and 26% are in conservation reserves. On the basis of these figures its conservation status seems relatively secure. However, the species should continue to be monitored since land clearance could change the situation greatly, particularly amongst its northern populations.

B. telmatiaea is typically a shrub up to 2 m in height growing in grey sandy soils which are seasonally wet. Surrounding vegetation is usually shrubland, sometimes with emergent moisture-loving trees such as Banksia littoralis and Melaleuca preissiana. Occasionally, trees may be sufficiently dense to form low open woodland. The main flowering period is April to August though flowers have also been recorded in November. New shoot growth appears to be mainly in summer though further data are needed to confirm this. The species is killed by fire and regenerates from seed.

BANKSIA TRICUSPIS Meissner (1855)

Lesueur Banksia, Pine Banksia

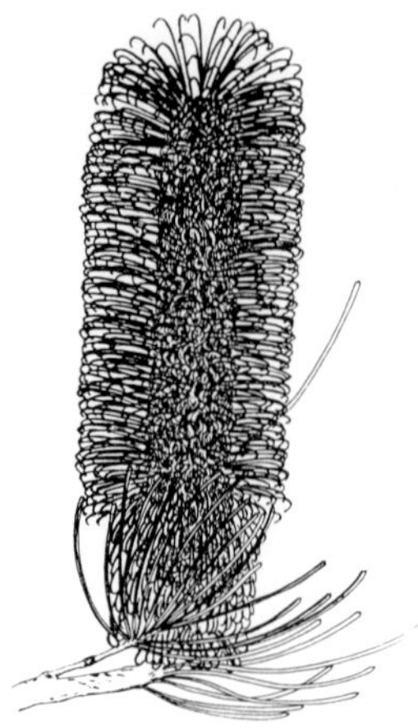

13 RECORDS: Jan (0) Feb (0) Mar (1) Apr (2) May (3) Jun (1) Jul (2) Aug (0) Sep (1) Oct (1) Nov (0) Dec (2).

Population Size: 1-10 (5) 10-100 (5) >100 (3) Unspecified (0).

Conservation Status: Restricted to road verge (0%) Not (92%) Unspecified (8%). In conservation reserve (0%) Not (92%) Unspecified (8%).

Tree/Shrub: Tree form (15%) Shrub form (85%) Unspecified (0%).

Height (metres)

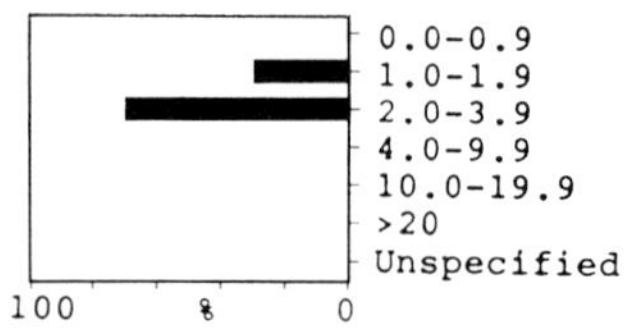

New Shoot Growth

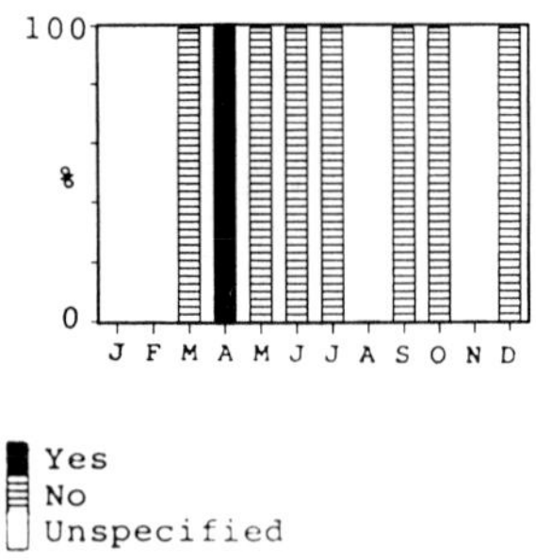

HABITAT

Within 2 km of coast: Yes (0%) No (100%) Unspecified (0%).

Altitude (metres)

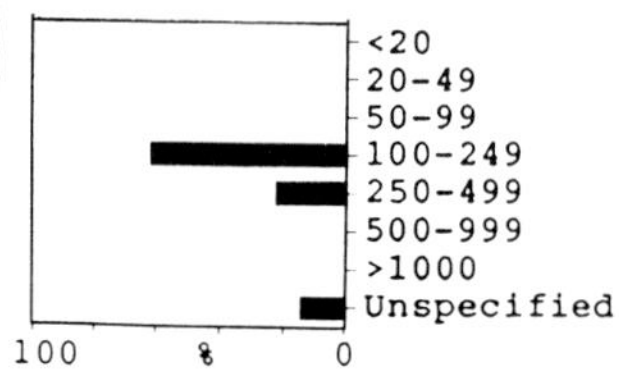

Vegetation

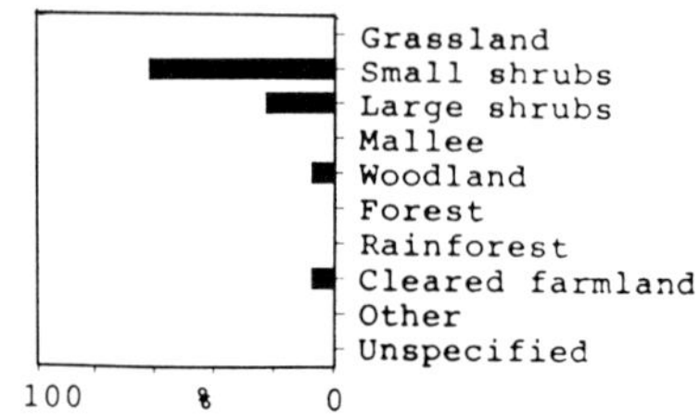

Soil Type

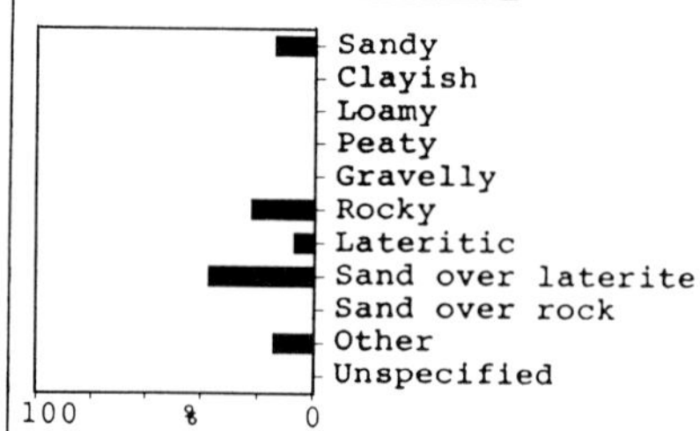

Landform

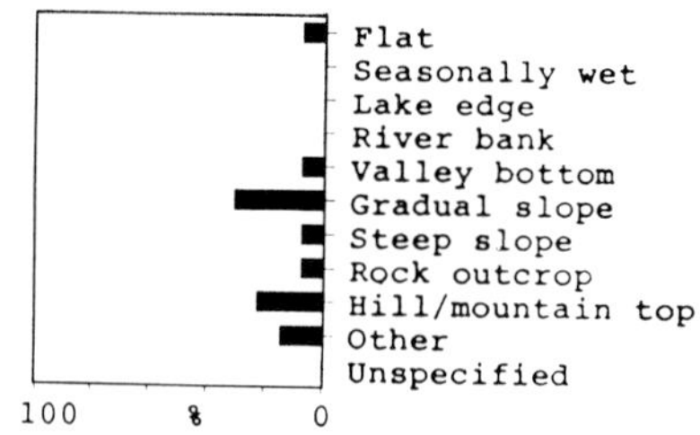

Response to Fire: 4 records

Growth response: Killed, new seedlings and trunk resprout (75%) trunk resprout (25%).

Flowering response: Number of records where flowering had occurred after fire (2). Median reported time to flowering (36-37 months); minimum reported time to flowering (1-12 months). Number of records where flowering had not occurred after fire (1). Maximum reported time to no flowering (<1 month).

Flowering

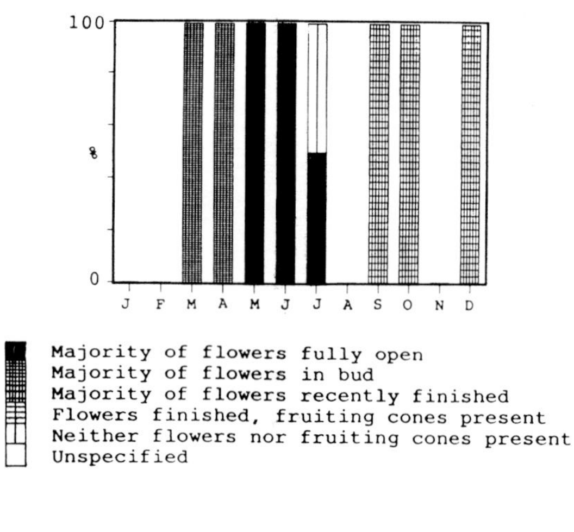

Possible pollinators

Pollinator type 7 records: bird (29%) bees, wasps, ants (29%) mammal (14%) no pollinator observed (29%).

Specific pollinator 4 records: Brown Honeyeater (1) Western Spinebill (1) European honey bee (1) Honey-possum (1)

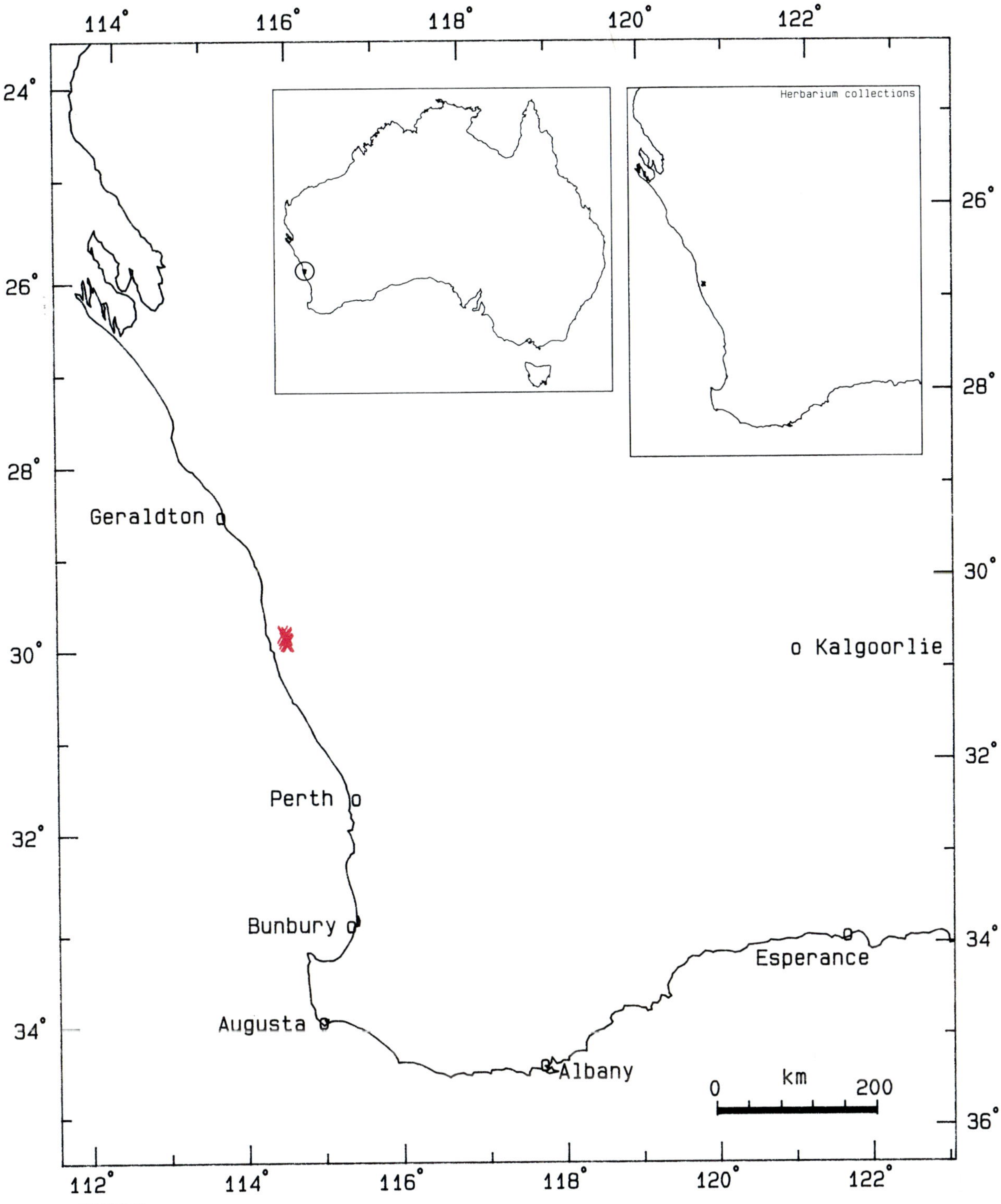

DISCUSSION

B. tricuspis is a declared rare plant with a maximum geographic range of only ca. 15 km. It is entirely restricted to an area near Jurien. A survey by Van Leeuwen in recent years has located 65 populations with a total of ca. 15 000 plants. Population size varied between 1 and 4 000 plants, but most contained less than 100. Currently, none of these populations is in conservation reserves. Almost all, except the most northern ones, would be included within the proposed Mt Lesueur Nature Reserve. The area in which the species occurs is the subject of many conflicting land-use interests including mining, agriculture and apiculture. The species must be considered at risk unless reserve status is granted.

B. tricuspis is typically a shrub or tree up to 4 m in height growing on rocky hill tops, slopes, gullies and breakaways. Soils are generally derived from laterite or sandstone. Sometimes, the plants grow almost straight out of rock cracks and crevices where very shallow soil has accumulated. In its typical habitat B. tricuspis is emergent amongst low or tall shrubland often itself forming an open low woodland. It is often associated with the rare Hakea neurophylla and may frequently co-exist with Banksia grossa and B. micrantha. A single population has a different habitat, occurring on flat sandplain country with shrubby forms of B. attenuata and B. menziesii along with Eucalyptus todtiana forming the dominant plants. Atlas contributors recorded the species flowering from May-July, though Van Leeuwen (1985) defined the flowering period as March-August. The same author stated that the species is probably pollinated by birds, mammals and insects with the Brown honeyeater and Honey-possum being the most important. Further data on new shoot growth are needed before reliable conclusions can be drawn. Following fire, the species resprouts from its lignotuber and from epicormic buds. Observations of seedlings suggests that they are killed by fire and do not tolerate burning until at least 20 years of age (Van Leeuwen 1985).

BANKSIA VERTICILLATA R.Brown (1810)

Granite Banksia, Albany Banksia

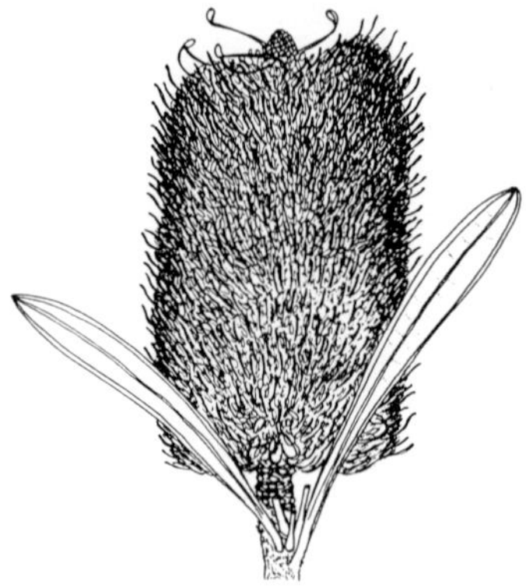

21 RECORDS: Jan (0) Feb (0) Mar (5) Apr (1) May (2) Jun (3) Jul (3) Aug (2) Sep (0) Oct (3) Nov (1) Dec (1).

Population Size: 1-10 (7) 10-100 (9) >100 (5) Unspecified (0).

Conservation Status: Restricted to road verge (0%) Not (81%) Unspecified (19%). In conservation reserve (81%) Not (19%) Unspecified (0%).

Tree/Shrub: Tree form (38%) Shrub form (62%) Unspecified (0%).

Height (metres)

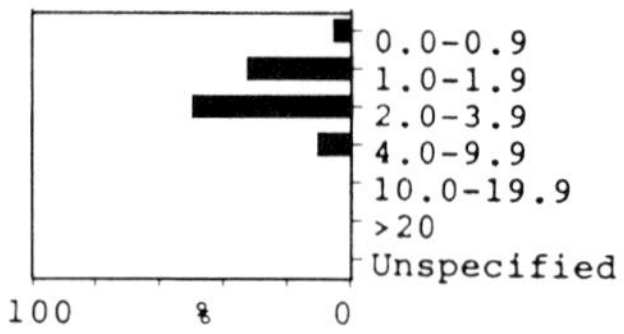

New Shoot Growth

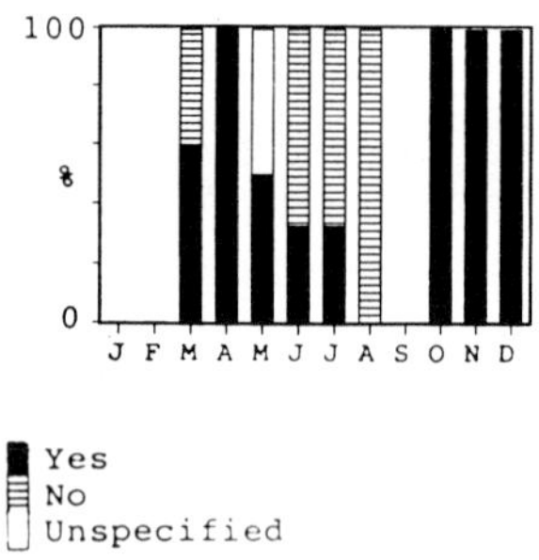

HABITAT

Within 2 km of coast: Yes (95%) No (0%) Unspecified (5%).

Altitude (metres)

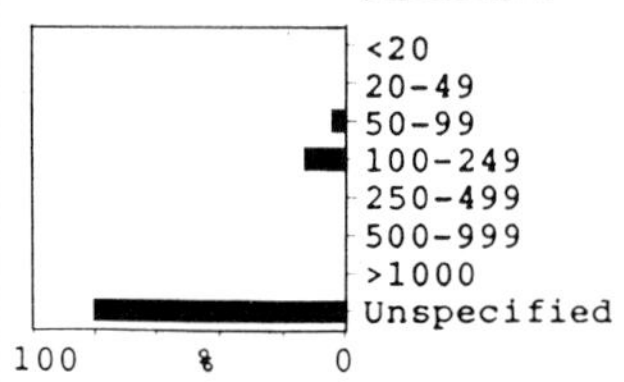

Vegetation

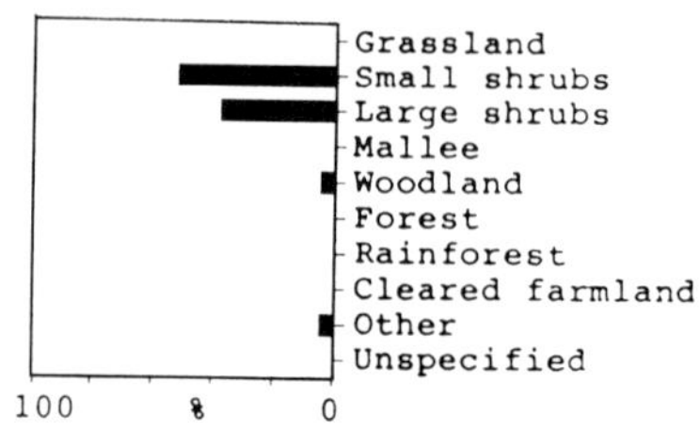

Soil Type

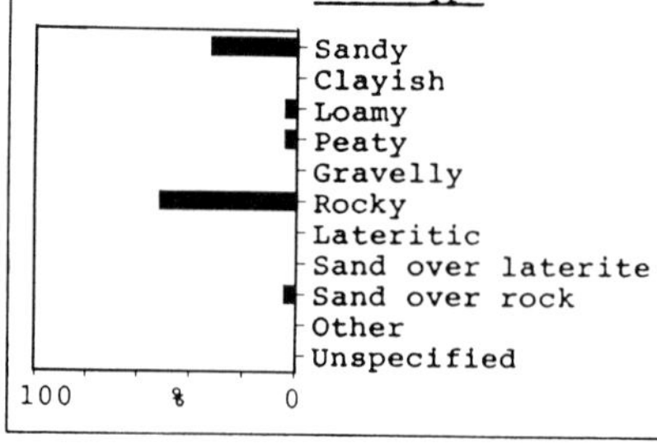

Landform

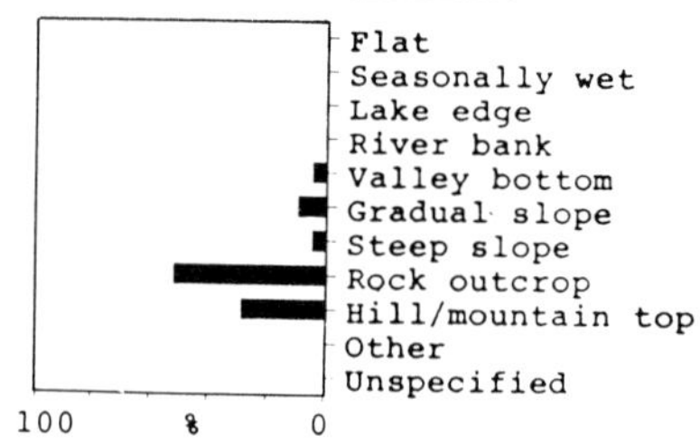

Response to Fire: 3 records

Growth response: Ground resprout and trunk resprout (33%) unspecified (67%).

Flowering response: Number of records where flowering had occurred after fire (2). Median reported time to flowering (12-13 months); minimum reported time to flowering (1-12 months). Number of records where flowering had not occurred after fire (1). Maximum reported time to no flowering (25-36 months).

Flowering

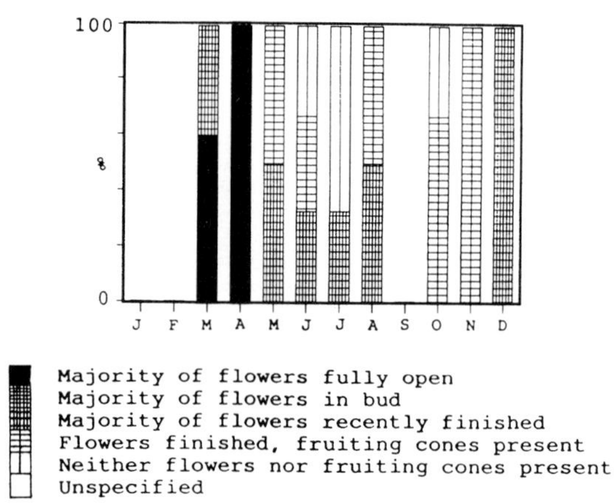

Possible pollinators

Pollinator type 3 records: bird (33%) no pollinator observed (67%).

Specific pollinator 1 records: New Holland Honeyeater (1)

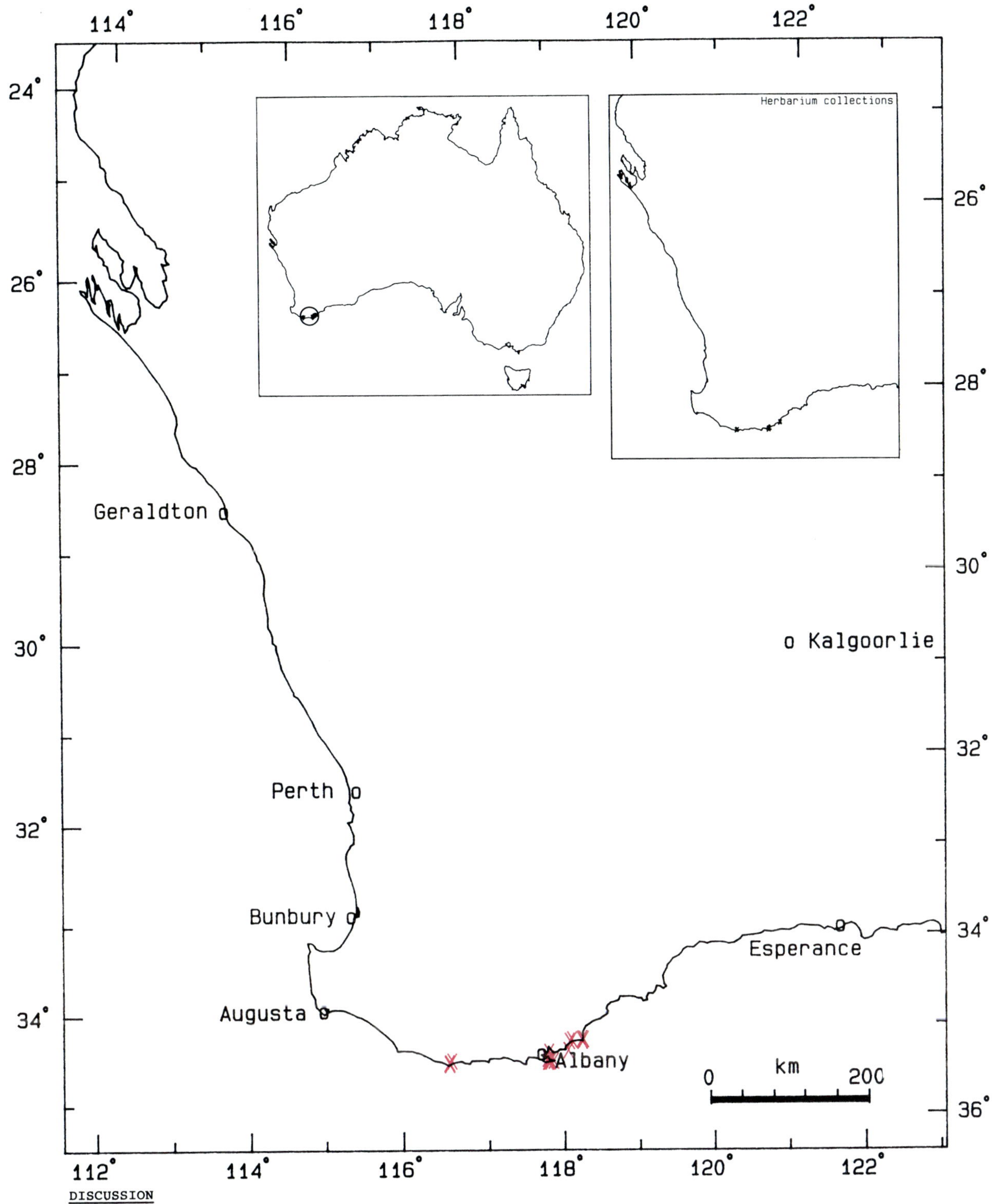

DISCUSSION

Atlas data have confirmed that B. verticillata occurs in two disjunct areas close to the south coast of Western Australia. B. verticillata is a rare species with only 10 or 11 relatively small populations known. Possibly only 3 populations consist of more than 100 plants and the maximum number of individuals at one site is c. 250. Its total range is c. 160 km. Although it is well represented in conservation reserves, the species is considered to be at risk. It is susceptible to dieback disease and may be killed by fire. It has therefore been recommended for inclusion on the list of declared rare and endangered plants in Western Australia.

B. verticillata occurs as a large bushy shrub or tree up to 5 m in height. It grows on or near exposed granite outcrops surrounded by low or tall shrubland. Soil is either rocky or with a shallow layer of sand over underlying rock. George (1981) stated the flowering period to be January to April. There were Atlas records of it flowering in March and April but no records at all from January and February. New shoot growth may be in late spring though further data are needed to confirm this. The species' response to fire also requires further investigation. At a location east of Albany, some plants were reported to have been killed by fire. Others were resprouting from both epicormic buds and lignotubers.

BANKSIA VICTORIAE Meissner (1855)

Woolly Orange Banksia

23 RECORDS: Jan (2) Feb (0) Mar (0) Apr (0) May (0) Jun (2) Jul (1) Aug (7) Sep (10) Oct (1) Nov (0) Dec (0).

Population Size: 1-10 (6) 10-100 (5) >100 (11) Unspecified (1).

Conservation Status: Restricted to road verge (17%) Not (78%) Unspecified (4%). In conservation reserve (48%) Not (48%) Unspecified (4%).

Tree/Shrub: Tree form (4%) Shrub form (91%) Unspecified (4%).

Height (metres)

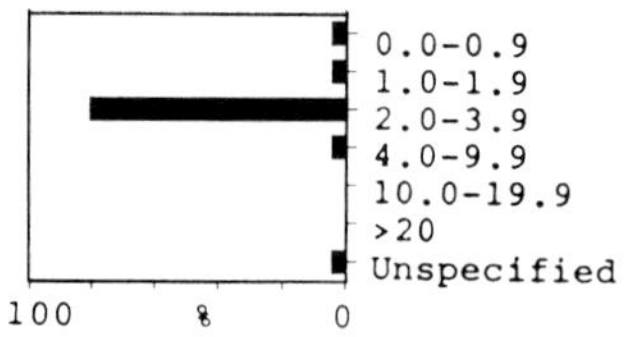

New Shoot Growth

HABITAT

Within 2 km of coast: Yes (4%) No (96%) Unspecified (0%).

Altitude (metres)

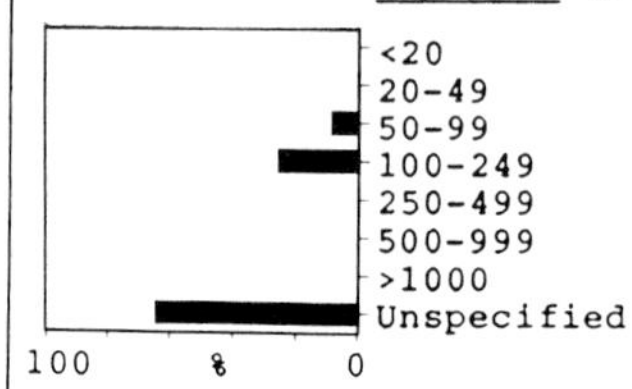

Vegetation

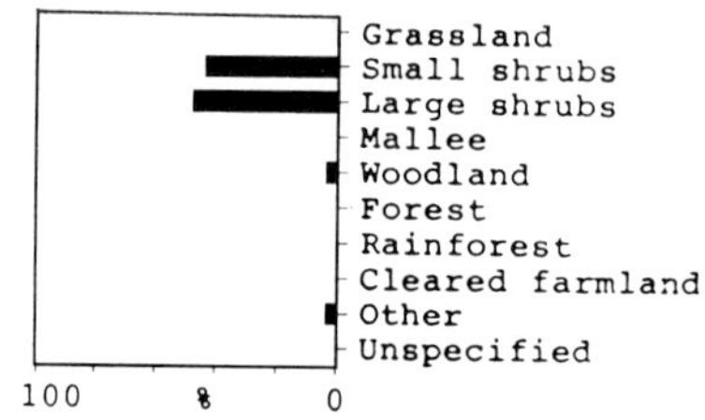

Soil Type

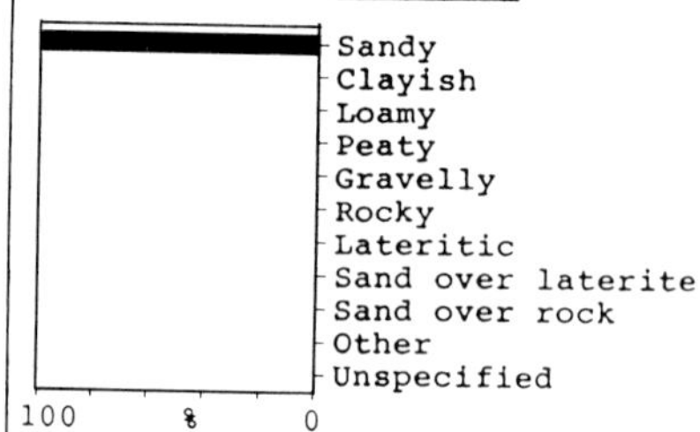

Landform

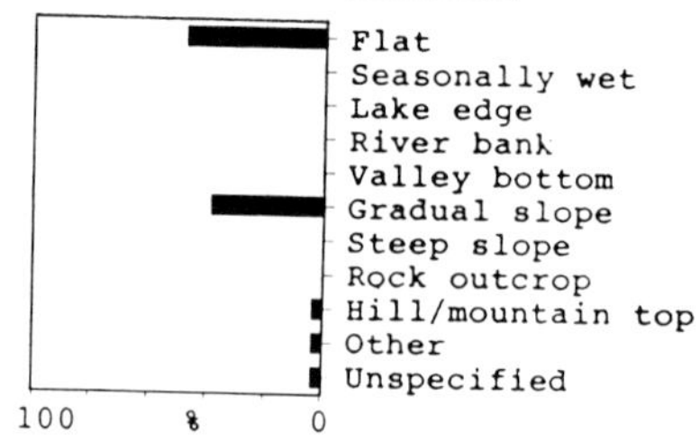

Response to Fire: 1 record

Growth response: Killed, new seedlings (100%).

Flowering response: Number of records where flowering had occurred after fire (0). Number of records where flowering had not occurred after fire (0).

Flowering

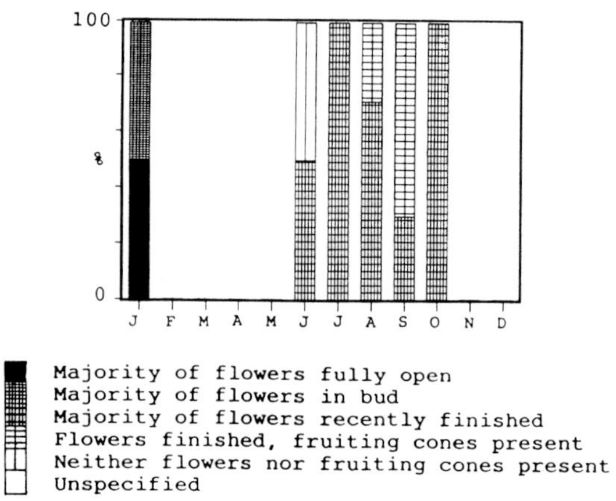

Possible pollinators

Pollinator type 2 records: no pollinator observed (100%).

Specific pollinator 0 records.

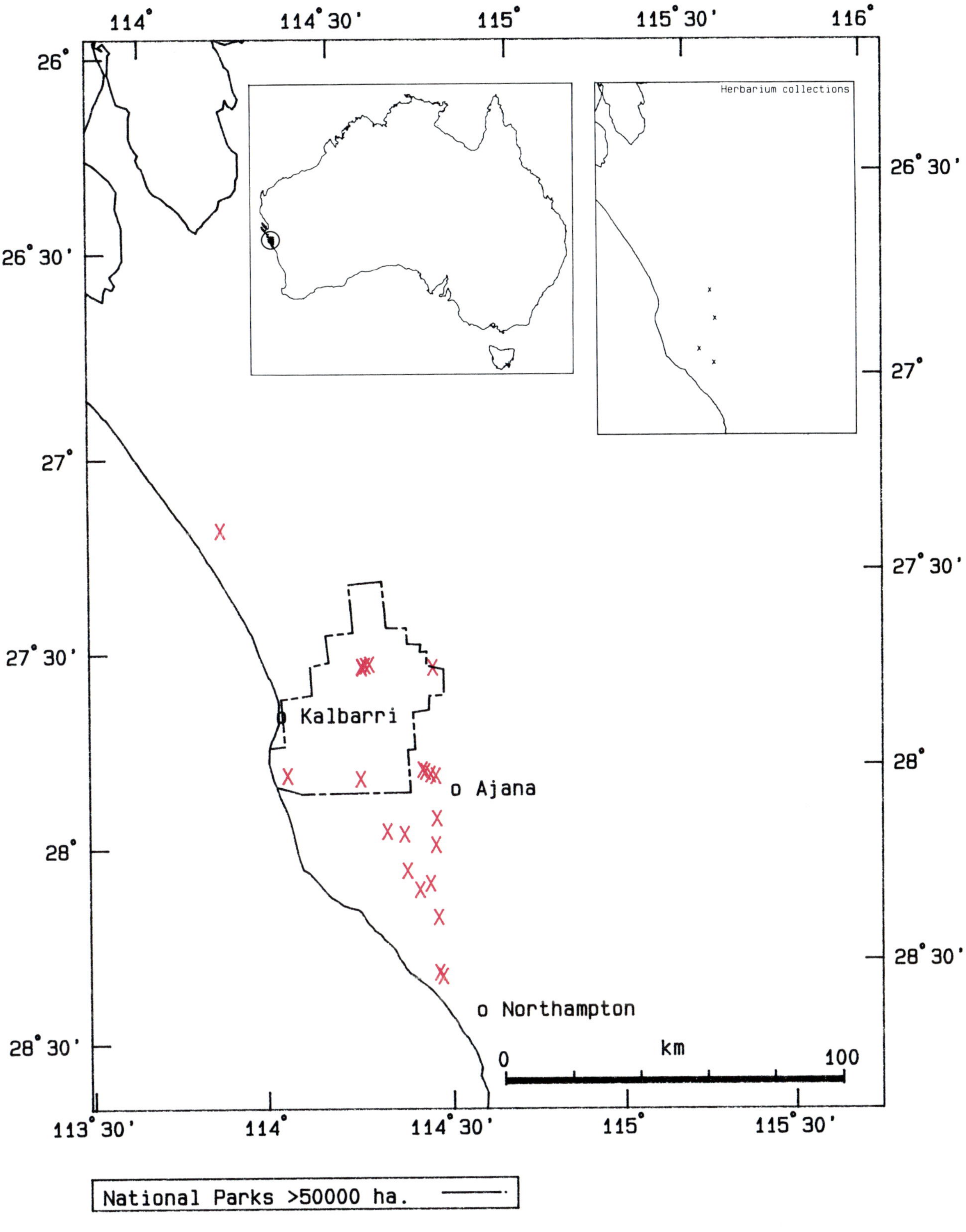

DISCUSSION

In the Western Australian Herbarium in 1984 there were only four collections of *B. victoriae* extending from the north-east boundary of Kalbarri National Park southwards to between Hutt River and Northampton. A single Atlas record has extended this range northwards to Zuytdorp National Park. At this location and at another in the south-west corner of Kalbarri National Park, populations occur within a few kilometres of the coast, whereas George (1981) described the distribution as always 20-30 km inland.

Banksia victoriae is a relatively rare species with few populations known and a total geographic range of *ca.* 140 km. Although the species does occur in conservation reserves (Kalbarri and Zuytdorp National Parks) the figure of 48% compiled from Atlas data is an overstatement since there are several duplicate records from a single locality along the lower Murchison River. The species is harvested for the commercial cut flower trade and is also very susceptible to dieback disease. It is killed by fire and regenerates from seedling regrowth so that successive fires at intervals too short to allow young plants to reach maturity, flower and set seed, could pose a significant threat. For all these reasons, the conservation status of the species should be monitored over future years.

B. victoriae is typically a large shrub up to 4 m in height, occasionally developing into taller tree-like specimens. Deep yellow sands are strongly preferred and surrounding vegetation is usually shrubland often including *B. attenuata* and *B. sceptrum*. *B. victoriae* is sometimes locally common and forms the dominant plant. Atlas records show the species to be flowering in January though lack of data from adjoining months prevent an assessment of the entire flowering period. However, George (1981) has defined it as January and February. The main period of new shoot growth appears to be summer though further data are needed to confirm this.

BANKSIA VIOLACEA C.Gardner (1928)

Violet Banksia

166 RECORDS: Jan (9) Feb (19) Mar (19) Apr (6) May (3) Jun (7) Jul (11) Aug (13) Sep (18) Oct (25) Nov (22) Dec (12).

Population Size: 1-10 (46) 10-100 (77) >100 (37) Unspecified (6).

Conservation Status: Restricted to road verge (19%) Not (78%) Unspecified (3%). In conservation reserve (48%) Not (46%) Unspecified (5%).

Tree/Shrub: Tree form (0%) Shrub form (100%) Unspecified (0%).

Height (metres)

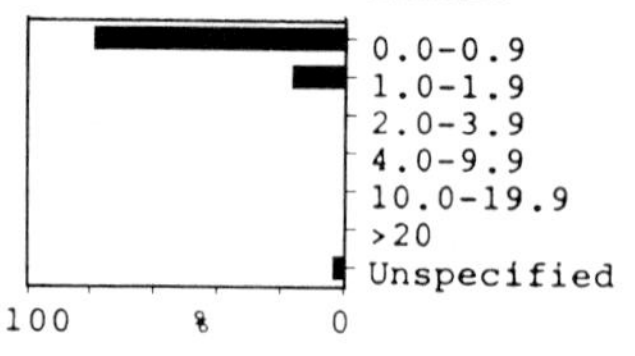

New Shoot Growth

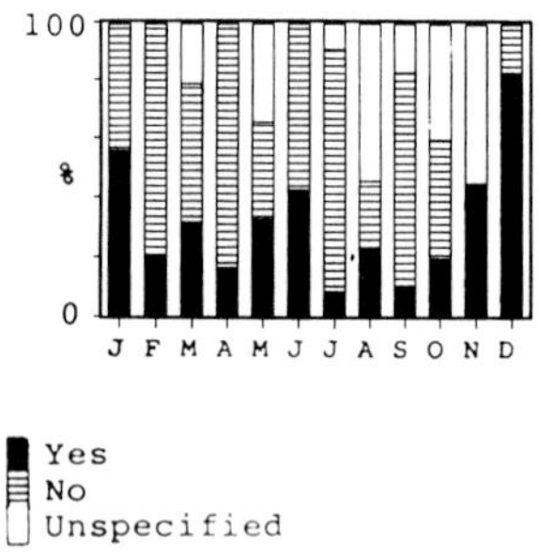

HABITAT

Within 2 km of coast: Yes (10%) No (90%) Unspecified (0%).

Altitude (metres)

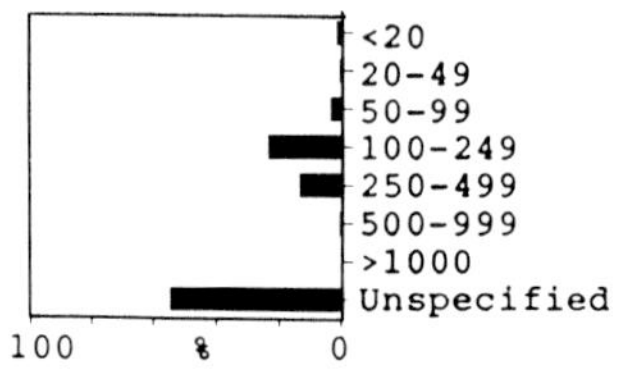

Vegetation

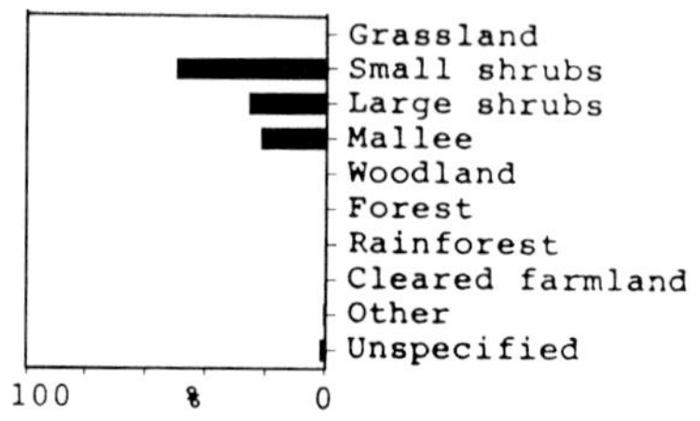

Soil Type

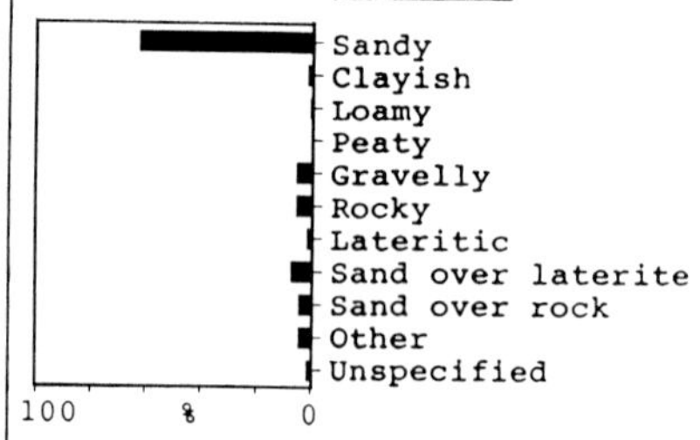

Landform

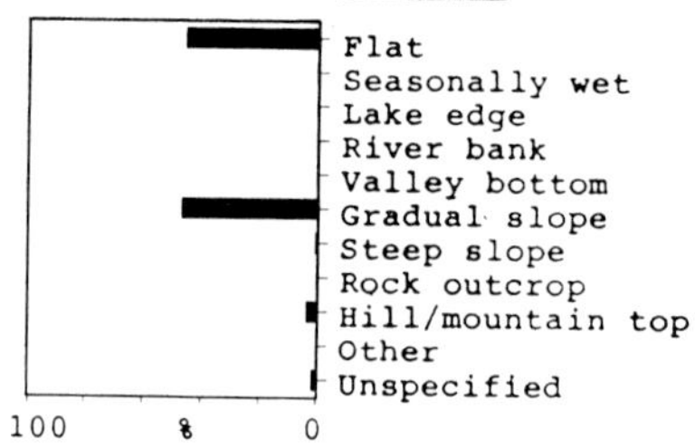

Response to Fire: 12 records

Growth response: Killed, new seedlings (58%) killed, no seedlings (25%) unspecified (17%).

Flowering response: Number of records where flowering had occurred after fire (4). Median reported time to flowering (49-60 months); minimum reported time to flowering (37-48 months). Number of records where flowering had not occurred after fire (4). Median reported time to no flowering (24-25 months); maximum reported time to no flowering (49-60 months).

Flowering

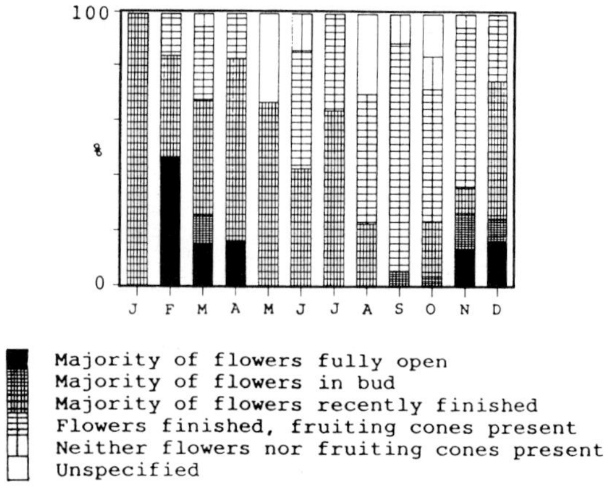

Possible pollinators

Pollinator type 36 records: bees, wasps, ants (11%) flies (3%) no pollinator observed (86%).

Specific pollinator 3 records: ant (3)

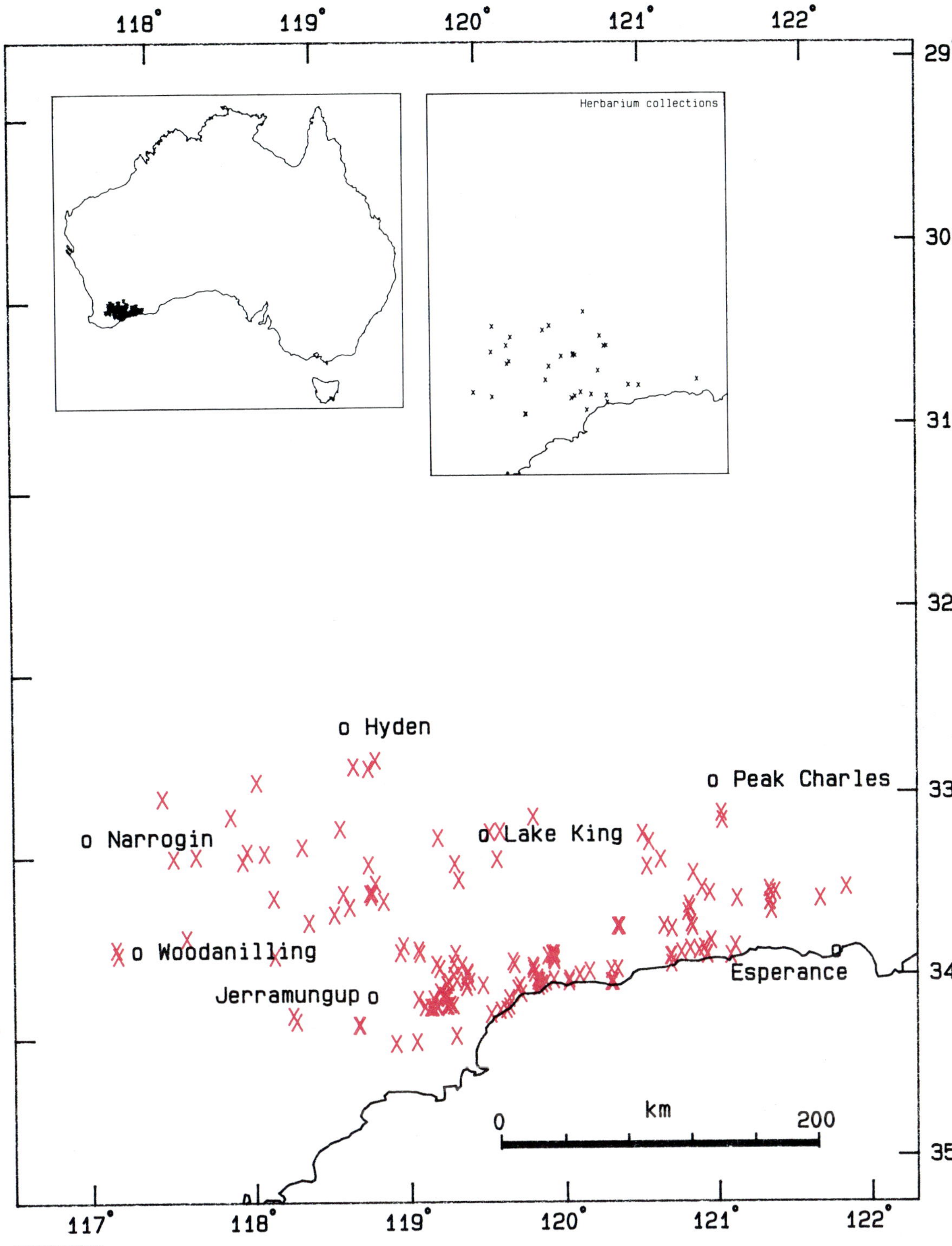

DISCUSSION

B. violacea occurs in the southern wheatbelt and adjoining coastal areas of Western Australia from Wickepin and Woodanilling eastwards to 20 km south of Peak Charles and 40 km north of Esperance. It is apparently absent from an area between Fitzgerald, Lake Magenta, Frank Hann National Park and Jerdacuttup. Western Australian Herbarium collections had indicated a similar distribution in the western part of its range but in the east there were no collections from beyond Lake King and only a few from near the coast between Ravensthorpe and Esperance. The records from south of Peak Charles therefore represent an 80-90 km range extension. With almost 50% of its recorded population occurring in conservation reserves, B. violacea appears relatively secure.

The species is typically a small shrub, less than one metre in height, though occasionally it may be larger. It grows predominantly in white sandy soils, sometimes overlying laterite. Surrounding vegetation is usually shrubland or mallee eucalypts, the two often intermixed. Atlas data suggest a low and possibly erratic rate of flowering. The flowering period appears to be from November to April, but even in the most floriferous month (February) less than 50% of recorded populations were in full flower. In January, not one of the nine records was of plants in full flower. The main period of new shoot growth is probably in summer. The species is generally non-lignotuberous, being killed by fire and regenerating from seedling germination and regrowth. However, in the vicinity of Woodanilling populations of lignotuberous plants were recorded.